T0328219

MODERNIZING GLOBAL HEALTH SECURITY TO PREVENT, DETECT, AND RESPOND

MODERNIZING GLOBAL HEALTH SECURITY TO PREVENT, DETECT, AND RESPOND

Edited by

SCOTT J.N. MCNABB

Hubert Department of Global Health, Rollins School of Public Health, Emory University, Atlanta, GA, United States

AFFAN T. SHAIKH

Hubert Department of Global Health, Rollins School of Public Health, Emory University, Atlanta, GA, United States; Yale University, New Haven, CT, United States

CAROL J. HALEY

Hubert Department of Global Health, Rollins School of Public Health, Emory University, Atlanta, GA, United States

Academic Press is an imprint of Elsevier
125 London Wall, London EC2Y 5AS, United Kingdom
525 B Street, Suite 1650, San Diego, CA 92101, United States
50 Hampshire Street, 5th Floor, Cambridge, MA 02139, United States
The Boulevard, Langford Lane, Kidlington, Oxford OX5 1GB, United Kingdom

ISBN: 978-0-323-90945-7

For information on all Academic Press publications visit our website at
https://www.elsevier.com/books-and-journals

Publisher: Stacy Masucci
Acquisitions Editor: Kattie Washington
Editorial Project Manager: Barbara Makinster
Production Project Manager: Punithavathy Govindaradjane
Cover Designer: Matthew Limbert

Typeset by TNQ Technologies

Contents

Section II

Global One Health to address pandemics - ecological and biological challenges in the dynamic planet

Wondwossen A. Gebreyes, Carol J. Haley, Jonna A.K. Mazet and Ermias Belay

6. (Re-)emerging viral zoonotic diseases at the human—animal—environment interface

Amanda M. Berrian, Zelalem Mekuria, Laura E. Binkley, Chima J. Ohuabunwo, Samantha Swisher, Kaylee Errecaborde, Stephane de la Rocque and Carol J. Haley

7. Emergence and dissemination of antimicrobial resistance at the interface of humans, animals, and the environment

Shu-Hua Wang, Senait Kebede, Ebba Abate, Afreenish Amir, Ericka Calderon, Armando E. Hoet, Aamer Ikram, Jeffrey T. LeJeune, Zelalem Mekuria, Satoru Suzuki, Susan Vaughn Grooters, Getnet Yimer and Wondwossen A. Gebreyes

8. Toxic and environmentally ubiquitous chemical agents

Michael Bisesi and Jiyoung Lee

9. Global climate change impacts on vector ecology and vector-borne diseases

Rafael F.C. Vieira, Sebastián Muñoz-Leal, Grace Faulkner, Tatiana Sulesco, Marcos R. André and Risa Pesapane

10. Assessment of critical gaps in prevention, control, and response to major bacterial, viral, and protozoal infectious diseases at the human, animal, and environmental interface

Muhammed S. Muyyarikkandy, Kalmia Kniel, William A. Bower, Antonio R. Vieira, María E. Negrón and Siddhartha Thakur

11. Urbanization, human societies, and pandemic preparedness and mitigation

Gonzalo M. Vazquez-Prokopec, Laura E. Binkley, Hector Gomez Dantes, Amanda M. Berrian, Valerie A. Paz Soldan, Pablo C. Manrique-Saide and Thomas R. Gillespie

Section III

Movement of people and things: The challenge of pandemic spread

Ann Marie Kimball and Wondimagegnehu Alemu

12. The interconnected world of trade, travel, and transportation networks

Mohamed Moussif, Marissa Morales and Ryan Rego

13. Mitigating negative economic impacts of pandemics

Jarjieh Fang, Erin Holsted, Krishna Patel, Ryan Rego and Thomas Kingsley

14. Health measures at points of entry as prevention tools

Nyri Safiya Wells, Charuttaporn Jitpeera, Mohamed Moussif, Peter S. Mabula and Sopon Iamsirithaworn

15. Rights-based global health security through all-hazard risk management

Qudsia Huda, Erin L. Downey, Ali Ardalan, Shuhei Nomura and Ankur Rakesh

Section IV

Tools and techniques to modernize prevention, detection, and response to epidemics

Joseph N. Fair, Aamer Ikram, Rana Jawad Asghar and Marjorie P. Pollack

16. Global laboratory systems

Lucy A. Perrone, Francois-Xavier Babin, Sebastien Cognat, Juliane Gebelin, Emmanuelle Boussieres, Allegra Molkenthin, Barbara Jauregui, Koren Wolman-Tardy, Hanvit Oh and Allison Watson

17. Modernizing public health surveillance

Louise Gresham, Wondimagegnehu Alemu, Nomita Divi, Noara Alhusseini, Oluwafunbi Awoniyi, Adnan Bashir, Affan T. Shaikh and Scott J.N. McNabb

18. Data for public health action: Creating informatics-savvy health organizations to support integrated disease surveillance and response

Nancy Puttkammer, Phiona Vumbugwa, Neranga Liyanaarachchige, Tadesse Wuhib, Dereje Habte, Eman Mukhtar Nasr Salih, Legesse Dibaba, Terence R. Zagar and Bill Brand

19. Analytics and intelligence for public health surveillance

Brian E. Dixon, David Barros Sierra Cordera, Mauricio Hernández Ávila, Xiaochun Wang, Lanyue Zhang, Waldo Vieyra Romero and Rodrigo Zepeda Tello

20. Navigating a rapidly changing information and communication landscape amidst "infodemics"

Jacqueline Cuyvers, Aly Passanante, Ed Pertwee, Pauline Paterson, Leesa Lin and Heidi J. Larson

21. Countering vaccine hesitancy in the context of global health

James O. Ayodele, Joann Kekeisen-Chen, Leesa Lin, Ahmed Haji Said, Heidi J. Larson and Ferdinand Mukumbang

Section V

Moving to the best-protected global community

Mohannad Al-Nsour, Chima J. Ohuabunwo, Rana Jawad Asghar and Randa K. Saad

22. Science and political leadership in global health security

Natalie Mayet, Eliot England, Benjamin Djoudalbaye, Ebere Okereke and Wondimagegnehu Alemu

List of editors

Editors

Scott J.N. McNabb, PhD, MS
Research Professor | Emory University, Rollins School of Public Health
Managing Partner | Public Health Practice, LLC
Highlands, NC, United States

Affan T. Shaikh, MPH, MBA
Adjunct Instructor | Emory University, Rollins School of Public Health
Atlanta, GA, United States

Carol J. Haley, PhD, MS
Adjunct Associate Professor | Emory University, Rollins School of Public Health
Atlanta, GA, United States

Section editors

Stella Chungong, MD, MS
Director | Health Security Preparedness Department
World Health Organization, Health Emergencies Programme
Geneva, Switzerland

Marjorie P. Pollack, MD
Former Deputy Editor & Epidemiology and Surveillance Moderator, ProMED
Independent Infectious Disease Epidemiology Consultant, Global
Brooklyn, NY, United States

Aamer Ikram, HI(M), SI(M)
Chief Executive Officer | National Institutes of Health, Pakistan
Executive Board Member | International Association of National Public Health Institutes
Chair, Advisory Board | Training Programs in Epidemiology and Public Health Interventions Network
Chair, Board | International Federation of Biosafety Associations
Executive Board Member | Gavi The Vaccine Alliance
Islamabad, Pakistan

Ann Marie Kimball, MD, MPH
President | Rotary Club of Bainbridge Island
Senior Consulting Fellow | Chatham House
Professor Emerita | University of Washington, School of Public Health
Seattle, WA, United States

Rana Jawad Asghar, MBBS, MPH
Adjunct Professor | University of Nebraska Medical Center, College of Public Health
Chief Executive Officer | Global Health Strategists and Implementers
Islamabad, Pakistan

Wondwossen A. Gebreyes, DVM, PhD, DACVPM
Hazel C Youngberg Distinguished Professor, Molecular Epidemiology | Ohio State University
Executive Director | Global One Health initiative
Columbus, OH, United States

Wondimagegnehu Alemu, MD, MPH and TM, FAIPH
Chief Technical Advisor | International Health Consultancy, LLC
Adjunct Associate Professor | Emory University, Rollins School of Public Health
Atlanta, GA, United States

Chima J. Ohuabunwo, MD, MPH, FWACP, MACE
Professor | Office of Global Health Equity, Morehouse School of Medicine
Adjunct Professor | Emory University, Rollins School of Public Health
Professor | University of Port Harcourt School of Public Health, Choba-Port Harcourt, Rivers, Nigeria
Atlanta, GA, United States

Chikwe Ihekweazu, MD, MPH, FFPH
Assistant Director-General | Division of Health Emergency Intelligence and Surveillance Systems, Incorporating the WHO Hub for Pandemic and Epidemic Intelligence
World Health Organization, Health Emergencies Programme
Geneva, Switzerland

Mohannad Al-Nsour, MD, MSc, PhD, FFPH
Executive Director | Eastern Mediterranean Public Health Network
Amman, Jordan

Joseph N. Fair, PhD, MSPH
Principal | TriCorder Health
Tyler, TX, United States

Tajudeen Raji, MD, MPH, FWACP
Head of Public Health Institutes and Research | Africa Centres for Disease Control and Prevention
Addis Ababa, Ethiopia

Ahmed Ogwell Ouma, MD, MPH, MPhil
Deputy Director General | Africa Centres for Disease Control and Prevention
Addis Ababa, Ethiopia

List of contributors

Ebba Abate Ohio State-Global One Health, Addis Ababa, Ethiopia

David Addiss Focus Area for Compassion and Ethics (FACE), The Task Force for Global Health, Decatur, GA, United States

Emmanuel Agogo Resolve to Save Lives, New York, NY, United States

Robert Agyarko Outbreaks and Epidemics: African Risk Capacity (ARC), Johannesburg, Gauteng, South Africa

Fatima Al Slail Public Health Agency, Ministry of Health, Riyadh, Saudi Arabia

Mohannad Al-Nsour Eastern Mediterranean Public Health Network (EMPHNET), Amman, Jordan

Wondimagegnehu Alemu International Health Consultancy, LLC, Decatur, GA, United States; Rollins School of Public Health, Emory University, Decatur, GA, United States

Noara Alhusseini Department of Biostatistics, Epidemiology, and Public Health, College of Medicine, Alfaisal University, Riyadh, Kingdom of Saudi Arabia

Afreenish Amir National Institutes of Health, Islamabad, Pakistan

Duaa Al Ammari College of Public Health and Health Informatics, King Saud Bin Abdulaziz University for Health Sciences, King Abdullah International Medical Research Center, Riyadh, Saudi Arabia

Marcos R. André Laboratório de Imunoparasitologia, Departamento de Patologia, Reprodução e Saúde Única, Faculdade de Ciências Agrárias e Veterinárias, Universidade Estadual Paulista (FCAV/UNESP), Jaboticabal, Brazil

Lawrence N. Anyanwu University of Glasgow, Glasgow, United Kingdom

Ali Ardalan WHO Regional Office for the Eastern Mediterranean, World Health Organization, Cairo, Egypt

Alex Riolexus Ario Uganda National Institute of Public Health, Kampala, Uganda

Rana Jawad Asghar Global Health Strategists & Implementers, Islamabad, Pakistan; University of Nebraska Medical Center, College of Public Health, Omaha, NE, United States

Oluwafunbi Awoniyi Hubert Department of Global Health, Rollins School of Public Health, Emory University, Atlanta, GA, United States

James O. Ayodele WHO Office for the Eastern Mediterranean Region, Cairo, Egypt

Francois-Xavier Babin Fondation Mérieux, Lyon, France

Caroline Baer Public Health Practice, LLC, Atlanta, GA, United States

Adnan Bashir Health Information Systems Program, Islamabad, Pakistan

David Barros Sierra Cordera Mexican Institute for Social Security, Mexico City, Mexico

Ermias Belay Division of High-Consequence Pathogens and Pathology, National Center for Emerging and Zoonotic Infectious Diseases, Centers for Disease Control and Prevention, Atlanta, GA, United States

Amanda M. Berrian Department of Veterinary Preventive Medicine, College of Veterinary Medicine, The Ohio State University, Columbus, OH, United States; Global One Health initiative (GOHi), The Ohio State University, Columbus, OH, United States

Barbara E. Bierer Brigham and Women's Hospital and Harvard Medical School, Boston, MA, United States

Laura E. Binkley Department of Veterinary Preventive Medicine, College of Veterinary Medicine, The Ohio State University, Columbus, OH, United States; Global One Health initiative (GOHi), The Ohio State University, Columbus, OH, United States

Michael Bisesi Global One Health Initiative (GOHi), The Ohio State University College of Public Health, Columbus, OH, United States

Emmanuelle Boussieres Fondation Mérieux, Lyon, France

William A. Bower Bacterial Special Pathogens Branch, Division of High-Consequence Pathogens and Pathology, National Center for Emerging and Zoonotic Infectious Diseases, Centers for Disease Control and Prevention, Atlanta, GA, United States

Bill Brand The Task Force for Global Health/ Public Health Informatics Institute, Decatur, GA, United States

Garrett Wallace Brown School of Politics and International Studies, University of Leeds, Leeds, United Kingdom

Ericka Calderon Inter-American Institute for Technical Cooperation (IICA), San Jose, Costa Rica

Vicky Cardenas Gibson, Dunn & Crutcher LLP, San Francisco, CA, United States

Parker Choplin Rollins School of Public Health, Emory University, Atlanta, GA, United States

Luc Bertrand Tsachoua Choupe Evidence and Analytics for Health Security, Geneva, Switzerland

Yashwant Chunduru Yale University, New Haven, CT, United States

Stella Chungong Health Security Preparedness Department, WHO Health Emergencies Programme, World Health Organization, Geneva, Switzerland

Sebastien Cognat World Health Organization, Lyon, France

Francis P. Crawley Good Clinical Practice Alliance – Europe (GCPA) & Strategic Initiative for Developing Capacity in Ethical Review (SIDCER), Brussels, Belgium

Jacqueline Cuyvers Department of Infectious Disease Epidemiology, London School of Hygiene & Tropical Medicine, London, United Kingdom

Stephane de la Rocque Health Security Preparedness Department, World Health Organization, Geneva, Switzerland

Lisa M. DeTora Writing Studies and Rhetoric, Hofstra University, Hempstead, NY, United States

Legesse Dibaba Ministry of Health, Addis Ababa, Ethiopia

Nomita Divi Ending Pandemics, San Francisco, CA, United States

Brian E. Dixon Richard M. Fairbanks School of Public Health, Indiana University, Indianapolis, IN, United States; Center for Biomedical Informatics, Regenstrief Institute, Indianapolis, IN, United States

Benjamin Djoudalbaye Africa Centres for Disease Control and Prevention (Africa CDC), Addis Ababa, Ethiopia

Erin L. Downey Harvard Humanitarian Initiative, Harvard T.H. Chan School of Public Health, Cambridge, MA, United States

Perihan Elif Ekmekci TOBB University of Economics and Technology, International Chair in Bioethics/WMA Cooperation Center, Ankara, Turkey

Eliot England Rollins School of Public Health, Emory University, Atlanta, GA, United States

Ngozi Erondu Chatham House, London, United Kingdom; O'Neil Institute, Georgetown University, Washington, DC, United States

Kaylee Errecaborde Health Security Preparedness Department, World Health Organization, Geneva, Switzerland

Joseph N. Fair TriCorder Health, Tyler, TX, United States

Jarjieh Fang ICAP at Columbia University, New York City, NY, United States

Grace Faulkner Center for Computational Intelligence to Predict Health and Environmental Risks (CIPHER), The University of North Carolina at Charlotte, Charlotte, NC, United States

Denise O. Garrett Applied Epidemiology Team, Sabin Vaccine Institute, Washington, DC, United States

Juliane Gebelin Fondation Mérieux, Lyon, France

Wondwossen A. Gebreyes Global One Health initiative (GOHi), The Ohio State University, Columbus, OH, United States; Department of Veterinary Preventive Medicine, College of Veterinary Medicine, The Ohio State University, Columbus, OH, United States

Brittany Gentry Hubert Department of Global Health, Rollins School of Public Health, Emory University, Atlanta, GA, United States

Thomas R. Gillespie Department of Environmental Sciences, Emory University, Atlanta, GA, United States

Kara Goldstone Rollins School of Public Health, Emory University, Atlanta, GA, United States

Hector Gomez Dantes Center for Health System Research, National Institute of Public Health, Cuernavaca, Mexico

Louise Gresham PAX sapiens and School of Public Health, San Diego State University, Washington, DC, United States

Susan Vaughn Grooters Department of Veterinary Preventive Medicine, College of Veterinary Medicine, The Ohio State University, Columbus, OH, United States

Dereje Habte US Centers for Disease Control and Prevention – Ethiopia, Addis Ababa, Ethiopia

Ahmed Haji Said Rollins School of Public Health, Emory University, Atlanta, GA, United States

Carol J. Haley Hubert Department of Global Health, Rollins School of Public Health, Emory University, Atlanta, GA, United States

Mauricio Hernández Ávila Mexican Institute for Social Security, Mexico City, Mexico

Marc Ho Ministry of Health, Singapore

Armando E. Hoet Department of Veterinary Preventive Medicine, College of Veterinary Medicine, The Ohio State University, Columbus, OH, United States

Erin Holsted Rollins School of Public Health, Emory University, Atlanta, GA, United States

Qudsia Huda WHO Headquarters, World Health Organization, Geneva, Switzerland

Sopon Iamsirithaworn Department of Disease Control, Ministry of Public Health, Bangkok, Thailand

Aamer Ikram National Institutes of Health, Islamabad, Pakistan

Barbara Jauregui Mérieux Foundation USA, Washington, DC, United States

Charuttaporn Jitpeera Hubert Department of Global Health, Rollins School of Public Health, Emory University, Atlanta, GA, United States

Nirmal Kandel Evidence and Analytics for Health Security, Geneva, Switzerland

Senait Kebede Hubert Department of Global Health, Rollins School of Public Health, Emory University, Atlanta, GA, United States

Joann Kekeisen-Chen Rollins School of Public Health, Emory University, Atlanta, GA, United States

Ann Marie Kimball Department of Epidemiology, University of Washington, Bainbridge Island, WA, United States; Chatham House, London, United Kingdom

Thomas Kingsley American Action Forum, Washington, DC, United States

Kalmia Kniel Department of Animal and Food Sciences, University of Delaware, Newark, DE, United States

Marika L. Kromberg Underwood Centre for Sustainable Healthcare Education, University of Oslo, Oslo, Norway

Heidi J. Larson Department of Infectious Disease Epidemiology, London School of Hygiene & Tropical Medicine, London, United Kingdom; Centre for the Evaluation of Vaccination, Vaccine & Infectious Disease Institute, University of

Antwerp, Antwerp, Belgium; Department of Health Metrics Sciences, University of Washington, Seattle, WA, United States

Jiyoung Lee College of Food, Agriculture, and Environmental Science, Global One Health Initiative (GOHi), The Ohio State University College of Public Health, Columbus, OH, United States

Jeffrey T. LeJeune The Food and Agriculture Organization of United Nations (FAO), Rome, Italy

Leesa Lin Department of Infectious Disease Epidemiology, London School of Hygiene & Tropical Medicine, London, United Kingdom; Laboratory of Data Discovery for Health (D24H), Hong Kong Science Park, Hong Kong SAR, China; WHO Collaborating Centre for Infectious Disease Epidemiology and Control, School of Public Health, LKS Faculty of Medicine, The University of Hong Kong, Hong Kong SAR, China

Neranga Liyanaarachchige Division of Global HIV and TB, Global Health Center, United States Centers for Disease Control and Prevention, Atlanta, GA, United States

Peter S. Mabula Emergency Medicine Department, Disaster Management, Safari City HealthCare Limited, Arusha, Tanzania

Sarah B. Macfarlane Epidemiology and Biostatistics, and Institute for Global Health Sciences, University of California San Francisco, San Francisco, CA, United States

Mabel K.M. Magowe School of Nursing, University of Botswana, Gaborone, Botswana

Peter Mala World Health Organization, Geneva, Switzerland

Pablo C. Manrique-Saide Department of Zoology, School of Biological Sciences, Autonomous University of Yucatan, Merida, Yucatan, Mexico

Natalie Mayet National Institute for Communicable Diseases, A Division of the National Health Laboratory Service, Johannesburg, South Africa

Jonna A.K. Mazet Grand Challenges, University of California, Davis, CA, United States

Scott J.N. McNabb Hubert Department of Global Health, Rollins School of Public Health, Emory University, Atlanta, GA, United States

Zelalem Mekuria Global One Health initiative (GOHi), The Ohio State University, Columbus, OH, United States; Department of Veterinary Preventive Medicine, College of Veterinary Medicine, The Ohio State University, Columbus, OH, United States

Allegra Molkenthin Mérieux Foundation USA, Washington, DC, United States

Marissa Morales Hubert Department of Global Health, Rollins School of Public Health, Emory University, Atlanta, GA, United States

Mohamed Moussif Casablanca International Airport, Ministry of Health, Casablanca, Morocco

Ferdinand Mukumbang Ingham Institute for Applied Medical Research, Liverpool, NSW, Australia; Sydney Institute for Women, Children and Their Families, Sydney, NSW, Australia

Sebastián Muñoz-Leal Departamento de Ciencia Animal, Facultad de Ciencias Veterinarias, Universidad de Concepción, Chillán, Chile

Brittany L. Murray Division of Pediatric Emergency Medicine, Global Health Office of Pediatrics at Emory, Emory University School of Medicine, Atlanta, GA, United States

Muhammed S. Muyyarikkandy Department of Population Health and Pathobiology, North Carolina State University, Raleigh, NC, United States

María E. Negrón Bacterial Special Pathogens Branch, Division of High-Consequence Pathogens and Pathology, National Center for Emerging and Zoonotic Infectious Diseases, Centers for Disease Control and Prevention, Atlanta, GA, United States

Shuhei Nomura School of Medicine, Keio University, Tokyo, Japan

Hanvit Oh Emory University, Atlanta, GA, United States

Chima J. Ohuabunwo Department of Medicine and the Office of Global Health Equity,

Morehouse School of Medicine, Atlanta, GA, United States; Hubert Department of Global Health, Rollins School of Public Health, Emory University, Atlanta, GA, United States; Department of Epidemiology, University of Port Harcourt School of Public Health, Rivers State, Nigeria

Ebere Okereke Africa Centres for Disease Control and Prevention (Africa CDC), Addis Ababa, Ethiopia; Tony Blair Institute for Global Change, London, United Kingdom

Saheedat Olatinwo Rollins School of Public Health, Emory University, Atlanta, GA, United States

Abbas Omaar World Health Organization (WHO), Geneva, Switzerland

Bernard Owusu Agyare Center for Global Health Science and Security, Georgetown University, Washington, DC, United States

Oyeronke Oyebanji Coalition for Epidemic Preparedness Innovations, London, United Kingdom

Aly Passanante Department of Infectious Disease Epidemiology, London School of Hygiene & Tropical Medicine, London, United Kingdom

Krishna Patel Rollins School of Public Health, Emory University, Atlanta, GA, United States

Pauline Paterson Department of Infectious Disease Epidemiology, London School of Hygiene & Tropical Medicine, London, United Kingdom; Health Protection Research Unit in Vaccines and Immunisation, London School of Hygiene & Tropical Medicine, London, United Kingdom; Grantham Institute - Climate Change and the Environment, Imperial College London, London, United Kingdom

Lucy A. Perrone University of British Columbia, Vancouver, BC, Canada; University of Washington, Seattle, WA, United States

Ed Pertwee Department of Infectious Disease Epidemiology, London School of Hygiene & Tropical Medicine, London, United Kingdom

Risa Pesapane Department of Veterinary Preventive Medicine, College of Veterinary Medicine, The Ohio State University, Columbus, OH, United States; School of Environment and Natural Resources, College of Food Agricultural and Environmental Science, The Ohio State University, Columbus, OH, United States

Marjorie P. Pollack PROMED, Brooklyn, NY, United States

Nancy Puttkammer Department of Global Health, University of Washington, Seattle, WA, United States

Ankur Rakesh WHO Headquarters, World Health Organization, Geneva, Switzerland

Ryan Rego Center for Global Health Equity, University of Michigan, Ann Arbor, MI, United States

Natalie Rhodes School of Politics and International Studies, University of Leeds, Leeds, United Kingdom

Waldo Vieyra Romero Mexican Institute for Social Security, Mexico City, Mexico

Randa K. Saad Center for Excellence for Applied Epidemiology, Eastern Mediterranean Public Health Network (EMPHNET), Amman, Jordan

Eman Mukhtar Nasr Salih Hubert H. Humphrey Fellowship Program, Rollins School of Public Health, Emory University, Atlanta, GA, United States

Julian Salim Hubert Department of Global Health, Rollins School of Public Health, Emory University, Atlanta, GA, United States

Affan T. Shaikh Hubert Department of Global Health, Rollins School of Public Health, Emory University, Atlanta, GA, United States; Yale University, New Haven, CT, United States

Valerie A. Paz Soldan Department of Tropical Medicine, Tulane School of Public Health and Tropical Medicine, New Orleans, LA, United States

Romina Stelter World Health Organization (WHO), Geneva, Switzerland

Laura C. Streichert Public Health Consultant, San Diego, CA, United States

Tatiana Sulesco Laboratory of Entomology, Institute of Zoology, Chisinau, Moldova

Rana Sulieman Rollins School of Public Health, Emory University, Atlanta, GA, United States

Ludy Suryantoro World Health Organization (WHO), Geneva, Switzerland

Satoru Suzuki Center for Marine Environmental Studies, Ehime University, Matsuyama, Ehime, Japan

Samantha Swisher Department of Veterinary Preventive Medicine, College of Veterinary Medicine, The Ohio State University, Columbus, OH, United States

Siddhartha Thakur Department of Population Health and Pathobiology, North Carolina State University, Raleigh, NC, United States

Gonzalo M. Vazquez-Prokopec Department of Environmental Sciences, Emory University, Atlanta, GA, United States

Rafael F.C. Vieira Department of Public Health Sciences, The University of North Carolina at Charlotte, Charlotte, NC, United States; Center for Computational Intelligence to Predict Health and Environmental Risks (CIPHER), The University of North Carolina at Charlotte, Charlotte, NC, United States

Antonio R. Vieira Bacterial Special Pathogens Branch, Division of High-Consequence Pathogens and Pathology, National Center for Emerging and Zoonotic Infectious Diseases, Centers for Disease Control and Prevention, Atlanta, GA, United States

Phiona Vumbugwa Department of Global Health, University of Washington, Seattle, WA, United States

Sejal Waghray Hubert Department of Global Health, Rollins School of Public Health, Emory University, Atlanta, GA, United States

Shu-Hua Wang Global One Health initiative (GOHi), The Ohio State University, Columbus, OH, United States; Division of Infectious Diseases, Department of Internal Medicine, The Ohio State University, Columbus, OH, United States

Xiaochun Wang Center for Global Public Health, Chinese Center for Disease Control and Prevention, Beijing, China

Tewodros A. Wassie International Health Consultancy, LLC, Atlanta, GA, United States

Allison Watson Emory University, Atlanta, GA, United States

Nyri Safiya Wells Hubert Department of Global Health, Rollins School of Public Health, Emory University, Atlanta, GA, United States

Koren Wolman-Tardy Mérieux Foundation USA, Washington, DC, United States

Tadesse Wuhib Division of Global HIV and TB, Global Health Center, United States Centers for Disease Control and Prevention, Atlanta, GA, United States

Getnet Yimer Ohio State-Global One Health, Addis Ababa, Ethiopia; Department of Genetics, Perelman School of Medicine, and Penn Center for Global Genomics & Health Equity, Perelman School of Medicine, University of Pennsylvania, Philadelphia, PA, United States

Terence R. Zagar Independent Contractor seconded to the Division of Global HIV and TB, Global Health Center, United States Centers for Disease Control and Prevention, Atlanta, GA, United States

Rodrigo Zepeda Tello Mexican Institute for Social Security, Mexico City, Mexico

Lanyue Zhang Rollins School of Public Health, Emory University, Atlanta, GA, United States

Foreword

"It has been a pleasure to read this book, which is a fantastic collection of evidence and insights from an impressive list of global experts on strengthening global health security. We now have an opportunity as a public health family to seize the moment and support new global health initiatives which capitalize on all we have learned through the COVID-19 pandemic. We must now act to strengthen prevention, detection, and response to public health hazards, transforming public health and health security.

The COVID-19 pandemic has, as I write this, claimed the lives of nearly six million people. Its catastrophic impact on the public's health over the past 2 years has highlighted the need for transformative change of global, national, and local public health systems for both transmissible and non-transmissible diseases. While we had for many years been anticipating and preparing for a global pandemic, there were huge gaps in our preparedness, and the impact has been massive for all countries. There has been significant learning as a result of our experience. We have had the opportunity to identify the strengths and weaknesses of existing policies, regulations, and approaches and we have compared our different responses and seen and better understood the impact of global and national inequalities. While we had insight into the potential consequences of a pandemic, with past large-scale epidemics of SARS, Ebola, and Zika reminding us of the economic, social, humanitarian, and political consequences of pandemics, the scale of disruption wrought by COVID-19 has opened our eyes wider to the risks we face, not only to infectious disease threats but to other hazards including other public health threats, infrastructure failings and related economic impacts.

In my role with IANPHI (the International Association of National Public Health Institutes), I have engaged with the leaders of National Public Health Institutes globally, listened to their concerns, and the global network has learnt from this collective experience. Many of the voices of those who have played a major role in the pandemic response are captured in this publication. The key partnerships that have been forged, with the leadership of the World Health Organization (WHO) and with an impressive range of partners from the public, private, academic sectors, and civil society organizations, are fundamental to this. Key elements of the knowledge gained are captured in the following pages, which explore the global health security architecture, the growing evidence base for more effective action, and describe the operational experience which must inform future action. It is through this process of collaboration and sharing of experience that we will build preparedness and epidemic and pandemic response capacity for the future.

We have important choices to make as a global community. We urgently need to recognize that until the entire world is safe, none of us are safe. We must now work together to recognize and understand the significant weaknesses in our global public health systems and seek to better our collective capacity to respond to future threats.

The COVID-19 response has prompted unprecedented international scientific collaboration in support of developing diagnostic tests for COVID-19, vaccines, and life-saving treatments for patients with COVID-19. We have seen valuable international collaboration and sharing of information and experiences between countries, which have provided a deeper understanding of the nature of the virus and its mutations and we must celebrate and build on the peer-to-peer exchanges that have made these achievements possible. No one country or region has the solution to the challenges we all face and every country needs to engage in learning from others and sharing experience. To build back our national systems better, and create inclusive, resilient health systems everywhere, we must have true global collaboration.

There have also been significant failures in national and international responses that we must acknowledge and learn from. The WHO has already warned the world of the consequences of the unequal global distribution of vaccines, an inequality that has contributed to the emergence of new COVID-19 variants that threaten the global recovery. The failure to share vaccines, vaccine patents, and production capacity cannot be allowed to persist. It leaves large groups of people, including health workers, vulnerable in many countries and delays the point at which the whole world can begin to feel more secure. We must work together through a fair and equitable system to support access to vaccines and production capacity. In the same vein, we must also collaborate to address inequity in access to data, research resources, and support for the development of diagnostics, treatments, medical goods, and personal protective equipment. We must work together as a global public health community to counter the misinformation that surrounds vaccine hesitancy and damages the public's trust in vaccination and public health institutes.

The link between COVID-19 morbidity and mortality and noncommunicable disease to obesity, hypertension, and diabetes highlights the need for global health security to also be linked to broader health system strengthening. The close link between human health and animal health and the environment also demands that our responses need to strengthen 'One Health' collaboration, ensuring we look beyond the health sector to the broader determinants of health in developing stronger global health security. We need an expanded and well-trained public health workforce and effective National Public Health Institutes capable of delivering the full range of public health services, we need strong academic and private sector partnerships, and we need more effective intersectoral collaboration.

The opportunity that we have now as a public health family is to embrace the energy, commitment, and ideas of the authors of this book and build a strong system from the new global health initiatives that have emerged. We need to advocate and win the argument for greater and sustained investment in public health and decisively break the historical cycle of "panic and forget" that has characterised government spending decisions. Perhaps best expressed by "show me your budget and I will know what you care about."

Investing in public health is the smartest thing we can do—good health underpins a strong economy. Up until now, efforts to invest in pandemic preparedness have fallen short. We need to be ready for the next health emergency. We are stronger together."

With best wishes,
Professor Duncan Selbie
President, International Association of National Public Health Institutes
Chief Adviser, Public Health Authority of Saudi Arabia

Preface

This book describes the gaps (or holes) in the global public health and healthcare fabrics revealed during the 2019–23 COVID-19 pandemic with a vision and practical actions on how to repair them. Highlighting academically solid, scientific guidance, in close collaboration and coauthorship with the World Health Organization (WHO), over 120 international authors—plus 20 Emory University, MPH student managers—generously offered their time and talent to chart a pathway to a modernized global health security framework. What went wrong? How do we fix it, so another pandemic doesn't happen?

We assimilate the background, analyses, case studies, vision, and practical actions to aid and assist WHO, guide global leaders, and help train the future public health and healthcare workforce. Covering a broad range of topics (chapters), stratified by health level and perspective, the authors capture the voice and experiences of frontline workers battling today's greatest public health and healthcare challenges.

The authors worked very, very hard during the 3-year COVID-19 pandemic to achieve this goal. And they faced extraordinary personal and professional hardships (e.g., sickness with COVID-19, heart attacks, family emergencies, professional challenges serving their communities). With this generous offering of time and talent, authors learned and practiced understanding, compassion, kindness, and faithfulness. The authors are inspiring; and science is honored by their efforts.

Scott J.N. McNabb

Acknowledgments

This book was inspired by the *WHO Vision of a World* in which all peoples attain the greatest possible level of health, while keeping the world safe and serving the vulnerable. Recent outbreaks of MERS-CoV, Zika, and Ebola showed more was needed to improve Global Health Security. From this, the book began to take shape. However, as the COVID-19 pandemic gripped the world and our authors, leading epidemiologists, laboratorians, front-line workers, and public health professionals faced unexpected and unprecedented workplace and personal challenges and the need to strengthen global health security intensified. The editorial team was reinvigorated to bring this book to fruition, while delicately balancing the commitments and availabilities of now highly, in-demand experts.

Throughout the world, COVID-19 challenged our capacity to prevent, detect, and respond to emerging disease outbreaks; it exposed significant weaknesses and major gaps in Global Health Security. Collectively, the scientific community was shaken. At the same time, the pandemic unveiled the beautiful hope in the community that lessons learned could improve Global Health Security. Answering the following two questions drove the vision and development of this book: What went wrong? And how do we fix this, so this doesn't happen again?

Critically understanding the COVID-19 pandemic while still amid this public health emergency was no easy task. The authors looked at historical and emerging evidence plus the hard-learned lessons from the recent past. They combined this review with their subject matter expertise to present recommendations to the global community. Each author team further built case studies that underscore valuable lessons.

The development and production of this book is only possible through the hard work, expertise, patience, support, and effort of many people, including authors, editors, publishers, family, and friends; far too many to list here. Please know that we are truly grateful to all of you. The generosity, commitment to the good of the community, and selflessness from over 100 scientific experts and students eager to make a difference are reflected in the pages that follow.

There were no monetary incentives; no funder provided support for the extraordinary, 3-year commitment of time and energy by the many teams of authors and editors. And this central grace makes this book genuine, pure, powerful, and uniquely extraordinary.

Written with the intent to fully support the mission of the World Health Organization and of any other entities that wish to make use of the vision and suggestions provided, all royalties from the sale of this book will go to charity.

We also acknowledge you—our readers—for considering the ideas presented here and for using them to seed your own efforts to meet your goals toward modernizing global health security to prevent, detect, and respond.

Warmly—Scott, Affan, and Carol

List of abbreviations

AAR	After Action Review
ACE2	human angiotensin-converting enzyme 2
ACT-A	Access to COVID-19 Tools Accelerator
AEs	Adverse Effects
Ae.	Aedes
Afrexim Bank	African Export–Import Bank
Africa CDC	Africa Centres for Disease Control and Prevention
AFROHUN	Africa One Health University Network
AFTCOR	African Task Force on Coronavirus
AI	Artificial Intelligence
AIDS	Acquired immunodeficiency syndrome
AMR	Antimicrobial Resistance
AMSP	Africa Medical Supplies Platform
AMU	Antimicrobial Use
ATL	Atlanta, Georgia, United States—Hartsfield International (Airport Code)
AU	African Union
AVAT	African Vaccine Acquisition Trust
AVATT	African Vaccine Acquisition Task Team
AVDA	African Vaccine Delivery Alliance
CCoP	Clinical Community of Practice
CDC	Centers for Disease Control and Prevention
CDPH	Chicago Department of Public Health
CEPI	Coalition for Epidemic Preparedness Innovations
CESCR	International Covenant on Economic, Social and Cultural Rights
CGH	Common Goods for Health
COCOA	COVID-19 Contact-Confirming Application
CONCVACT	Consortium for COVID-19 Clinical Vaccine Trials
CORDS	Connecting Organizations for Regional Disease Surveillance
COVAX	COVID-19 Vaccine Global Access Facility
COVAX AMC	COVAX-Advance Market Commitment
COVID-19	Coronavirus Disease (2019)
CRC	Coronavirus Resource Center
CSOs	Civil Society Organizations
DARWIN	Data Analysis and Real-World Interrogation Network
dd	Digital Droplet
DHIS2	District Health Information System
DPM	Dynamic Preparedness Matrix
DRC	Democratic Republic of Congo

DRR	Disaster Risk Reduction
EAR	Early Alerting and Reporting platform
ECA	United Nations Economic Commission for Africa
ECDC	European Centre for Disease Prevention and Control
EHDS	European Health Data Space
eHeath literacy	digital health literacy
EHEC	Enterohemorrhagic *E. coli*
EIDs	Emerging infectious diseases
EIEC	Enteroinvasive *E. coli*
EIOS	Epidemic Intelligence from Open Sources
EIR	Electronic Immunization registries
EMA	European Medicines Agency
EMM	European Media Monitor
EPEC	Enteropathogenic *E. coli*
EPHFs	Essential public health functions
ETEC	Enterotoxigenic *E. coli*
EU	European Union
EVD	Ebola Virus Disease
FAO	Food and Agriculture Organization of the United Nations
FCTC	Framework Convention on Tobacco Control
FDA	Food and Drug Administration
FETP	Field Epidemiology Training Program
Gavi	The Vaccine Alliance
GDP	Gross Domestic Product
GFATM	Global Fund to Fight AIDS, Tuberculosis and Malaria
GHC	Global Health Community
GHS	Global Health Security
GHSA	Global Health Security Agenda
GHSI	Global Health Security Initiative
GHSI	Global Health Security Index
GI	Gastrointestinal
GISAID	Global Initiative on Sharing Avian Influenza Data
GISRS	Global Influenza Surveillance and Response System
GLEWS	Global Early Warning and Response System
GloPID-R	Global Research Collaboration for Infectious Disease Research
GOARN	Global Outbreak Alert and Response Network
GPH	Global Public Health
GPHIN	Global Public Health Intelligence Network
GPS	Global positioning systems
H1N1	Novel influenza A
H1N1	species Influenza A virus of the genus Influenzavirus A
H1N1	"H1N1" Hemagglutinin 1, Neuraminidase 1, proteins found on the envelope of the influenza virus
HACCP	Hazard Analysis Critical Control Points
HCV	Hepatitis C virus

HEDRM	Health Emergency Disaster Risk Management
HHS	Health and Human Services
HIC	High-income country
HiC	Health information and communication
HICs	High Income Countries
HIS	Hospital Safety Index
HIV	Human Immunodeficiency Virus
HIV/AIDS	Human immunodeficiency virus infection/acquired immunodeficiency syndrome
HSforHS	Health Systems for Health Security framework
HSS	Health System Strengthening
HUS	Hemolytic Uremic Syndrome
IAEA	International Atomic Energy Agency
IANPHI	International Association of National Public Health Institutes
IARs	Intraaction Reviews
ICAO	International Civil Aviation Organization
ICD	International Classification of Disease
ICT	Information and Communication Technology
ICVP	International Certificate of Vaccination or Prophylaxis
IDSR	Integration Disease Surveillance and Response
IHR	International Health Regulations
IHR	International Health Regulations (IHR-2005)
IHR MEF	IHR Monitoring and Evaluation Framework
IHR-RC	IHR Review Committee
IIS	Immunization Information Systems
IL	Interleukin
IMO	International Maritime Organization
IMSS	Instituto Mexicano del Seguro Social
IOAC	Independent Oversight and Advisory Committee
IOM	International Organization for Migration
IoT	Internet of Things
IPPPR	Independent Panel for Pandemic Preparedness and Response
ISC	International Sanitary Convention
ISR	International Sanitary Regulations
IT	Information Technology
IWD	Institute for Workforce Development
JDN	Juventude ba Dezenvolvimentu Násional
JEE	Joint External Evaluation
JFK	New York, New York, United States—John F. Kennedy International Airport (Airport Code)
KSA	Kingdom of Saudi Arabia
LAC	Latin America and Caribbean
LAMP	Loop-mediated Amplification
LAX	Los Angeles, California, United States—Los Angeles International Airport (Airport Code)

LIC	Low-income country
LMIC	Low- and Middle-Income Country
LOINC	Logical Observations, Identifiers, Names and Codes
MADLI-TOF	Matrix-assisted Laser Desorption/Ionization with Time of Flight
MEDISYS	Medical Information System
MEF	Monitoring and Evaluation Framework
MERS	Middle East Respiratory Syndrome
MERS-CoV	Middle East respiratory syndrome coronavirus
MoAF	Ministry of Agriculture and Fishery
MoH	Ministry of Health
MPC	Multisectoral Preparedness Coordination
mRNA	Messenger Ribonucleic Acid
MS	Member States
NAPHS	National Action Plan for Health Security
NBW	National Bridging Workshop
NCDs	Noncommunicable Diseases
NCITs	Noncontact infrared thermometers
NFP	National Focal Point
NGO	Non-Governmental Organization
NIMS	National Immunization Management System
NOHP	National One Health Platform
NoV	Norovirus
NPC	National Pharmacovigilance Center
NPHI	National Public Health Institutes
NPI	Nonpharmaceutical intervention
OECD	Organization for Economic Cooperation and Development
OHHLEP	One Health High-Level Expert Panel
OIE	World Organization for Animal Health
ORD	Chicago, Illinois, United States—Chicago O'Hare International Airport (Airport Code)
PACT	Partnership to Accelerate COVID-19 Testing
PAHO	Pan American Health Organization
PAVM	Partnership for African Vaccine Manufacturing
PCR	Point-of-Care Tests
PCR	Polymerase Chain Reaction
PEI	Polio Eradication Initiative
PH	Public Health
PHC	Primary Health Care
PHEIC	Public Health Emergency of International Concern
PHI	Public Health Institute
PHS	Public health surveillance
PIP	Pandemic Influenza Preparedness
PoE	Point of Entry
PPE	Personal protective equipment

PRC	Prevention, Retention, and Contingency
ProMED	Program for Monitoring Emerging Diseases
PTF	Presidential Task Force
PVS	Performance of Veterinary Services
qPCR	Quantitative Polymerase Chain Reaction
RC	Review Committee
RCTs	Randomized controlled trials
RDTs	Rapid Diagnostic Tests
RWICA	Rural Water Initiative for Climate Action, Ltd.
SARS	Severe Acute Respiratory Syndrome
SARS-CoV-1	severe acute respiratory syndrome coronavirus 1
SARS-CoV-2	Severe Acute Respiratory Syndrome Coronavirus 2
SDGs	Sustainable Development Goals
SFDA	Saudi Food and Drug Authority
SFO	San Francisco, California, United States—San Francisco International Airport (Airport Code)
SimEx	Simulation Exercises
SMART	Spatial Monitoring and Reporting Tool
SMEs	Small to medium-sized enterprises
SPAR	State Party Annual Reporting
STAR	Strategic Tool for Assessing Risks
STDC	Short Term Disability Claims
STEC	Shiga Toxin-producing *Escherichia coli*
TAG	Technical Advisory Group
TB	Tuberculosis
TBRF	Tick-borne Relapsing Fever
TEPHINET	The Training Programs in Epidemiology and Public Health Interventions Network
TMP-SMX	Trimethoprim-sulfamethoxazole
TSCs	Thermal scanner cameras
TWG	Technical Working Group
TZG	Tripartite Zoonoses Guide
U.S.	United States
UHC	Universal Health Coverage
UHPR	Universal Health and Preparedness Review
UK	United Kingdom
UN	United Nations
UNEP	the United Nations Environment Programme
UNICEF	United Nations Children's Fund
UPPR	Universal Periodic Peer Review
US	United States
US CDC	United States Centers for Disease Control and Prevention
USA	United States of America
USAID	United States Agency for International Development
VAERS	Vaccine Adverse Event Reporting System

VBD	Vector-borne disease
VSD	Vaccine Safety Datalink
VUCA	Volatility, Uncertainty, Complexity, and Ambiguity
WHA	World Health Assembly
WHO	The World Health Organization
WNV	West Nile Virus
WOAH	World Organisation for Animal Health
WTO	World Trade Organization
ZCS	Zika Congenital Syndrome

CHAPTER

1

Vision guiding modernization of global health security

Scott J.N. McNabb[1], *Stella Chungong*[2], *Affan T. Shaikh*[1,3], *Mohamed Moussif*[4], *Ann Marie Kimball*[5,6], *Carol J. Haley*[1] *and Sejal Waghray*[1]

[1]Hubert Department of Global Health, Rollins School of Public Health, Emory University, Atlanta, GA, United States; [2]Health Security Preparedness Department, WHO Health Emergencies Programme, World Health Organization, Geneva, Switzerland; [3]Yale University, New Haven, CT, United States; [4]Casablanca International Airport, Ministry of Health, Casablanca, Morocco; [5]Department of Epidemiology, University of Washington, Bainbridge Island, WA, United States; [6]Chatham House, London, United Kingdom

In the 21st century, Global Health Security (GHS) has become not just confined to protecting human health, but also to guarding the human interface of animal health and environmental security. This interface—called One Health—is multisectoral and multidisciplinary and addresses natural and human activity in the diverse domains that influence public health. At the same time, recent globalization of laboratory capabilities has created risk. However, as recent history demonstrates, GHS must also include legislative, regulatory, political, and economic well-being; cultural and gender equity; active engagement by local communities; and ethical and uniform adaptation of information and communication technology (ICT).

As such, GHS optimizes public health and wellness in a **whole of society** approach. COVID-19 reinforced awareness of the critical requirement to coordinate preventive, preparatory, and responsive national and international policies and actions beyond just human health. Insight and harmonization across all national sectors are required. This book explores these critical linkages: where they exist and where and how they must be strengthened.

At the Member State (MS) level, the World Health Organization (WHO)—in its role as the United Nations (UN) agency for health—provides technical and assessment guidance for implementing the International Health Regulations (IHR) (2005). And the IHR are currently undergoing targeted amendments, taking into account lessons learned from the COVID-19

Modernizing Global Health Security to Prevent, Detect, and Respond
https://doi.org/10.1016/B978-0-323-90945-7.00027-0

pandemic and other recent epidemics. This ongoing work must include all sectors so MS can better understand and embrace how the **whole of society's** preparedness and response should be actualized. Harmonization of existing policies for emergencies and pandemics is essential. Such harmonization must include the private sector, civil society, and other key stakeholders. COVID-19 also revealed that manufacturing, shipping, transportation, finance, agriculture, and other industrial and commercial entities are critical in the overall global response to pandemic(s).

There are calls from those in many sectors for a new paradigm; one that demands a closer look at global, regional, and MS-level architecture for health emergency prevention, preparedness, response, and resilience. As part of the discussion on global architecture, there are ongoing intergovernmental negotiations on a pandemic accord and on revisions to the IHR (2005). These include the need for strong governance, resilient systems, plus sustainable and predictable financing; all underpinned by principles of equity, solidarity, and coherence. Community trust is key, with increased transparence and accountability.

Paradigm shifts must occur in the diverse settings of human endeavor—from local market regulation to global trade regimes; and from rural to urban settings. With the rapid rate of political and societal change over which COVID-19 settled and with expected continued rapid change over the next decades and beyond, consideration of the stability and robustness of norms, as well as emergent and innovative, influences present challenges to effective pandemic policy.

These approaches now become increasingly important as the world faces more disease outbreaks, pandemics, natural disasters, and humanitarian crises. For example, unless properly managed, wet markets create the perfect ecosystem for viruses to emerge, interact, persist, and mutate. The severe acute respiratory syndrome (SARS) coronavirus pandemic of 2002 is thought to have started from Himalayan palm civets (Paguma *larvata*) in such a market in Foshan, Guangdong, China.[1,2]

Wet markets play an important role in food security for many communities, and disrupting critical food supply chains can stimulate an unregulated black market for animal products. Yet wet markets represent just one node of zoonotic transmission. And as the global population of the world passes 8 billion (almost 30% living in cities)[a], other animal–human interfaces including those that have long existed (e.g., among city-dwelling animal populations such as rats and birds and newer interfaces brought about by factors such as human-habitat encroachment, hunting, hanrvesting, behavioral adaptations, and climate change) become increasingly important.

Modernizing GHS requires a new framework to fully achieve prevention, detection, and response capabilities to mitigate (re)emerging public health threats. This new framework posits GHS in its proper relation to political, cultural, One Health, economic, individual community, and ICT dimensions. These relations are reviewed and evaluated; and—with a proposed vision—needed steps are outlined in the chapters of the five sections of this book (Fig. 1.1).

This book explores the history of GHS and its links to ethical, governance, and global legal structures. It explores how GHS has performed (or not) in the past to prevent, detect, and respond to (re)emerging health threats. Of course GHS is only as effective as local public

[a] https://www.worldometers.info/world-population/.

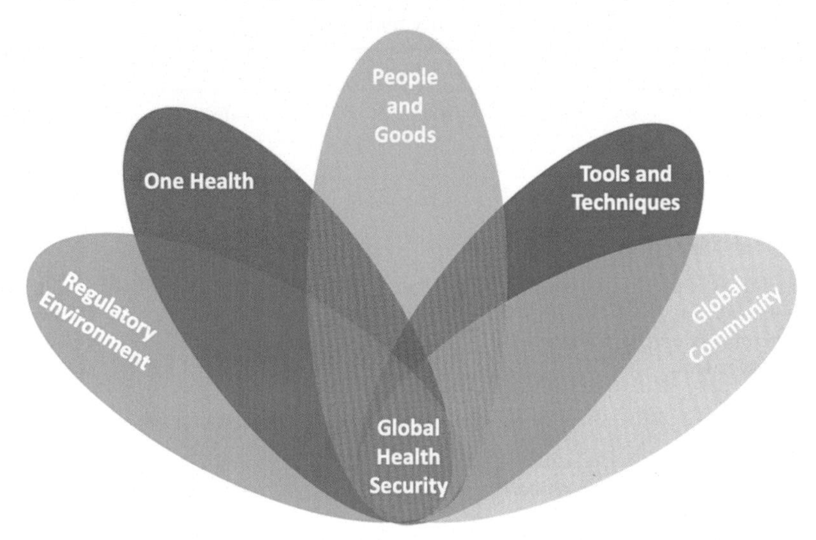

FIGURE 1.1 Global health security intersectionality framework.[b]

health; pandemics exploit the weakest of these linked systems to transmit and amplify negative health outcomes.

This book presents the state of current public health environments with the understanding that outbreaks are expected, but pandemics are preventable.[3] The authors highlight the reality (gaps) we now face, a vision we should achieve, and the challenges (actions needed) to move to a better, modernized GHS. This volume is the product of intense research to bring forward the most up to date and cutting-edge insights for pandemic preparedness and response. Case studies in each chapter illustrate lessons learned—what worked and didn't work—as well as recommendations.

Realizing together that we have both the obligation and opportunity to support the global health community (GHC) by providing a thoughtful, thorough, and diverse scientific review and analyses of the gaps and impediments to GHS, the authors give the background, vision, and practical actions to assist national and international public health institutions, guide global health leaders, plus train the future public health workforce. Our intent is to provide a solid, scientific review to modernize GHS while suggesting actionable goals at various governance tiers.

GHS can be influenced at many levels. Effective and efficient GHS is not just the responsibility of governments and international organizations; it involves everyone—from individuals to communities at all levels of society. Every person has a role to play to ensure we are best prepared to prevent, detect, and respond to global health threats. By working together and taking action at all levels, we can strengthen our collective ability to protect global health (Table 1.1).

[b] At the core of Fig. 1.1 is GHS. Each petal branching off and overlapping indicates the different subjects that must be a part of a new GHS framework. Furthermore, the lotuslike shape is indicative of the journey to modernizing GHS as the flower is a symbol of strength and resilience.

TABLE 1.1 Levels of global health security influence and theories of change.

Level	Definition	Stakeholders	Examples	Influence	Theory of change
Individual	Persons unique from others, possessing one's own needs, goals, rights, and responsibilities	Individual	Self	Lifestyle behaviors (e.g., mask-wearing, staying informed)	Build trust and transparency to ensure interventions are successful
Community	Social units with commonality (e.g., geography, norms, religion, values, customs, or identity)	Family and friends; close neighbors; local businesses	Religious and cultural groups; special interests; clubs; informal local institutions	Upholding GHS norms; staying informed; countering spread of misinformation	Identify key changemakers to help build trust and ensure local buy-in
District	Lower tiers of public administration within a member state	Local leaders; Health officers	Province; Township; Village; District; County	Tailoring strategies from national level to meet district needs	Understand specific needs and circumstances to help bridge global best practices to local successes
National	Highest tiers of public administration within a member state	National leaders	President; Prime minister; Heads of ministries (e.g., minister of health, minister of agriculture, ministry of education, secretary of state)	Implementing strategies from regional and international tiers	Invest in building critical GHS capacities to create robust systems and people to prevent, detect, and respond
Regional	Governance of a specific area and domain by institutions given authority by member states	Supranational organizations; Regional organizations	European union; World Health Organization regional offices; United nations regional offices (e.g., FAO and OIE); International Association of National Public Health Institutes	Adopting strategies on GHS implementation (from international tier)	Build cross-cutting collaborative networks to support knowledge sharing
International	Transnational organization of nations having similar goals and objectives	Intergovernmental organizations	United Nations World Health Organization	Global framework; setting strategies on GHS implementation	Revise global frameworks to become relevant to evolving societies and advancing technology to create guiding principles that can usher positive change

Why modernize GHS now to prevent, detect, and respond?

GHS is constantly under threat by a range of hazards including outbreaks of (re)emerging infectious diseases; natural disasters; and biological, chemical, or radiation events. Despite considerable measures to ensure prevention and strengthen national preparedness, health emergencies continue to negatively impact every region of the world. COVID-19 has shown that national governments, societies, and the global multilateral systems are currently ill-equipped to deal effectively with severe epidemics and pandemics. It is more critical now than ever to ensure that the world is prepared to prevent, detect, and respond to global health threats and safeguard the health and well-being of all people.

The post COVID-19 world is vulnerable to the next pandemic. In recent years, we have seen a rise in the frequency of global health threats. Deforestation with encroachment and the loss of animal habitat, urbanization with increased density of human populations, and climate change with attendant extreme weather and warming of biological systems contribute to the (re)emergence of disease. Health systems under COVID buckled, directing attention to how societies assure access to care. Further, the world is more interconnected than ever, with increased travel and trade among countries. Diseases can spread quickly and easily across borders.

Beyond saving lives, investing in effective GHS has significant economic benefits. Preventing and controlling outbreaks of infectious diseases can help avoid the economic costs of lost productivity, trade restrictions, and travel bans. Historically, such losses have figured in the billions of dollars. On the positive side, advances in technology, such as artificial intelligence (AI), big data analytics, and mobile health applications, have potential to transform how we prevent, detect, and respond to global health threats. Modernizing GHS will ensure that we take advantage of these technologies to ensure that countries are better prepared to detect and respond to (re)emerging diseases before they become global health emergencies.

The WHO first learned on December 31, 2019, of what would soon become the most significant global health crisis in a century.[4]

> WHO's Country Office in the People's Republic of China picked up a media statement by the Wuhan Municipal Health Commission from their website on cases of 'viral pneumonia' in Wuhan, People's Republic of China. The Country Office notified the International Health Regulations (IHR)[c] focal point in the WHO Western Pacific Regional Office about the Wuhan Municipal Health Commission media statement of the cases and provided a translation of it. WHO's Epidemic Intelligence from Open Sources (EIOS) platform also picked up a media report on ProMED[d] … about the same cluster of cases of 'pneumonia of unknown cause', in Wuhan.[4]

On January 30, 2020, the Director-General of the WHO declared the novel coronavirus outbreak a Public Health Emergency of International Concern (PHEIC), WHO's highest level of alarm.[5] This PHEIC pronouncement ended on May 5, 2023. To communicate the current status, the WHO's Coronavirus Dashboard provides a real-time update on the number of confirmed cases, deaths, and vaccines administered worldwide.[6]

[c] In 2005, the World Health Assembly adopted the International Health Regulations (IHR, 2005) as the primary international legally binding instrument coordinating the public health response to major multinational outbreaks.[4]

[d] https://promedmail.org/.

COVID-19 challenged our capacity to prevent, detect, and respond to emerging outbreaks. Collectively, the scientific community was shaken by this catastrophe, and answering the following two questions drove the vision and development of this book: *What went wrong? How do we fix this, so this doesn't happen again?*

The COVID-19 pandemic exposed significant weaknesses and major gaps in GHS. Covering a broad range of GHS areas (in 27 Chapters) by various health levels (local, district, national, and international) and perspectives, this book is organized into five Sections.

1. International Regulatory Environment to Prevent, Detect, and Respond and Importance of a Global Point of View
2. Global One Health to Address Pandemics—Ecological and Biological Challenges on Our Dynamic Planet
3. People and Goods on the Move
4. Tools and Techniques to Modernize Prevention, Detection, and Response
5. Moving to the Best-protected Global Community

Capturing the voice and experience of frontline workers plus global leadership battling today's greatest public health challenges, we present a practical vision of what the world could (should) do and be if we address current shortfalls.

We begin by understanding how the world has changed through ...

- rapid, human transportation and communication.
- urbanization.
- mass gatherings of people.
- immediacy of human, avian, and mammalian animal interactions due to expanding human development.
- climate change.

And we explore the increase in modern-day threats through ...

- (re)emerging diseases.
- environmental catastrophes.
- biosecurity.
- politics, conflict, and human rights.
- lifestyle choices.

We also acknowledge the broad-reaching global changes and formidable challenges in the wake of COVID-19 (e.g., impediments to supply chains; inequities in vaccine distribution; greater need for universal healthcare and health system strengthening; increased social isolation; exponential growth in e-commerce; socioeconomic divides growing wider; demands for health-related expertise; need for new ICT and artificial intelligence [AI] access and strategies; and concerns over data sharing and privacy).

Recalling an inspiring proverb, *It's always darkest before the dawn*[e], we know that things always seem the worst just before they improve.[7] Moreover, late American singer and social

[e] The first person to use this proverb was Thomas Fuller, an Englishman, in the year 1650. It appeared in his work titled *A Pisgah-Sight of Palestine and the Confines Thereof.*

activist, Peter Seeger, introduced powerful lyrics embedded in his song, Turn! Turn! Turn!, that encourages and reminds us that we must change (or turn) for the better.

> To everything (turn, turn, turn)
>
> There is a season (turn, turn, turn)
>
> And a time to every purpose, under heaven.[8]

This is the season for GHS to *turn* to the better. In fact, in the interest of future generations, it is imperative.

History of pandemics?

Throughout time, preventative GHS measures have fallen short of stopping major pandemics. And, in spite of modern improvements in public health and medicine we continue to fall short; the morbidity and mortality of recent pandemics tell this story (Fig. 1.2).

Recorded history of pandemics began in the 14th century with the bubonic plague, followed by smallpox in the 16th century, cholera in the 19th century, Spanish flu in 1918, the 1957 flu, 1967 Hong Kong flu, and AIDS in 1980s. We have chosen to limit public health history of

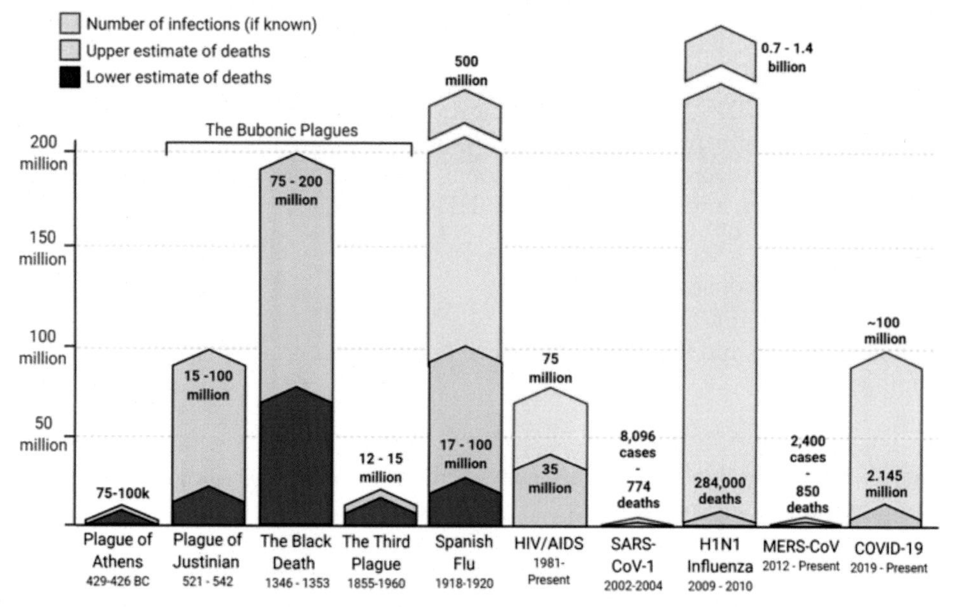

FIGURE 1.2 Historical timeline of major pandemics showing number of infections and upper and lower estimates of deaths.[9,f]

[f] An updated present record of COVID-19 deaths can be found at the WHO's Coronavirus Dashboard.

pandemics to the 1918 influenza pandemic, extending to the discovery of Ebola in 1976 and subsequent outbreaks: the 2003 SARS pandemic that stimulated the 2005 revision of the IHRs; the 2009 H1N1 pandemic and first global crisis formally designated a PHEIC; the 2014 Ebola pandemic; 2015 MERS-CoV pandemic; 2016 Zika crisis; and the pandemic of COVID-19. Each—in their own way—reflected failures in prevention, detection, and response.

With prior historic pandemics, the timeframe between them was in centuries. However, in recent major pandemics there is a much smaller time gap. These events drove the creation of new programs, with the goal to strengthen preparedness and public health surveillance (PHS) capacities.[10]

While *pandemic* is defined as the prevalence of an infectious disease around the world, *epidemic* is defined as a widespread occurrence (beyond the ordinary) of a disease in a community, at a particular time. *Outbreaks* are defined as the increase above the expected or a sudden or violent start of something unusual, such as war or disease.[11] In the past, novel strains of influenza caused pandemics of illness, death, and economic disruption. The Spanish flu of 1918, Asian flu of 1957, Hong Kong flu of 1968, and H1N1 of 2009 were pandemics with severe morbidity and mortality.[12] The introduction of penicillin and the advent of antibiotics ushered in a period of complacency in the 1970s and 1980s. These agents are not effective against viral threats, and the global pandemic of HIV/AIDS awakened a new appreciation of pandemic risk.[13]

We have witnessed three pandemics in this century associated with novel coronaviruses. Severe Acute Respiratory Syndrome (SARS), the Middle East respiratory syndrome (MERS), and COVID-19 were responsible for causing acute respiratory tract infections (ARTIs) and were highly contagious with significant mortality.[14] Additionally, these pandemics had substantial impacts on the economy of MS and the world.[15]

According to IHR (2005), a "PHEIC" means an extraordinary event that is determined to constitute a public health risk to other MS through the international spread of disease and potentially requires a coordinated international response.[16] Further, under the IHR, the WHO Director-General, after establishing an IHR Emergency Committee, requests its views on ...

(a) whether an event constitutes a PHEIC.
(b) termination of a PHEIC.
(c) proposed issuance, modification, extension, or termination of temporary recommendations.

Of the 10 IHR emergency committees convened by the WHO Director-General, nine have closed, and one is ongoing as of June 2, 2023. Updates and relevant details can be found at the WHO IHR Emergency Committees website.[17]

Among other factors, links to environmental changes, changes in land-use patterns such as deforestation and intensive agriculture have been suggested as causes. While most human infectious diseases are zoonotic (i.e., coming from animals), pathogens transfer from animal to human easily when humans and animals are in close proximity due to deforestation and the extinction of animal habitats.[18,19] Further, social media and ICT have evolved and information is spread rapidly significantly impacting consumer engagement.[20]

What is the future vision?

To strengthen GHS and prepare for future challenges, this book recognizes key factors that address ...

- a unified approach to GHS.
- synchronization of emergency regimens, preparations, and protocols among international and national stakeholders (e.g., in trade and transportation, finance, business, and labor).
- reforming emergency preparedness, readiness, response, and monitoring guidelines, informed by risk.
- greater insight into legal, political, social, and ethical forces at play and investment in health equity.

First and foremost, a unified approach to GHS is crucial to addressing current shortcomings. While GHS has increased in colloquial usage, there is "no universally agreed definition" or approach.[21] As a result, various organizations have chosen to emphasize different parts of a similar definition and focus on GHS in an uncoordinated manner. Definitions utilized by various organizations are summarized in Table 1.2.

Given that public health is interdisciplinary and health preparedness is multisectoral with little standardization, it is important that a universal understanding be established for GHS to, more properly, identify process and outcome indicators and harmonize funding.[25] In fact, in 2014, the Global Health Security Agenda (GHSA) was established to bring together different global health colleagues (ranging from countries to private sector companies) and acknowledge the *essential need for a multilateral and multi-sectoral approach to strengthen … capacity to prevent, detect, and respond to infectious threats.*[26] Further, in 2001, the World Health Assembly (WHA) through Resolution WHA54.14 adopted the Secretariat report (WHA54/9) on *Global Health Security: Epidemic Alert and Response* that established GHS as an action-based plan to address health preparedness internationally.[27,28]

TABLE 1.2 Various sources and defintions of global health security.

Source	GHS definition
World Health Organization	"Global public health security is defined as the activities required, both proactive and reactive, to minimize the danger and impact of acute public health events that endanger people's health across geographical regions and international boundaries."[22]
U.S. Center for Disease Control and Prevention	"Global health security is the existence of strong and resilient public health systems that can prevent, detect, and respond to infectious disease threats, wherever they occur in the world."[23]
Global Health Security Agenda	"… to build a world safe and secure from infectious disease threats whether naturally occurring, accidentally released, or intentional."[24]

International cooperation and regulation are crucial for public health sustainability, and a clear definition of GHS is needed for unified approval and delegation of tasks organized across borders, stakeholders, and donors.[29] This book poses successful organization and delegation for GHS actions as it reviews the distinct roles of various disciplines and sectors for GHS reform.

Second, it is important to revisit international governance instruments such as the IHRs. When IHR (2005) was adopted, Member states (MS) agreed to cooperate and collaborate for early detection, notification, verification, and emergency preparedness during situations of international concern.[30,31] While the IHRs addressed the need to regulate the international health system, recent global health crises have exposed gaps and demonstrated the urgency to update this legal instrument. The 1969 version of the IHR only addressed specific infectious diseases (i.e., cholera, plague, and yellow fever). Infectious disease outbreaks in the Democratic Republic of Congo (Ebola), China (SARs), and the Middle East (MERS-CoV) made it crucial to learn from other events and locations.[32] While the IHR (2005) expanded diseases addressed and enforced on a global level, there is need for additional progress. It is important now to reevaluate the IHR (2005) to address gaps seen by MS during the COVID-19 pandemic. Technologic advancements have also increased the *spread and speed of interconnectivity*. Moreover, the ability to cross national borders more efficiently has *accelerated flow of traded goods*.[33]

The IHRs must address globalization (e.g., data advancement in communication, rapid human transportation, mass gatherings of individuals, immediacy of interaction). A working group constituted by MS to the IHR was recently established and is discussing amendments to the regulations report of the Review Committee regarding amendments to the International Health Regulations (2005).[34] Support for low- and middle-income countries (LMICs); transparency in global health communication; and addressing urbanization are essential for GHS strengthening—a vision shared throughout this book as gaps are identified and addressed.

WHO established foci to invest in essential health functions and the foundations of primary healthcare. By prioritizing international attention to the roots of GHS, we address the limitations of inequities and current disproportionate impacts on marginalized or vulnerable populations. Historically, GHS has centered around an approach where high-income countries (HICs) support LMICs with necessities for controlling infectious diseases. During COVID-19, many LMICs with experience in managing epidemics and investment in essential public health functions were better positioned to manage the pandemic. HIC and LMIC health systems across the board were unable to scale up to deal with a pandemic of the magnitude and severity of COVID-19 and continue providing essential health services.

This calls for investment in health systems resilience allowing health systems to surge to meet the demands imposed by health emergencies.[35] The book addresses the global need to build LMIC health systems. While complex and costly, this is essential to successful GHS. The core value of trust among communities, governments, and health sectors has proven central to GHS. That trust is built through assuring the voice and will of communities.

To provide a unified vision for different sectors, gaps, and approaches to modernize GHS, we address eight key principles for practical, collective, future action.

1. Understand the long arc of history to improve public health prevention, detection, and response;
2. Need of global partnerships; especially including private sector actors and community actors
3. Effective interventions to curtail transmission through travel and international mass gathering events require strong relationships among countries at origin and destination;
4. One Health is rising in prominence but needs broader adoption and consistent inclusion of environmental concerns as well as other sectors beyond the human–animal interface through a multisectoral and whole-of-society approach;
5. Social partnerships, ranging from academia and private sector to parliaments and communities, are crucial;
6. Economic concerns underpin public health decision-making processes; balancing the two while preserving life is critical to modernize GHS;
7. Technologic advancements and innovations present new and exciting opportunities and can enhance GHS; laboratory biosafety and biosecurity is not yet universalized and must be.
8. People at community level with community action remain core to modernizing GHS; training and building networks are fundamental.

The 27 Chapters address these eight principles while recognizing current public health shortcomings and encouraging reflection on long-term strategies to improve GHS.

What's inside?

The book is divided into five sections.

Section I. **International Regulatory Environment to Prevent, Detect, and Respond and Importance of a Global Point of View** explores the necessity of a global perspective to prevent pandemics. It considers the IHRs: what are the goals, and how did they function during the recent pandemic crises of the 2007 H5N1, 2009 H1N1, 2014 Ebola, 2014 MERS-CoV, 2015 Zika, and 2019 COVID-19? Where did the IHR (2005) fail during the COVID-19 pandemic? What are the strengths and weaknesses, and what should be done to improve or update them?

How do we value and enhance ethics? What is the influence of discrimination (i.e., racism, ethnic minorities, extreme nationalism, age) and other inequities in public health practice and as well as consider how nationalism can be balanced with responsible global collaboration? How can the IHRs have an impact on this? How well did the IHRs function; what has been learned? Pandemics pose challenges to existing ethical guidelines. They highlight the limitations of international cooperative agreements (based on money and power) and show what goodwill or selflessness could accomplish.

Existing ethical frameworks are inadequate and may be completely disregarded during pandemic responses. Although the IHRs align to human rights principles, the term *ethics* does not appear in the documents. An entirely new awareness of global imbalances has shown the negative health outcomes based on lopsided money and power. The legal foundation for GHS and pandemic response must be grounded in a modern ethical framework.

Ethics and human rights approaches may share a common purpose, but they rely on different principles and spheres of influence.

What roles do infodemics, weak healthcare, and public health systems play in pandemic vulnerability, and how did these contribute to the most recent pandemic? What are examples of how universal healthcare can be a successful intervention? It explores the benefits and challenges of developing and sustaining multiorganizational and sectoral collaborations and the delicate balance required to maintain GHS.

Section II. **Global One Health to Address Pandemics—Ecological and Biological Challenges on Our Dynamic Planet** elevates the importance of integrating human, animal, environmental, and plant health systems for a more holistic approach to addressing pandemics. This section seeks to answer what are the interactions among humans, the natural environment, and animals that contribute to pandemics and GHS risk? What predictions can we make about how these actors will progress and how they may change dynamics of pandemics? What global ecological factors contribute to pandemics and increase GHS risk? How have changes in these factors increased GHS risk? What provisions in the IHR address the interactions of veterinary, ecological, and human health that contribute to human pandemics and increase GHS risks?

From exploring the increasing prevalence of zoonotic diseases contributing to the global burden of disease to the impact of antimicrobial resistance (AMR) on GHS, this section also expands One Health to consider environmental factors and builds on the environmental theme by exploring how climate change is impacting GHS. It dives deeply into the ecology of vectors and the impact of climate change on that ecology, and the ecology of water and the emergence of water-borne pathogens, and returns to the human element and considers how growing urbanization is contributing to public health emergencies.

Section III. **People and Goods on the Move** considers the movement of people and goods on GHS. Specifically, the section aims to answer how has the movement of people and goods have changed since IHR (2005) and what has been the impact on pandemics and GHS risk. China has published transmission studies on cold store trade pertinent to COVID transmission in their setting. These findings should be acknowleged and duplicated to be confirmed. What are the direct and indirect impacts of stopping traffic? How can emergency or pandemic regimes in the sectors of transportation, manufacturing, agriculture, and travel reinforce GHS? How have disruptions to supply chains that have resulted from constraints on movement contributed to GHS risk? Do current provisions in the IHRs adequately address this? How have they helped or not helped in the current pandemic? How can nations allow free exchange of goods, protect economies, and allow citizens to exercise personal freedom to travel while protecting GHS? How can nations both protect their citizens and support the GHS fabric? What changes to the IHR would support these needs? It also explores the complex connections and networks linking the world, and the GHS implications and how financial markets and supply chains can be protected in pandemics. Further, it discusses how points of entry can be strengthened against public health threats and demonstrates what the WHO is doing to help improve pandemic preparedness efforts.

Section IV. **Tools and Techniques to Modernize Prevention, Detection, and Response** examines tools and techniques employed to prevent, detect, and respond to public health threats. The section addresses what provisions in the IHRs deal with tools and techniques to improve GHS? What are the challenges? What are the gaps in each of the areas addressed in this section? What should be addressed in the ongoing IHR amendment discussions to improve the current situation? How do we better assess and monitor pandemic preparedness? And finally what tools and techniques can help address infodemics?

The section begins by exploring tools and techniques available to laboratories for preventing pandemics and delves deeply into how modern technologies are advancing public health surveillance (PHS) methodologies across the world. It underscores the important concepts around data needs for GHS and reveals the advanced analytics that are enabling the next generation of PHS, including predictive techniques for effective public health intelligence. It looks deeply into how new technologies can help contain infodemics from exacerbating public health emergencies and explores the latest innovations safeguarding populations from (re) emerging threats.

Section V. **Moving to the Best-protected Global Community** presents the vision to prevent future pandemics. It proposes political approaches or systems to enable, enhance, and empower public health action. What provisions in the IHRs support these recommendations? What changes in the IHRs in the context of global initiatives include the Global Architecture for Health Emergency Preparedness, Response, and Resilience (HEPR)?[g] IHR amendments and negotiation of the pandemic accord will help to remove impediments or close gaps. What other international regimes need to harmonize? How do we communicate with people using good information using the National Provider Information Standard (NPIS) to access healthcare and therapies and vaccine access? How can global regimes most effectively support societal aspirations for good health systems at national and local levels? How do we develop and promote agile leadership and management? How do we measure progress? How do we allocate resources? And how do we diversify global supply chains for readiness?

This final section explores how these two key areas are inextricably interlinked and dives deeply into understanding the capital to modernize GHS. It explores the relationship between institutions and the public and focuses on the public health professionals, scientists, and healthcare providers instrumental to our institutions. Additionally, it looks more closely into how institutions can be strengthened to prevent future pandemics and the metrics needed to ensure GHS goals are being met.

This book is the culmination of hard work by over 120 people from different parts of the world, with a wide range of experience and expertise, and with many different perspectives. Our contributors wrote this book while experiencing the personal and professional challenges of the COVID-19 pandemic; many developed COVID-19 themselves. In spite of all this, they engaged thoughtfully and collaboratively and with an innovative spirit to propose a vision and actions that can modernize GHS to prevent pandemics. We invite our readers to follow the contributors' example.

For the reader, we've structured the book in such a way so you can explore various topics most relevant for you. We hope that you find value and together we can move to modernize GHS.

[g] HEPR is a program established by the World Bank Group that "provides financing to low-income countries and to countries with low health emergency preparedness and response capabilities." Its goal is to "improve their capacities to prepare for, prevent, respond, and mitigate the impacts of epidemics on people".[36]

References

1. Lin B, Dietrich ML, Senior RA, Wilcove DS. A better classification of wet markets is key to safeguarding human health and Biodiversity. *Lancet Planet Health*. 2021;5(6). https://doi.org/10.1016/s2542-5196(21)00112-1.
2. A chronicle on the SARS epidemic. *Chin Law Govern*. 2014;36(4):12–15. https://doi.org/10.2753/clg0009-4609360412.
3. TED: Ideas Worth Spreading. *A Global Pandemic Calls for Global Solutions*. YouTube; 2020. Retrieved October 28, 2022, from https://www.youtube.com/watch?v=8bj0GR34XWc.
4. Listings of WHO's Response to Covid-19. World Health Organization. https://www.who.int/news/item/29-06-2020-covidtimeline. Published January 29, 2021. Accessed February 2022.
5. IHR Emergency Committee on Novel Coronavirus (2019-nCov). World Health Organization. https://www.who.int/director-general/speeches/detail/who-director-general-s-statement-on-ihr-emergency-committee-on-novel-coronavirus-(2019-ncov). Published January 30, 2020. Accessed February 2022.
6. WHO Coronavirus (COVID-19) Dashboard World Health Organization. https://covid19.who.int/. Published 2022. Accessed February 2022.
7. Fuller T. *A Pisgah-Sight of Palestine, and the Confines Thereof, with the History of the Old and New Testament Acted Thereon*. London: Williams; 1869.
8. Seeger P. *Turn! Turn! Turn! (To Everything There is a Season)*; 1962. https://www.youtube.com/watch?v=CGDoxnjCSLY. Accessed February , 2022.
9. Feehan J, Apostolopoulos V. Is COVID-19 the worst pandemic? *Maturitas*. 2021;149:56–58. https://doi.org/10.1016/j.maturitas.2021.02.001.
10. McNabb SJN, Magowe M, Shaw N, et al. Delivering modern global health learning requires new obligations and approaches. *Ann Glob Health*. 2021;87(1):68, 1–9. https://doi.org/10.5334/aogh.3261.
11. *Merriam-Webster's Collegiate Dictionary (10th ed.)*. Merriam-Webster Incorporated; 1999.
12. Akin L, Gözel MG. Understanding dynamics of pandemics. *Turk J Med Sci*. April 21, 2020;50(SI-1):515–519.
13. Aminov RI. A brief history of the antibiotic era: lessons learned and challenges for the future. *Front Microbiol*; December 8, 2010. Retrieved from https://www.ncbi.nlm.nih.gov/pmc/articles/PMC3109405/#!po=63.8889.
14. Khan M, Adil SF, Alkhathlan HZ, et al. COVID-19: a global challenge with old history, epidemiology and progress so far. *Molecules*. January 2021;26(1):39.
15. Chernogor N, Zemlin A, Kholikov I, Mamedova I. Impact of the spread of epidemics, pandemics and mass diseases on economic security of transport. *E3S Web Conf*. 2020;203, 05019.
16. International Health Regulations (2005) Third Edition. Available from: https://www.who.int/publications-detail-redirect/9789241580496.
17. IHR Emergency Committees. Available from: https://www.who.int/teams/ihr/ihr-emergency-committees.
18. Smith KF, Goldberg M, Rosenthal S, et al. Global rise in human infectious disease outbreaks. *J R Soc Interface*. 2014;11(101):20140950. https://doi.org/10.1098/rsif.2014.0950.
19. Bloomfield LS, McIntosh TL, Lambin EF. Habitat fragmentation, livelihood behaviors, and contact between people and Nonhuman Primates in Africa. *Landsc Ecol*. 2020;35(4):985–1000. https://doi.org/10.1007/s10980-020-00995-w.
20. Volkmer I, Thompson W. *Social Media and COVID-19: A Global Study of Digital Crisis Interaction Among Gen Z and Millennials*. World Health Organization; 2021. Retrieved July 6, 2022, from https://arts.unimelb.edu.au/__data/assets/pdf_file/0007/3958684/Volkmer-Social-Media-and-COVID.pdf.
21. Aldis W. Health security as a public health concept: a critical analysis. *J Health Policy Syst Res*. 2008;23(6):369–375. https://doi.org/10.1093/heapol/czn030.
22. World Health Organization. Health Topics: Health Security. World Health Organization. n.d. Retrieved 2023, from https://www.who.int/health-topics/health-security#tab=tab_1.
23. US Centers for Disease Control and Prevention. What is Global Health Security? *US Centers for Disease Control and Prevention: Global Health Security Agenda*; May 18, 2022. Retrieved 2023, from https://www.cdc.gov/globalhealth/security/what.htm.
24. U.S. Department of State/Global Health Security Agenda. Building Global Health Security - United States Department of State. *Office of International Health and Biodefense*; January 5, 2021. Retrieved 2023, from https://www.state.gov/key-topics-office-of-international-health-and-biodefense/building-global-health-security/.

25. Kickbusch I, Heymann D, Ihekweazu C, Khor SK. *Global Health Security Demands National as Well as Global Responses. Health Policy Watch; January 29,* 2022. Retrieved July 6, 2022, from https://healthpolicy-watch.news/global-health-security-national-global/.
26. About the GHSA. *Global Health Security Agenda*; 2014. Retrieved July 6, 2022, from https://ghsagenda.org/about-the-ghsa/.
27. World Health Assembly Report, 54/9. *Global Health Security - Epidemic Alert and Response: Report by the Secretariat.* World Health Organization; 2001. https://apps.who.int/iris/handle/10665/78718.
28. World Health Assembly Resolution, WHA54.14. *Global Health Security - Epidemic Alert and Response*; 2001. https://apps.who.int/gb/ebwha/pdf_files/WHA54/ea54r14.pdf.
29. Salm M, Ali M, Minihane M, Conrad P. Defining global health: findings from a systematic review and thematic analysis of the literature. *BMJ Glob Health*. 2021;6(6). https://doi.org/10.1136/bmjgh-2021-005292.
30. Annex 2 of the International Health Regulations. World Health Organization. https://www.who.int/publications/m/item/annex-2-of-the-international-health-regulations-(2005). Published 2005. Accessed February 2022.
31. Ginsbach KF, Monahan JT, Gottschalk K. Beyond Covid-19: Reimagining the Role of International Health Regulations in the Global Health Law Landscape. Health Affairs Forefront. https://www.healthaffairs.org/do/10.1377/forefront.20211027.605372. Published November 1, 2021. Accessed February 23, 2022.
32. Lee T. Vanderbilt Journal of Transnational Law. n.d. Retrieved March 15, 2022, from https://scholarship.law.vanderbilt.edu/cgi/viewcontent.cgi?article=1174& context=vjtl.
33. Labonté R. The growing impact of globalization for health and public health practice. Annual Reviews. n.d. Published December 21, 2010. Retrieved March 14, 2022, from https://www.annualreviews.org/doi/10.1146/annurev-publhealth-031210-101225.
34. World Health Organization. *Working Group on Amendments to the International Health Regulations (2005)*. World Health Organization - Governance; 2023. Retrieved from https://apps.who.int/gb/wgihr/.
35. Fukuda-Parr S, Buss P, Yamin AE. Pandemic treaty needs to start with rethinking the paradigm of Global Health Security. *BMJ Glob Health*; June 1, 2021. Retrieved March 14, 2022, from https://gh.bmj.com/content/6/6/e006392.
36. The World Bank. *Health Emergency Preparedness and Response (HEPR) Umbrella Program.* The World Bank | IBRD . IDA; July 19, 2021. Retrieved May 13, 2023, from http://worldbank.org/en/topic/health/brief/health-emergency-preparedness-and-response-hepr-umbrella-program.

SECTION I

International regulatory environment to prevent, detect, and respond and importance of a global point of view

Scott J.N. McNabb[1], *Affan T. Shaikh*[1,2] *and Carol J. Haley*[1]

[1]Hubert Department of Global Health, Rollins School of Public Health, Emory University, Atlanta, GA, United States; [2]Yale University, New Haven, CT, United States

In the past few years, the world has witnessed increasing and far-impacting cross-border outbreaks, culminating in the COVID-19 pandemic. These crises highlight the urgent need to modernize and strengthen global health security (GHS) to effectively prevent, detect, and respond to public health threats on a global scale. Underscoring any effort is the need to understand governance and collaboration across geographies and stakeholders. But global health is embedded in the historical need of developed, western countries to protect their borders from diseases of far-off lands. This colonial framing continued as Western organizations brought both capital and expertise to drive agendas of their own making. Understanding this history and establishing a need for equity sets the stage for the rest of the conversation to understand governance structures and collaborations needed to modernize GHS.

The chapter **Ethics and Global Health Security** discusses the need for an explicit ethical framework in GHS, and within the International Health Regulations (IHR). While the IHR aligns with human rights principles and is legally binding in many countries, it lacks a clear ethical framework. The chapter emphasizes the importance of better incorporating ethics and human rights principles into the IHR to improve global readiness and response to pandemics. It suggests the use of existing models for balancing ethics during public health emergencies as a guide for developing an ethical framework for GHS.

The chapter **National Interagency Collaboration for Public Health** focuses on the importance of national interagency collaboration to address public health threats. It highlights the challenges in implementing and sustaining these collaborations and emphasizes the need for sustained coordination, communication, collaboration, and capacity building across multiple sectors within countries. The text provides practical ideas and recommendations for closing the gaps in national interagency collaboration and increasing countries' capacity for effective coordination and collaboration.

The chapter **The Imperative for Global Cooperation to Prevent and Control Pandemics** discusses the need for global cooperation in preventing and controlling pandemics. It criticizes the lack of solidarity and urgency in expanding cooperative initiatives to counter COVID-19 and future pandemics. The text emphasizes the need for multiple stakeholders to work together to build core capacities for public health surveillance, knowledge generation, and fair distribution of essential resources. It examines global governance mechanisms and evaluates the COVAX initiative as a case study. The text considers the benefits of a legally binding pandemic treaty and provides actions that stakeholders can take to enhance cooperation and build trust.

The last chapter in this section, **International Legal Issues of National Sovereignty and Authority Impacting Global Health Security**, examines the international legal issues surrounding national sovereignty and authority in the context of GHS. It highlights the critical role of the International Health Regulations (IHR) in promoting an effective global response to public health threats. However, it also acknowledges criticisms that the IHR have not fully fulfilled their mandate. The text emphasizes the need to strengthen GHS by amending the IHR while maintaining a balance between national sovereignty and global authority. It explores the intersection between the IHR and domestic health systems and emphasizes the importance of a synergistic relationship to enhance GHS.

CHAPTER

2

Ethics and global health security

Senait Kebede[1], *Lisa M. DeTora*[2,a], *Perihan Elif Ekmekci*[3], *Tewodros A. Wassie*[4], *Caroline Baer*[5], *David Addiss*[6], *Francis P. Crawley*[7] *and Barbara E. Bierer*[8]

[1]Hubert Department of Global Health, Rollins School of Public Health, Emory University, Atlanta, GA, United States; [2]Writing Studies and Rhetoric, Hofstra University, Hempstead, NY, United States; [3]TOBB University of Economics and Technology, International Chair in Bioethics/WMA Cooperation Center, Ankara, Turkey; [4]International Health Consultancy, LLC, Atlanta, GA, United States; [5]Public Health Practice, LLC, Atlanta, GA, United States; [6]Focus Area for Compassion and Ethics (FACE), The Task Force for Global Health, Decatur, GA, United States; [7]Good Clinical Practice Alliance – Europe (GCPA) & Strategic Initiative for Developing Capacity in Ethical Review (SIDCER), Brussels, Belgium; [8]Brigham and Women's Hospital and Harvard Medical School, Boston, MA, United States

Introduction

The WHO defines global health security (GHS) as the activities needed to prevent, detect, and respond to "acute public health events" that pose dangers across national boundaries.[1,2] The International Health Regulations (IHR), which is legally binding in 194 WHO member states and 2 additional countries, was designed to coordinate, report, and respond to potential international health threats of significant concern. Following a severe acute respiratory syndrome coronavirus (SARS-CoV-1) outbreak in 2003, an IHR update emphasized the need to address emerging and reemerging infections detected by national health authorities without too much interference in international travel or trade.[3] During the COVID-19 pandemic, greater IHR adherence was associated with lower disease incidence and mortality in a cross-sectional analysis of 114 countries.[4] However, both the IHR and an associated GHS Index that ranks IHR compliance have certain limitations,[5,6] including the lack of an explicit ethics framework to guide decision-making.[7]

[a] Co-first author.

Modernizing Global Health Security to Prevent, Detect, and Respond
https://doi.org/10.1016/B978-0-323-90945-7.00018-X

As with other international and national documents rooted in human rights or international law, the IHR mentions ethical principles like respect for persons, dignity, and human rights. However, the IHR is largely silent regarding the concept of ethics more generally or how to address ethical concerns during public health emergencies. The words "ethics" and "values" are not even mentioned in the IHR. However, ethical frameworks exist for public health, for research, and for healthcare delivery. Existing regulations, by not explaining how to balance these frameworks, seem to presume that individual actors and fields will manage their own ethical decision-making. In practice, specific ethical guidelines are developed for individual outbreaks or pandemic events; however, each pandemic or epidemic has had new features that required ethical guidance. The complexities of GHS and the IHR require an overall framework to balance potentially competing ethical frameworks, an implementation guide, and a mechanism for ongoing review, discussion, and revision. Considerations outlined below will be important for formulating this framework (Table 2.1).

One reason a framework is needed to balance existing ethical frameworks is a persistent global imbalance of power and access to resources. Existing public health and research structures are governed by policies and interests that can support or undermine the independence needed for local health security. For instance, public health interventions, healthcare provision, and medical research may rely on external sources of funding and evaluation.[8,9] Power asymmetries often result in certain people, institutions, and organizations being perceived as objects of charity rather than full partners. This circumstance can affect groups within communities or entire nations. Understanding power relationships will be vital to ensure long-term sustainable GHS.

Ethical considerations during pandemics impact many activities, including public health interventions, research, and healthcare delivery. Each of these activities has existing ethical codes and/or guidelines. In emergency situations, conflicts may arise between individual rights and collective benefits. Existing ethical guidelines for public health emergencies often presume preparation time, the availability of outside aid, and good quality research data.[10–12] COVID-19 revealed a lack of comprehensive guidance for maintaining ethical standards in situations of uncertainty or while preparing for outbreak response. Responses to COVID-19 worsened existing social and economic inequalities and left public health officials susceptible to criticism.[13] While ethical obligations remain unchanged (except, perhaps in gaining importance) during pandemics or epidemics,[14] recent public health emergencies have thrown into sharp relief gaps and inconsistencies within and across existing ethical frameworks—exposing tensions between fundamental human rights and the need to control disease transmission and curtail morbidity and mortality.[15] No unified guideline explains how to balance ethics in research and public health decision-making during health emergencies in the context of GHS.[9,16,17] We outline some gaps in the published literature and then suggest a process to develop a GHS ethics framework modeled on existing work in public health emergencies[12] that integrates existing ethics frameworks for clinical research, public health, health emergencies, and global health.

TABLE 2.1 Selected ethical guidelines and frameworks.

Ethical frameworks/Guidelines	Organization	Link
International Health Regulations 2005 (IHR)	WHO	https://apps.who.int/iris/bitstream/handle/10665/246107/9789241580496-eng.pdf
Human genome editing: a framework for governance	WHO	https://www.who.int/publications/i/item/9789240030060
Public health guidance for community-level preparedness and response to severe acute respiratory syndrome (SARS)	US CDC	https://www.cdc.gov/sars/guidance/core/index.html
Genomic sequencing of SARS-CoV-2: a guide to implementation for maximum impact on public health	WHO	https://www.who.int/publications/i/item/9789240018440
Ethical standards for research during public health emergencies: Distilling existing guidance to support COVID-19 R&D	WHO	https://www.who.int/blueprint/priority-diseases/key-action/liverecovery-save-of-ethical-standards-for-research-during-public-health-emergencies.pdf
Guidance for managing ethical issues in infectious disease outbreaks	WHO	https://apps.who.int/iris/handle/10665/250580
Guidelines on ethical issues in public health surveillance	WHO	https://apps.who.int/iris/bitstream/handle/10665/255721/9789241512657-eng.pdf?sequence=1
Ethical considerations in developing a public health response to pandemic influenza	WHO	https://www.who.int/csr/resources/publications/WHO_CDS_EPR_GIP_2007_2c.pdf
Ethical considerations to guide the use of digital proximity tracking technologies for COVID-19 contact tracing	WHO	https://www.who.int/publications/i/item/WHO-2019-nCoV-Ethics_Contact_tracing_apps-2020.1
Ethical considerations for use of unregistered interventions for Ebola virus disease (EVD)	WHO	https://apps.who.int/mediacentre/news/statements/2014/ebola-ethical-review-summary/en/index.html
International ethical guidelines for health-related research involving humans	Council for International Organizations of Medical Sciences (CIOMS) in collaboration with WHO	https://www.who.int/docs/default-source/ethics/web-cioms-ethicalguidelines.pdf?sfvrsn=f62ee074_0
Guidance for research ethics committees for rapid review of research during public health emergencies	WHO	https://www.who.int/publications/i/item/9789240006218

(*Continued*)

TABLE 2.1 Selected ethical guidelines and frameworks.—cont'd

Ethical frameworks/Guidelines	Organization	Link
The Belmont report: Ethical principles and guidelines for the protection of human subjects of research	National Commission for the Protection of Human Subjects of Biomedical and Behavioral Research	https://www.hhs.gov/ohrp/regulations-and-policy/belmont-report/read-the-belmont-report/index.html
Nagoya protocol on access and benefit-sharing	Secretary-General of the United Nations	https://www.cbd.int/abs/
World Medical Association Declaration of Helsinki: ethical principles for medical research involving human subjects	World Medical Association	https://pubmed.ncbi.nlm.nih.gov/24141714/

For an integrated example, see: Emergency Ethics: Public Health Preparedness and Response. *Oxford UP. https://doi.org/10.1093/med/9780190270742.001.0001.*

Literature review and gap analysis

No ethics framework for GHS has to date been published

While many articles indexed in PubMed and PMC discuss ethics in connection with GHS, no overarching framework or consolidated guidelines for GHS ethics have appeared. This literature is dominated by papers that borrow the research principles of beneficence, autonomy, and justice outlined in the *Belmont Report*, rather than referring to published literature for public health ethics or health emergency ethics.[18] Among the most common topics are the ethical design of research protocols, equitable access to healthcare, fair allocation of scarce resources, effective responses among healthcare workers, community engagement, and the need for better definitions of GHS.[19,20] Research ethics is reviewed in connection informed consent,[21] randomized controlled trials,[22–24] and attitudes toward dual-use research during health emergencies.[25] Other papers discuss ethical issues regarding vaccines,[26] laboratory research, big data, vulnerable populations,[27,28] data and sample sharing, biosecurity, risk perception, privacy and confidentiality, fear, One Health, the handling of infectious disease isolates, the GHS Index, and mass gatherings.[29] Many papers consider specific prior outbreaks or epidemics, especially in Africa, and may be overlooked as a source of general advice.

Like the literature on GHS and ethics, the indexed literature on global health ethics lacks a consistent, coherent overall framework. Instead, this literature mainly discusses specific topics such as vaccine distribution or travel restrictions, largely from the perspective of authors in high-income countries and in the context of the principles outlined in the *Belmont Report*.[30,31] Definitions of global health also vary across this literature.[32]

The existing indexed literature before identifying gaps and suggesting a future vision will be reviewed in the following sections.

The indexed literature discusses moral failings regarding GHS

Some indexed literature on GHS and ethics reveals concerns about moral failure, as evident during the 2014 Ebola outbreak and COVID-19.[19,32,33] Emergencies may prompt individuals and public health authorities to forget their ethical and moral responsibilities or to adopt inappropriately paternalistic behavior, thus undermining the autonomy of the less powerful.[12,34] Physicians, researchers, and public health officials retain ethical obligations across all conditions and urgencies under which they practice.[35] Physicians working in pandemics and other emergency situations may lack supplies, protective equipment, or other resources required to adhere to current best practices for patient care.[12] While guides to health emergency ethics exist,[12] each emergency situation presents unanticipated challenges in managing conflicts between the need to protect individuals and communities, to provide care, and to protect health workers.

During the 2014 Ebola outbreaks, the international community showed a limited capacity to prevent or mitigate the public health emergency, a problem only compounded by failures of transparency and community engagement.[33] A similar pattern of failure was repeated during COVID-19, revealing divisions between national and regional resources as well as between communities and individuals that prevented proper protections and impeded access to needed healthcare interventions.[32] Privileged persons and communities may resist even limited personal inconveniences to protect public health, and less privileged persons and groups may be put in life-threatening danger by measures to protect the preferences of larger or more affluent communities.[19] Some public health measures, such as shelter-in-place orders, may excessively burden vulnerable persons while other protective measures, such as social distancing, may not be effectively implemented for certain groups (e.g., those living in detention centers, prisons, multigenerational housing, or refugee camps).[36] As shown during COVID-19, a hasty response, however robust, once an emerging infectious agent is detected does not resolve underlying public health issues. Hasty responses may even exacerbate preexisting inequities in access to healthcare and likely will not overcome a lack of capacity for early public health surveillance.[37] Any community lacking adequate investment in healthcare endangers GHS, reinforcing a moral argument against deprioritizing the protection and care of any region or group.[19]

Karan calls for a new concept of global health beneficence, one that expands the research principle of doing good beyond research and across national borders, to ensure GHS and overcome past moral failings.[32] Karan also suggests adapting the concept of beneficence to address preexisting disparities within and between nations. This paper suggests practical measures: supporting stronger health systems; improved surveillance and laboratory capacities to enable early warning and response; and better funding.[34] Investment in primary and routine healthcare infrastructure is needed to ensure trust between governments, public health workers and facilities, and local communities. Permanent healthcare infrastructure systems that can mitigate disease outbreaks before they endanger other communities are urgently needed.[32]

Existing ethics frameworks conflict in a GHS context

Ethics guidelines and/or frameworks exist for research, clinical care, public health, and public health emergencies. The ethical concepts and principles in these existing frameworks should form the basis for a balanced ethics framework and guidance for GHS similar to frameworks for emergency ethics.[12] Well-recognized principles of clinical research ethics (such as autonomy, social justice, beneficence, and nonmaleficence)[31,32] emphasized in the indexed literature on GHS and ethics were not originally conceived in terms of public health concerns or health security.[32] Foundational bioethical documents, including the Nuremberg Code,[39] the Declaration of Helsinki,[40] and the *Belmont Report*,[31,38] responded to research abuses of human subjects,[41] focusing, therefore, on protecting individual research participants.[42] In healthcare settings, patients' rights and the right to health found in the WHO Constitution[43] focus on the right of individuals not to be exploited or harmed, and even to be helped, by the medical community and/or the state. Members of the medical community should be protected as well. Prior public health emergencies like the 2014 Ebola outbreaks revealed the limitations of these individualistic values, particularly in the context of ongoing failures to address existing healthcare disparities.[32]

Public health measures often leverage government intervention to achieve future-oriented outcomes, which can create ethical conflicts and tensions[44] between individual rights and the responsibility to protect communities and nations.[45] Upshur[41,46] and Childress et al.[47] outline principles and offer heuristics for decision-making that balance individual and public health concerns. Upshur also suggests principles such as transparency of decision-making before undertaking public health interventions, identifying the least restrictive means of protecting public health, determining which harms to individuals are justified (and why) in this pursuit, and providing reciprocity or reasonable compensation for individuals who experience such harms.[41,46] Public health interventions should not infringe on fundamental human rights as expressed in international constitutional laws. Ideally, the interests of all groups will be considered before public health measures are implemented.

Childress and colleagues emphasize that no algorithmic solution to public health ethics will ever exist. Ongoing evaluation and intellectual engagement will always be necessary to maintain morally, and ethically, justifiable public health practices.[47] These authors emphasize transparent communication and the need to consider both community and individual values, engaging with stakeholders rather than delivering dicta. Harms should be avoided actively. Distributive justice and transparency are necessary in the allocation of resources, as is procedural justice in determining the methods for decision-making. Individual rights must be respected, including those of autonomy, privacy, and confidentiality. Trust must be developed and maintained through honesty, transparency in science and decision-making, as well as in the promises and commitments made to the public.

In the setting of healthcare delivery, clinical and public health ethics should be at the core of decision-making during emergencies, whether by physicians, researchers, public health workers, or policymakers.[9,16] Unger et al.[16] see challenges to this kind of thoughtful decision-making as arising in part from corporate health insurance algorithms that ignore patient interests and limit physician autonomy in healthcare delivery. Drawing on published ethical tenets for medical care and public health, the authors suggest expanding existing ethical concepts, such as beneficence, medical, and public health engagement, to better promote health equality as well as community and individual participation. The authors encourage physicians to remain aware of public health ethics but focus on their duties to patients and communities through professional excellence, nonmaleficence, and healthcare delivery informed by scientific excellence as opposed to algorithms.[9,16] During pandemics, healthcare workers often work in unhealthy conditions, exposed to disease risk and other personal harms, which can be worsened by poor policy decisions or misinformation spread through the media. Ethics guidance for GHS and pandemic responses must address these concerns, not only for individual healthcare providers but also proactively for health systems and communities.

Emergency Ethics: Public Health Preparedness and Response provides a helpful guide to ethics frameworks for preparedness, social justice, stockpiling, protecting vulnerable populations, engagement with the public, and research, as well as civic, personal, and professional ethical obligations.[12] The authors also provide an overall framework for decision-making during public health emergencies. Specific advice includes recommendations to be concrete in ethical thinking, to enumerate ethical values affected in a given situation, to evaluate the response after the emergency has abated, and to resist unwarranted urgency.[12] A successful implementation of ethics in all settings, from the global to the individual health practitioner, is needed to ensure GHS. Ethics frameworks must also account for differences across national boundaries and navigate foreign health policies that might emphasize national security concerns rather than support global health initiatives.[15] The *Nagoya Protocol on Access to Genetic Resources and the Fair and Equitable Sharing of Benefits Arising from their Utilization to the Convention on Biological Diversity*,[48] which entered into force for its signatories in 2014, by its very existence hints at a need for greater attention to ethics in decision-making, benefit sharing, and social justice in the context of global disease surveillance (see Boxes 2.1 and 2.2). Surveillance burdens in countries already facing serious public health challenges also may create tensions between healthcare delivery to residents and the protection of health that can affect GHS.[49] Being explicit about the frameworks that undergird international regulatory and legal agreements of relevance to GHS can provide a crucial bridge for more effective, cooperative, and ethical responses to global health threats.

The legacy of colonialism: impact on GHS and ethics

GHS requires cooperation among nations and international organizations and also from communities and individuals. However, cooperation implies the interaction of equal partners. An unfortunate legacy of past colonialism and military conflict informs current power imbalances and other inequities, which negatively impacts the spirit of trust necessary to ensure GHS.[8,9,50] COVAX, a pillar of the Access to COVID-19 Tools Accelerator, should have ensured access to COVID-19 vaccines to all people without reference to wealth, but vaccine distribution instead favored the affluent.[51] COVID-19 responses illuminate a persistent legacy of colonialism that undermines agreements by governments, other organizations, and communities.[52] With the growing call for equity in global health, existing health structures and professionals must examine power asymmetry and its impact on health practice and GHS, to determine appropriate moral and ethics frameworks. Thus, calls have been made to "decolonize" global health.[8]

BOX 2.1

Case study 1

Case study of Asian Avian Influenza: Trust in international collaboration

Background

The Asian Avian Influenza A H5N1 virus is contagious and deadly among domestic poultry, other birds (and some animals). H5N1 is endemic in some areas, most commonly Southeast and East Asia. While rare, cross-species transmission to humans can result in severe disease, pneumonia, respiratory failure, and death, most typically in younger people. The first human case was reported in 1997, and several outbreaks have occurred since.

In July 2005, Indonesia reported its first case of human H5N1 and, consistent with prevailing surveillance practices, sent clinical samples to two World Health Organization (WHO) reference laboratories for confirmation and risk assessment. From the first isolation until December 2007, 116 cases were reported, some clustered, with an 81% case fatality rate. In accordance with the 2005 International Health Regulations (IHR) and in deference to public health concerns, Indonesia routinely shared early biological specimens, particularly because it initially did not have the capacity to detect H5N1 virus. Indonesia understood that neither the samples nor the viral genetic sequence would be shared without their express permission, and that there would be fair and equitable benefit sharing, consistent with Article 19[63] of the principles of the multilateral Convention on Biological Diversity treaty.

Trust began to erode in April 2006. Data from Indonesian samples were presented publicly with little advance notice to the country, and coauthorship was offered only late in manuscript preparation. While Indonesia deposited the then-available viral sequences in a public databank, it decided to test subsequent biospecimens itself and not to participate in the WHO system for case confirmation (although it continued to provide viral isolates for risk assessment). Later that year, an Australian company announced its intention to develop and commercialize a vaccine against H5N1 virus, using an Indonesian viral strain. In January 2007, in the absence of any international commitment to benefit sharing, Indonesia decided to withhold further specimens.[64]

The sovereign decision to withhold specimens from the global community that feared an emerging pandemic prompted international attention to structuring a resolution. On the one hand, data and specimen sharing of emerging viral threats is necessary to enable pandemic preparedness, and on the other, a system of fair reciprocity that results in health equity and global health security must also result.[65]

Observations

Resource poor countries have few mechanisms to ensure fair and equitable access to medicines and vaccines for their populations.

The breakdown of trust in international collaboration can lead to significant harm.

BOX 2.1 *(cont'd)*

Key issues

Whether viral isolates, capable of global transmission, are owned by any country is a topic of continued debate.

A transparent system that endorses data and biospecimen sharing for public health must be balanced by fair and equitable benefit sharing for contributors.

Global health security depends upon timeliness and trust.

Interventions

Multiple international meetings of designated representatives were necessary to advance any resolution to enable progress.

Lessons learned

What worked?

Unilateral action by a resource poor country led to change and international recognition

Meetings and negotiation resulted in tentative agreements

What didn't work?

The commitment to consent and permission of the contributing country for use, distribution, and commercialization of viral sequences was abrogated.

High resource countries and manufacturers failed to consider the needs of resource poor countries to partake in access to medicine and vaccines.

Recommendations

An international agreement, based on foundational ethical principles and addressing global health security, should be developed, signed by all parties, and implemented.

A defined process for timely resolution of conflicts, misunderstandings, and challenges should be developed.

The concept of GHS emerged from definitions of global health and tropical medicine established by groups whose membership came disproportionately from developed nations and powerful institutions.[9,16] Access to medical care and public health benefits arising from biomedical discoveries have been controlled and limited by those in power, frequently at the expense of vulnerable populations. Dramatic inequities in power and wealth threaten the goal of GHS. Indeed, existing ethics frameworks and agreements can perpetuate these imbalances, even when their stated aim is to address inequity.[8,53] Imbalances like these also undermine attempts to develop an ethics framework upon which global health, and GHS, should rest. Even the evidence base for international health projects is dominated by funding and research from a limited number of powerful government and nongovernment organizations,[8] which may drown out or seek to control the voices of those often most affected by public health emergencies. The need exists to recognize the effects of past colonialism and decolonize global health at all levels, from the ethical to the pragmatic, if GHS is to be achieved.[8] Thus, any ethics framework for GHS must seek to include and protect vulnerable and marginalized populations. Equity matters in pandemic preparedness and control; we have an ethical obligation to work toward greater equity worldwide.

BOX 2.2

Case study 2

Successful surveillance implementation in low-resource setting[66]

Background

The IHR core surveillance and response capacity requirement entail that member states develop and maintain capabilities to detect, assess, and report disease events to WHO within 48 h of confirmation. The WHO consultation on the pandemic influenza A H1N1 2009 that affected over 214 countries and caused over 18,449 deaths concluded that the world was unprepared to respond to pandemics partly due to inadequate global surveillance and response capacity. Technical and financial support was provided for Strengthening Influenza Sentinel Surveillance in Africa (SISA) project, which took place in 2011.[67]

Observations

Of the eight countries included in the project, Sierra Leone was unique in having no influenza surveillance activity already in place. Support was provided to implement influenza sentinel surveillance (ISS) and improve epidemiological and virological data collection and reporting through global databases. The health system in Sierra Leone had only a small, trained workforce and additional burdens to overcome, including a recently ended civil war and a lack of basic resources and infrastructure.

Interventions

The process began with engagement at national level to ensure the participation, trust, and confidence in collaboration of decision-makers and staff involved in surveillance activities at various levels. The capacity of current surveillance systems was assessed, potential synergies were identified, and roles were clearly defined. Following the assessment, national action plan and relevant tools were developed. Epidemiological and laboratory surveillance data were collected, processed, and reported to national and global levels in line with the WHO quality standards. To ensure sustainability, influenza surveillance activity was implemented using Integrated Disease Surveillance and Response (IDSR) infrastructure that promotes integration of resources through the coordination and streamlining of common surveillance activities.

Advisory and technical support were focused and provided with deliberate effort to promote national leadership and full engagement of local professionals for specific interim targets. Stable funding sources were identified, with external support, to cover core functions. Communication was a critical element for success. Review meetings, budget updates, and resource assessments took place on an ongoing basis. Updates were provided regularly to all stakeholders. Staff were motivated through recognition and nonmonetary incentives. National ownership was encouraged through early engagement of and routine communication with decision-makers.

Lessons learned

The successful implementation of an ISS system in Sierra Leone was the result of many factors, including the following:

BOX 2.2 *(cont'd)*

- A systematic approach, national ownership, appropriate timing, and external support are essential for rapid implementation of surveillance activities for early detection and data generation during pandemics.
- Monitoring with regular communication, training/capacity building, and staff recognition promotes motivation.
- Improved relationship between decision-makers, public health agencies, technical experts, academics, and other stakeholders is critical for pandemics response.

Recommendations

- Partners engagement that recognizes local leadership, expertise, and knowledge yields better outcome.
- Clear communication and understanding by participants of each other's capabilities, roles, and responsibilities are critical for efficient implementation of activities.

Vision for the future

A need to strengthen healthcare ethics always exists. Although ethics guidelines exist for the responsible conduct of health-related research,[31,40] public health measures,[16,46,47] and resource allocation in pandemic or epidemic situations,[12] these existing guidelines have limitations. Existing guidelines should be evaluated and amended by appropriate groups to account for situations of uncertainty like those encountered during recent epidemics and COVID-19. In addition to the ethical considerations reviewed above as related to public health and research ethics, guidelines for GHS ethics should consider nonpharmaceutical interventions; the ability of countries to manufacture and distribute needed medical supplies, vaccines, and medicines; equity and access; resource allocation; funding; and protection of healthcare workers and the public. Data and biospecimen sharing, publication ethics, access to the published literature, and effective communication with various stakeholders also should be included in this new ethics framework. Attention should be paid to addressing the needs of marginalized groups and ensuring inclusivity by seeking out input from marginalized groups rather than simply delivering information. We believe that the WHO should exhibit leadership in this area.

A successful GHS ethics framework must balance multiple existing ethics frameworks and value systems. As suggested by Childress et al.,[47] no algorithmic solutions for ethics in complex situations like public health will ever exist. The same is true for GHS: ongoing evaluation and intellectual engagement will always be necessary, not only because ethics must be considered in different situations but also because inconsistencies between the ethics frameworks of different disciplinary and practical fields must be balanced.[12,47] An understanding of power relationships will be necessary and may be complicated by factors such as differences in language, values, religious beliefs, gender, citizenship status, or even access to education and health information.[54] Past histories of conflict and colonization also may affect GHS. Initiatives like One Health,[55] which broadens the frame of ethical considerations to address the interface of human health, animal health, and the environment, include the humane treatment of animals, food resources, safety and security, and the environmental hazards including chemicals and air pollution. Various good practice guidelines for clinical research,[56] laboratory practice (especially regarding data and biospecimen sharing and the humane treatment of animals),[57] and publications of such research[58] should be better integrated into responses to health emergencies. Ideally, in any proposed model for GHS ethics, the more vulnerable or marginalized the population, the more attention should be paid to ensuring beneficence and limiting harms.

An ethics framework for GHS can contribute to ongoing efforts to decolonize global health, or begin to undo the damage of past colonialism. Such efforts will require assessing current power dynamics and moving toward a more balanced, ethical, and equitable approach to healthcare, research, and public health interventions.[8] An effective ethics framework for GHS must be compatible with current efforts to decolonize global health[8] and to move from unbalanced systems and dynamics toward an equitable approach for global health. Funding and aid often foster dependencies that also operate to undermine the good. A prevailing prescriptive top-down approach, in which funders dictate priorities, does not advance health security (or GHS) or foster independence and self-reliance in local communities. A robust ethics framework for GHS should include requirements for sustained and meaningful community engagement that also supports independence.

We envision a future in which a transparent and explicit discussion of ethical principles informs the actions taken and decisions made in the various domains of GHS and global health, such as (but not limited to): clinical care, public health interventions, resource allocation, and research. Communities will be meaningfully engaged in these endeavors and marginalized groups will be included. Adjustments to interventions and other plans will be made in response to feedback and changing situations, in order to limit harms to individuals, communities, and larger groups. Prior political histories will be examined to identify the causes of health inequalities and better address them. The public will have access to scientifically sound publications as well as research results and data, and health information will be made clear and accessible for audiences with differing access to health and scientific literacy. Technological advancements in communication and collaboration at distance can help foster communication and inclusivity.

Governments and regulators should transparently explain their decision-making, enumerating the ethical principles that informed any decision; such transparency should promote feedback and allow for revision of existing ethical principles when necessary. Scientific information and data should also be made available. The public posting of clinical trial results is legally required in the European Union, using language that is accessible to the public. This trend should continue more broadly, as should the use of plain language summaries of manuscripts in biomedical journals. Plan S,[59] which was implemented by some granting bodies to ensure that research results are freely available to the public, should encourage all researchers to use open access options. Anonymized and appropriately redacted data should be available for all marketed medicinal products, or products in widespread emergency use, for all indications, so that scientists, patients, and communities have access to the research that informed the introduction of the product into the public health space. Greater investment in equitable access to the technology needed to access medical and scientific information is also needed.

Guidelines for transparency, publication management, peer review, editing, and even developing plain language summaries should present a coherent ethics framework that reflects public health needs and interests. To help the public understand this information, appropriate working groups should be identified to develop accessible communications for various audiences, including the public. To encourage manufacturers to consider the broader ethical impacts of their work, it may be helpful if responsible health authorities began to require a more considered ethics statement, in the spirit of the benefits and risks conclusions of the Common Technical Document,[60] that outlines the specific ethics framework(s) used in development program as part of regulatory submissions. Communication strategies should follow culture-centered approaches, which begin with authentic stakeholder engagement and consider how access to healthcare and personal agency can be promoted to overcome existing inequities.[61–63]

GHS will be impossible unless international organizations, regions, countries, and communities address the expectations of their people regarding healthcare infrastructure and delivery of care. Independent decision-making is required to meet the needs of diverse populations and individual citizens as well as to build healthcare infrastructure, develop appropriate science and research, and garner and distribute resources appropriately. Further, investments are needed in ethics education and infrastructures, including for ethics committees, to assure respect for regional, national, and local values and perspectives. These perspectives must be acknowledged in decisions on health-related science, research, and medical interventions. Attention to training and education is needed at various relevant tiers such as preservice and in-service training for health professionals.

Different sectors and groups will naturally prioritize different activities. For instance, frontline healthcare workers retain a duty of care even in uncertain circumstances while research ethics committees must ensure that research studies are justified and well designed, however urgent the medical situation is. Our vision is that working groups of affected stakeholders will set reasonable ethical priorities for these sectors and groups, which will then be communicated honestly and clearly, debated openly, and subject to potential reconsideration and revision, if necessary. The working groups should be inclusive and representative of the structures and the heterogeneity of the populations affected. As many stakeholders as possible should be included and consulted to ensure that marginalized and previously colonized groups are not overlooked. Working groups should seek input broadly before making suggestions and include marginalized groups in their consultations. The areas to be addressed will include, but are not limited to: pandemic preparedness, nonpharmaceutical public health interventions, supply chain security, pharmaceutical and medical device manufacturing, access to healthcare, protection of healthcare workers, epidemiologic surveillance, access to vaccines and medicines, research conduct and integrity, corporate profits on medicinal products during public health emergencies, and impacts on day-to-day life. In practice, gaps will be inevitable during health emergencies; for this reason, it will be vital for local governments and communities to have an opportunity to contextualize, communicate, and seek adjustments to public health interventions to avoid any compounding of harms.

Our vision will not be possible unless all people responsible for promoting GHS, from healthcare workers to corporations to national health authorities to the public, make a commitment to ongoing improvements, evaluation, and transparency in ethical thinking and decision-making. A good starting point will be the inclusion of a specific, considered, and evidence-based ethics framework in all healthcare delivery and public health policy statements as well as other forms of health documentation. Heuristics should be in place to help decision-makers consider ethical issues, including the short- and long-term benefits and harms of suggested interventions and applicable ethical principles. The development of GHS policies, planning, and implementation of activities for pandemic preparedness and response should be based on values and ethical principles. These ethical principles should be derived from research ethics, medical ethics, public health ethics, and the ethics of public health emergencies while also adapted for GHS. Outcomes should support a just, inclusive, and comprehensive framework that embodies the principle of "health for all." All existing GHS working groups and associated organizations should develop and add an ethics framework to their core documents, a framework aligned with IHR ethics recommendations as they are developed.

An honest and objective assessment—involving the relevant stakeholders—is needed to identify shortcomings and refine approaches to achieve GHS. The following recommendations should be considered.

Recommended actions

- The IHR requires a single, consolidated, coherent framework for ethics in GHS, epidemics, pandemics, and international public health emergencies. Associated guidance should provide specific examples and sufficient nuance to guide practical application. The international community should identify standing working groups to the WHO to develop and continually evaluate such ethics frameworks and guidance in general and adapted to specific situations. These frameworks should explain how to balance and apply existing ethics guidelines within the context of GHS.
- The working group(s) should include various stakeholders, including groups from frontline healthcare providers to persons at health authorities, local governments, and community members, and patients. Representation from marginalized and vulnerable groups is necessary within the working groups.
- The working groups should identify and design a specific approach to GHS ethics that balances existing ethical frameworks for research, public health, and health emergencies to create frameworks for ethical conduct and decision-making that are suitable for unpredictable or changing circumstances. Frameworks should be sufficiently flexible to limit harms in changing circumstances and should follow the "living document" model of the Declaration of Helsinki and be subject to ongoing evaluation and revision.

Considerations for the suggested actions

- Data and biospecimen sharing, principles of reciprocity, and commitments to integrity and the responsible conduct of research and transparency should be considered. Socio-economic vulnerabilities should be accounted for to limit harmful consequences, like impacts on livelihood, which also have important health consequences.
- The ethics framework suggested by the working groups should indicate how various actors (such as healthcare providers, national laboratory managers, and regulators) should prioritize actions, including resource allocation, with specific reference to existing ethics and codes of conduct. Attention must be paid to the role of various corporate entities, including healthcare insurance companies and pharmaceutical manufacturers, as well as to the government sector.
- All frameworks developed by these groups should be reviewed and revised on a regular, ongoing basis and not only in response to emergencies. A standing analysis and adjudication board, able to do posthoc assessments of specific interventions and events, would also be advisable.
- Measures for transparent communication, information sharing, and access are also needed for all the actions suggested. These communication measures might take a culture-centered approach to address existing imbalances of power and access to different kinds of resources that can impact day-to-day health as well as health emergencies.

References

1. International Health Regulations. 2nd ed. 2005. https://www.who.int/publications/i/item/9789241580410. (Accessed 06.06.2021).
2. WHO Health security. https://www.who.int/health-topics/health-security/#tab=tab_1. (Accessed 06.06.2021).
3. US CDC. *International Health Regulations (IHR) last reviewed*; 2019 (Accessed 06.06.2021) https://www.cdc.gov/globalhealth/healthprotection/ghs/ihr/index.html.
4. Wong MC, Huang J, Wong SH, Yuen-Chun Teoh J. The potential effectiveness of the WHO International Health Regulations capacity requirements on control of the COVID-19 pandemic: a cross-sectional study of 114 countries. *J R Soc Med*. 2021;114(3):121–131. https://doi.org/10.1177/0141076821992453.
5. Tabbaa D. International Health Regulations and COVID-19 pandemic, challenges and gaps. *Vet Ital*. 2020;56(4):237–244. https://doi.org/10.12834/VetIt.2335.13296.1.
6. Ravi SJ, Meyer D, Cameron E, Nalabandian M, Pervaiz B, Nuzzo JB. Establishing a theoretical foundation for measuring global health security: a scoping review. *BMC Publ Health*. 2019;19(1):954. https://doi.org/10.1186/s12889-019-7216-0.
7. Global Health Security Index. 2021. https://www.ghsindex.org. (Accessed 06.06.2021).
8. Eichbaum QG, Adams LV, Evert J, Ho MJ, Semali IA, van Schalkwyk SC. Decolonizing global health education: rethinking institutional partnerships and approaches. *Acad Med*. 2021;96(3):329–335. https://doi.org/10.1097/ACM.0000000000003473.
9. Holst J. Global Health – emergence, hegemonic trends and biomedical reductionism. *Glob Health*. 2020;16(1):42. https://doi.org/10.1186/s12992-020-00573-4.
10. WHO Research ethics in international epidemic response: WHO technical consultation, Geneva, Switzerland, 2009: meeting report. https://apps.who.int/iris/handle/10665/70739. (Accessed 06.06.2021).
11. WHO Ethical issues in pandemic influenza planning. https://www.who.int/teams/health-ethics-governance/emergencies-and-outbreaks/pandemic-influenza. (Accessed 06.06.2021).
12. Jennings B, Arras JD, Barrett DH, Ellis BA. *Emergency Ethics: Public Health Preparedness and Response*. Oxford University Press; 2016.

13. Gopalan PD, Joubert IA, Paruk F, et al. The Critical Care Society of Southern Africa guidelines on the allocation of scarce critical care resources during the COVID-19 public health emergency in South Africa. *S Afr Med J*. 2020;110(8):700–703.
14. Wynia MK. Ethics and public health emergencies: encouraging responsibility. *Am J Bioeth*. 2007;7(4):1–4. https://doi.org/10.1080/15265160701307613.
15. Lencucha R. Cosmopolitanism and foreign policy for health: ethics for and beyond the state. *BMC Int Health Hum Right*. 2013;13:29. https://doi.org/10.1186/1472-698X-13-29.
16. Unger JP, Morales I, De Paepe P, Roland M. In defence of a single body of clinical and public health, medical ethics. *BMC Health Serv Res*. 2020;20(Suppl 2):1070. https://doi.org/10.1186/s12913-020-05887-y.
17. Ho A, Dascalu I. Global disparity and solidarity in a pandemic. *Hastings Cent Rep*. 2020;50(3):65–67. https://doi.org/10.1002/hast.1138.
18. Robson G, Gibson N, Thompson A, Benatar S, Denburg A. Global health ethics: critical reflections on the contours of an emerging field, 1977–2015. *BMC Med Ethics*. 2019;20(1):53. https://doi.org/10.1186/s12910-019-0391-9.
19. Yimer B, Ashebir W, Wolde A, Teshome M. COVID-19 and global health security: overview of the global health security alliance, COVID-19 response, African countries' approaches, and ethics. *Disaster Med Public Health Prep*. 2020:1–5. https://doi.org/10.1017/dmp.2020.360.
20. Straetemans M, Buchholz U, Reiter S, Haas W, Krause G. Prioritization strategies for pandemic influenza vaccine in 27 countries of the European Union and the Global Health Security Action Group: a review. *BMC Publ Health*. 2007;7:236. https://doi.org/10.1186/1471-2458-7-236.
21. Walsh K. Case reports on dangerous infectious diseases: a review of patient consent. *BMJ Mil Health*. 2020;166(3):179–180. https://doi.org/10.1136/jramc-2018-000949.
22. Adebamowo C, Bah-Sow O, Binka F, et al. Randomised controlled trials for Ebola: practical and ethical issues. *Lancet*. 2014;384(9952):1423–1424. https://doi.org/10.1016/S0140-6736(14)61734-7.
23. Lanini S, Ioannidis JPA, Vairo F, et al. Non-inferiority versus superiority trial design for new antibiotics in an era of high antimicrobial resistance: the case for post-marketing, adaptive randomized controlled trials. *Lancet Infect Dis*. 2019;19(12):e444–e451. https://doi.org/10.1016/S1473-3099(19)30284-1.
24. North CM, Dougan ML, Sacks CA. Improving clinical trial enrollment – in the Covid-19 era and beyond. *N Engl J Med*. 2020;383(15):1406–1408. https://doi.org/10.1056/NEJMp2019989.
25. Sarwar S, Ilyas S, Khan BA, et al. Awareness and attitudes of research students toward dual-use research of concern in Pakistan: a cross-sectional questionnaire. *Health Secur*. 2019;17(3):229–239. https://doi.org/10.1089/hs.2019.0002.
26. Ismail SJ, Hardy K, Tunis MC, Young K, Sicard N, Quach C. A framework for the systematic consideration of ethics, equity, feasibility, and acceptability in vaccine program recommendations. *Vaccine*. 2020;38(36):5861–5876. https://doi.org/10.1016/j.vaccine.2020.05.051.
27. Krubiner CB, Faden RR, Karron RA, et al. Pregnant women & vaccines against emerging epidemic threats: ethics guidance for preparedness, research, and response. *Vaccine*. 2021;39(1):85–120. https://doi.org/10.1016/j.vaccine.2019.01.011.
28. Matlin SA, Depoux A, Schütte S, Flahault A, Saso L. Migrants' and refugees' health: towards an agenda of solutions. *Publ Health Rev*. 2018;39(1). https://doi.org/10.1186/s40985-018-0104-9.
29. Memish ZA, Al-Rabeeah AA. Public health management of mass gatherings: the Saudi Arabian experience with MERS-CoV. *Bull World Health Organ*. 2013;91(12). https://doi.org/10.2471/BLT.13.132266, 899–899A.
30. Salm M, Ali M, Minihane M, Conrad P. Defining global health: findings from a systematic review and thematic analysis of the literature. *BMJ Glob Health*. 2021;6(6), e005292. https://doi.org/10.1136/bmjgh-2021-005292.
31. Adashi EY, Walters LB, Menikoff JA. The Belmont report at 40: reckoning with time. *Am J Publ Health*. 2018;108(10):1345–1348. https://doi.org/10.2105/AJPH.2018.304580.
32. Karan A. How should global health security priorities be set in the global north and west? *AMA J Ethics*. 2020;22(1):E50–E54. https://doi.org/10.1001/amajethics.2020.50.
33. Smith MJ, Upshur REG. Ebola and learning lessons from moral failures: who cares about ethics? *Publ Health Ethics*. 2015;8(3):305–318. https://doi.org/10.1093/phe/phv028.
34. Thomson S, Ip EC. COVID-19 emergency measures and the impending authoritarian pandemic. *J Law Biosci*. 2020;7(1):lsaa064. https://doi.org/10.1093/jlb/lsaa064.

35. Lillywhite L. Medical services policy in respect of detainees: evolution and outstanding issues. *BMJ Mil Health*. 2021;167(1):23–26. https://doi.org/10.1136/jramc-2019-001156.
36. Erfani P, Uppal N, Lee CH, Mishori R, Peeler KR. COVID-19 testing and cases in immigration detention centers, April–August 2020. *JAMA*. 2021;325(2):182–184. https://doi.org/10.1001/jama.2020.21473.
37. Hellewell J, Abbott S, Gimma A, et al. Feasibility of controlling COVID-19 outbreaks by isolation of cases and contacts. *Lancet Global Health*. 2020;8(4):e488–e496. https://doi.org/10.1016/S2214-109X(20)30074-7.
38. Beauchamp T, Childress J. Principles of biomedical ethics: marking its fortieth anniversary. *Am J Bioeth*. 2019;19(11):9–12. https://doi.org/10.1080/15265161.2019.1665402.
39. Moreno JD, Schmidt U, Joffe S. The Nuremberg code 70 years later. *JAMA*. 2017;318(9):795–796. https://doi.org/10.1001/jama.2017.10265.
40. World Medical Association. World Medical Association Declaration of Helsinki: ethical principles for medical research involving human subjects. *JAMA*. November 27, 2013;310(20):2191–2194. https://doi.org/10.1001/jama.2013.281053. PMID: 24141714.
41. *International Ethical Guidelines for Health-Related Research Involving Humans*. 4th ed. Geneva: Council for International Organizations of Medical Sciences (CIOMS); 2016.
42. van Delden JJM, van der Graaf R. Revised CIOMS international ethical guidelines for health-related research involving humans. *JAMA*. 2017;317(2):135–136. https://doi.org/10.1001/jama.2016.18977.
43. International Health Conference. Constitution of the World Health Organization. 1946. *Bull World Health Organ*. 2002;80(12):983–984.
44. Faden RR, Shebaya S, Siegel AW. Distinctive challenges of public health ethics. In: *The Oxford Handbook of Public Health Ethics*. 2019.
45. Meier BM, Evans DP, Phelan A. Rights-based approaches to preventing, detecting, and responding to infectious disease. *Infect Dis N Millennium*. 2020:217–253. https://doi.org/10.1007/978-3-030-39819-4_10.
46. Upshur RE. Principles for the justification of public health intervention. *Can J Public Health*. 2002;93(2):101–103. https://doi.org/10.1007/BF03404547.
47. Childress JF, Faden RR, Gaare RD, et al. Public health ethics: mapping the terrain. *J Law Med Ethics*. 2002;30(2):170–178. https://doi.org/10.1111/j.1748-720x.2002.tb00384.x.
48. The Nagoya Protocol on Access and Benefit-sharing. https://www.cbd.int/abs/. (Accessed 07.06.2021).
49. Youde J. *Biopolitical Surveillance and Public Health in International Politics*. Springer; 2010 (Accessed 07.06.2021) https://link.springer.com/chapter/10.1057%2F9780230104785_7.
50. Juran L, Trivedi J, Kolivras KN. Considering the "public" in public health: popular resistance to the Smallpox Eradication Programme in India. *Indian J Med Ethics*. 2017;2(2):104–111. https://doi.org/10.20529/ijme.2017.025.
51. Gavi. *COVAX Explained*; 2020 (Accessed 11.01.2021) https://www.gavi.org/vaccineswork/covax-explained.
52. Time, COVAX Was a Great Idea, But Is Now 500 Million Doses Short of Its Vaccine Distribution Goals. What Exactly Went Wrong? 2021. https://time.com/6096172/covax-vaccines-what-went-wrong/. (Accessed 11.01.2021).
53. Foucault M. *The Birth of the Clinic*. 3rd ed. Routledge; 2003. https://doi.org/10.4324/9780203715109.
54. Witter S, Govender V, Ravindran TKS, Yates R. Minding the gaps: health financing, universal health coverage and gender. *Health Policy Plan*. 2017;32(suppl_5):v4–v12. https://doi.org/10.1093/heapol/czx063.
55. WHO. *One Health*; 2007 (Accessed 06.12.2021) https://www.who.int/news-room/q-a-detail/one-health.
56. Corneli A, Forrest A, Swezey T, Lin L, Tenaerts P. Stakeholders' recommendations for revising good clinical practice. *Contemp Clin Trials Commun*. 2021;22, 100776. https://doi.org/10.1016/j.conctc.2021.100776.
57. Bolon B, Baze W, Shilling CJ, Keatley KL, Patrick DJ, Schafer KA. Good laboratory practice in the academic setting: fundamental principles for nonclinical safety assessment and GLP-compliant pathology support when developing innovative biomedical products. *ILAR J*. 2018;59(1):18–28. https://doi.org/10.1093/ilar/ily008.
58. Battisti WP, Wager E, Baltzer L, et al. Good publication practice for communicating company-sponsored medical research: GPP3. *Ann Intern Med*. 2015;163(6):461–464. https://doi.org/10.7326/M15-0288.
59. Plan S Making full and immediate Open Access a reality. https://www.coalition-s.org. (Accessed 08.09.2021).
60. a ICH. ICH M4E(R2). *Revision of M4E Guideline on Enhancing the Format and Structure of Benefit-Risk Information in ICH Efficacy M4E*; 2016. https://database.ich.org/sites/default/files/M4E_R2__Guideline.pdf. Accessed August 21, 2022;

    b Dutta MJ. *Communicating Health: A Culture-Centered Approach*. Polity; 2008.

61. Dutta MJ. Decolonizing communication for social change: a culture-centered approach. *Commun Theor.* 2015;25(2):123–143.
62. Dutta MJ. Culture-centered approach in addressing health disparities: communication infrastructures for subaltern voices. *Commun Methods Meas.* 2018;12(4):239–259.
63. Convention on Biological Diversity. *Article 19. Handling of Biotechnology and Distribution of its Benefits*; 2006 (Accessed 08.01.2021) https://www.cbd.int/convention/articles/?a=cbd-19.
64. Sedyaningsih ER, Isfandari S, Soendoro T, Supari SF. Towards mutual trust, transparency and equity in virus sharing mechanism: the avian influenza case of Indonesia. *Ann Acad Med Singapore.* 2008;37(6):482–488.
65. CIDRAP. *Indonesia details reasons for withholding H5N1 viruses*; July 15, 2008 (Accessed 08.01.2021) https://www.cidrap.umn.edu/news-perspective/2008/07/indonesia-details-reasons-withholding-h5n1-viruses.
66. Kebede S, Conteh IN, Steffen CA, et al. Establishing a national influenza sentinel surveillance system in a limited resource setting, experience of Sierra Leone. *Health Res Pol Syst.* 2013;11(22). https://doi.org/10.1186/1478-4505-11-22.
67. Steffen C, Debellut F, Gessner BD, et al. Improving influenza surveillance in sub-Saharan Africa. *Bull World Health Organ.* 2012;90(4):301–305. https://doi.org/10.2471/BLT.11.098244.

CHAPTER

3

National interagency collaboration for public health

Alex Riolexus Ario[1], *Benjamin Djoudalbaye*[2], *Saheedat Olatinwo*[3], *Abbas Omaar*[4], *Romina Stelter*[4] *and Ludy Suryantoro*[4]

[1]Uganda National Institute of Public Health, Kampala, Uganda; [2]Africa Centres for Disease Control and Prevention (Africa CDC), Addis Ababa, Ethiopia; [3]Rollins School of Public Health, Emory University, Atlanta, GA, United States; [4]World Health Organization (WHO), Geneva, Switzerland

Introduction: The need for collaboration

The International Health Regulations [IHR (2005)] and the Global Health Security Agenda (GSHA) encourage commitment and collaborative efforts between countries, organizations, and sectors to address and strengthen public health preparedness and response.[1,2] Despite this, effective interagency and multisectoral collaborations remain a challenge. In the wake of the COVID-19 pandemic, the gaps in interagency collaboration have become more apparent. Some of these gaps include information and communication challenges, administrative and logistical challenges, political challenges, tension, and unfavorable organizational cultures.[3,4] All of these may impact the ability to establish and maintain effective partnerships (e.g., through loss of power, the strain on resources, and delayed responses).

The emergence and re-emergence of zoonotic diseases from increased interaction between humans and wildlife/domestic and humans led to the field of the One Health approach. The One Health approach recognizes that the health of humans, animals, and ecosystems are interconnected. Health risks can emerge from various sources, and a multidisciplinary and cross-sectoral approach is necessary to address these risks.[4] In addition, the spread and impact of infectious disease are exacerbated by the social and economic conditions of the population. Thus, public health should be represented in activities, data, and policies developed across multiple sectors of government and stakeholders, utilizing a whole-of-society approach.

Modernizing Global Health Security to Prevent, Detect, and Respond
https://doi.org/10.1016/B978-0-323-90945-7.00006-3

The core responsibility of protecting public health lies with a nation's government. Public health security requires an appropriate and timely response, which relies on effective communication, and collaboration among all necessary agencies at all levels (local, state, national, and international).[5–7] During emergencies, it becomes essential for agencies (governmental and nongovernmental) and organizations across multiple sectors to work together to protect the public health.

In this chapter, interagency collaboration is defined as a process in which organizations come together and engage critical stakeholders in a coordinated and integrated effort—through joint decision-making, exchange of information and resources, and enhancing each other's capacity for mutual benefit—to improve health security globally. This involves organizations from different levels in the nation and across multiple systems and sectors. Interagency collaboration promotes program implementation and positive outcomes, sound policy development, information and resource sharing, effective monitoring and evaluation, and disease surveillance.

Failure to collaborate may result in duplication of efforts and waste of resources. More so, it may lead to more severe consequences in the context of a public health emergency due to the uncertain and disrupted environment. In different health areas, collaborative efforts have been seen to help make considerable progress in achieving health goals.[8] A notable example is the Access to COVID-19 Tools Accelerator (ACT-A)—a global collaboration of public health agencies to accelerate the development, production, and access to COVID-19 vaccines, diagnostics, and therapeutics.[5] Another is the African Vaccine Acquisition Trust (AVAT), a partnership among Africa Centers for Disease Control and Prevention (Africa CDC) and the African Union (AU), African Export-Import Bank (Afreximbank), and the United Nations Economic Commission for Africa (ECA) to safeguard access to COVID-19 vaccines across Africa.[6] At the international level, the Global Health Security Agenda (GHSA) serves as a good example of collaboration between countries and organizations. GHSA leverages and complements the strengths and resources of multisectoral and multilateral partners to address priorities and gaps in efforts to build and improve countries' capacity and leadership in the prevention and early detection of, and effective response to, infectious disease threats.[9]

Countries will only achieve progress against complex problems such as achieving health security through a collective practical approach across sectors. The purpose of this chapter is to expand the understanding of national interagency collaboration and provide guidance to enhance and sustain effective collaboration for public health preparedness and response. Our focus is at the national and local levels.

Literature review and gap analysis

Overview

National health systems and public health programs are developed based on local and national contexts, needs, and conditions. Several public health events, such as the Ebola virus pandemic in 2014 and the COVID-19 pandemic in 2020, have revealed that health security by and large is determined by extensive collaboration across multiple sectors based on their shared vision and collective actions. This requires a wide range of actors—government agencies, businesses or private sector, nongovernmental and nonprofit organizations, and international agencies—to effectively communicate, coordinate, collaborate, and build capacity at all public health emergency preparedness and response stages.[2,10]

The concept of interagency or intersectoral collaboration is not new and has been pushed forward since the early 2000s during the drive for integrated surveillance for various diseases,[11] One Health approach to address potential threats from infectious and chemical hazards at the interface of human health, animal health, and the environment,[10] Health in All Policies for population health promotion[10] and the revised International Health Regulations (IHR 2005).[12] More recently, Universal Health Coverage (UHC), the Sustainable Development Goals (SDGs), the Global Health Security Agenda (GHSA), and other efforts to prevent and protect public health have emphasized the importance and interdependence of various sectors, such as human and animal health, agriculture, education, security, law enforcement, social welfare, foreign affairs, trade, and finance, among others—in achieving its goals and objectives.[1,2,13–15]

Interagency collaboration here represents a fundamental aspect of preparing and responding to public health emergencies.[2] It incorporates the health sector and nonhealth sectors and requires action at several levels (local, national, regional, and global) by different organizations to potentially improve public health.[2] This includes collaborations that directly deliver health benefits and produce ripple effects for public health.

This chapter examined peer-reviewed literature and evidence from technical reports and guidelines[16] to assess interagency collaboration. We summarize the benefits, risks, existing challenges, and opportunities of interagency cooperation at the national level and describe key components or factors for effective collaboration. This is then followed by a case study of the One Health Platform in Uganda. Our analysis highlights key components that fostered collaboration, successes, and challenges, which allowed us to make recommendations to improve collaborative functioning.

Forms of collaboration

Interagency collaboration is integral to managing public health events and has the potential to promote a holistic approach to needs across different sectors. These collaborations occur in various forms and between different groups. Russel Dynes in his book, Organized Behavior in Disaster, categorized organizations into four (4) groups based on the nature of the task and organizational structure—established, expanding, extending, and emergent.[17]

Established organizations (type I) are those that exist and carry out their regular tasks during emergencies.[17] For example, in health emergencies, public health agencies have a role to protect the public and will continue activities such as case investigation and disease surveillance. Expanding organizations (type II) carry out their regular tasks and spread out their capacities and resources with a new structure within their organizations.[17] This may include the establishment of advisory groups made up of academic or professional associations, for example, the Technical Advisory Group (TAG) for Pan American Health Organization (PAHO) that provides technical recommendations on vaccine-preventable diseases for the WHO Region of the Americas and identifies research needs to strengthen immunization programs.[18] Extending organizations (type III) are existing organizations from other sectors that provide their expertise and resources for nonregular tasks in the emergency response.[17]

For example, the complex realities of border closures and mitigation measures during the COVID-19 pandemic resulted in unintended consequences such as a negative impact on the supply of essential goods and services, economic activities, and migrant communities.[19] This resulted in partnerships between the International Organization for Migration (IOM) and national and regional immigration and border authorities to support data collection, disease surveillance, risk communication, and many other interventions.[20] Lastly, emergent organizations (type IV) are groups of individuals or representatives of government agencies, newly formed to perform tasks that were not specified prior to the emergency.[17] The ACT-A is a great example of an emergent organization.[5]

Typically, these organizations include government agencies, intergovernmental organizations, nongovernmental organizations, the private sector, civil society organizations, and international development organizations, among others, all of which are crucial in emergencies.[2] In public health, different agencies interact and collaborate to complement and/or share skills, knowledge, and resources to achieve their desired goals and objectives. Interagency collaboration may serve various purposes including, but not limited to policy development, program implementation, technical assistance, research and development, advocacy, and resource mobilization. For example, the establishment of the Presidential Task Force (PTF) on COVID-19 in 2020 in Nigeria to provide coordination, strategy and policy guidance, including oversight of the different health (e.g., Nigeria Center for Disease Control; National Primary Health Care Development Agency; National Institute for Pharmaceutical Research and Development) and nonhealth (e.g., Federal Ministry of Humanitarian Affairs, Disaster Management and Social Development; Ministry for Information and Culture; National Emergency Management Agency) sectors.[21] Similarly, in 2004, the Ministry of Public Health in Thailand established the Committee on Trade in Health and Services to study the implication of trade liberalization policies on the Thai health sector.[22] The committee consists of representatives from the ministries of public health, commerce, food and agriculture, the Private Hospitals' Association, and various professional councils and sectors.[22]

One Health approach

The COVID-19 pandemic has underscored the need for multisectoral and multidisciplinary action using the One Health approach for research, decision-making, preparedness, and response to current and future pandemics.[23] Operationalizing the One Health approach at the global level followed long-term active collaboration and engagement among intergovernmental organizations: WHO and Food and Agriculture Organization of the United Nations (FAO), which joined together in 1962 to create the commission to support and operationalize the Codex Alimentarius.[4] Codex Alimentarius is a compendium of reference standards that address food production. That alliance was financially supported and provided a basis for interagency collaboration, which eventually evolved into "One Health."[24]

One Health is defined by the United Nations Quadripartite—FAO, the World Organization for Animal Health (OIE), the World Health Organization (WHO), and the United Nations Environment Program (UNEP) as an "integrated, unifying approach that aims to sustainably balance and optimize the health of people, animals and ecosystems."[25] The One Health approach recognizes the interdependence of humans, animals, plants, and the wider environment. This involves coordination, communication, collaboration, and capacity building across multiple disciplines and sectors to promote and protect health.[26]

The IHR 2005, GHSA 2030, Performance of Veterinary Services, and One Health approach among others have called for collaboration to respond to public health events. However, most countries have inadequate mechanisms for effective collaboration among the human, animal, and environmental sectors.[27,28] Some of the impediments to effective collaboration include the lack of joint preparation, coordination, and information sharing across relevant sectors that may lead to confusion, obstructed, or delayed response and failure to meet goals and objectives.[28] More recently, the UN Tripartite of FAO, OIE, and WHO, developed the Tripartite Zoonoses Guide (TZG), which considers the need for countries to adapt to their local contexts by providing guidance and tools for implementing the One Health approach to address public health threats at the interface of human health, animal health, and environmental systems. This will allow expanding surveillance systems and ensure effective coordination, communication, monitoring and evaluation, governance, and sustainable financing.[28]

The whole-of-society approach

A whole-of-society approach is critical in public health emergency preparedness and response.[29] Preventing or addressing public health threats requires health systems to interact with other government sectors (nonhealth sectors) at all levels and to collaborate and develop partnerships with the private sector, civil society, nongovernmental organizations, and international organizations.[29] The focus of the whole-of-society approach is on building and maintaining relationships and communication within and between health and nonhealth sectors and having a coordinated plan for public health emergency preparedness and response. This helps reduce the negative consequences of public health events not only for health but also for well-being and livelihood.

The WHO Multisectoral Preparedness Coordination Framework, developed in 2020, aims to improve collaboration and coordination between the health sector and all other relevant sectors using a whole-of-government, whole-of-society approach to advance health emergency preparedness and response. The framework provides states parties, ministries, and relevant sectors and stakeholders with an overview of the key elements for overarching, all-hazard, multisectoral coordination with a focus on emergency preparedness to result in a better response. It aims to improve coordination among relevant public stakeholders, particularly actors beyond the traditional health sector, such as finance, foreign affairs, interior and defense ministries, national parliaments, nonstate actors, and the private sector, including travel, trade, transport, and tourism, even before an emergency response is required.[2]

The guidance document discusses key elements for effective multisectoral coordination for health security preparedness, including high-level political commitment, country ownership and leadership, and formalizing mechanisms that contribute to multisectoral preparedness coordination. Transparency, trust, accountability, communication, resources, and monitoring will be required for the proper functioning of such mechanisms. Importantly, context-specific priorities need to be considered in developing coordination mechanisms that reflect the different needs, incentives, and contributions of the multisector stakeholders for engaging in health security preparedness.[2]

Sustaining collaboration for public health

Effective interagency collaborations offer an avenue to mobilize and pull resources, knowledge, and funds to create systems better equipped to address current and future public health challenges. For several systems, agencies, or sectors to work together, it is important to understand their individual roles and the ways they interact with each other through shared visions and policies. Understanding this relationship is crucial when building systems that support working together. Four key mechanisms—communication, coordination, collaboration, and capacity building are vital for different agencies to work together.[26,30,31]

Coordinated mechanisms are put in place for organizations or agencies to achieve common goals. This is done collectively involving representatives from the agencies involved. Coordinated systems have policies, mechanisms, committees, and terms to assign tasks, leveraging organizations' strengths and resources to ensure the overall goal is met efficiently. Coordination involves "aligning one's actions with those of other relevant actors and organizations to achieve a shared goal."[30] It requires interaction and cooperation between the actors to ensure they work together without interference while they achieve their own goals and overarching vision. Coordination forms the backbone for organizations across sectors or countries to work together, with collaboration being at the high end.[2] Many emergencies are coordinated through the thematic cluster approach, with the Cluster Coordinator or Cluster Lead facilitating the overall management of the cluster. The cluster lead or coordinator lays down structures, processes, principles, and commitments to coordinate support for responses based on the national government's request or international standards.[32]

Communication involves the exchange of data, information, knowledge, and experiences between agencies, organizations, and countries for mutual benefit. The communication process in emergency preparedness and response involves integrated and interoperable systems, transparency, timeliness, and frequency of information (information on risk, decisions, and resources).[2,32] The Integrated Surveillance System for Antimicrobial Resistance (AMR) and Antimicrobial Use (AMU) in Canada is an example of a surveillance system that fosters communication through assembled data on resistant nosocomial pathogens, zoonotic and food-borne pathogens, and elements of AMU in humans, animals, and horticulture across One Health sectors.[33]

Collaboration is about the different, but complementary and integrated roles, skills, and knowledge agencies bring together to achieve a shared vision. At this stage, different agencies explore and leverage their information and resources and are able to come together to exchange or merge these information and resources to achieve the desired goals. Here, decisions are made together, and efficiency is prioritized. Fundamentally, interagency collaboration requires a shared vision with aligned goals and a matrix to track progress, a strategy or roadmap for action, and effective governance to facilitate collective decision-making and ensure accountability.[2,31,34,35]

Capacity building will enhance sustainable collaboration by addressing capacity and knowledge gaps among the collaborating agencies or sectors. It is necessary to identify the opportunities, capabilities, and needs of each multisectoral agency in order to harmonize knowledge and expectations among all agencies. Capacity building occurs at different levels, including the individual level (human resource development), organizational level (organizational development), and system level (institutional and legal framework development).[2,26,36]

Enabling factors for collaboration

There are several factors that enable and prevent effective interagency collaboration for public health. Some of the major motivating factors for interagency collaboration are political will at all levels of government to support commitment to collaborative efforts.[2] This includes focus on policy development or changes, commitment to national and international strategic plans or goals, and availability of sustainable resources (financial, technical, human, and infrastructure) across the various sectors for collaborative implementation of activities.

Another important factor is a mutual understanding of the issues, vision, and actions. This builds on trust and understanding and ensures that partners, stakeholders, or agencies have coordinated actions, monitoring, evaluation, and public health outcomes. As agencies formalize their collaborative relationships, they strengthen their commitment to and accountability for the agreed goals and policies.[2,16,36]

Effective communication is required to achieve operational and technical goals. Regular communication between collaborating members allows the exchange of information among members to discuss collaboration goals, mechanisms, and issues. Communication through interoperable and integrated systems, joint evaluation, and the establishment of risk communication channels or protocols increases transparency and credibility and helps meet the technical goals of the collaboration.[2,16]

A clear leadership structure is required to institutionalize collaboration and facilitate technical efforts. Within the country, leadership may be in a single agency or institution, shared between agencies, or can rotate on an agreed schedule. The efficient functioning of collaborative agencies remains the responsibility of the lead and all participating members and requires active participation by all members.[2,36]

Challenges and issues

There have been multiple successful national interagency collaborations to protect health across the globe. Nonetheless, interagency collaboration is faced with several challenges and risks that have made its implementation or sustenance difficult. A major limitation is inadequate adoption of collaboration guidance and mechanisms at the national level and failure to adapt standard guidelines at a global level to fit local context.[37] Another major challenge is the lack of effective communication among agencies (public and private and health and non-health). Ku, Han, and Lee discussed the limited sharing of information and knowledge which resulted in delays in diagnosis, isolation, understanding, and knowledge of the virus during the Middle East respiratory syndrome coronavirus (MERS-CoV) outbreak in South Korea in 2015.[38] Similarly, the lack of an integrated or interoperable system also contributes to poor communication and delayed response to emergencies.[39]

Since the national government has the prime responsibility for protecting the public, political will has a strong influence on interagency collaboration. Underutilization and underfunding of agencies may result in shortage of essential resources, lack of governance capacity, and trust in government.[39] Cultural differences between different agencies can also cause a strain on collaborative efforts.[39,40] Some publications and reports have explained that challenges such as physical barriers (including working internationally) and differences in salaries for the same job type have an impact on maintaining communication, and building relationships and trust.[2,39,40] Some other challenges to effective interagency collaboration include difficulty synchronizing work activities; inflexible organizational structures; unclear vision, goal, and responsibilities.[2,40–43]

CASE STUDY

Uganda One Health Platform

Background

Uganda has been described as one of the Congo basin's "hot spots" for emerging and re-emerging infections where over 70% of the diseases are of zoonotic nature.[44,45] Since 2000, numerous outbreaks of viral hemorrhagic fevers and zoonotic influenza have occurred. In addition, endemic zoonotic diseases like rabies, brucellosis, anthrax, cysticercosis, tuberculosis, and trypanosomiasis continue to cause illness and death among humans, domestic animals, and wildlife leading to direct and indirect economic losses.[46] The huge burden and threat of zoonotic diseases prompted the government to adopt a One Health approach to strengthen and institute formal structures for collaboration between ministries and sectors responsible for human, livestock, wildlife, and environmental health to address these challenges in a coordinated and sustainable way.[47] This approach was to promote multisectoral and multidisciplinary collaboration to strengthen the prevention and management of zoonotic diseases and other public health and environmental challenges.[44,47]

Interventions

- A National One Health Platform (NOHP) was established in 2016 to strengthen and formalize multisectoral and multidisciplinary collaboration by having in place organizational structures, frameworks, and resources to address zoonotic disease challenges.[44]
- NOHP consisted of four government sectors—Ministry of Health, Ministry of Water and Environment, Ministry of Agriculture, Animal Industry and Fisheries, and Uganda Wildlife Authority—and was led by the Ministry of Health. The Ministry of Defense and Veteran Affairs has been an active participant since the formation of the NOHP.
- The One Health Technical Working Group (TWG) developed the One Health Framework with a Memorandum of Understanding.[48]
- The One Health TWG composed of 38 technical experts and policymakers from government and other stakeholders with clear Terms of Reference was operationalized. This TWG which comprises high level technical officers in various ministries and agencies, developed all the technical documents of the One Health Platform for Uganda.
- The One Health Coordination Office was formed and located within the Uganda National Institute of Public Health, Ministry of Health.
- A 5-year One Health Strategic Plan, 2018–22 was crafted and launched.[44]

Observations

Since the launch of the NOHP in 2016, implementation of the plan has been going on with support from the government and partners. The One Health TWG meets every quarter, and the chair rotates among the four

CASE STUDY *(cont'd)*

consortium ministries every 6 months. The TWG also calls for adhoc meetings during urgent situations. The availability of technical officers trained by the Africa One Health University Network (AFROHUN) (formerly One Health East and Central Africa) has strengthened the implementation process.[49]

Key issues

The process by which ministries, departments, agencies, and stakeholders collaborate to address zoonotic disease challenges have developed well. The regular meetings of the TWG and Directors of the different ministries with critical partners like World Health Organization, Food and Agriculture Organization (FAO), and United States Centers for Disease Control and Prevention (US CDC) have provided valuable working points for the implementation of priority diseases programs.

Lessons learned

- It is possible to bring together different groups focused on different disciplines using a common binding factor, in this case One Health
- Establishing a common goal allows for effective implementation across ministries
- Using Directors to chair the meetings has led to the building of trust between ministries and agencies
- A strong leadership as exhibited by the Ministry of Health is essential in interagency collaboration

What worked

- Implementation of the One Health Framework
- The establishment of a One Health Coordination Office improved the implementation of programs and supported collaboration
- Frequent and effective meetings ensured collective decision-making and resource allocation
- Funding of the components of the strategic plan which is in tandem with the National Plan for Health Security through partner's support

What challenges were observed?

- Funding of the One Health Coordination Office would benefit from greater resourcing, for example, staffing, capacity building, infrastructure, etc.
- Incorporating the One Health Strategic Plan in the sector ministries and agencies plans needs to be fully achieved. Roll out of One Health to sub-national levels has been challenged with only 27 of 135 districts having District One Health Teams.
- Sharing of information is limited, only antimicrobial resistance data are shared regularly through the Data Information Sharing Portal.
- Greater funding of the OH strategic plan is needed

Recommendations

- Strengthen appropriation of the One Health program at the country level by allocating greater resources and specific budget lines from nationally available resources to accelerate the implementation of the One Health strategic plan
- Incorporate One Health strategic plan in the sector, ministries, and agency plan

(Continued)

CASE STUDY *(cont'd)*

- Strengthen surveillance systems at national and district levels to allow interoperable system for electronic reporting from human health, animal health, and environmental systems across ministries and agencies
- Advocate for funding, staffing, and equipping the One Health Coordination Office and Strategic Plan
- Establish a public health emergency financing mechanism for easy and timely access to funds to respond to emergencies

Vision for the future

The mechanisms for and types of collaboration in public health are diverse, as can be seen by examples in this chapter. The case study on the National One Health Platform in Uganda reflects a single type of collaboration—an integrated One Health approach—for health security.

Interagency collaboration promotes functional working relationships among multiple agencies to improve process outcomes while maximizing resources. No single institution or agency has sufficient knowledge and resources to handle it alone. Therefore, interagency collaboration is a sustainable and practical approach to be put into practice for successful health security from the central to the field level in every country. This also extends beyond COVID-19 to all aspects of public health and thus, to successfully address public health challenges, we need to work together: collaborate.

Though each collaboration has unique benefits and dynamics, the concepts, and recommendations in this chapter layout a conceptual vision that can be considered and adapted. This chapter envisions a driven multisectoral and multidisciplinary system that seeks to promote and protect the public against public health threats. (a) A system where public health agencies will form partnerships and networks across sectors with shared or complementary goals and shared indicators to evaluate performance against those goals to hold each other accountable. The future, powered at the local level, will enable cross-sector, real-time data-sharing. (b) A system where public health agencies are integrated with health and nonhealth partners from private and nonprofit sectors to effectively design and implement policies or programs and respond efficiently to tackle public health events.

Actions

All recent health emergencies including COVID-19 have underscored the urgency and importance of interagency collaboration as a vital means for ensuring robust preparedness and response at national and sub-national levels. It is crucial that interagency collaboration is scaled-up to advance preparedness and response, and it is vital that individual agencies prioritize the strengthening of their capacity to effectively participate in joint action for health security with other stakeholders.[2] Below are some key actions/recommendations for consideration that are proposed to support achieving this.

1. Strengthen the IHR (2005) through technical amendments to the IHR (2005) in order to foster the multisectoral and cross-sectoral coordination and implementation of capacity building activities, detection and response capacities, addressing nontraditional health stakeholders such as national parliaments, security authorities including the militaries, and local authorities in cities and urban settings.
2. Cross-agency engagement for health security should be set against a robust strategy that guides the nature, scope, and approach for collaboration among partners. The strategy should be developed from the outset of collaborative efforts (where possible) and regularly reviewed to ensure that the implementation of cross-agency collaboration is in line with stated aims and objectives and that the desired expectations are met. Interagency collaboration within a country should also be carried out under the authority of the highest possible levels of political—and organizational leadership to facilitate the greatest level of commitment among partners and achieve a whole-of-government contribution to preparedness efforts. This will support continued alignment among stakeholders as collaboration efforts are carried out, and it can broaden the level of engagement to a wide range of agencies and stakeholders from across government ministries and sectors (e.g., health, finance, agriculture, travel and trade, parliaments, technology, etc.). Although the incentives and contributions that different agencies can have may differ, diversity among collaborative actors can lead to new and unique perspectives being introduced, which can ultimately be highly valuable for complementary action in building preparedness.
3. At the practical level, interagency collaboration should be guided by comprehensive workplan activities that indicate the full scope of roles and responsibilities among partners/stakeholders. Although it may not be always possible to establish and adhere to joint work plans—especially during times of health emergencies where the nature of interagency collaboration must frequently adapt as needed—the presence of a well-established work plan can enable effective engagement among partner agencies. Clear work plans with identifiable milestones can also facilitate stakeholders to monitor and evaluate the progress of joint action for preparedness and anticipate bottlenecks ahead of time.
4. National and international agencies that are involved in health security strengthening should establish and maintain collaborative linkages before a global health emergency occurs. Often, interagency coordination and collaboration in the context of health emergencies is only set up during a response action when a crisis occurs, and it is likely that this is carried out in light of specific diseases or hazards. It is also common to see that such collaboration lasts only for the duration of the health emergency itself and those interagency linkages, which are formed often disappear once a public health event is controlled. However, proactively seeking where potential linkages between national and international agencies can be made before a health emergency occurs should facilitate

the collaborative approach that is taken at all stages including during a time of crisis. Much progress has been made in advancing interagency collaboration during COVID-19, and it is imperative that this must be carried forward beyond the pandemic. Although much of the recent interagency collaboration has been based on addressing the risks and impact of the pandemic specifically, it would be useful for this to be adapted toward addressing other health emergency threats through an all-hazards approach. This would better enable the impact of existing interagency collaboration for health security to extend to other public health threats.

5. Global guidance such as the WHO Multisectoral Preparedness Coordination (MPC) Framework, the WHO National Collaboration Framework for Civil Military Collaboration, or IPU-WHO Handbook for parliamentarians on Strengthening Health Security can support countries when planning and implementing sustainable interagency collaboration. Countries need to contextualize and adapt their interagency and multisectoral collaboration. There is no "one size fits all" for multisectoral collaboration.
6. There should be regular information sharing between agencies to enhance opportunities for collaboration. It is clear that agencies will not always be able to provide the full amount of information that is available however, sharing knowledge, information, and experiences between agencies can support continued alignment among partners in ways that facilitate engagement toward common action for preparedness. Any barriers or boundaries for information sharing—including those relating to potential conflicts of interest—should be openly discussed at the outset of the establishment of interagency collaboration in order and during meetings as other issues arise to avoid and resolve potential challenges.
7. Interagency collaboration should be based on principles of shared financial and technical resources where possible, including shared human resources. This will support the sustainability of interagency collaborative action and partnerships and provide a basis of reliability that stakeholders can build upon to feel confident in committing significantly toward working with each other. In particular, it is useful to establish costing needs early on and to carry out regular resource mobilization to ensure that the availability of funding and other resources is not a challenge.
8. Establishing a steering committee to make strategic goals and decisions to strengthen domestic health security. The committee convenes political representatives, members of the collaboration agencies (including the leadership), as well as invited public and private stakeholders. This will enhance political and stakeholder buy-in and accountability. The committee facilitates the building of a collaborative environment for information and knowledge sharing among participating agencies and members. In addition, smaller groups or subcommittees can be established to facilitate prompt decision-making or to achieve specific objectives.
9. All instances of interagency collaboration, including at the sub-national, regional, national, and global levels, should regularly document and disseminate knowledge about their experiences. This may be developed by the technical groups or subcommittees based on their knowledge and experiences and commissioned by the steering committee. Identifying successes and challenges is highly valuable to enable others to understand what has worked well and what measures should potentially be avoided in other ventures of interagency collaboration. Clear documentation and wide dissemination of

best practices also limit the scope for potential duplication, and it also presents stakeholders with an understanding of opportunities where they may be able to provide complementary contributions that can build upon existing progress.

References

1. Katz R, Sorrell EM, Kornblet SA, Fischer JE. Global health security agenda and the international health regulations: moving forward. *Biosecur Bioterror*. Sep-Oct 2014;12(5):231–238. https://doi.org/10.1089/bsp.2014.0038.
2. World Health Organization. *Multisectoral Preparedness Coordination Framework: Best Practices, Case Studies and Key Elements of Advancing Multisectoral Coordination for Health Emergency Preparedness and Health Security*. 2020.
3. Errecaborde KM, Macy KW, Pekol A, et al. Factors that enable effective One Health collaborations - a scoping review of the literature. *PLoS One*. 2019;14(12):e0224660. https://doi.org/10.1371/journal.pone.0224660.
4. Mackenzie JS, Jeggo M, Daszak P, Richt JA. In: Compans RW, ed. *Food Safety and Security, and International and National Plans for Implementation of One Health Activities*. One Health: The Human–Animal–Environment Interfaces in Emerging Infectious Diseases.
5. World Health Organization. What is the ACT-Accelerator. World Health Organization. https://www.who.int/initiatives/act-accelerator/about. Accessed January, 2022.
6. Africa Centres for Disease Control and Prevention. A Partnership for COVID-19 Vaccination in Africa. Africa Centres for Disease Control and Prevention. https://africacdc.org/news-item/mastercard-foundation-and-africa-cdcs-saving-lives-and-livelihoods-initiative-delivers-first-tranche-of-over-15-million-vaccines/. Accessed July, 2022.
7. Organization WH. *The World Health Report 2007: A Safer Future: Global Public Health Security in the 21st Century*. World Health Organization; 2007.
8. Fidler DP. Public health and national security in the global age: infectious diseases, bioterrorism, and realpolitik. *Geo Wash Int'l L Rev*. 2003;35:787.
9. Global Health Security Agenda (GHSA). *Global Health Security Agenda (GHSA) 2024 Framework*; 2018. https://ghsagenda.org/wp-content/uploads/2020/06/ghsa2024-framework.pdf.
10. Zinsstag J, Schelling E, Waltner-Toews D, Tanner M. From "one medicine" to "one health" and systemic approaches to health and well-being. *Prev Vet Med*. September 1, 2011;101(3–4):148–156. https://doi.org/10.1016/j.prevetmed.2010.07.003.
11. World Health Organization, Centres for Disease Control and Prevention, United States Agency for International Development, Support for Analysis and Research in Africa, United Nations Foundation Interagency Partnership. *The Implementation of Integrated Disease Surveillance and Response in the African and Eastern Mediterranean Regions: Synthesis Report*; 2003:26. https://pdf.usaid.gov/pdf_docs/Pnads693.pdf. Accessed March, 2022.
12. Institute of Medicine (US) Forum on Microbial Threats. *Global Public Health Governance and the Revised International Health Regulations. Infectious Disease Movement in a Borderless World: Workshop Summary*. The National Academies Press; 2010. https://www.ncbi.nlm.nih.gov/books/NBK45725/.
13. Wolicki SB, Nuzzo JB, Blazes DL, Pitts DL, Iskander JK, Tappero JW. Public health surveillance: at the core of the global health security agenda. *Health Secur*. May-Jun 2016;14(3):185–188. https://doi.org/10.1089/hs.2016.0002.
14. World Health Organization. *Thirteenth General Programme of Work 2019–2023*; 2018. https://apps.who.int/iris/bitstream/handle/10665/324775/WHO-PRP-18.1-eng.pdf.
15. Hinton R, Armstrong C, Asri E, et al. Specific considerations for research on the effectiveness of multisectoral collaboration: methods and lessons from 12 country case studies. *Glob Health*. 2021;17(1):18. https://doi.org/10.1186/s12992-021-00664-w.
16. Olatinwo S. *Interagency Collaboration in Public Health Emergencies*. Emory University: Hubert Department of Global Health; 2022.
17. Dynes R. *Organized Behavior in Disaster*. Health and Company; 1970.
18. Pan American Health Organization. *XXVI Meeting of PAHO's Technical Advisory Group (TAG) on Vaccine-Preventable Diseases*; 2021. https://iris.paho.org/bitstream/handle/10665.2/54833/PAHOFPLIMCOVID-19210038_eng.pdf?sequence=1&isAllowed=y. Accessed March, 2022.

19. Organisation for Economic Co-operation and Development. *The Territorial Impact of COVID-19: Managing the Crisis across Levels of Government. OECD Policy Responses to Coronavirus (COVID-19)*; 2020. https://read.oecd-ilibrary.org/view/?ref=128_128287-5agkkojaaa&title=The-territorial-impact-of-covid-19-managing-the-crisis-across-levels-of-government.
20. International Organization for Migration. *COVID-19: Immigration and Border Management Response*; 2020. https://www.iom.int/sites/g/files/tmzbdl486/files/documents/en_covid-19ibmresponseinfosheet_3pages.pdf.
21. *National COVID-19 Pandemic Multi-Sectoral Response Plan 47*. 2020.
22. Smith RD, Lee K, Drager N. Trade and health: an agenda for action. *Lancet*. 2009;373(9665):768–773. https://doi.org/10.1016/S0140-6736(08)61780-8.
23. Häsler B, Bazeyo W, Byrne AW, et al. Reflecting on one health in action during the COVID-19 response. Perspective. *Front Vet Sci*. 2020-October-30;7. https://doi.org/10.3389/fvets.2020.578649.
24. Lee K, Brumme ZL. Operationalizing the One Health approach: the global governance challenges. *Health Pol Plann*. 2012;28(7):778–785. https://doi.org/10.1093/heapol/czs127.
25. World Health Organization. Tripartite and UNEP support OHHLEP's definition of "One Health". World Health Organization. https://www.who.int/news/item/01-12-2021-tripartite-and-unep-support-ohhlep-s-definition-of-one-health. Accessed March, 2022.
26. World Health Organization. *Strengthening WHO Preparedness for and Response to Health Emergencies: Strengthening Collaboration on One Health*; 2022. https://apps.who.int/gb/ebwha/pdf_files/WHA75/A75_19-en.pdf. Accessed August 16, 2022.
27. *WHO Regional Office for the Eastern Mediterranean (WHO EMRO). Zoonotic Disease: Emerging Public Health Threats in the Region*; 2011. http://www.emro.who.int/fr/pdf/about-who/rc61/zoonotic-diseases.pdf?ua=1. Accessed April, 2022.
28. World Health Organization, Food Agriculture Organization of the United Nations, World Organisation for Animal Health. *Taking a Multisectoral, One Health Approach: A Tripartite Guide to Addressing Zoonotic Diseases in Countries*. World Health Organization; 2019.
29. Mosselmans M, Waldman R, Cisek C, Hankin E, Arciaga C. *Beyond Pandemics: A Whole-Of-Society Approach to Disaster Preparedness*. 2011.
30. Comfort LK. Crisis management in hindsight: cognition, communication, coordination, and control. *Publ Adm Rev*. 2007;67(s1):189–197. https://doi.org/10.1111/j.1540-6210.2007.00827.x.
31. Axelsson R, Axelsson SB. Integration and collaboration in public health–a conceptual framework. *Int J Health Plann Manage. Jan-Mar*. 2006;21(1):75–88. https://doi.org/10.1002/hpm.826.
32. Clarke PK, Campbell L. *Exploring Coordination in Humanitarian Clusters*; 2015. https://reliefweb.int/sites/reliefweb.int/files/resources/study-coordination-humanitarian-clusters-alnap-2015.pdf.
33. Otto SJG, Haworth-Brockman M, Miazga-Rodriguez M, Wierzbowski A, Saxinger LM. Integrated surveillance of antimicrobial resistance and antimicrobial use: evaluation of the status in Canada (2014-2019). *Can J Public Health = Revue canadienne de sante publique*. 2022;113(1):11–22. https://doi.org/10.17269/s41997-021-00600-w.
34. Feiock RC. The institutional collective action framework. *Pol Stud J*. 2013;41(3):397–425. https://doi.org/10.1111/psj.12023.
35. Emerson K, Nabatchi T. Evaluating the productivity of collaborative governance regimes: a performance matrix. *Publ Perform Manag Rev*. 2015;38(4):717–747. https://doi.org/10.1080/15309576.2015.1031016.
36. Organization WH. *Stronger Collaboration for an Equitable and Resilient Recovery towards the Health-Related Sustainable Development Goals, Incentivizing Collaboration: 2022 Progress Report on the Global Action Plan for Healthy Lives and Well-Being for All*. 2022.
37. Abbas SS, Shorten T, Rushton J. Meanings and mechanisms of One Health partnerships: insights from a critical review of literature on cross-government collaborations. *Health Pol Plann*. 2021;37(3):385–399. https://doi.org/10.1093/heapol/czab134.
38. Ku M, Han A, Lee KH. The dynamics of cross-sector collaboration in centralized disaster governance: a network study of interorganizational collaborations during the Mers epidemic in South Korea. *Int J Environ Res Publ Health*. December 21, 2021;19(1). https://doi.org/10.3390/ijerph19010018.
39. Kapucu N, Hu Q. An old puzzle and unprecedented challenges: coordination in response to the COVID-19 pandemic in the US. *Publ Perform Manag Rev*. 2022:1–26. https://doi.org/10.1080/15309576.2022.2040039.
40. Raftery P, Hossain M, Palmer J. A conceptual framework for analysing partnership and synergy in a global health alliance: case of the UK public health rapid support team. *Health Policy Plan*. Mar 4 2022;37(3):322–336. https://doi.org/10.1093/heapol/czab150.

41. Bistaraki A, Georgiadis K, Pyrros DG. Organizing health care services for the 2017 "Athens Marathon, the authentic:" perspectives on collaboration among health and safety personnel in the marathon command center. *Prehospital Disaster Med.* October 2019;34(5):467–472. https://doi.org/10.1017/s1049023x19004722.
42. Antonio CAT, Bermudez ANC, Cochon KL, et al. Recommendations for intersectoral collaboration for the prevention and control of vector-borne diseases: results from a modified delphi process. *J Infect Dis.* October 29, 2020;222(Suppl 8):S726–s731. https://doi.org/10.1093/infdis/jiaa404.
43. Torti C, Zazzi M, Abenavoli L, et al. Future research and collaboration: the "SINERGIE" project on HCV (South Italian network for rational guidelines and international epidemiology). *BMC Infect Dis.* 2012;12(2):S9. https://doi.org/10.1186/1471-2334-12-S2-S9.
44. *Uganda One Health Strategic Plan 2018-2022 52.* 2018.
45. Wang L, Crameri G. Emerging zoonotic viral diseases. *Rev Sci Tech.* 2014;33(2):569–581.
46. Centers for Disease Control and Prevention (CDC). *Workshop Summary: One Health Zoonotic Disease Prioritization for Multi-Sectoral Engagement in Uganda;* 2017:24. https://www.cdc.gov/onehealth/pdfs/uganda-one-health-zoonotic-disease-prioritization-report-508.pdf. Accessed March, 2022.
47. Sekamatte M, Krishnasamy V, Bulage L, et al. Multisectoral prioritization of zoonotic diseases in Uganda, 2017: a One Health perspective. *PLoS One.* 2018;13(5):e0196799. https://doi.org/10.1371/journal.pone.0196799.
48. *Framework to Strengthen One Health Approach to Prevent and Control Diseases in Uganda.* 2016.
49. Buregyeya E, Atusingwize E, Nsamba P, et al. Operationalizing the one health approach in Uganda: challenges and opportunities. *Journal of epidemiology and global health.* 2020;10(4):250–257. https://doi.org/10.2991/jegh.k.200825.001.

CHAPTER

4

The imperative for global cooperation to prevent and control pandemics

Robert Agyarko[1], *Fatima Al Slail*[2], *Denise O. Garrett*[3], *Brittany Gentry*[4], *Louise Gresham*[5], *Marika L. Kromberg Underwood*[6], *Sarah B. Macfarlane*[7] *and Mohamed Moussif*[8]

[1]Outbreaks and Epidemics: African Risk Capacity (ARC), Johannesburg, Gauteng, South Africa; [2]Public Health Agency, Ministry of Health, Riyadh, Saudi Arabia; [3]Applied Epidemiology Team, Sabin Vaccine Institute, Washington, DC, United States; [4]Hubert Department of Global Health, Rollins School of Public Health, Emory University, Atlanta, GA, United States; [5]PAX sapiens and School of Public Health, San Diego State University, Washington, DC, United States; [6]Centre for Sustainable Healthcare Education, University of Oslo, Oslo, Norway; [7]Epidemiology and Biostatistics, and Institute for Global Health Sciences, University of California San Francisco, San Francisco, CA, United States; [8]Casablanca International Airport, Ministry of Health, Casablanca, Morocco

Introduction

The [COVID-19] pandemic has demonstrated that highly infectious pathogens cannot be contained by any single sovereign state. We can only control them by working together in solidarity and with a one health approach that addresses the links between human, animal and planetary health. ***Tedros Adhanom Gebreyesus, Executive Director, World Health Organization, October 2020.***[1]

The COVID-19 pandemic reminds us that "unless everyone is safe from the virus, no one is safe."[2] The world's communal safety depends not only on effective global structures, but also on regional and cross-border cooperation to prevent, treat, and control this virus and the

https://doi.org/10.1016/B978-0-323-90945-7.00019-1

next. Many stakeholders including governments, nongovernmental organizations, the private sector, philanthropy, the media, and communities have to learn to work together to build preparedness and response capacities, generate knowledge, and fairly distribute scarce resources. These cooperative functions are challenging during routine operations and grow exponentially in complexity when stress-tested by outbreaks.

COVID-19 revealed the fragility of global cooperation. The pandemic demonstrated the need for the world to act collectively and with clarity and that countries cannot fight global health threats independently. Successful detection, validation, and response to potential pandemics call for informal, formal, and codified cooperation among leaders and organizations.

Two years into the pandemic, there were still extreme shortages of ventilators, oxygen, diagnostics, and personal protective equipment especially in low- and lower-middle-income countries (LMICs), and worldwide disparities in the distribution of COVID-19 vaccines. At the end of 2021, 75% of people living in high-income countries (HICs) had received at least one dose of vaccine compared to 7.5% living in low-income countries (LICs).[3] Many have spoken out about the "shocking imbalance"[4] and "horrifying inequity"[5] as being "immoral"[6] but these worldwide disparities remain.

Communal safety would seem to be a clear incentive for the world to cooperate. However, the COVID-19 pandemic exposed a lack of solidarity among nations; countries struggled to formulate shared aims, even in HICs and rising hypernationalism undermines the World Health Organization (WHO) and other global frameworks. In 2021, a Pew Research Center poll of 14 advanced economies indicated "a prevailing view that international cooperation would have reduced coronavirus cases".[7]

The world has had varied successes in cooperating to control health emergencies. The HIV/AIDS pandemic emerged in the early 1980s and was followed by pandemic threats such as SARS in 2003, Swine flu (H1N1) in 2009–10, MERS in 2012, the Ebola outbreaks in West Africa between 2014–16, and then in Central Africa between 2018–19, followed by Zika in 2015–16 and COVID-19 from 2019.[8,9] We review the mechanisms of cooperation set up to tackle these health emergencies or that were created to respond to the COVID-19 pandemic and we make recommendations to strengthen these mechanisms to enhance the global response to future pandemics. Global cooperation arose to deal with the challenge but stakeholders hold mixed opinions on the outcomes.

Literature review and gap analysis

Using peer-reviewed literature and contemporaneous reports and communiques,[10] we reviewed key mechanisms for international cooperation to address health emergencies such as COVID-19 and identify some of the obstacles that COVID-19 highlights.

Our literature review revealed multiple forms of cooperation by type and level of stakeholder. While governments are responsible for protecting their own populations from public health emergencies and for meeting internationally agreed regulations, other stakeholders have important roles in the complex mosaic of global cooperation, including international bodies, academic institutions, philanthropists, not-for-profit and community/civil society organizations, individuals, the media, and several others. These players come together—globally, regionally, and nationally—as formal and informal networks, networks of networks, alliances, public–private partnerships, and social media platforms.

Cooperative activities around health emergencies generally aim to address one or more of the following functions: (1) building national capacity to prevent, detect, report, and respond effectively to an outbreak, and to monitor and control its progress; (2) maintaining a global early warning and disease surveillance system to prevent, detect, report, and respond effectively to an outbreak and to monitor and control its progress; (3) generating knowledge to prevent and treat the infective agent virus with global dissemination of publicly accessible knowledge; and/or (4) producing and fairly distributing products such as vaccines.

We start this chapter by reviewing the governance of pandemics and the WHO International Health Regulations (IHR). The IHR and other collaborative modalities we discuss below provide a prism through which we can analyze existing collaboration or lack of, and how this may have impacted the global COVID-19 pandemic response. We then identify the major cooperative initiatives that have been active in responding to COVID-19. Our case study on the COVID-19 Vaccines Global Access Facility (COVAX), the vaccine pillar of ACT-A,[11] provides an analysis of the distribution of vaccines as the pandemic unfolded, detailing successes, challenges, and shortfalls. We conclude the chapter by contrasting the normative and institutional power of WHO to implement the IHR against its lack of structural power to enforce them; and discuss the opportunities presented by the proposed International Treaty on Pandemics.[12]

Global governance of pandemics

The global community established WHO in 1948 to respond to global health emergencies such as the 1918 influenza epidemic, and to prevent cholera, yellow fever, tuberculosis, and typhoid spreading across international borders. Governed by Member States through the World Health Assembly (WHA), WHO is the only global body responsible for coordinating global health policy, preparedness, and response. Through its six regional offices and its headquarters in Geneva, WHO works with its Member States to implement the resolutions of the WHA.[13]

Early in the COVID-19 pandemic, recognizing that pathogens cannot be contained by any single sovereign state, the WHA proposed a one health approach.[14] The Food and Agriculture Organization of the United Nations (FAO), the World Organization for Animal Health (OIE), the United Nations Environment Program (UNEP), and WHO developed the following operational definition: "One Health is an integrated, unifying approach that aims to sustainably balance and optimize the health of people, animals and ecosystems."[15] In 2021, WHO prepared a COVID-19 Strategic Preparedness and Response Plan (SPRP 2021) to guide "national, regional, and global levels to overcome the ongoing challenges in the response to COVID-19, address inequities, and plot a course out of the pandemic."[16]

The 58th WHA adopted the WHO IHR (2005)[17] based on the support and cooperation of intergovernmental organizations and international bodies. The purpose and scope of the IHR are "to prevent, protect against, control and provide a public health response to the international spread of disease in ways that are commensurate with and restricted to public health risks, and which avoid unnecessary interference with international traffic and trade."[17] The IHR (2005) entered into force on June 15, 2007.

As a legally binding instrument (but one that has no enforcement capability), the IHR require Member States to establish core capacities to early detect, assess, notify, and better respond to potential public health events of international concern. The IHR refer to international collaboration in several ways: with the provision of technical cooperation, logistical support, financial resources, and formulation of proposed laws; with special treaties or arrangements for direct and rapid exchange of public health information between neighboring countries; and with bilateral or multilateral agreements to prevent or control the international transmission of disease at ground crossings.

The limited adherence to the IHR 2005 was not enough to assure capacity at the country level nor cooperation at a global level to prevent early pneumonia clusters from becoming the COVID-19 pandemic. International coordination and collaboration can be greatly expanded by IHR Focal Point organizations.[18] These focal points are critical to the timely, secure sharing of information but their current lack of authority significantly limits and delays notifications. Political support and resources for focal points are needed to support fuller implementation of the IHR.

Mechanisms of cooperation to prevent and mitigate health emergencies

Global and regional surveillance

One of WHO's earliest activities, in 1952, was to form the *Global Influenza Surveillance Network* "to provide early warning of changes in influenza viruses circulating in the global population to help mitigate the consequences of such a pandemic and maintain the efficacy of seasonal influenza vaccines."[19] Now known as the Global Influenza Surveillance and Response System (GISRS), this WHO-led network coordinates 150 laboratories in 114 countries[20] through WHO-designated National Influenza Centers and WHO Collaborating Centers on influenza. In 2011, largely as a response to the emergence of Avian Influenza (H5N1), GISRS was adjusted and expanded under the Pandemic Influenza Preparedness (PIP) framework.[21]

PIP included the establishment of the Global Initiative on Sharing Avian Influenza Data (GISAID), an open-access database of genomic data of influenza viruses which has been invaluable during the COVID-19 pandemic. By April 2021, scientists collaborating with GISAID had uploaded and shared more than 1.2 million coronavirus genome sequences from 172 countries and territories on its online data platform. This was facilitated by terms of agreement for publication that avoided exploitation of other researchers' data. According to Tulio de Oliveira, the director of the KwaZulu-Natal Research Innovation and Sequencing Platform in Durban, South Africa. "This is the first time I've seen people sharing so much data before publication."[22]

In 2000, WHO set up the *Global Outbreak Alert and Response Network (GOARN)* to "strengthen and coordinate rapid mobilization of experts in responding to international outbreaks and to overcome the sometimes chaotic and fragmented operations characterizing previous responses."[23] GOARN is now WHO's vehicle to provide technical assistance to Member States in operationalizing the IHR during outbreaks. It consists of over 250 collaborating technical international organizations and networks that have responded to major outbreaks of cholera, dengue, encephalitis, influenza, meningitis, Nipah, plague, severe acute respiratory syndrome (SARS), viral hemorrhagic fevers (VHF), yellow fever, and other emerging and epidemic-prone pathogens.[23] GOARN has integrated COVID-19 into its activities coordinating new research and expertise from its partners and networks, and providing guidance.

While WHO, OIE, and FAO administer activities through their regional offices, other less formal networks complement the tripartite efforts. Regional surveillance networks in the Middle East, Southeast Asia, and Africa, for example—coordinated by the *Connecting Regional Disease Surveillance Networks (CORDS)* formed in 2012—share information on cross-border disease events. Their cooperation and coordination are made possible by trust and by well-exercised national and regional pandemic preparedness policies.[24,25] Cooperation with highly pathogenic influenza in the countries of Southeast Asia and the Middle East has been practiced in real outbreaks and tested in simulation settings. The CORDS networks thereby model the kind of transnational cooperation that can mount the needed flexible and coordinated response to the spread of COVID-19 and future pandemic threats.[26] The consortium illustrates the value of regional disease surveillance networks in shaping and managing cohesive policies on past, current, and future threats.[26,27] The partnerships also exemplify to other parts of the world that finding common ground is imperative to promoting health security and cooperation where it is most lacking and needed and that building trust across the most difficult boundaries in the world translates into the ability to counter unscripted threats.

To undertake effective disease surveillance and investigation, ministries of health require qualified field epidemiologists. *The Training Programs in Epidemiology and Public Health Interventions Network (TEPHINET)* manages a global network of 75 Field Epidemiology Training Programs (FETP) across more than 100 countries.[28,29] These FETPs work collaboratively within ministries of health, universities, national public health institutes, and other public health organizations, to fight new public health threats such as COVID-19. TEPHINET builds capacity by training epidemiologists using standardized methodologies and tested outbreak and laboratory guidelines to detect and respond quickly to outbreaks.[30]

Internet-based surveillance

Launched in 1994, the *Program for Monitoring Emerging Diseases (ProMED)* is an example of internet-based, unstructured public health surveillance. Expert moderators in 32 countries cooperate to constantly scan the web to identify emerging and reemerging infectious diseases and toxins affecting humans, animals, and plants and to share data. Community members, public health leaders, clinicians, and government officials can access the data and commentary on ProMED 24 h a day. Key to agile and efficient cooperation, ProMED has established surge capacity in LMICs, often the geographic locations at the highest risk of diseases with pandemic potential. ProMED was early to report numerous disease outbreaks including SARS, MERS, Ebola, and Zika.[31] In April 2020, WHO disclosed that it first learned of the COVID-19 outbreak from a ProMED Editor in Brooklyn, New York—not by formal reporting as required by IHR.[32]

Complementing the work of ProMED, the *Global Public Health Intelligence Network (GPHIN)* practices big data surveillance by analyzing information on public health threats reported through global media sources.[33] Set up in the 1990s as a free electronic public health early warning system by the Government of Canada, GPHIN provides more than 40% of WHO's early warning outbreak information.[34] GPHIN also identified early cases of

SARS-CoV-2 and in late December 2019, GPHIN proposed a prototype model for an individual-level pandemic notification system.[35]

Research and development

Fondation Mérieux founded *GloPID-R* in 2013 to fill a gap in global partnerships that plan and invest in research and innovation *before* a pandemic. This global alliance of research funding organizations facilitates effective and rapid research on a significant outbreak of a new or reemerging infectious disease with epidemic or pandemic potential. The need for the rapid development of essential diagnostics, vaccines, and therapeutics at the outset of an emerging infectious disease outbreak was highlighted when Ebola struck West Africa in 2014. Likewise, the importance of planning and investing in research and innovation before a health crisis occurs was seen with SARS-CoV-2.[36] Soon after WHO declared COVID-19 a pandemic, GloPID-R assessed existing research funding and launched emergency calls for funding.[36]

The *Coalition for Epidemic Preparedness Innovations (CEPI)*, launched in 2017, is a partnership between public, private, philanthropic, and civil society organizations with the primary objective to develop vaccines to halt epidemics. Through this partnership, CEPI finances independent research projects to develop vaccines mainly for emerging infectious diseases and enable access to people when needed during outbreaks.[37,38] Before COVID-19, CEPI focused on Ebola, Lassa Virus, Rift valley Fever, and others.[37,38] CEPI aims to compress the development of "pandemic-busting" vaccines from six plus years to 100 days and specifically supports partnerships to develop COVID-19 vaccines.[37]

Mechanisms established during the COVID-19 pandemic

Very early in the COVID-19 pandemic, global health leaders and institutions feared that any available therapeutics and vaccines to curb the spread of COVID-19 might be subjected to the same scramble as was experienced during the H1N1 influenza pandemic in 2009.[39] To address this concern, early in 2020, WHO, the European Union (EU), France, and the Bill and Melinda Gates Foundation created the *Access to COVID-19 Tools Accelerator (ACT-A)* to fast-track research for the development, production, and equitable access to tests, treatments, and vaccines for COVID-19.[38]

Within the ACT-A framework, CEPI, Gavi the Vaccine Alliance and WHO formed *COVAX* to focus on vaccine needs, particularly in fast-tracking research and development and facilitating market investment and demand to encourage vaccine producers to create the conditions vaccine manufacturers would need to advance vaccine production. With vaccines produced at scale, it was hoped that this would decrease inequitable access to COVID-19 vaccines when they became available.[39,40] COVAX is the focus of our case study.

The WHO Epidemic and Pandemic Intelligence Hub was inaugurated in September 2021 by Germany's Chancellor Angela Merkel and Dr. Tedros Adhanom Ghebreyesus, Director-General of WHO. The Hub brings together experts in a state-of-the-art building in Berlin Germany, creating cooperative momentum at the country and global level to prepare, detect, and respond to health emergencies.[41] The Hub is a mechanism to generate collaborative intelligence, trusted interactions, and science solutions and responds to the need for equity in global decision-making. Open data and analytics are the hallmarks of readiness, timeliness, and appropriate response to epidemic and pandemic risk. It is envisioned that further hubs will be created to cooperatively link national and global data to improve intelligence and equitable access to data for decision-makers.

Deepening global health cooperation

COVID-19 has demonstrated the restricted scope of WHO's authoritative institutional power, and its lack of legitimacy to speak for all. WHO's limited institutional and normative powers have played out amid the structural powers that framed the pandemic by exerting imbalanced economic, political, and social influences.[42,43] There is no global social contract. For example, Italy could not force the EU to share health supplies; the United States could not force China to share data.[44] LICs have no independent wealth upon which to lean and subsequently turn to international financial institutions such as multilateral, regional, and bilateral development banks for emergency support. This is not a failure of global health governance but illustrates how extreme imbalance can undermine global cooperation.

The Independent Panel for Pandemic Preparedness and Response (IPPPR) Report *COVID 19: Make it the Last Pandemic 12* May 2021[45] even a year and a half into the pandemic—spoke concisely to the panel's deep grasp of normative, institutional, and structural powers that shape the ability to cooperate in health emergencies. The IPPPR asks the WHA (the body that requested WHO Director General to initiate the review in 2020) to give WHO both the explicit authority to publish information about outbreaks with pandemic potential without prior approval of national governments and the power to investigate.[46] These expanded authorities would circumvent full reliance on unpredictable sharing of data.

WHO has the responsibility to declare if and when a health emergency becomes a Public Health Event of International Concern (PHEIC)—as it did for COVID-19. Under the IHR, Member States are obliged to respond in a coordinated fashion. The IPPPR recognized Member States' concerns about the broad design of the PHEIC[17] ranging from regional outbreaks to pandemics. The IPPPR report ponders whether or not a PHEIC declaration is amply clear to trigger action[45] and recommends that WHO base the PHEIC upon precautionary principles, clear objectives, and published criteria. Precautionary principles are a mainstay of the practice of public health, especially given we increasingly act under uncertainty. The precautionary principle is stringently but creatively used as a basis for preventive actions while avoiding unintended consequences. As pointed out by Nuzzo, WHO's process for declaring a global public health emergency was called into question during the delayed Ebola 2014 PHEIC, and again, more troubling, delayed even after COVID-19 had spread through the Chinese city of Wuhan December 2019.[47] It is no wonder WHO was called into question about political influence from Beijing upon the PHEIC declaration of January 31, 2020, a process intended to be science driven.[47]

An analysis of IHR Emergency Committees and PHEIC designations published in 2020 summarized a chronology of declarations for the past 11 years to include H1N1, MERS, Polio, Ebola Virus Disease, Zika, Yellow Fever, and then COVID-19, finding significant inconsistencies.[48] Clear criteria used to declare a PHEIC declaration would infuse confidence in the declaration process. Notwithstanding, the analysis bluntly confronts the IHR as pitted against institutional and structural powers and confronts the questionable consistency and transparency of WHO's application of PHEICs. Member States and experts support the IHR as a cornerstone of international public health and health security law.[49] And, yet, the non-enforceable IHR are of lesser value than a treaty that would enforce IHR along with global coordination and national laws.

While the IHR have been instrumental in helping the world better organize and inform itself on the dangers of health emergencies, they have not built the cooperation required for effective action. A more enforceable arrangement is needed.

Proposal to create an international treaty on pandemics

The global community was caught unprepared and slow to marshal a global response to COVID-19. There was no globally agreed upon pandemic blueprint to guide roles and responsibilities, needs, and actions required at a global scale. Global health leaders have called for a global pandemic treaty[45,47,48] with the primary purpose of protecting the world from the threats of pandemics. The IHR do not serve this purpose. Although its constitution gives it authority to develop international law for public health good, WHO has been reluctant to use this power.

In the last 70 years, the 2003 Framework Convention on Tobacco Control (FCTC)[50] has been the only treaty to be ratified through accession, acceptance, or approval, only then bestowing a high level of enforceability. Conversely, the adoption of regulations like the IHR or WHO nomenclature or recommendations under PIP does not need ratification, opening the way for looser adherence due to the lack of enforceability.

During the 15th G20 Leaders' Summit on Nov 21, 2020, the EU Council president announced that "An international Treaty on Pandemics could help prevent future pandemics and help us respond more quickly and in a more coordinated manner. It should be negotiated with all nations, United Nations (UN) organizations and agencies, particularly WHO."[12] This call attracted widespread support of other international bodies and organizations. In September 2021, the third meeting of the Member States Working Group on Strengthening WHO Preparedness for and Response to Health Emergencies (WGPR) considered the benefits of a pandemic treaty focused on governance, financing, systems and tools and equity.[51]

Some countries oppose the idea of a new pandemic treaty and propose instead opening up the IHR for amendments.[52] There is also discussion of WHO's suitability to negotiate the Treaty—questions were raised about WHO's lack of mandate in areas such as finance, trade, cross-border travel, law enforcement, and supply chain management—calling for consideration to negotiate the treaty within the broader mandate of the United Nations.[53] However, rather than creating parallel institutions, others have opined that it is better to build on existing platforms such as the IHR and strengthen WHO's mandate and capacity to prepare for and respond to global pandemic threats.[54] A treaty would provide WHO the legal authority, and resources to work with its Member States to prevent, detect, and respond to pandemic threats, by developing innovative norms, governance, and compliance mechanisms needed to prepare for novel outbreaks with pandemic potential.[55]

The proposed treaty would ensure that its signatories adhere to timely sharing of disease outbreak information and of the science and technologies needed to increase global capacity and supply of vaccines, therapeutics, and other medical countermeasures for an adequate response to pandemics. COVID-19 has proven that our collective security and health depend on working together, in solidarity and not in nationalism and isolationism.[55]

Lastly, treaty negotiations take time to agree. According to Frieden and Buissonnière, the "FCTC took 8 years to negotiate, another 3 years to ratify and many more years to agree on protocols for the different components."[53] The negotiators for a pandemic treaty would have to avoid such a prolonged process. Given the complexity and time delays of treaty negotiation, development must be conducted in tandem with investments to strengthen and capacitate current global health mechanisms to respond to any dangers that may arise. Time is a luxury the world no longer has.

As part of the revisions in its operations in relation to health emergencies, and to strengthen the IHR, WHO is introducing a Universal Health Preparedness Review,[56] a complement to the current IHR to build mutual trust and accountability for health, by bringing nations together as neighbors to support a whole-of-government approach to strengthening national capacities.

CASE STUDY

International cooperation to distribute vaccines fairly

Background

As COVID-19 intensified into pandemic status, the world waited anxiously for scientists to develop vaccines to protect everyone from the virus. In preparation, Gavi, WHO, and CEPI established COVAX as the vaccine pillar of the ACT-A. COVAX was to provide a "global framework that puts international cooperation above vaccine nationalism in stopping the pandemic."[57]

Observations

COVAX's initial goal was, by the end of 2021, to make 2 billion doses of vaccine available to protect high-risk/vulnerable populations and frontline healthcare workers in every country, and to expand distribution fairly as vaccines became available.[58] In October 2021, WHO raised the global vaccination targets to 40% of the world's population by the end of 2021 and 70% by mid-2022.[58]

Interventions

COVAX's mechanism of cooperation was to invite countries to sign up to buy vaccines through COVAX so that it could procure them on a large scale at a fair price.[40] It also developed an advanced purchasing mechanism—COVAX-Advance Market Commitment (AMC)—through which governments, the private sector, philanthropies, and the general public could donate for COVAX to purchase vaccines to send to eligible LMICs.[58]

Key issue

COVAX's plan was to employ creative funding mechanisms that primarily incentivize development with a diversified portfolio—investing in vaccine candidates in various countries and using different vaccine development methods. This was a pioneering experiment in centralizing and thus equitizing the vaccine supply chain to avoid the type of competition that would outprice most LMICs as experienced during the 2009 Swine Flu Pandemic.[40,59]

What worked

- By mid-2020, almost all countries had signed up for COVAX, of which 80+

Continued

CASE STUDY (cont'd)

were self-financing and 92 were AMC eligible.[60]

- By mid-2021, donors had pledged sufficient funding for COVAX AMC to secure 1.8 billion fully subsidised doses to deliver to AMC countries in 2021 and early 2022[61]although there were significant delays in subsequent donation and delivery.
- In mid-January 2022, COVAX delivered its billionth dose worldwide, almost a third of these delivered in December 2021.[62]

What did not work

- By the end of 2021, 7.5% of people living in LICs had received at least one vaccine dose compared to only 75% living in HICs.[3]
- Even as COVAX started up, wealthy countries secured the most available doses at undisclosed prices and some pharmaceutical companies chose not to sell to COVAX, preferring lucrative bilateral deals.[63,64]
- When scarcity was no longer a problem, governments chose not to share surplus vaccines and when they did, the doses they shared had short shelf life affecting uptake in LMICs.[65,66]
- COVAX relied largely on deliveries from the Indian Serum Institute which halted supplies when the Delta variant hit India.[67]

Lessons learned

- Solidarity mechanisms function best if they preexist an emergency, which was not the case with COVAX.[68] The challenge in ensuring buy-in and follow-through is not a lack of humanitarian sentiments or an understanding of global interconnectedness among world leaders, but instead one of prioritizing these motivations in the midst of a crisis. Nationalistic sentiments often prevail, and WHO lacks authority to demand potential cooperative solutions, such as removing optionality or establishing a completely centralized distribution hub.[68]
- Equitable access, globally, to diagnostics, therapeutics, and vaccines is imperative in times of global health crisis, while that was largely lacking for COVID-19, the pandemic encouraged steps toward this idea—such as the World Trade Organization Agreement on Trade-Related Aspects of Intellectual Property Rights (TRIPS) waiver -calling for the temporary suspension on intellectual property rights to make COVID-19 vaccines and new technologies available to LMICs[69,70] and the mRNA vaccine technology transfer hub in South Africa.

Recommendations

- Create and implement binding commitments in noncrisis periods to allow leaders to mitigate the potential for nationalism and lack of cooperation.

CASE STUDY *(cont'd)*

- Cooperatively share vaccine technology as a solution to production and distribution problems, ensuring that manufacturing capacity is fully used. Ideologically, it follows in the spirit of shared value, equity, and collective well-being.[69] Practically, it eliminates for-profit motivation and removes decision-making from the hands of those motivated by profit.
- Collaboratively strengthen healthcare systems in LMICs at all levels, perhaps the only ethical and practical way to end any pandemic.

Vision for the future

The field of global health aims to unify the wide-ranging constellation of global actors in or adjacent to healthcare under one unified system, be it centralized to a single health organization or multiple organizations with an agreed division of labor.[71] For decades, effective cooperation has been a key missing component of global health, a flaw exacerbated during the COVID-19 pandemic. Simply put, our vision for the future is a fully flourishing materialization of global health cooperation.

As we have outlined in the literature review section of this chapter, progress has certainly been made. The implementation of the IHR was the first step in a paradigm shift to require rapid identification, reporting, and response to potential public health emergencies, notwithstanding issues with compliance and scope. On a diffuse level, enterprising minds have created important entities of cooperation from ProMED to CORDS to ACT-A and beyond, which serve to counterbalance increasing fragmentation. But not much has changed in the way the international community collaborates. With advancements in technology, the world is experiencing unparalleled amounts of innovation and information collection, but without a similar expansion in the development of strategies for unification on an international level.

In the true sense of solidarity and universal preparedness, our perception of the future involves the recovery of multilateralism. At a policy level, we will only succeed if the Member States reach an agreement for a strong international pandemic preparedness and response treaty with binding and nonbinding instruments of compliance, while at the same time navigating geopolitical tensions. This new pandemic treaty would be coordinated by WHO and the UN, involving a wide representation of countries, organizations, and communities. Cooperation and flexibility among negotiating bodies or willing nations are essential and will only come if they meet the interests of all involved parties. For countries to engage and adhere to the agreement, the treaty will need to consider the national interests of its members, combining flexibility to accommodate some unilateral decisions, representativeness, and the financial resources to be implemented. Any concept for the future must take into consideration reasonable changes in national interests over time. A vision additionally includes strengthening the flow of information and experience between informal stakeholders, gaining unprecedented importance with the rise of modern technology.

In our view for the future, we learn from our mistakes. We adopt a science-, equity- and solidarity-based approach in the creation of a new pandemic treaty encompassing the solutions needed to protect lives and the well-being of everyone, everywhere. We live at a time when aggressive transnational threats—with COVID-19 likely not being the last—require international cooperation above all. High performing cooperation will involve specific actions at the global, regional, and national levels to explore partnerships and institutions that can create scientific consensus, international trust, and knowledge to forge a common perspective of a problem and collective action. In this manner, scientific, political, and societal dialogue is distributive and strongest where it should be, at the source of infection.

Actions

At the global level

1. *Listen equitably to the voices of WHO Member State*s (from low-, middle-, and high-income countries) on how to strengthen the existing global preparedness and response framework. No single sovereign state or single sector can contain pathogens. Early in the COVID-19 pandemic, Member States brought forward suggestions to work together cooperatively, with a one health approach that crosses human, animal, and planetary health sectors. Their voices on one health serve to mitigate the weaknesses and shortfalls in current global health governance and management.
2. *Create an integrated global early warning platform.* As recognized across high-level review panels including IPPPR and the Group of Seven (G7), the creation of a global early warning platform has urgency. Technology and goods alone will not solve preparedness goals but the role of public health surveillance is elevated by transparency, timely data, equitable data sharing, and trust. As noted in the literature review, the tension between countries and international organizations can limit data sharing and drown out public health priorities, especially in emergencies, GISRS was one example we reviewed that shows how countries can share open-access genomic data on circulating influenza. For 3 decades, GPHIN has laced together global cooperation through multilanguage, worldwide alerts of global health threats. The WHO Berlin Epidemic and Pandemic Intelligence Hub is a contemporary example of a mechanism to generate collaborative intelligence, accessible data, and cooperative solutions to pandemic threats.
3. *Improve global coordination and leadership.* All stakeholders must cooperate to strengthen the role of WHO as the global health coordinating body. The WHA should agree upon an enforceable legally binding instrument, such as a global pandemic instrument/treaty, to strengthen global architecture, surveillance, financing, and medical countermeasures to prevent pandemics. The instrument must address the challenges of global health by introducing mechanisms that will: ensure predictable funding for global health needs; promote equity through sharing of knowledge and technologies and waiver patents where required; ensure respect for human rights; and provide equitable access to healthcare services and medical countermeasures, including vaccines.

At the regional and country levels

4. *Create platforms for leaders* to speak to key stakeholders about national interests that often converge with wider international agendas. We cited CORDS as an example of trust, based upon tested communication and disease surveillance activities that constitute committed relationships among key stakeholders within universities, ministries, and philanthropies. Building trust between countries (between health authorities in multiple sectors) promotes cooperation and transparency.[72] The relationships are intended to improve service delivery and investments in capacities for the production of medicines, therapeutics, diagnostics, and vaccines in LMICs, thereby minimizing over reliance on a few countries not meeting global needs in times of great crisis. GloPID-R was another illustration that cooperative agreements that planned and invested in research *before* the Ebola 2014 outbreak facilitated more rapid development of therapeutics, diagnostics, and vaccines.

At the community level

5. *Strengthen the community's role in the dissemination of information* and the building of cooperative atmospheres that address misinformation, dispel false rumor, and uphold population acceptance of public health measures and directives. National and subnational action plans can include an evaluation of institutional capacity to cooperate and deliver timely and effective public health performance. Often, success or failure in response to health emergencies is contingent on local and national dynamics, ranging from vaccine hesitancy to insecurities. Community engagement with networks of professionals was highlighted by ProMED's use of informal data to create one health threat notifications of emerging and reemerging disease threats, all the while giving communities, clinicians, and governments global free access to curated and timely information. With the recognition that COVID-19 has caused illness and deaths among multiple nonhuman animal species, the question of cooperative programs for animal vaccines, mainly in zoos, is open for rich discussion. Zoos recognize that animal infections can spread back to humans and they are also compelled to protect rare and high-value animals. Zoological staff, as a network of experts, can work cooperatively on diagnostics and vaccinations.

References

1. World Health Organization. *WHO Director-General's Opening remarks at Executive Board Meeting*. World Health Organization; October 5, 2020. https://www.who.int/director-general/speeches/detail/who-director-general-s-opening-remarks-at-executive-board-meeting. Accessed June 28, 2023.
2. United Nations Children's Fund. *No-one Is safe until everyone Is safe – Why we need a global response to COVID-19*. United Nations Children's Fund; May 24, 2021. https://www.unicef.org/press-releases/no-one-safe-until-everyone-safe-why-we-need-global-response-covid-19. Accessed June 28, 2023.
3. Our World in Data. *Share of people who received at least one dose of COVID-19 vaccine, Dec 31 2021*. Our World in Data; 2021. https://ourworldindata.org/grapher/share-people-vaccinated-covid?time=2021-12-31&country=High+income~Low+income. Accessed June 28, 2023.
4. Ghebreyesus TA. *WHO Director-General's Opening remarks at the WTO - WHO High Level Dialogue: Expanding COVID-19 vaccine manufacture to promote equitable access*. World Health Organization; July 21, 2021.

https://www.who.int/director-general/speeches/detail/who-director-general-s-opening-remarks-at-the-wto—who-high-level-dialogue-expanding-covid-19-vaccine-manufacture-to-promote-equitable-access. Accessed June 28, 2023.

5. Online News Editor. *Guterres condemns "Vaccine Nationalism" in fight against Covid-19*. La Prensa Latina Bilingual Media; October 7, 2021. https://www.laprensalatina.com/guterres-condemns-vaccine-nationalism-in-fight-against-covid-19/. Accessed June 28, 2023.
6. Jerving S. *Winnie Byanyima: The world needs a COVID-19 plan*. devex; September 23, 2021. https://www.devex.com/news/winnie-byanyima-the-world-needs-a-covid-19-plan-101689. Accessed June 28, 2023.
7. Devlin K, Connaughton A. *Most approve of national response to COVID-19 in 14 advanced economies*. Pew Research Center; August 27, 2020. https://www.pewresearch.org/global/2020/08/27/most-approve-of-national-response-to-covid-19-in-14-advanced-economies/. Accessed June 28, 2023.
8. Huremović D. *Brief History of Pandemics (Pandemics throughout History)*. New York City, NY: Springer International Publishing; 2019:7–35.
9. Staff KM. *Pandemics in recent history*. knowable Magazine; July 16, 2020. https://knowablemagazine.org/article/health-disease/2020/pandemics-recent-history. Accessed June 28, 2023.
10. Gentry B. Global collaboration needed to prevent pandemics. In: *Hubert Department of Global Health*. Atlanta, GA: Emory University; 2021.
11. World Health Organization. What is the ACT-Accelerator. World Health Organization. https://www.who.int/initiatives/act-accelerator/about. Accessed 28 June 2023.
12. G20 Summit. *G20 leaders united to address major global pandemic and economic challenges*. European Council; November 22, 2020. https://www.consilium.europa.eu/en/press/press-releases/2020/11/22/g20-summit-g20-leaders-united-to-address-major-global-pandemic-and-economic-challenges/. Accessed June 28, 2023.
13. World Health Organization. Where we work: WHO organizational structure. World Health Organization. https://www.who.int/about/structure. Accessed 28 June 2023.
14. World Health Organization. *The Seventy-Fourth World Health Assembly closes*. World Health Organization; May 31, 2021. https://www.who.int/news/item/31-05-2021-the-seventy-fourth-world-health-assembly-closes. Accessed June 28, 2023.
15. Food and Agriculture Organization of the United Nations. One health. Food and Agriculture Organization of the United Nations. https://www.fao.org/one-health/en/. Accessed 28 June 2023.
16. World Health Organization. *COVID-19 Strategic Preparedness and Response Plan (SPRP 2021)*. World Health Organization; 24 February 2021. https://www.who.int/publications/i/item/WHO-WHE-2021.02. Accessed June 28, 2023.
17. World Health Organization. *International Health Regulations (2005)*. Third. Geneva: World Health Organization; 2016. https://www.who.int/publications/i/item/9789241580496. Accessed June 28, 2023.
18. World Health Organization. National IHR focal points. World Health Organization. https://www.who.int/teams/ihr/national-focal-points. Accessed 28 June 2023.
19. Hay AJ, McCauley JW. The WHO global influenza surveillance and response system (GISRS)-A future perspective. *Influenza and Other Respiratory Viruses*. 2018;12(5):551–557. https://doi.org/10.1111/irv.12565.
20. Broor S, Campbell H, Hirve S, et al. Leveraging the Global Influenza Surveillance and Response System for global respiratory syncytial virus surveillance—opportunities and challenges. *Influenza and Other Respiratory Viruses*. 2020;14(6):622–629. https://doi.org/10.1111/irv.12672.
21. World Health Organization. Pandemic Influenza Preparedness (PIP) Framework. World Health Organization 2022. https://www.who.int/initiatives/pandemic-influenza-preparedness-framework. Accessed 28 June 2023.
22. Maxmen A. One million coronavirus sequences: popular genome site hits mega milestone. *Nature*. 2021;593(7857):21–21. https://doi.org/10.1038/d41586-021-01069-w.
23. Mackenzie JS, Drury P, Arthur RR, et al. The global outbreak alert and response network. *Global Publ Health*. 2014;9(9):1023–1039. https://doi.org/10.1080/17441692.2014.951870.
24. Pushpa RW, Ofrin RH, Bhola AK, Inbanathan FY, Bezbaruah S. Pandemic influenza preparedness in the WHO South-East Asia Region: a model for planning regional preparedness for other priority high-threat pathogens. *WHO South-East Asia Journal of Public Health*. 2020;9(1):43–49. https://doi.org/10.4103/2224-3151.282995.
25. Gresham L, Ramlawi A, Briski J, Richardson M, Taylor T. Trust across borders: responding to 2009 H1N1 influenza in the Middle East. *Biosecur Bioterrorism Biodefense Strategy, Pract Sci*. 2009;7(4):399–404. https://doi.org/10.1089/bsp.2009.0034.

26. Gresham LS, Smolinski MS, Suphanchaimat R, Kimball AM, Wibulpolprasert S. Creating a global dialogue on infectious disease surveillance: connecting organizations for regional disease surveillance (CORDS). *Emerg Health Threats J*. 2013;6(1):19912. https://doi.org/10.3402/ehtj.v6i0.19912.
27. Bond KC, Macfarlane SB, Burke C, Ungchusak K, Wibulpolprasert S. The evolution and expansion of regional disease surveillance networks and their role in mitigating the threat of infectious disease outbreaks. *Emerg Health Threats J*. 2013;6(1):19913. https://doi.org/10.3402/ehtj.v6i0.19913.
28. Jones DS, Dicker RC, Fontaine RE, et al. Building global epidemiology and response capacity with field Epidemiology training programs. *Emerg Infect Dis*. 2017;23(13). https://doi.org/10.3201/eid2313.170509.
29. André AM, Lopez A, Perkins S, et al. Frontline field epidemiology training programs as a strategy to improve disease surveillance and response. *Emerg Infect Dis*. 2017;23(13). https://doi.org/10.3201/eid2313.170803.
30. Hilmers A. On the first World Field Epidemiology Day, honoring our disease detectives on the front lines. *Int J Infect Dis*. 2021;110:S1–S2. https://doi.org/10.1016/j.ijid.2021.08.070.
31. ProMED. About ProMED. ProMED. https://promedmail.org/about-promed/. [28 June 2023].
32. Meek J. *The Health Transformation Army*. London Review of Books; 2 July 2020, 42 (13) https://www.lrb.co.uk/the-paper/v42/n13/james-meek/the-health-transformation-army. Accessed June 28, 2023.
33. Dion M, Abdelmalik P, Mawudeku A. Big data and the global public health intelligence network (GPHIN). *Can Comm Dis Rep*. 2015;41(9):209–214. https://doi.org/10.14745/ccdr.v41i09a02.
34. Mykhalovskiy E, Weir L. The Global Public Health Intelligence Network and early warning outbreak detection: a Canadian contribution to global public health. *Can J Public Health*. 2006;97(1):42–44. https://doi.org/10.1007/BF03405213.
35. Sakib MN, Butt ZA, Morita PP, Oremus M, Fong GT, Hall PA. Considerations for an individual-level population notification system for pandemic response: a review and prototype. *J Med Internet Res*. 2020;22(6):e19930. https://doi.org/10.2196/19930.
36. COVID-19 Research GloPID-R Synergies Meeting Working Group; Meeting Co-chairs. Ending COVID-19: progress and gaps in research—highlights of the july 2020 GloPID-R COVID-19 research synergies meetings. *BMC Med*. 2020;18(1). https://doi.org/10.1186/s12916-020-01807-3.
37. Coalition for Epidemic Preparedness Innovations. *New vaccines for a safer world*. CEPI; 2021. https://cepi.net/. Accessed June 28, 2023.
38. World Health Organization. *COVAX Announces Additional Deals to Access Promising COVID-19 Vaccine Candidates: Plans Global Rollout starting Q1 2021*. World Health Organization; 18th December 2020. https://www.who.int/news/item/18-12-2020-covax-announces-additional-deals-to-access-promising-covid-19-vaccine-candidates-plans-global-rollout-starting-q1-2021. Accessed 28 June 2023.
39. Eccleston-Turner M, Upton H. International collaboration to ensure equitable access to vaccines for COVID-19: the ACT-accelerator and the COVAX facility. *Milbank Q*. 2021;99(2):426–449. https://doi.org/10.1111/1468-0009.12503.
40. Berkley S. *COVAX explained*. Gavi; September 3, 2020. https://www.gavi.org/vaccineswork/covax-explained. Accessed June 28, 2023.
41. World Health Organization. *WHO HUB for Pandemic and Epidemic Intelligence*. World Health Organization; 2021. https://cdn.who.int/media/docs/default-source/2021-dha-docs/who_hub.pdf?sfvrsn=8dc28ab6_5. Accessed June 28, 2023.
42. Patterson A, Clark MA. COVID-19 and power in global health. *Int J Health Pol Manag*. 2020;Oct 9(10):429–431. https://doi.org/10.34172/ijhpm.2020.72.
43. Chen D. *Coronavirus battle in China: Process and prospects*. Shanghai Institutes for International Studies (SIIS); February 2020. http://en.people.cn/n3/2020/0202/c90000-9653225.html. Accessed June 28, 2023.
44. Wright T, Kahl C. *Aftershocks: Pandemic Politics and the End of the Old International Order*. St. Martin's Press; 2021.
45. Independent Panel for Pandemic Preparedness and Response. COVID 19: Make it the last pandemic. Independent Panel for Pandemic Preparedness and Response: May 2021. https://theindependentpanel.org/wp-content/uploads/2021/05/COVID-19-Make-it-the-Last-Pandemic_final.pdf. Accessed 28 June 2023.
46. Sirleaf EJ, Clark H. Report of the independent panel for pandemic preparedness and response: making COVID-19 the last pandemic. *Lancet*. 2021;398(10295):101–103. https://doi.org/10.1016/S0140-6736(21)01095-3.
47. Nuzzo J. *To stop a pandemic: A better approach to global health security*. Foreign Affairs; December 8 2021. https://www.foreignaffairs.com/articles/china/2020-12-08/stop-pandemic. Accessed June 28, 2023.

48. Mullen L, Potter C, Gostin LO, Cicero A, Nuzzo JB. An analysis of international health regulations emergency committees and public health emergency of international concern designations. *BMJ Glob Health*. 2020;5(6):e002502. https://doi.org/10.1136/bmjgh-2020-002502.
49. Wieler LH. *Statement to the 148th Executive Board by the Chair of the Review Committee on the functioning of the International Health Regulations (2005) during the COVID-19 Response*. World Health Organization; January 19, 2021. https://www.who.int/news/item/19-01-2021-statement-to-the-148th-executive-board-by-the-chair-of-the-review-committee-on-the-functioning-of-the-international-health-regulations-(2005)-during-the-covid-19-response. Accessed June 28, 2023.
50. De Luca M, Ramirez ML. A pandemic treaty: learning from the framework convention on tobacco control. *Health Secur*. 2023;Mar-Apr;21(2):105–112. https://doi.org/10.1089/hs.2022.0135.
51. World Health Organization. *Member States Working Group on Strengthening WHO Preparedness for and Response to Health Emergencies (WGPR)*. World Health Organization; September 30, 2021. https://apps.who.int/gb/wgpr/pdf_files/wgpr3/A_WGPR3_6-en.pdf. Accessed June 28, 2023.
52. Labonté R, Wiktorowicz M, Packer C, Ruckert A, Wilson K, Halabi S. A pandemic treaty, revised international health regulations, or both? *Glob Health*. 2021;17(1). https://doi.org/10.1186/s12992-021-00779-0.
53. Frieden TR, Buissonnière M. Will a global preparedness treaty help or hinder pandemic preparedness? *BMJ Glob Health*. 2021;6(5):e006297. https://doi.org/10.1136/bmjgh-2021-006297.
54. Velásquez G, Syam N. *A new WHO International Treaty on Pandemic Preparedness and Response: Can it address the needs of the Global South?* Infojustice; May 12, 2021. http://infojustice.org/archives/43154. Accessed June 28, 2023.
55. Gostin LO, Halabi SF, Klock KA. An international agreement on pandemic prevention and preparedness. *JAMA*. 2021;326(13):1257. https://doi.org/10.1001/jama.2021.16104.
56. World Health Organization. *Universal Health and Preparedness Review (UHPR): Member States information session*. World Health Organization; 23 December 2020. https://apps.who.int/gb/COVID-19/pdf_files/23_12/UHPR.pdf. Accessed June 28, 2023.
57. Berkley S, Hatchett R, Swaminathan S. *A moment of truth in the pandemic*. Project Syndicate; September 16, 2021. https://www.project-syndicate.org/commentary/covax-vaccine-global-access-binding-commitments-by-seth-berkley-et-al-2020-09?barrier=accesspaylog. Accessed June 28, 2023.
58. Hatchett R, Berkley S, Ghebreyesus TA, Fore H. *COVAX Joint Statement: Call to action to equip COVAX to deliver 2 billion doses in 2021*. United Nations Children's Fund; May 27, 2021. https://www.unicef.org/press-releases/covax-joint-statement-call-action-equip-covax-deliver-2-billion-doses-2021-0. Accessed June 28, 2023.
59. Fidler DP. Negotiating equitable access to influenza vaccines: global health diplomacy and the controversies surrounding avian influenza H5N1 and pandemic influenza H1N1. *PLoS Med*. 2010;May 4;7(5). https://doi.org/10.1371/journal.pmed.1000247.
60. COVAX. *Self-Financing Participants*. COVAX; May 12, 2021. https://www.gavi.org/sites/default/files/covid/pr/COVAX_CA_COIP_List_COVAX_PR_12-05-21.pdf. Accessed June 28, 2023.
61. Gavi. *World leaders unite to commit to global equitable access for COVID-19 vaccines*. Gavi; June 2 2021. https://www.gavi.org/news/media-room/world-leaders-unite-commit-global-equitable-access-covid-19-vaccines. Accessed March 31, 2022.
62. United Nations Children's Fund. *COVAX: 1 billion vaccines delivered*. United Nations Children's Fund; January 17, 2022. https://www.unicef.org/supply/stories/covax-1-billion-vaccines-delivered. Accessed June 28, 2023.
63. Reuters. COVAX struggling as some nations bid more for scant vaccines, says WTO. https://www.reuters.com/business/healthcare-pharmaceuticals/covax-struggling-some-nations-bid-more-scant-vaccines-says-wto-2021-07-13/. Accessed June 28, 2023.
64. Apuzzo M, Gebrekidan S. *Governments sign secret vaccine deals. Here's what they hide*. New York Times; January 28, 2021. https://www.nytimes.com/2021/01/28/world/europe/vaccine-secret-contracts-prices.html. Accessed June 28, 2023.
65. Cheng M. *WHO: Rich countries should donate vaccines, not use boosters*. AP News; July 12, 2021. https://apnews.com/article/europe-business-health-government-and-politics-coronavirus-pandemic-02b1157d4f0def0460c9c4548f1c7679. Accessed June 28, 2023.
66. Goldhill O, Furneaux R, Davies M. *'Naively Ambitious': How COVAX failed on its promise to vaccinate the world*. STAT; October 8, 2021. https://www.statnews.com/2021/10/08/how-covax-failed-on-its-promise-to-vaccinate-the-world/. Accessed June 28, 2023.

67. Bostock B. *COVAX has received zero shipments - It's 140 million doses short - since March due to India's outbreak*. Insider; May 17, 2021. https://www.businessinsider.co.za/covax-missing-coronavirus-doses-india-outbreak-2021-5. Accessed June 28, 2023.
68. Rahman-Shepherd A, Clift C, Ross E, et al. *Solidarity in response to the COVID-19 Pandemic*. Chatham House; July 14, 2021. https://www.chathamhouse.org/2021/07/solidarity-response-covid-19-pandemic/03-case-study-1-covax-vaccines-and-solidarity. Accessed June 28, 2023.
69. Agyarko R. *Opinion: Solidarity can overcome this pandemic*. devex; November 24, 2021. https://www.devex.com/news/opinion-solidarity-can-overcome-this-pandemic-102078. Accessed June 28, 2023.
70. Usher AD. South Africa and India push for COVID-19 patents ban. *Lancet*. 2020;396(10265):1790–1791. https://doi.org/10.1016/S0140-6736(20)32581-2.
71. Lee K. Shaping the future of global health cooperation: where can we go from here? *Lancet*. 1998; 351(9106):899–902. https://doi.org/10.1016/S0140-6736(97)11455-6.
72. McCloskey B, Dar O, Zumla A, Heymann D. Emerging infectious diseases and pandemic potential: status quo and reducing risk of global spread. *Lancet Infect Dis*. 2014;14(10):1001–1010. https://doi.org/10.1016/S1473-3099(14)70846-1.

CHAPTER

5

International legal issues of national sovereignty and authority impacting global health security

Rana Sulieman[1], *Lawrence N. Anyanwu*[2], *Vicky Cardenas*[3], *Mohamed Moussif*[4], *Ebere Okereke*[5,6] *and Oyeronke Oyebanji*[7]

[1]Rollins School of Public Health, Emory University, Atlanta, GA, United States; [2]University of Glasgow, Glasgow, United Kingdom; [3]Gibson, Dunn & Crutcher LLP, San Francisco, CA, United States; [4]Casablanca International Airport, Ministry of Health, Casablanca, Morocco; [5]Tony Blair Institute for Global Change, London, United Kingdom; [6]Africa Centres for Disease Control and Prevention (Africa CDC), Addis Ababa, Ethiopia; [7]Coalition for Epidemic Preparedness Innovations, London, United Kingdom

Introduction

Significant challenges exist in balancing global health security (GHS) and national sovereignty. We must consider the benefits of global governance while maintaining the delicate legal balance of legitimate national sovereignty and authority that benefits all. Promoting reform that prioritizes transparency and accountability while preserving sovereignty and authority is the path forward.

This chapter explores the historic context in which the International Health Regulations (IHR) were established and evolved and assessed the functionality of current IHR (2005) guidelines during major international crises. This review aims to examine the intersection between the IHR (2005) and the global health architecture, while determining how to establish functioning GHS. The objective was to identify key issues prompting amendments to the current legal infrastructure and reframe the IHR in the context of the global health law.

Modernizing Global Health Security to Prevent, Detect, and Respond
https://doi.org/10.1016/B978-0-323-90945-7.00015-4

Emergence of the International Health Regulations

The IHR serve as an instrument of international agreement and provide an overarching governing framework that respects and defines national rights and obligations during public health crises. The IHR represent a legally binding (refer to "Relevant Legal Instruments and Processes" section for more information) global agreement among 196 countries, including the 194 World Health Organization (WHO) Member States (MSs) *"to prevent, protect against, control, and provide a public health response to the international spread of disease in ways that are commensurate with and restricted to public health risks and that avoid unnecessary interference with international traffic and trade."*[1]

The origin of the IHR dates back to the first International Sanitary Convention (ISC), held in Paris in 1851 to address the European cholera epidemics.[2] Perhaps more importantly, the IHR were originally developed to prevent cross-border transmission of this disease in a way that minimized interference with international trade and travel. Subsequently (from 1851 to 1938), 13 additional ISCs were held to standardize international quarantine regulations against the spread of cholera, plague, and yellow fever.[3]

WHO was established on April 7, 1948, by the United Nations to fulfill the mandate of protecting GHS.[4] WHO administered the International Sanitary Regulations (ISR) by a central international health organization with near universal membership.[5] At the Fourth World Health Assembly (WHA) in 1951, 100 years after the first ISC, the ISR were adopted by WHO member states with few changes, including the expansion of infectious diseases under regulation to include cholera, plague, yellow fever, smallpox, relapsing fevers, and typhus and the integration of WHO into its operation.[5] In 1969, the ISR were renamed the IHR.[1]

Evolution of the IHR

The IHR (1969) focused on control of a short list of diseases (cholera, plague, yellow fever, smallpox, relapsing fevers, and typhus) and obliged MS to report these specific diseases and maintain minimal public health capabilities at ports and borders.[6] Following public health threats and outbreaks outside the regulation—including the Ebola hemorrhagic fever in Zaire in 1976—and the nonapplicability of the IHR at the time in addressing the growing threat of other (re-) emerging infectious diseases (e.g., HIV/AIDS, tuberculosis, and malaria), the WHA and other WHO governance structures recognized the need to revise IHR (1969).[7]

The revision process commenced in 1995,[1] preceded by an initial amendment in 1981 that removed smallpox from the list due to its global eradication in the late 1970s.[5] This revision process culminated in the IHR (1998) draft.

About 5 years after the IHR (1998) draft, the global spread of Severe Acute Respiratory Syndrome (SARS) in 2003 revealed the urgent need for a new set of rules to prevent, control, and provide a public health response to international disease threats.[8] This accelerated another IHR revision process and within 18 months a new set of regulations was agreed upon.[7] On May 23, 2005, the WHA adopted the IHR (2005) and concluded the decade-long effort led by WHO to revise the IHR (1969) to make them more effective against global disease threats. IHR (2005) entered into force on June 15, 2007.[9]

The IHR (2005) constitute the most radical and far-reaching change in international public health law since the beginning of international health cooperation in the mid-19th century.[5]

Under the IHR (2005), there is an improved comprehensive governance strategy that applies to significant international threats emanating from biological, chemical, or radiological sources that are naturally occurring, accidental, or intentionally caused.[5]

While bi- and multilateral collaborative efforts are the basis for the global management of infectious disease epidemics, these efforts are strengthened by the IHR. Since IHR (2005), there have been several global disease outbreaks including H1N1 (commonly known as swine flu), Zika, Ebola, and COVID.[6] There has been a noticeable shift toward global governance, defined as a purposeful order that leans on a mutual arrangement among global institutions to synchronize action for the greater good.

Within the context of GHS, this can manifest in formal or informal sets of policies and recommendations transcending national boundaries; this is rooted in a set of rules that are driven by a combination of economic and moral incentives. This shift toward global governance is highly dependent on reputable institutions to negotiate, adopt, and evaluate normative rules among sovereign nations. Currently, WHO has the constitutional authority to negotiate and evaluate normative instruments, making it the central driver guiding stakeholders toward collaborative efforts.[1] The constitution of the WHO articulates the universal value of the right to health, a nearly universally accepted international legal right. To understand the issues surrounding GHS response, it is important to discuss relevant legal instruments and processes.[10]

Relevant legal instruments and processes

Global health law is not organized as a standardized legal body, but rather unified by a network of "hard" and "soft" laws. Hard laws denote legally binding obligations that can be legally enforced before a court of law (e.g., treaties).[11] Soft laws refer to agreements, principles, and declarations that are not legal binding (e.g., codes of practice); soft law instruments are primarily utilized in the international sphere.[11] The combination of hard and soft legal instruments greatly impacts the realm of GHS and has come under the auspices of the WHO. Legal norms are not binary, but are rooted in normativity, ranging from binding to nonbinding instruments. While the WHA defines the powers that the WHO can apply to develop global health law, only two treaties have been established since the WHA's inception: the Framework Convention on Tobacco Control (FCTC) and the International Health Regulations.[10]

WHO has extraordinary access to treaty-making powers; the processes for negotiating agreements, conventions, and regulations are well-established. MSs have 18 months to accept or reject a convention after its adoption by the WHA. This is an influential mechanism that obliges MS to consider a given treaty in accordance with their respective national, constitutional practices. Nevertheless, WHO lacks power to enforce compliance at a national level, which leads to a heavy reliance on MS involvement to implement conventions through domestic policy.[12] WHO has the authority to negotiate a wide range of health topics, including sanitation and quarantine, disease nomenclature, and standards of safety of pharmaceuticals.[13] Regulations are set into motion after the adoption by the WHA; the process of adoption occurs within an 18-month period. If an MS does not opt out, they are bound to the regulation by default.[12]

Revisions of the IHR establish a more-modern framework for addressing GHS. Because the IHR are legally binding, they establish a centralized process for coordinating the detection of and response to "public health emergencies of international concern" (PHEICs).[14] IHR (2005) require a minimum set of "core capacities" (in relation to health governance) from its signatories, detailed from Annexes 1 and 2.

- WHO shall collect information regarding events through its surveillance activities and assess their potential to cause international disease to spread and possible interference with international traffic.
- Each State Party shall assess events occurring within its territory by using the decision instrument in Annex 2. Each State Party shall notify WHO, by the most efficient means of communication available, by way of the National IHR Focal Point, and within 24 h of assessment of public health information, of all events which may constitute a public health emergency of international concern within its territory in accordance with the decision instrument, as well as any health measure implemented in response to those events. If the notification received by WHO involves the competency of the International Atomic Energy Agency (IAEA), WHO shall immediately notify the IAEA.
- Following a notification, a State Party shall continue to communicate to WHO timely, accurate and sufficiently detailed public health information available to it on the notified event, where possible including case definitions, laboratory results, source and type of the risk, number of cases and deaths, conditions affecting the spread of the disease and the health measures employed; and report, when necessary, the difficulties faced and support needed in responding to the potential public health emergency of international concern.
- When WHO receives information of an event that may constitute a PHEIC, it shall offer to collaborate with the State Party concerned in assessing the potential for international disease spread, possible interference with international traffic and the adequacy of control measures. Such activities may include collaboration with other standard-setting organizations and the offer to mobilize international assistance to support the national authorities in conducting and coordinating on-site assessments. When requested by the State Party, WHO shall provide information supporting such an offer.[1]

The competencies support a process that involves

1. Detecting
2. Identifying
3. Reporting
4. Verifying and responding

In addition, MSs are expected to coordinate support in the event of an international public health crisis. Once an event is reported, it is reviewed by the WHO to assess if the event should be declared a PHEIC (i.e., constitutes an international public health risk through the international spread of disease prompting a highly coordinated international response plan).[14]

There have been seven PHEIC declarations: the 2009 H1N1 (swine flu) pandemic, the 2014 polio declaration, the 2014 outbreak of Ebola in Western Africa, the 2015–16 Zika virus epidemic, the 2018–20 Ebola epidemic, the ongoing COVID-19 pandemic, and the 2022 monkeypox outbreak.[14] In the past 5 years, >100 countries have gone through the latest monitoring and evaluation practices developed by the WHO—the Joint External Evaluation

(JEE). This framework identifies gaps to develop national plans of action through multisectoral approaches.

Strategies such as the US Center for Disease Control and Prevention's (CDC) One Health guideline were born from these evaluation efforts, identifying a need to engage in collaborative, multisectoral, and transdisciplinary approaches at the local, regional, national, and global efforts. Strategies that emphasize the necessity of a coordinated and collaborative effort to prepare and prevent infectious disease outbreaks and for mitigate harm.[15]

Literature review and gap analysis

IHR's underpinnings for global pandemic responses

In today's interconnected and symbiotic world, it is more important than ever to ensure all countries can respond to and contain public health threats. The IHR (2005) creates a potential avenue toward achieving this through its purpose and scope. It consists of 66 articles, including communication and coordination mechanisms and activities among WHO and state parties, roles, and responsibilities of WHO and IHR National Focal Points (NFPs) within MS, and public health surveillance (PHS) and response activities required in-country and at points of entry. It is legally binding on the 196 MSs.[16]

However, concerns and challenges have been observed since IHR (2005). The world has changed considerably in the past 17 years, prompting the need to consider revising IHR (2005) to adapt to a shifting global landscape and to become better equipped for future crises. In the past 13 years, there have been unprecedented crises, each providing a stronger case for legislative reform.[7]

2009 H1N1 Influenza

The 2009 H1N1 influenza pandemic was the first litmus test of IHR (2005); it quickly revealed strengths and weaknesses. Strengths included the prompt notification of the emergence of a novel influenza strain by Mexico to the Pan American Health Organization; and the leadership of WHO in coordinating the response shown through the appointment of an emergency committee and eventual official determination of the first PHEIC[14] WHO initiated the use of epidemic intelligence functions to strengthen timely detection and monitoring of the pandemic, and the structured response activities leading up to its declaration all took place within the framework of the revised International Health Regulations [IHR (2005)].[9] Additionally, through the establishment of National Focal Points (NFPs), the IHR (2005) enabled communication between WHO and all member states, which ultimately served as a guiding framework for the coordination of response efforts.

However, the H1N1 pandemic also highlighted weaknesses in IHR (2005) that could derail its successful future implementation and effectiveness. These included varying disease detection and response capabilities among MSs; violations of IHR rules; and sovereign issues as nations made unilateral decisions outside the governance structure of the IHR.[14]

In 2009, during the aftermath of the H1N1 influenza pandemic, WHO's Executive Board assembled an independent review of the efficacy of IHR (2005).[17] The review highlighted strengths, but ultimately concluded much more needed to be done to establish a feasible and sustainable contingency plan for future public health emergencies. This served as a major lesson and initiated a series of recommendations as underpinnings for future PHEIC mitigation plans.

2016 Zika virus

The 2016 Zika put IHR (2005) under scrutiny, emphasizing the importance of proficient public health surveillance. The Zika virus outbreak response between 2015 and 2016 was magnified because of the criticisms of the preceding 2014–15 Ebola pandemic.[18] WHO and other key players were harshly scrutinized for the delay in decisive action at the outset of the Ebola pandemic. There was a significant delay in declaring Ebola as a PHEIC, resulting in a delay of the call-to-action from the international community to provide financial and technical resources in response to the GHS threat posed by Ebola.

The lesson was learned and in the early stages of the Zika epidemic, Director-General Margaret Chan declared clusters of microcephaly and neurological syndromes associated with Zika virus infection a PHEIC.[18] The rationale for this declaration was driven by concerns surrounding the virus' propensity to lead to congenital Zika syndrome.[18] The IHR's emergency committee was able to quickly respond and prompt an immediate international coordinated effort. There was heightened concern about the sexual transmission of Zika and how it could lead to the introduction of the virus in other countries. To mitigate spread, the WHO and CDC established travel advisories, specifically for individuals traveling from Zika-endemic countries to avoid pregnancy, including Brazil. Several countries issued travel warnings and there was a sharp decline in tourism to countries that experienced outbreaks; this impacted the call to cancel the Olympics in Rio de Janeiro.[17]

While the response was better, there were PHS shortcomings. Zika responses were reliant on case-based PHS, which had significant validity issues. Health departments became the conveyors of PHS, using case reports to track infections in state and local areas, as well as evaluating the efficacy of prevention and control programs while guiding public health action.

This process created what is known as "PHS artifacts," or erroneous markers that increase the number of cases that are not due to an increase in disease.[19] PHS artifacts can provide misleading information about the spread of a disease in each population. This can have significant implications on public health practice. On February 16, 2016, WHO released the "Zika Strategic Response Framework and Joint Operations Plan."[20] This was an imperative step toward an effective response to the Zika epidemic, taking a leap toward reestablishing WHO's global health integrity. This strategy focused on the mobilization and coordination of global partners, while distributing resources to help countries enhance their PHS of Zika.[19]

COVID-19 pandemic

COVID-19 (described as the biggest challenge that human society has faced since WWII)[8] has tested the scope of IHR (2005). This pandemic has necessitated the employment of unconventional measures including timely triage, referral of suspected cases, provision of designated isolation facilities, institution of socioeconomic support to promote widespread uptake of public health measures, transparent communications, and development of multi-level partnerships across sectors.

The implementation of these, all within IHR (2005) and other policies at all levels of government, has had a positive role in limiting the spread of disease.[15] However, to objectively assess the performance of IHR (2005) during COVID-19, WHO set up an IHR expert review committee. Key findings from this review revealed insufficient preparedness efforts across many MSs, including a lack of multisectoral coordination and leadership as well as under-funding of pandemic activities.[16] These were particularly regrettable since the world had witnessed the devastating impact of novel viruses in recent times, including 2003 SARS, 2009 H1N1 influenza, 2012 MERS-CoV, and 2014 Ebola pandemics.[18] COVID-19 has brought into sharp focus the limitations of IHR (2005). These include China's delays on notifying the WHO of public health risks; WHO's delays in declaring a PHEIC, as neither the timing of the threat nor the actual international spread of disease is constitutive elements of a PHEIC stated in the IHR; and many MSs shirking global solidarity for infectious disease prevention, detection, and response activities.[20]

While the IHR are binding on MSs, they contain no enforcement. As a result, WHO has been unable to hold MSs to their obligations—or discipline those that have failed to meet them.[21]

The gaps identified in IHR (2005) emphasize a need for whole-of-government and society approach to public health, including governance and financing and empowering the WHO in coordinating public health responses, through ensuring mandatory reporting and discipline of MSs that do not comply with WHO guidance. Over the last 30 years, GHS crises have resulted in pointed criticisms of the international health community's ability to deal with such threats. These crises offer opportunities growth and improvement.

An important outcome has been an incremental strengthening of international resolve and know-how to promote and improve GHS.[22] The Joint External Evaluation (JEE)—a component of the IHR Monitoring and Evaluation Framework—has been leveraged by various countries to improve GHS.[23,24] While the IHR (2005) are not perfect, they help the world prepare to cope with public-health emergencies and significantly advance the protection of GHS.

Sovereignty and GHS

Among the biggest challenges that exist within GHS is the concept of national sovereignty. In the global context, sovereignty is often regarded as a central aspect to the modern international system, meaning it serves as a code of conduct in the international arena.[25] However, as policymakers continue to unravel the concept of sovereignty, they find it is rather complicated and it becomes difficult to disentangle as an idealized conception of the world and as a reality.

Security-development nexus

Security in the current context has been conventionally defined as the *protection of territorial integrity, stability, and vital interests of states through the use of political, legal, or coercive instruments at the state or international level.*[25] This definition was broadened in the mid-1990s to include nonmilitary threats that may result in violent conflict affecting the security of individuals or states; these range from events like mass migration, resource conflicts, and civil war.[25]

Development refers to the processes through which communities and societies, by-and-large, seek a certain standard of equitable living. This could be achieved through activities that promote socio-economic growth, provision of healthcare services, establishment of sound educational systems, and overall improvements in infrastructure.[24,25]

The concept of the security-development nexus highlights the importance of a mutually reinforced relationship between maintaining security and promoting development.[26] Tackling the nexus involves identifying how security strategies can become better aligned with development objectives in our increasingly unstable geopolitical landscape. Within the context of the promotion of health on an international scale, it is imperative to set contingency plans that emphasize equal development and reduction of harm, particularly among the most vulnerable populations.

Viral sovereignty

Viral sovereignty is the concept that viruses isolated from within the territorial boundaries of a nation state are the sovereign property of that state. The history of viral sovereignty as a concept in international law is nuanced, resting on the intersection between politics and ethics. Access to biological specimens is obviously necessary for scientific research and the processes that provide diagnostics, therapeutics, and vaccines in response to (re)emerging diseases.[27] Information from research allows for the identification of crucial information about the mechanism of replication and infection and serves as the first step toward developing effective vaccines.

Historically, obtaining access to biological specimens has been a challenge perpetuated by viral sovereignty. The argument that sovereignty is a form of "organized hypocrite," which is defined when states rhetorically sanction the normative values or rules associated with sovereignty, but their policies and actions violate these same rules.[32] This is not only true in the context of the international system and its structures, but it also points to a fruitful conversation regarding the dangers of such unrealistic concepts of sovereignty.[26,27] The desire to gatekeep important biological specimens or information in times of potential pandemic damages an interconnected, global system.

A primary reason that access to biological specimens has waned in the context of GHS is the ascendance of the material transfer agreement (MTA) as the medium of ownership and transference of information among researchers (academic, commercial, or nonprofit).[28] MTAs are contracts protected by law. Therefore, if one provision is broken, the contract is considered breached, and the parties involved may be brought into a lawsuit; this needs reform. Clauses that protect intellectual property while prioritizing aid during PHEICs should be added. Sovereignty as an absolute principle deteriorates under international norms of humanitarian intervention, responsibility to protect, and overall liberal ways of order.

CASE STUDY

The synergy of global health security and healthcare: the role of health systems in global health preparedness and security

Background

In the contemporary global landscape, the dialogue surrounding GHS is often in tandem with universal healthcare (UHC). Proponents of both UHC and GHS emphasize key global agendas that aim to strengthen the health and well-being of the global population. The linkages between health systems and health security are historically existent, but there is a need to address the gaps that focus on the importance of health system strengthening. According to the WHO, health security relates to '*the activities required, both proactive and reactive, to minimize vulnerability to acute public health events that endanger the collective health of populations living across geographical regions and international boundaries.*'[33] There are several efforts to emphasize the synergistic relationship between UHC and GHS, namely in the context of public health crises. The COVID-19 pandemic has served as a timely reminder of global unpreparedness in the wake of (re)emerging infectious diseases and PHEICs. This relationship has been gradually become emphasized in global, political discourses, with an ultimate goal of establishing effective GHS to respond to these crises. For this to happen, there must be a concerted effort at local, national, and global levels.

A scoping review was conducted by a group of researchers (Brown, G.W., Bridge, G., Martini, J. et al.) to identify literature on the linkage between health systems and health security.[34] Given the minimal synthesized evidentiary content associated between health systems and health security, the review implemented an augmented methodology as developed by Arksey and O'Malley[35] and further refined by Levac et al.[36] and the Joanna Briggs Institute.[37] The aim of the review was to help the WHO "1) locate and clarify key conceptual and practical linkages between health systems and health security in the peer reviewed academic literature; 2) identify key characteristics or factors related to these linkages; 3) act as a precursor to a systematic review, and; 4) identify and analyze knowledge gaps."[34]

Methods

- Seven electronic databases: Jstor, PubMed, ProQuest, Scopus, ScienceDirect, MedLine, SAGE, two institutional library repositories: USYD library, DOAJ Directory of Open Access Journals, and one search engine: Google Scholar were searched.
- The authors used existing policy framework to produce a list of key search terms. These include resilience, emergency preparedness, WHO benchmarks for IHR capacities, health emergency and disaster risk management, disaster risk reduction, disaster management, common goods for health, and disease control. Variants and combinations of these search terms were used to identify literature.[34]
- The data was imported into a spreadsheet and assessed based on the following criteria:
  - High: The publication makes direct links between health systems and health security (primary RQ) and

(Continued)

CASE STUDY *(cont'd)*

provides lessons on elements and characteristics of a "strengthened" health system for health security (subquestion A) with the presentation of case examples (subquestion B).

- Medium: The publication presents links between health systems for health security (primary RQ) but provides a more general refection with few concrete lessons (subquestion A) and no case studies/examples (subquestion B).
- Low: The publication presents few if any links between health systems for health security (primary RQ).

Observations

The results are presented in two parts. The first part described the overall implications of the study findings, underlining "how in much of the existing literature health security is prioritized over health system strengthening, how health security is considered in terms of exceptionalism and that there is a focus on acute emergencies rather than day to day challenges".[34] In the second part of the results, a series of practical examples were identified as "building blocks" for health systems and health security and what key links can be derived from these examples. It was observed that.

- Of the 343 articles that were initially identified, 204 discussed health systems and health security (high and medium relevance), 101 discussed just health systems and 47 discussed only health security (low relevance).
- Within the high and medium relevance articles, several concepts emerged, including "the prioritization of health security over health systems, the tendency to treat health security as exceptionalism focusing on acute health emergencies, and a conceptualization of security as 'state security' not 'human security' or population health."[34]
- "Much of the literature discussed health security in terms of being an exceptional form of response versus a concept embedded within the wider public health continuum and/or health system strengthening approach."[34]
- The overall analysis of the PHEICs addressed showed a clear propensity in the literature to focus on acute health emergencies and much less on preparedness.

Key issues

There is an urgent need to employ more resilient health care strategies to strengthen the capacities of MSs and local government structures to be better equipped to deal with PHEICs. The need to address this gap gains relevance in light of recent crisis, such as the COVID pandemic, as it confirmed a critical relationship between health system competences and applicable health security response.

Lessons learned

- ► There is a synergy between resilient healthcare systems and GHS, therefore promoting healthcare strategies to support GHS (e.g., high immunization coverage to prevent outbreaks; providing UHC for patients to establish earlier detection systems; improved case management and health system capacity to improve response).

CASE STUDY (cont'd)

- There is a verified need to embed IHR (2005) core capacities into the health system architecture. This must be done across all six health system functions and must include a conscious effort to involve governments in local, national, and global sectors.
- Strong global governance is needed to invest in a more robust and interconnected GHS system, this includes the implementation of healthcare strategies to bolster the effectiveness of IHR (2005) and improve adherence.

Recommendations

In conclusion, the results of this study indicate inadequate capacities to facilitate prevention, detection, and response to PHEICs or other security threats at national and regional levels. Furthermore, the ambition must be realized in tandem with healthcare resilience while maintaining legal sovereignty. This requires.

- Active planning by global institutions to improve the effectiveness of crisis preparedness agendas and ease the tensions between the two concepts.
- Implementation of strategic, integrated implementation approaches.
- Reducing the research gaps by incorporating a whole-of-society approach for GHS through the amalgamation of health security capacities and mechanisms from health systems and other sectors that work in synergy to meet the demands inflicted by health emergencies.

Vision for the future

With the ongoing COVID-19 pandemic, there has been tremendous loss; a tragedy with an overwhelming impact on peace and security across the globe resulting in crumbling economies, political tension, and undermining of social cohesion. All of these exposed major cracks in the GHS architecture. At the core is IHR (2005). Moving forward requires reflection on the current global, legal infrastructure and the process of another revision of the IHR.

Lessons from the COVID-19 pandemic emphasize the need for global leaders to marshal resources and collaborative strategies to mend the GHS architecture. Change takes time, but major progress can be made by leveraging the synergy between GHS and other strategies, including universal healthcare (UHC).[29] Revising a standing legal framework and creating new international instruments are not mutually exclusive.

An important step involves strengthening the IHR (2005) through early warning systems; centralized information and data sharing; and amended pandemic tracing efforts. These efforts are reinforced when made through collaborative strategies, as Secretary-General of the United Nations (UN) states, *The world faces security challenges that no single country or organization can address alone.*[30]

Currently, WHO is proactively supporting IHR (2005) training and capacity development in countries to promote the effectiveness of proficient information-sharing systems. An emphasis on supporting timely communication through a global network is essential for ensuring a strong operational capacity. To this end, the JEE under the IHR Monitoring and Evaluation Framework should be complemented with other programs having overlapping frameworks (e.g., UHC, the Sustainable Development Goals, Essential Public Health Functions/Operations).[31] Working in tandem with similarly oriented approaches help bolster WHO's efficient operational role in emergency preparedness while establishing baseline accountability.

Change is required on a multilevel matrix, through the core of WHO programmatic efforts. The goal should be to embed IHR (2005) requirements into health systems, at every level, to promote strong and stable GHS. One way this may be achieved is if the framework is laid across all public health functions and this is dependent on the strength of communication within inter-regional clusters. Moreover, these efforts must be morally sound and aim to minimize inequalities by promoting health equity and aiding the most vulnerable communities. The international community must grasp the momentum around the current pandemic to ensure preparedness for the next.

Actions

All recent health emergencies including COVID-19 have underscored the urgency and importance of interagency collaboration as a vital means for ensuring robust preparedness and response at national and subnational levels. It is crucial that interagency collaboration is scaled-up to advance preparedness and response, and it is vital that individual agencies prioritize the strengthening of their capacity to effectively participate in joint action for health security with other stakeholders.[2] Below are some key actions/recommendations for consideration that are proposed to support achieving this.

1. Cross-agency engagement for health security should be set against a robust strategy that guides the nature, scope, and approach for collaboration among partners. The strategy should be developed from the outset of a crisis (where possible) and regularly reviewed to ensure that the implementation of cross-agency collaboration is in line with stated aims and objectives and that the desired expectations are met. Interagency collaboration within a country should also be carried out under the authority of the highest possible levels of political—and organizational leadership to facilitate the greatest level of commitment among partners and achieve a whole-of-government contribution to preparedness efforts. This will support continued alignment among stakeholders as collaboration efforts are carried out, and it can broaden the level of engagement to a wide range of agencies and stakeholders from across government ministries and sectors

(e.g., health, finance, agriculture, travel and trade, parliaments, technology). Although the incentives and contributions from different agencies can have may differ, diversity among collaborative actors can lead to the introduction of new and unique perspectives which is valuable for complementary action in building preparedness.

2. At the practical level, interagency collaboration should be guided by comprehensive work plan activities that indicate the full scope of roles and responsibilities among partners/stakeholders. Although it may not be always possible to establish and adhere to joint work plans—especially during times of health emergencies where the nature of interagency collaboration must frequently adapt as needed—the presence of a well-established work plan can enable effective engagement among partner agencies. Clear work plans with identifiable milestones can also facilitate stakeholders to monitor and evaluate the progress of joint action for preparedness and anticipate bottlenecks ahead of time.
3. National and international agencies that are involved in health security strengthening should establish collaborative linkages before a global health emergency occurs. Often, interagency coordination and collaboration in the context of health emergencies are only set up during a response action when a crisis occurs, and it is likely that this is carried out in light of specific diseases or hazards. It is also common to see that such collaboration lasts only for the duration of the health emergency itself and those interagency linkages which are formed often disappear once a public health event is controlled. However, proactively seeking where potential linkages between national and international agencies can be made before a health emergency occurs should facilitate the collaborative approach that is taken at all stages including during a time of crisis. Much progress has been made in advancing interagency collaboration during COVID-19, and it is imperative that this must be carried forward beyond the pandemic. Although much of the recent interagency collaboration has been based on addressing the risks and impact of the pandemic specifically, it would be useful for this to be adapted toward addressing other health emergency threats through an all-hazards approach. This would better enable the impact of existing interagency collaboration for health security to extend to other public health threats.
4. Global guidance such as the WHO Multisectoral Preparedness Coordination (MPC) Framework can support countries when planning and implementing interagency collaboration. There is no "one size fits all" for multisectoral collaboration, therefore, countries need to contextualize and adapt their interagency and multisectoral collaboration.
5. There should be regular information sharing between agencies to enhance opportunities for collaboration. Agencies may not always be able to provide the full amount of information that is available however sharing knowledge, information, and experiences between agencies can support continued alignment among partners in ways that facilitate engagement toward common action for preparedness. Any barriers or boundaries for information sharing—including those relating to potential conflicts of interest—should be openly discussed at the outset of interagency collaboration to avoid and resolve potential challenges.
6. Interagency collaboration should be based on principles of shared financial and technical resources where possible, including shared human resources. This will support the sustainability of interagency collaborative action and partnerships and provide a basis

of reliability that stakeholders can build upon to feel confident in committing significantly toward working with each other. It is useful to establish costing needs early on and to carry out regular resource mobilization to ensure that the availability of funding and other resources is not a challenge.

7. All instances of interagency collaboration including at the subnational, regional, national, and global levels should regularly document and disseminate knowledge about their experiences. Identifying successes and challenges is highly valuable to enable others to understand what has worked well and what measures should potentially be avoided in other ventures of interagency collaboration. Clear documentation and wide dissemination of best practices also limit the scope for potential duplication, and it also presents stakeholders with an understanding of opportunities where they may be able to provide complementary contributions that can build upon existing progress.

References

1. World Health Organization. *International Health Regulations*. World Health Organization; 2005. Retrieved June 18, 2022, from https://www.who.int/publications/i/item/9789241580410.
2. Gostin L 0. World health law: toward a new conception of global health governance for the 21st century. *Yale J Health Policy Law Ethics [Internet]*. 2005;1:413–424. http://165.158.1.110/english/pro-salute/contents.htm;.
3. Howard-Jones N. The scientific background of the international sanitary conferences 1851-1938. *WHO Chron [Internet]*. 1974;28(11):495–508. https://pubmed.ncbi.nlm.nih.gov/4607249/.
4. World Health Organization. *World Health Organization | Encyclopedia.Com [Internet]*; 2018. Retrieved June 19, 2022, Available from https://www.encyclopedia.com/social-sciences-and-law/political-science-and-government/united-nations/world-health-organization.
5. Fidler DP. From international sanitary conventions to global health security: the new international health regulations. *Chin J Int Law*. January 1, 2005;4(2):325–392.
6. Youde J. *The International Health Regulations. Biopolitical Surveillance and Public Health in International Politics*; 2010:147–175. Retrieved June 20, 2022. Available from https://link.springer.com/chapter/10.1057/9780230104785_7.
7. Gostin DPF and LO. The new international health regulations: an historic development for international law and public health. *J Law Med Ethics*. 2006;34(1):85–94.
8. Gostin LO. International infectious disease law revision of the world health organization's international health regulations. *J Am Med Assoc Med Assoc [Internet]*. 2004;291(21):2623–2627. http://jama.jamanetwork.com/.
9. Katz R, Fischer J. The Revised International Health Regulations: A Framework for Global Pandemic Response (Internet). http://www.ghgj.org.
10. Baker MG, Fidler DP. Global public health surveillance under new international health regulations. *Emerg Infect Dis*. 2006;12(7):1058–1065. https://doi.org/10.3201/eid1207.051497.
11. Abbott KW, Snidal D. Hard and soft law in international governance. *Int Organ*. 2000;54(3):421–456. http://www.jstor.org/stable/2601340.
12. World Health Organization. *Trade and Health: Towards Building a National Strategy*; 2015. http://apps.who.int/iris/bitstream/10665/183934/1/9789241565035_eng.pdf?ua=1.
13. World Health Organization. *Everybody's Business: Strengthening Health Systems to Improve Health Outcomes. WHO's Strategic Framework for Action*; 2007. http://www.who.int/healthsystems/strategy/everybodys_business.pdf.
14. Mujica OJ, Brown CE, Victora CG, Goldblatt PO, da Silva JB. Health inequity focus in pandemic preparedness and response plans. *Bull World Health Organ [Internet]*. 2022;100(2), 91-91A https://www.who.int/publications/journals/bulletin/.
15. Mackenzie JS, McKinnon M, Jeggo M. One health: from concept to practice. *Confronting Emerging Zoonoses*. 2014:163–189. https://doi.org/10.1007/978-4-431-55120-1_8.

16. Kluge H, Martín-Moreno JM, Emiroglu N, et al. Strengthening global health security by embedding the International Health Regulations requirements into national health systems. *BMJ Glob Health*. 2018;3(Suppl 1):e000656. https://doi.org/10.1136/bmjgh-2017-000656.
17. Calain P. From the field side of the binoculars: a different view on global public health surveillance. *Health Policy Plan [Internet]*. 2007;22(1):13–20. https://academic.oup.com/heapol/article/22/1/13/674982.
18. Wilder-Smith A, Osman S. Public health emergencies of international concern: a historic overview. *J Trav Med*. 2020;27(8):taaa227. https://doi.org/10.1093/jtm/taaa227.
19. Nsubuga P, White ME, Thacker SB, et al. Public Health Surveillance: A Tool for Targeting and Monitoring Interventions. In: Jamison DT, Breman JG, Measham AR, et al., eds. *Disease Control Priorities in Developing Countries*. 2nd ed. Washington (DC): The International Bank for Reconstruction and Development/The World Bank; 2006 (Chapter 53). Available from https://www.ncbi.nlm.nih.gov/books/NBK11770/Co-published by Oxford University Press. https://apps.who.int/iris/handle/10665/204420. New York. World Health Organization. (2016). Zika strategic response framework and joint operations plan, January-June 2016. World Health Organization.
20. Tabbaa D. International Health Regulations and COVID-19 pandemic, challenges and gaps. *Vet Ital*. 2020;56(4):237–244.
21. OH & A Phillips-Robins. *COVID-19 and International Law Series: WHO's Pandemic Response and the International Health Regulations*. 2020.
22. Gostin LO, Habibi R, Meier BM. Has global health law risen to meet the COVID-19 challenge? Revisiting the international health regulations to prepare for future threats. *J Law Med Ethics*. 2020;48(2):376–381.
23. World Health IMA. *Stopping Ebola in its Tracks: Maximizing a Health Systems Approach for Improved Epidemic Response*. Washington DC: IMA World Health; 2015.
24. Gostin LO, DeBartolo MC, Friedman EA. The international health regulations 10 years on: the governing framework for global health security. *Lancet*. 2015;386:2222–2226.
25. MAKINDA SM. Sovereignty and global security. *Secur Dialog*. 1998;29(3):281–292. https://doi.org/10.1177/0967010698029003003.
26. Clark D, Berson T, Lin HS. Committee on developing a cybersecurity primer: leveraging two decades of national academies work; computer science and telecommunications board; national research council. In: *At the Nexus of Cybersecurity and Public Policy: Some Basic Concepts and Issues*. Washington (DC): National Academies Press (US); June 16, 2014.
27. Sam F. Halabi, viral sovereignty, intellectual property, and the changing global system for sharing pathogens for infectious disease research. *28 Ann Health Law*. 2019;101. Available at https://scholarship.law.missouri.edu/facpubs/743.
28. Rourke MF. Restricting access to pathogen samples and epidemiological data: a not-so-brief history of "viral sovereignty" and the mark it left on the world. In: Eccleston-Turner M, Brassington I, eds. *Infectious Diseases in the New Millennium*. Cham: Springer; 2020:. International Library of Ethics, Law, and the New Medicine; Vol 82.
29. Frenk J, Moon S. Governance challenges in global health. *N Engl J Med*. 2013;368:936–942.
30. Beyond COVID-19: reimagining the role of international health regulations in the global health law landscape, Health Affairs Blog, November 1, 2021.
31. Volk C. *The Problem of Sovereignty in Globalized Times. Law, Culture and the Humanities*. February 2019. https://doi.org/10.1177/1743872119828010.
32. Krasner SD. *Sovereignty: Organized Hypocrite*. Princeton, NJ: Princeton University Press; 1999. https://doi.org/10.1515/9781400823260.
33. WHO. *The World Health Report 2007 - the World Health Report 2007 : A Safer Future : Global Public Health Security in the 21st Century*. World Health Organization; 2007. https://apps.who.int/iris/bitstream/handle/10665/43713/9789241563444_eng.pdf?sequence=1&isAllowed=y.
34. Brown GW, Bridge G, Martini J, et al. The role of health systems for health security: a scoping review revealing the need for improved conceptual and practical linkages. *Glob Health*. 2022;18:51. https://doi.org/10.1186/s12992-022-00840-6.
35. Arksey H, O'Malley L. Scoping studies: towards a methodological framework. *Int J Soc Res Methodol*. 2005;8:19–32.
36. Levac D, Colquhoun H, O'Brien K. Scoping studies: advancing the methodology. *Implement Sci*. 2010;5. https://doi.org/10.1186/1748-5908-5-69. https://implementationscience.biomedcentral.com/articles/.
37. Peters M, Godfrey C, McInerney P, et al. Scoping reviews. In: *Joanna Briggs Institute Reviewer's Manual*. Adelaide, Australia: Joanna Briggs Inst; 2017. https://reviewersmanual.joannabriggs.org/JBI.

SECTION II

Global One Health to address pandemics - ecological and biological challenges in the dynamic planet*

Wondwossen A. Gebreyes[1], Carol J. Haley[2], Jonna A.K. Mazet[3] and Ermias Belay[4]

[1]Global One Health initiative (GOHi), The Ohio State University, Columbus, OH, United States; [2]Hubert Department of Global Health, Rollins School of Public Health, Emory University, Atlanta, GA, United States; [3]Grand Challenges, University of California, Davis, CA, United States; [4]Division of High-Consequence Pathogens and Pathology, National Center for Emerging and Zoonotic Infectious Diseases, Centers for Disease Control and Prevention, Atlanta, GA, United States

Introduction

About 75% of emerging and reemerging infectious diseases in humans are of animal and ecosystem origin. Coronavirus including SARS-CoV-2 is among these pathogens. The world is facing rapidly accelerating change that impacts human and animal health and disrupts our ecosystems, the environment, and overall planetary health. The increasing world population that has currently reached 8 billion is expected to reach nearly 9.7 billion by 2050 (UN Report).[a] Regions, such as Africa and Asia, are expected to have the fastest population growth. This intense population pressure has several potentially irreversible consequences. The land-use pattern is changing

*Contribution of author Ermias Belay is subject to public domain.

[a]https://www.un.org/en/global-issues/population.

with accelerating urbanization; the current 55% urban-dwelling population is expected to reach 70% by 2050 (UN 75 Report).[b] The rate of deforestation to expand human habitats and food production is resulting in massive changes to our atmosphere. Associated industrialization and expansion of intensive animal production to meet human protein demand increase greenhouse gas emissions that result in thinning of the ozone layer. This sustained climate change, in turn, will result in further ecosystem changes with altered distributions of insect vectors that carry potentially deadly pathogens, resulting in emergence of zoonotic vector-borne disease epidemics.

The interconnection of humans, animals, plants, and the environment is changing in ways that require us to rethink how we prevent, detect, and respond to epidemics and pandemics. The One Health paradigm recognizes the health of humans, domestic and wild animals, plants, and the wider environment (including ecosystems) are closely linked and interdependent. This approach with the terminology of "One Health" was described as early as 1984 by Calvin Schwabe of the University of California at Davis. The approach mobilizes multiple sectors, disciplines, and communities at varying levels of society to work together to foster well-being and tackle threats to health and ecosystems, while addressing the collective need for clean water, energy and air, safe and nutritious food, taking action on climate change, and contributing to sustainable development.[1] The Global One Health approach is a proactive approach that further expands in this recent One Health definition and facilitates building institutional capacity with a comprehensive (global) integrated system addressing communicable and noncommunicable diseases to preemptively promote global health security and prepare for and prevent the emergence and occurrence of biological and chemical hazards at the subnational, national, and international (geographic global) levels. Such a proactive approach is known to be cost-effective and could save a large number of human lives each year. For example, the World Bank estimates that it costs $4 billion over 10 years to build and operate One Health systems for effective control of emerging infectious diseases in developing countries.[2] This is estimated to be 10% of the cost of a large epidemic.[c] By investing this amount, we could help to prevent catastrophic epidemics and pandemics, similar to COVID-19 by focusing on preparedness and building workforce and infrastructure capacity. The impact of COVID-19 surpassed $9 trillion in less than 2 years.[3] While we recognize there are inevitable consequences of ever-changing global dynamics, the implementation of the One Health approach should be accelerated. It is encouraging to recognize the recent activities by the quadripartite (WHO, FAO, WOAH, and UNEP), G20 and others are calling for coordination and implementation of One Health globally to prevent

[b]https://www.un.org/en/un75/finalreport.

[c]https://www.science.org/doi/10.1126/science.aap7463.

epidemics/pandemics. Recently, the quadripartite released the One Health Joint Plan of Action (2022–26) that articulated six priority tracks.[d]

Efforts on predicting and preventing potential pandemics have, so far, been very limited. One example is the US Agency for International Development Emerging Pandemic Threats (EPT) program. EPT networks attempted to strengthen systems to identify, prevent, respond to, and mitigate pandemic threats. The strengthened systems were instrumental in helping partner countries to respond quickly to the threat of zoonotic spillovers, including identifying the first introduced cases of COVID-19, into countries neighboring China, allowing the public health systems to be activated early, even before the full sequence of SARS-CoV-2 was known.[4] Such investment needs to be more diverse across the US federal agencies, as well as other academic and nongovernmental implementing partners.

It is clear that without broader application of the One Health approach and significant resource commitment, we will continue to struggle to deal with complex public health problems that require a holistic problem-solving approach to prevent and mitigate them. We must act now to fully consider all of the human behaviors that increase our risk of the next epidemic and another pandemic. Top among these is human interactions with domestic (food and companion) as well as and wild animals (in the wild as well as in captivity), since the majority of emerging health threats arise from pathogens that are shared with animals, especially wildlife.[5] The One Health approach allows us to consider how people are increasing their risk of exposure through the development of growing industries (e.g., bat guano farming and cave tourism), our thirst for new products that take us deep into rainforests and other unexplored territories, and our appetite for "gourmet" meat and products that fuel wildlife trade and trafficking.

The Global Health Security Agenda (GHSA) was launched in 2014 by the US federal government and partner country governments to address the global threat of infectious diseases in an increasingly interconnected world.[6–8] It recognizes that health security is a shared responsibility and depends on a multisectoral approach to strengthen capacity in the animal, human, and environmental health sectors to prevent, detect, and respond to infectious disease threats.[9] The key objectives of GHSA include enhancing country capacities to meet International Health Regulation (IHR) requirements, promote multisectoral engagement and collaboration, and focus on common, measurable targets.[10] Efforts are ongoing to incorporate lessons learned from the COVID-19 pandemic to IHR and Joint External Evaluation (JEE) assessment tools and processes. In some countries, capacities established as part of GHSA implementation have been instrumental in addressing some of the surveillance and laboratory challenges faced in responding to the COVID-19 pandemic.

[d]https://apps.who.int/iris/rest/bitstreams/1473175/retrieve.

The world is at a pivotal point. This section of the book has been developed with the aim of revisiting several aspects of the existing Global One Health paradigm[11] in light of the recent pandemic and proposing how the paradigm can best be integrated into policies to improve prevention and detection of and response to pandemic threats. As indicated earlier, Global One Health is not limited to infectious agents. Chemical hazards associated with noncommunicable diseases are important components of Global One Health. The chapters address critical issues that, if not addressed, may lead to further health security concerns worldwide. These issues necessitate consistent attention and focus under the IHR guidelines in the future. The topics covered in this section are diverse including zoonotic viral diseases and their emergence and reemergence as epidemic and pandemic concerns in discussed in Chapter 6 and antimicrobial resistance (Chapter 7 that continues to impact the world in profound ways. Chapter 8 addresses toxic chemical challenges and hazards associated with environmental matrices: water, soil, and air. This chapter highlights the critical role the environment plays in the global disease burden. It also emphasizes the need for increased attention to chemical hazards in addition to the key emphasis on communicable diseases. The increasing need for the One Health approach to inherently embrace environmental and chemical hazards is among the key messages in this chapter aimed at the global scientific community and policy makers. Chapter 9 focuses on vector ecology and vector-borne diseases and their emergence and reemergence, primarily in association with climate change. It provides recommendations on how we can build and strengthen partnerships. The section then transitions to bacterial and protozoan diseases of zoonotic, food, and water-borne significance in Chapter 10. The last chapter of the section (Chapter 11 focuses on urban ecosystems and how the increasingly urbanized planet can effectively address One Health issues at the interface of humans, animals, and plants and their shared environment. This chapter discusses, in depth, the urban ecosystem and how it shapes infectious disease transmission dynamics.

The main goal of this book is to inform the global stakeholders of the critical importance of global health security and broader approaches to inform the future of IHR. This section's particular emphasis is to highlight the major role One Health plays at the interface of humans, animals, plants, and the environment. As in all chapters of this book, the contributors provide visions and actions for improvement. Whether we are working to identify the sources of future pandemics, attempting to stem the tide of climate change, or trying to improve our food and water systems, lasting solutions can be found through transdisciplinary and multisectoral collaborations and a holistic approach that is enshrined in the One Health approach.

The findings and conclusions in this report are those of the authors and do not necessarily represent the official position of the Centers for Disease Control and Prevention.

References

1. One Health High-Level Expert Panel (OHHLEP), Adisasmito WB, Almuhairi S, Behravesh CB, Bilivogui P, Bukachi SA, Casas N, Cediel Becerra N, Charron DF, Chaudhary A, Ciacci Zanella JR, Cunningham AA, Dar O, Debnath N, Dungu B, Farag E, Gao GF, Hayman DTS, Khaitsa M, Koopmans MPG, Machalaba C, Mackenzie JS, Markotter W, Mettenleiter TC, Morand S, Smolenskiy V, Zhou L. One Health: A new definition for a sustainable and healthy future. *PLoS Pathog*. 2022;18(6):e1010537. https://doi.org/10.1371/journal.ppat.1010537. PMID: 35737670; PMCID: PMC9223325.
2. World Bank. *People, pathogens and our planet volume 2 the economics of one health*. 2012.
3. World Economic Forum. Global Risks Report 2022. https://www.weforum.org/reports/global-risks-report-2022/.
4. PREDICT Consortium. *Advancing Global Health Security at the Frontiers of Disease Emergence*. Davis: One Health Institute, University of California; December 2020:596. Hard copies are available, and it can be downloaded at https://ohi.vetmed.ucdavis.edu/programs-projects/predict-project.
5. Jones K, Patel N, Levy M, et al. Global trends in emerging infectious diseases. *Nature*. 2008;451:990–993. https://doi.org/10.1038/nature06536.
6. Inglesby T, Fischer JE. Moving ahead on the global health security agenda. *Biosecur Bioterror*. 2014;12(2):63–65. https://doi.org/10.1089/bsp.2014.3314. Epub 2014 Mar 18. PMID: 24641445.
7. Kluge H, Azzopardi-Muscat N, Figueras J, McKee M. Trust and transformation: an agenda for creating resilient and sustainable health systems. *BMJ*. 2023;380:651. https://doi.org/10.1136/bmj.p651. PMID: 36940935.
8. Wolicki SB, Nuzzo JB, Blazes DL, Pitts DL, Iskander JK, Tappero JW. Public Health Surveillance: At the Core of the Global Health Security Agenda. *Health Secur*. 2016;14(3):185–188. https://doi.org/10.1089/hs.2016.0002. PMID: 27314658; PMCID: PMC6937158.
9. Gronvall GK, Wang L, McGrath PF, Cicero AJ, Yuan Y, Parker MI, Zhang W, Sun Y, Xue Y, Zhang J, Zhang X, Yu L, Song J, Trotochaud M. The Biological Weapons Convention should endorse the Tianjin Biosecurity Guidelines for Codes of Conduct. *Trends Microbiol*. 2022;30(12):1119–1120. https://doi.org/10.1016/j.tim.2022.09.014. Epub 2022 Oct 10. PMID: 36229380.
10. Katz R, Sorrell EM, Kornblet SA, Fischer JE. Global health security agenda and the international health regulations: moving forward. *Biosecur Bioterror*. 2014 Sep-Oct;12(5):231–238. https://doi.org/10.1089/bsp.2014.0038. PMID: 25254911.
11. Gebreyes WA, Dupouy-Camet J, Newport MJ, Oliveira CJ, Schlesinger LS, Saif YM, Kariuki S, Saif LJ, Saville W, Wittum T, Hoet A, Quessy S, Kazwala R, Tekola B, Shryock T, Bisesi M, Patchanee P, Boonmar S, King LJ. The global one health paradigm: challenges and opportunities for tackling infectious diseases at the human, animal, and environment interface in low-resource settings. *PLoS Negl Trop Dis*. 2014;8(11):e3257. https://doi.org/10.1371/journal.pntd.0003257. PMID: 25393303; PMCID: PMC4230840.

CHAPTER

6

(Re-)emerging viral zoonotic diseases at the human–animal–environment interface

Amanda M. Berrian[1,2], *Zelalem Mekuria*[1,2], *Laura E. Binkley*[1,2], *Chima J. Ohuabunwo*[3,4,5], *Samantha Swisher*[1], *Kaylee Errecaborde*[6], *Stephane de la Rocque*[6] *and Carol J. Haley*[4]

[1]Department of Veterinary Preventive Medicine, College of Veterinary Medicine, The Ohio State University, Columbus, OH, United States; [2]Global One Health initiative (GOHi), The Ohio State University, Columbus, OH, United States; [3]Department of Medicine and the Office of Global Health Equity, Morehouse School of Medicine, Atlanta, GA, United States; [4]Hubert Department of Global Health, Rollins School of Public Health, Emory University, Atlanta, GA, United States; [5]Department of Epidemiology, University of Port Harcourt School of Public Health, Rivers State, Nigeria; [6]Health Security Preparedness Department, World Health Organization, Geneva, Switzerland

Introduction

The legacy of zoonotic diseases

Of the five largest pandemics in human history, two (the 1918 H1N1 influenza and AIDS, caused by the human immunodeficiency virus (HIV) that was first described in 1981) emerged in the 20th century and both had zoonotic origins. The source of the 1918 influenza outbreak remains controversial; some analytic methods suggest that it derived from an avian strain, while others suggest that it was more likely related to a mammalian (human or swine) strain.[1] The virus proved unusually virulent among young, healthy adults; by the time the pandemic was over, it had killed 40–50 million people globally.[2] While the pandemic eventually ended

https://doi.org/10.1016/B978-0-323-90945-7.00012-9

after 3–4 successive waves of infection, the genetic material of the 1918 H1N1 pandemic strain is still present in many of the influenza A strains that infect humans and pigs today.[1,3]

In the decades after the 1918 influenza pandemic, great strides were made in science and medicine, and many believed that infectious diseases would soon be obsolete. An epidemiologist from Johns Hopkins, Aidan Cockburn, went so far as to assert in 1962 that "it seems reasonable to anticipate that within some measurable time, such as 100 years, all the major infections will have disappeared."[4] It was around this time that HIV, a zoonotic virus that likely spilled over from chimpanzees (simian immunodeficiency virus) into humans in the early 1900s, was starting to make its way undetected from Africa to Haiti and from there to the United States. Since then, the virus has spread throughout the world, killing an estimated 40.1 million people and infecting 38.4 million more who are still living with the disease.[5] HIV has had crushing economic and development consequences for countries where infection rates are high. In addition to the costs of treatment and prevention (estimated to be $29 billion for low- and middle-income countries in 2020 alone), HIV can have a significant impact on family livelihoods and the composition of the workforce.[5] In many parts of the world, HIV has had a disproportionate impact on women and girls; in sub-Saharan Africa, six out of seven newly infected adolescents are girls.[5] Beyond the measurable health and economic consequences, the stigma often associated with HIV infection has had serious social consequences for the individuals and communities most severely affected.[6–8]

It is tempting to believe, as Aidan Cockburn did in 1962, that recent advances in diagnostic techniques and vaccine development have made humans less vulnerable to pandemics. However, the recent COVID-19 pandemic has demonstrated that even in a world where we can quickly identify the etiologic agent and develop vaccines with unprecedented speed, emerging viruses can still have devastating impacts on human lives and the global community (Fig. 6.1).

The rise of global health security

Beginning in 1997 and fueled by the anxieties of pandemic-potential avian influenzas, the term "health security" was used to describe the need to mitigate the threats of infectious diseases, recognizing that in a globalized world, health risks emerging in one country pose an immediate economic and health security risk to other countries.[10,11] A health security approach looked to international capacity development and a focus on building response capacity in disease "hotspots." This spurred a series of global programs focused on disease mitigation at the human–animal–environment interface. Beginning in 2010 and formalized in 2017, the World Health Organization (WHO), together with the World Organization for Animal Health (WOAH, founded as OIE) and the Food and Agriculture Organization of the United Nations (FAO), formed the "Tripartite," providing the standards that protect public health worldwide and allow for safe trade of products that impact human, animal, and environmental health.[12] The global movement toward "health security" was made official by 2014, when the concept garnered commitments from over 64 nations, international organizations, and nongovernmental stakeholders to establish the Global Health Security Agenda (GHSA). As a multilateral, multinational policy initiative, GHSA supports collaborative, capacity-building efforts to achieve targets around biological threats, while accelerating the achievement of national core capacities required by international institutions.[13] Importantly,

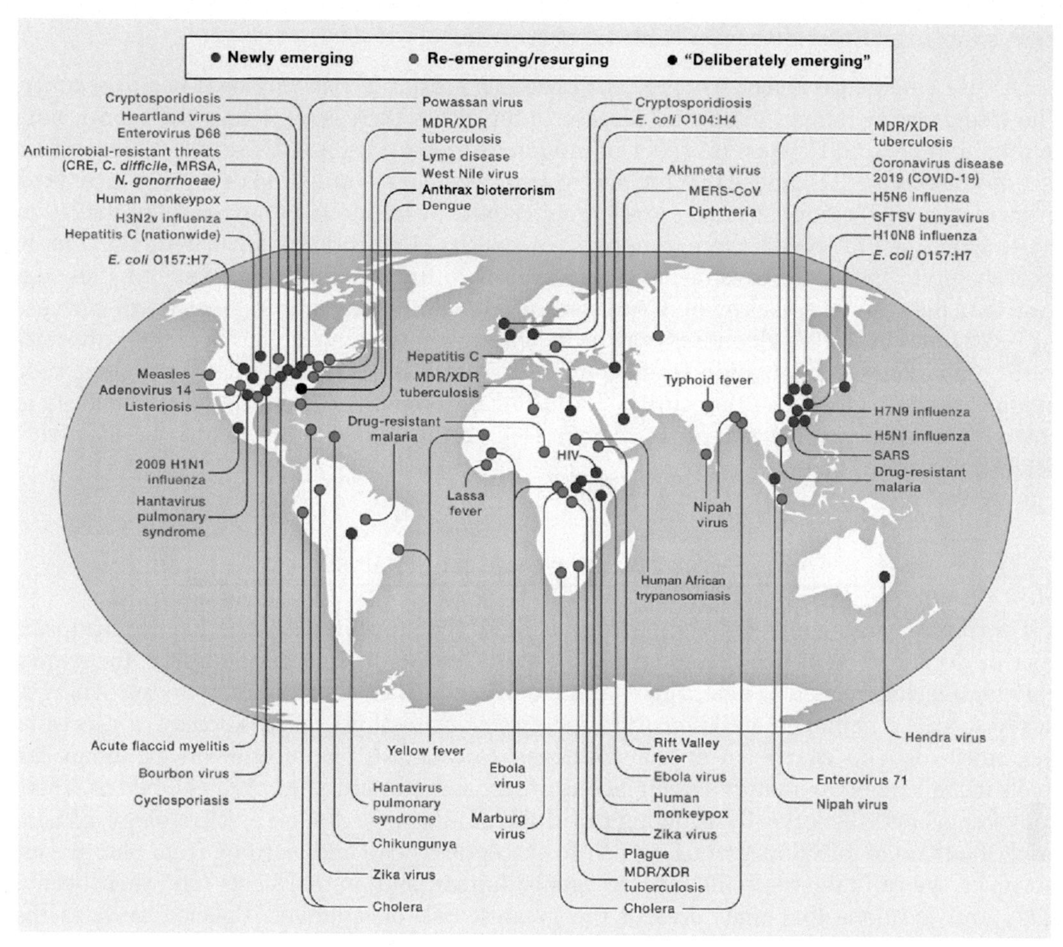

FIGURE 6.1 The global extent of newly emerging, reemerging, and "deliberately emerging" infectious disease from 1981 to the present (2020) by Morens, Fauci.[9] *Picture Copyright 2020: reproduced with permission.*

this initiative has linked GHSA milestones and evaluations with the International Health Regulations (2005)[14] (IHR) and the 2030 Sustainable Development Goals that were adopted in 2015.[15] This multisectoral approach via the Tripartite was further expanded to strengthen the links with the environmental component of One Health through the engagement with the United Nations Environment Programme.[16] This collaboration, formalized in 2022, is now known as the Quadripartite and is well-positioned to fill this "whole-of-society" response coordination for health security. International strategies for global health security continue to evolve, particularly in response to gaps and shortcoming revealed by the COVID-19 pandemic. At the 75th World Health Assembly (WHA75), countries formed a working group to guide discussions on proposed changes to the IHR (2005) as well as develop the draft One Health Joint Plan of Action to be "better able to prevent, predict, detect, and respond to health threats and improve the health of humans, animals, plants, and the environment, while contributing to sustainable development."[17]

Emerging infectious diseases: The next frontier

Disease emergence is increasingly recognized as a priority risk to global health security. The frequency of human infectious disease outbreaks is increasing across the world, both in total number and richness (number of unique diseases) of causal diseases.[18] Emerging infectious diseases (EIDs), those that are newly recognized in a population or rapidly increasing in incidence or geographic range, are largely caused by zoonotic pathogens. From 1940 to 2004, 60.3% of EID events were zoonoses, of which, 71.8% originated in wildlife[19]; more recently, up to 75% of EID events have been attributed to zoonotic pathogens.[20] Models suggest that the rate of spillover of viruses from animals to humans in Africa will increase 1.75–3.2 times by 2070.[21] Zoonoses may be increasingly more novel in terms of their diversity in the global human population when compared to human-specific diseases and, thus, pose unique threats to global health security.[18] Additionally, in many cases, EIDs are most likely to emerge and subsequently spread in regions lacking the resources to be able to effectively detect and control them.[19]

Literature review and gap analysis

For most emerging or reemerging diseases, those that appear after a period of disappearance or decline, there is not a single driver or event leading to emergence; rather, there are a variety of factors related to host, agent, and environmental characteristics, working synergistically to cause pathogen spillover (cross-species transmission).[22] The successive processes that must align to enable an animal pathogen to establish an infection in a human can make it challenging to identify which factors are most important or to predict when or where spillover events will occur. By applying principles of infectious disease epidemiology, considering the chain of infection (from reservoir to susceptible host) and learning from past events, however, we can target specific disease agents, human and animal hosts (and their behaviors), and locations that may present the greatest risk of spillover. Policies, such as the IHR, can help ensure that these targeted surveillance strategies (and response activities in the event of an outbreak) are implemented and coordinated, regardless of where they originate (e.g., people, domestic animals, wildlife, environment).

Factors that increase spillover and pandemic potential

Host factors: Reservoirs, susceptible hosts, and the chain of infection

The transmission dynamics resulting in spillover from wild animal to human populations are complex and often caused by numerous factors, some of which operate simultaneously. The process does not occur in isolation but rather involves ecological interactions at multiple scales (e.g., individual, species, community, global).[23] Pathogens must either become accommodated to or be accommodated by their reservoir hosts (populations or environments in which the pathogen can be permanently maintained and from which infection is transmitted to the target population).[24] The success of pathogen transmission in the reservoir depends on multiple factors, including molecular interactions between the pathogen and host cell receptors, the distribution and diversity of the reservoir species, the distribution of the pathogen within the host species' geographical range, and the connectivity to other reservoir populations.[23]

One simplified approach to illustrate this process defines two stages that must exist for emergence to occur and two additional stages for a pandemic emergence to result. In this approach, prerequisites for emergence include (1) contact between an infectious reservoir host and a susceptible individual from a secondary host species; and (2) successful cross-species transmission, or spillover, in which infection is able to be independently maintained within the secondary host.[23] To facilitate cross-species transmission, an intermediary host such as an arthropod vector or vertebrate host may also be required. For a pandemic to occur from this spillover event, two additional stages are required: (1) sustained transmission of the pathogen between members of the secondary host species in the absence of new spillover events; and (2) genetic and phenotypic changes of the pathogen that facilitate large-scale transmission.[23] This framework is similar to the "infect—shed—spill—spread cascade" that has been proposed.[25] Modifying factors, which can be abiotic (nonbiotic), intrinsic biotic (evolutionarily/biologically driven), and/or extrinsic biotic (external) factors, determine the extent of morbidity and mortality resulting from such pathogen emergence.[23]

Abiotic factors can reduce the potential for contact between reservoir host and secondary host species or their infectious intermediates. Intrinsic biotic factors, such as host characteristics, include those that either improve the ability of the pathogen to cross species barriers or improve the susceptibility of the host to infection (see *Agent Factors: Viral Evolution, Emergence, and Tropism*).[23] Conversely, the genetics and pathogen exposure history of the host will determine whether or not a pathogen is able to effectively establish infection within the host.[26] Extrinsic biotic factors are those factors working outside of natural systems that can expand opportunities for disease transmission. Nearly all extrinsic factors are the result of anthropogenic changes such as habitat modification/encroachment, modern agricultural and medicinal practices, domestication of animals, and international travel, among others.[23,27] In most cases, emergence is influenced by a combination of modifying factors. The level and distribution of zoonotic infections in wildlife (e.g., specific population dynamics), the degree to which wildlife are shedding the pathogen (e.g., influenced by stress), and patterns of human—wildlife interaction (e.g., contact rates) all interact to influence pathogen spillover.[25]

One example of modifying factors working synergistically comes from the emergence of Nipah virus in Malaysia. In this case, intensive pig farming resulted in human encroachment (extrinsic factor) that brought fruit bats (genus *Pteropus*) into close contact with humans and domestic animals. Susceptible pigs (intrinsic factor) were then able to acquire the virus through the consumption of fruit contaminated by the surrounding bat population (extrinsic factor), causing respiratory infection in the pigs. Close contact of pig handlers with infected pigs due to farming practices (extrinsic factor) resulted in transmission to humans, which caused a fatal encephalitis in infected humans.[28] Each factor can act as a potential stopping point or "gate" within the chain of infection where interventions can be implemented or public health surveillance initiated to prevent, detect and mitigate such outbreaks.[28] A more thorough understanding of modifying factors for priority species and geographies for zoonotic spillover can guide public health surveillance and inform interventions to protect public health.

In the case of the Nipah virus, for example, bat reproductive cycles and ages (intrinsic factor) can inform high-risk periods for viral shedding.[29] In these periods, contact with bats could be avoided or minimized through the implementation of outreach and policies to impede public access to forested areas inhabited by bats (extrinsic factor), disrupting the chain of transmission and the spillover of pathogens with pandemic potential.

A One Health surveillance strategy to address gaps at the host level

Zoonotic spillover, and the emergence of infectious diseases, is largely inevitable; furthermore, the likelihood of these events reaching pandemic potential increases as barriers diminish and vulnerabilities (e.g., underfunded public health systems, comorbidities, aging populations, socio-cultural practices) intensify. Given the global vulnerability to zoonoses, a coordinated public health surveillance system must be in place to detect priority pathogens in the environment and in human and animal hosts to respond to zoonotic disease events across all relevant sectors. Human surveillance activities may benefit from an approach that focuses on those individuals with the most constant and intimate contact with animals and are, thus, most likely to be "patient zero" so that rapid intervention can take place before a virus develops effective human-to-human transmission. This approach was used to identify a novel canine coronavirus in hospitalized pneumonia patients in Malaysia coming from areas where domestic animal and wildlife exposure is common.[30]

This approach should be coupled with wildlife disease monitoring to accomplish a robust One Health surveillance strategy[23]; however, global public health surveillance of the wildlife sector remains challenging. In a recent review of major health security reports published between 2007 and 2019, 57.9% (62/107) of reporting countries did not provide evidence of a functional wildlife health surveillance program.[31] Additionally, there are no existing capacity assessment tools for national-level wildlife or environmental services that parallel available public health and veterinary services evaluations, such as the WHO Joint External Evaluation (JEE) and WOAH's Performance of Veterinary Services (PVS) Pathway, respectively.[31] This lack of evaluation tools leads to an omission of wildlife health and environmental considerations from public health surveillance systems, risk assessments, health security plans, training, financing, and implementation efforts.

The Tripartite, reflecting their joint efforts to operationalize the One Health approach, mapped the IHR (2005) and the PVS Pathway. Bridging the two frameworks was put into practice rapidly after international communities called for more concrete actions to improve compliance with the IHR (2005), particularly regarding capacities at the human—animal—environmental interface.[32] This effort led to the IHR-PVS National Bridging Workshop (NBW),[33] a 3-day event held at the national level to improve collaboration between the human and animal health sectors by providing them with a method to "bridge" their sector-specific efforts for improved One Health operationalization.

The Global Early Warning and Response System (GLEWS), an FAO-WOAH-WHO collaborative system enables immediate exchange of information and rapid risk assessment for major animal diseases including zoonotic events.[34] Given the nature of EIDs, countries will not only require coordination and communication with stakeholders across internal sectors, but also the linkage of regional and international networks.

Upon conducting One Health surveillance, we must have procedures and data management tools to coordinate, integrate and share resources to act on results. Open-source monitoring and reporting tools such as the SpillOver: Viral Risk Ranking tool (https://spillover.global/), the Spatial Monitoring and Reporting Tool (SMART) platform[35] and its associated Wildlife Health

Information Platform database[36] (https://oneworldonehealth.wcs.org/Initiatives/WildHealthNet.aspx), are highly valuable tools that are clearly capable of detecting and sharing information on novel pathogens with the global community.[37,38] Not only are more tools becoming available, but metrics to measure the progress toward pandemic prevention and impact of One Health measures have also been established through the Salzburg Timeliness Metrics for One Health Surveillance.[39] To be effective, however, the appropriate networks and systems must be in place and ownership of the data by nations, particularly their local communities, is critical. Established systems must also have some level of standardization across countries and require public reporting and accountability. Currently, priority diseases for surveillance and public health surveillance systems are not standardized across countries.[31] Pandemics such as COVID-19 have proven that gathering these data is not only well worth the investment, but critical to response and recovery efforts. Further development of One Health surveillance programs and assessment tools will allow for the identification of potential stopping gates for zoonotic spillover, reducing the risk of future pandemics.[40]

Agent factors: Viral evolution, emergence, and tropism

A pathogen's ability to spill over and infect a new species (human or animal) is a key factor for diversification, however, such ability is not universal to all viruses.[41] The higher rate of genetic plasticity in RNA viruses compared to DNA viruses is a differential factor that drives evolutionary mechanisms to niche expansion and pathogen-host shifts.[27] In nature, viruses must survive by infection and ongoing cycles of replication and, during this process, RNA genomes exhibit considerably higher error rates (due to lack of error-corrective function from the RNA polymerases), generating mutations randomly across the viral genome. As a result, many of the most problematic EIDs are of viral origin, and more specifically, RNA viruses (e.g., influenza A, HIV, SARS-CoV-2).[40] Understanding how viral populations evolve within a host, across hosts, and during species jumps can elucidate the mechanisms of emergence, at least in part.

Viral evolution determines major events in the biology of viruses.[42] Recently, with the emergence of SARS-CoV-2, the seventh coronavirus known to infect humans,[43] a notable feature facilitating the host jump was found in structural studies and biochemical experiments that showed binding to the human angiotensin-converting enzyme 2 (ACE2) receptor.[44] An additional adaptive feature includes a functional polybasic cleavage (furin) site, which has been shown to be important for expanding host range, transmission, and pathogenesis in SARS-CoV-2 and other viruses such as SARS-CoV-1, MERS, and avian influenza.[45–48]

Since its emergence reported in December 2019, SARS-CoV-2 has managed to evolve further and generate variants with altered transmission potential[49,50] and capacity for immune escape, gaining the ability to infect the same patient multiple times as a result of acquired mutations that evade natural or vaccine-induced immunity.[51–54] Due to global surveillance efforts, such as the genomic consortium in the UK and South Africa (NGS-SA), we have witnessed the emergence of concerning SARS-CoV-2 variants in unprecedented detail.[55] Real-time genomic surveillance of variants is being used to monitor changes in variant frequency relative to other circulating lineages and to identify variants of concern, enabling countries to respond more proactively to newly emerging strains.[55]

Beyond the dominant picture of SARS-CoV-2 as a highly human-adapted virus, it has occasionally shown the capacity to spill back to animals.[56] Human-to-animal infections of captive mink populations in Europe caused respiratory diseases with increased mortality. This event raised concerns that mink might become a source of human infection and resulted in the mass culling of farmed mink in Denmark.[57] Infections of other animals including dogs, cats, tigers, lions, ferrets, and gorillas were demonstrated in natural, captive and experimental scenarios.[44,58,59]

Understanding the molecular mechanisms and evolution of viruses has great practical importance. At the beginning of the COVID-19 pandemic, amid political tensions between countries, public anxiety, and widespread conspiracy theories over the origins of the virus, scientists used viral genomic data to suggest the natural origins of COVID-19 emergence.[45] In these initial comparative genomic studies, SARS-CoV-2 showed concordance with natural evolutionary progression, which supported the concept of direct zoonotic spillover (with or without an intermediate host) into humans. Knowing how COVID-19 emerged is critical for informing global strategies to mitigate the risk of future outbreaks.

The COVID-19 pandemic has also been an instructive example of the implications of viral evolution. Countries have implemented specific policies and public health recommendations to mitigate virus introduction and spread. The WHO, US Centers for Disease Control and Prevention (CDC), and many other public health agencies now have specific guidelines to monitor and classify SARS-CoV-2 variants. To be classified as a variant of high consequence in the United States, a variant should demonstrate impact on medical countermeasures and cause the failure of available diagnostics.[60] Emergence of such variants requires notification to WHO under the IHR (2005), reporting to US CDC, an announcement of strategies to prevent or contain transmission, and recommendations to update treatments and vaccines.[60] The information generated from studying viral genomes and evolution have further guided scientific innovation and, consequently, the United States has now witnessed the second iteration of currently authorized vaccines modified to include antigens to boost immunity against potential escape variants.[61]

Researchers across the globe are working to assimilate their knowledge to assess spillover risk and pandemic potential of different known and unknown viruses. Tools like SpillOver: Viral Risk Ranking supplement and support the various viral discovery projects using deep sequencing approaches, which have revealed the expansive inventory of viruses in animals with poorly defined host ranges and transmission potentials.[62]

An applied One Health research strategy to address gaps at the agent level

In the era of genomics, we have seen a dramatic increase in viral discovery; however, the enormity of the virosphere requires regular sampling across geographic scales to track viruses with pandemic potential.[63] One study estimated 320,000 mammalian viruses awaiting discovery.[64] Using patterns of host sharing among virus species, another, more recent, study estimated 40,000 viruses in mammals (including 10,000 with zoonotic potential).[65] While viral discovery is an important aspect of pandemic preparedness, we still lack the capacity to explore the huge viral diversity in wildlife. Additionally, there is a growing interest in utilizing viral genome data for the assessment of zoonotic spillover risk, specifically for RNA viruses. While such tools are useful, they are currently biased toward a few species of socioeconomic importance, with less than 1% of possible vertebrate viruses being represented.[65] Metagenomic studies may also lead to incorrect host associations as viruses detected from fecal, cloacal, or gut samples may be present in the host diet and microbiome, rather than those actively replicating in the animal of interest.[63]

Viral sequence analysis alone may not be sufficient to understand viral evolution, zoonotic spillover, and emergence. Identification of coexisting microbes is also an important consideration while studying the virosphere because of the potential positive and negative influences that may occur between them. Similarly, how viruses interact with the new hosts and whether cross-protective immunity plays a role in viral spillover and establishment in new hosts are critical gaps, especially given the diversity and abundance of related viruses within a single host species.[63]

Enhancing and expanding our understanding of animal-origin viruses, and continuing to refine risk assessments, can increase prevention and preparedness efforts for the riskiest disease agents at the most likely locations and interfaces.[62] Given the importance of certain viruses and viral families for emergence and future pandemic risk, research efforts can prioritize the development of targeted broad-spectrum antiviral therapies, vaccines, and diagnostics for use at the site of an outbreak, saving time, money, and lives.[62] As such, a One Health surveillance strategy should be used to inform EID preparedness research and development.

Environmental factors: The nexus between nonhuman- and human-dominated landscapes

Activities at the human–animal–environment interface increase risk of zoonotic pathogen transmission and may drive emergence by altering the distribution of animal populations and increasing human exposure to novel pathogens.[66] It is widely believed that such extrinsic, anthropogenic factors are contributing to the increasing incidence of EIDs. On a broad scale, three major drivers include land-use change, decreased wildlife biodiversity, and increased human and livestock population density.[38] Human agricultural, dietary, cultural, and economic practices can all play a role both in increasing the probability of spillover from an animal host into humans and in allowing a small, local outbreak to propagate and become a pandemic (see Chapter 11). Spillover is also facilitated by the increasing demand for protein, both from domestic and wild animal sources. Farming practices and land conversion can bring domestic animals into closer proximity to wildlife, facilitating the transmission of infectious agents. Farming and consumption of wild animals may have contributed to the emergence of SARS-CoV-2 in China.[67,68] Hunting and trade of free-ranging wildlife, both for subsistence and for luxury consumption, has been associated with numerous zoonotic disease spillover events including HIV and Ebola.[69] The presence of live wildlife in markets may be especially risky, as demonstrated by the suspected role of civet cats in the SARS-CoV-1 outbreak in 2002–03.[69]

Another point of increasing convergence between humans and wildlife is extractive industries, such as forestry and mining. In addition to habitat disruption, these industries often bring workers into remote areas where they are more likely to contact wildlife. For example, forestry workers in South America often work in areas inhabited by monkeys that may be supporting the sylvatic cycles of Yellow Fever or Zika. If infected, workers may return to urban areas for treatment, moving the pathogens with them and increasing the chances of an urban disease cycle developing.[70]

Spillover of disease from wildlife to humans is not a new phenomenon but is increasingly associated with outbreaks that spread on a regional or global scale. Molecular clock analyses suggest that the M group of HIV-1 first emerged in the Democratic Republic of Congo in the early 1900s, well before it developed into the global disease threat that it is today. The disease ultimately spread on a much larger scale because of the expansion of cities like Kinshasa and the development of more extensive trade networks in the area.[71] More recently, case numbers in several Ebola outbreaks increased dramatically when infected individuals traveled from rural areas to urban centers.[72] The increasing mobility of human populations allows outbreaks that may previously have remained contained to reach pandemic proportions, threatening global health security. SARS-CoV-2 has demonstrated the interconnectedness of communities and how rapidly certain pathogens can spread across the globe. The propagation of infectious disease events through urbanization is explored further in Chapter 11.

Risk factors for disease emergence are often examined as broad categories such as climate change, human population growth, urbanization, or destruction of habitat.[73] Similarly, many approaches only look at one scale and may miss drivers at other scales (e.g., temporal, spatial). Though the link between land use and wildlife disease are well known, the specific causes have rarely been investigated from a systematic, broad-scale perspective incorporating multiple different landscapes.[25] Consequently, there is neither an established framework to manage land use to minimize zoonotic disease emergence, nor sufficient data to put the framework into action to address spillover risk.[25]

One health policymaking to address gaps at the environment level

Policy can be an effective tool to combat environmental changes that intensify zoonotic spillover risk. Historically, health policies and programs, including the IHR (2005) have focused more on disease prevention, preparedness, and response, without fully addressing many of the known drivers for disease emergence.[74] Broader prevention can be achieved by addressing the upstream stressors and root causes of ecological disruption.[25,75] Though significant progress has been made on illegal or unsustainable wildlife trade through the Convention on the International Trade of Endangered Species of Wild Fauna and Flora (CITES) and the Wildlife Trade Monitoring Network (known as TRAFFIC), buy-in from local governments and investment of resources proves to be a challenge.[76] There is a need for policy that can operationalize land-use planning and place protected area initiatives in a biosecurity context.[25] Enforced policy to regulate wildlife movement across borders is also lacking.

The Quadripartite partners are major players in the international health system and numerous tools and frameworks exist for national and international preparedness, including those that support zoonotic disease prioritization at the intergovernmental and country levels. For example, each country's animal health sector contributes to the implementation of the IHR through public health surveillance, reporting, and response to disease events that emerge at the human–animal interface. While legally binding for Member States, the IHR does not provide the WHO Secretariat with the authority to impose sanctions on countries for noncompliance. The integration of results from the JEE or the State Party Self-Assessment Annual Report (SPAR) (both part of the IHR Monitoring and Evaluation framework), and PVS Pathway through the NBWs, as well as the National Action Plan for Health Security (NAPHS), support the accelerated implementation of IHR core capacities. However, these assessments are "snapshots" in time, and there is a need for repeated assessments to determine ongoing functionality and systems improvements. Additionally, with the exception of SPAR, assessments are voluntary, and a single assessment cannot address all the core capacities needed to ensure health security across multiple sectors and disciplines, as demonstrated during the COVID-19 pandemic when health systems were overwhelmed. As a result, many (high-income) countries with higher scores have shown no correlation with lower COVID-19 deaths.[77] Despite the importance of zoonoses to pandemics (past and future), wildlife and environmental considerations remain disadvantaged in health security planning, assessments, and policymaking.[31]

Each of these factors (e.g., host, agent, environment) that increase spillover and pandemic potential of zoonotic pathogens is critical to long-term solutions planning using a One Health approach. The implementation of complementary activities that address these factors (e.g., public health, animal health and environmental health surveillance, research, policymaking) can be the difference between rapid detection and containment and a prolonged epidemic with significantly higher economic costs and lives lost. A comparison of strategies implemented in West Africa in response to the 2014 Ebola virus disease epidemic is explored in the case study, which can be highly instructive for highlighting the drivers of zoonotic spillover, gaps in response management, and factors that can lead to the propagation of an epidemic.

The 2014 West Africa Ebola virus disease (EVD) epidemic—one outbreak in two settings

Background

Ebola virus disease (EVD) is a highly fatal zoonosis caused by viruses of the genus *Filoviridae*. The virus spills over into human populations through direct contact with infected animals (e.g., fruit bats, chimpanzees, gorillas) and is sustained in humans through person-to-person contact.[78] Since its discovery in Zaire in 1976, EVD epidemics have been rare but repeated among communities of central and east African countries.[79] The largest EVD epidemic to date occurred in West Africa, beginning in Guinea in December 2013 and spreading to Sierra Leone, Liberia (Setting A) and Nigeria[80] (Setting B). On March 25, 2014, WHO announced the outbreak of EVD in Guinea with spread to neighboring Sierra Leone and Liberia; 86 cases (49 confirmed, 37 suspected) and 29 deaths were already recorded. The index case had occurred 3 months prior in a 2-year-old male residing near a large colony of Angolian free-tailed bats in rural Guinea.[81] On July 20, 2014, a diplomat exposed to an EVD-infected family member in Liberia arrived in Lagos, Nigeria by air with symptoms and was admitted to Hospital in Setting B. The doctor at Hospital B suspected EVD and promptly collected a blood sample; on July 23, EVD was confirmed via PCR.[82] Hospital B immediately reported the case to the local health department and the State Ministry of Health, which notified the Federal Ministry of Health (FMOH). The FMOH declared an EVD outbreak the same day and commenced contact tracing with existing One Health Field Epidemiology and Laboratory Training Program (FELTP) officers and partner-agency staff.[82] A multidisciplinary Emergency Operations Center (EOC) was activated with five working group units: (1) Strategic coordination, (2) Epidemiology, surveillance and laboratory, (3) Social mobilization and communication, (4) Points of entry, and (5) Case management and infection prevention and control (IPC). The outbreak, which affected two large Nigerian cities, was contained within 2 months with a total of 20 cases and 8 deaths.[82] Following the extension of the epidemic, WHO declared the West Africa EVD outbreak a public health emergency of international concern (PHEIC) on August 8, 2014. By June 9, 2016, when the outbreak was over, there were a total of 28,646 cases and 11,325 deaths affecting nine countries across three continents and an estimated 4.3B USD in total cost.

The 2014 West Africa Ebola virus disease (EVD) epidemic—one outbreak in two settings *(cont'd)*

Observations

	Outbreak setting A	Outbreak setting B
1	Delayed detection due to low index of suspicion and absence of surveillance at human–animal interface (e.g., wildlife sentinel surveillance, human surveillance in high-risk populations)	Early case detection due to high index of suspicion and vigilance given the urban setting and heightened risk of epidemic propagation
2	Challenged human health surveillance system with lack of capacity to initiate a One Health approach; minimal and ad hoc communication between human health, veterinary and environmental health services	Functional human health surveillance system with environmental health officers serving as disease surveillance officers; existing coordination/ collaboration with veterinary services
3	Notable IPC cultural challenges that began before risk communication reached communities (e.g., challenges with safe burial, poor EVD risk perception)	Reduced IPC cultural challenges with early public health education and community-based risk communication (e.g., safe burial practices, high EVD risk perception)
4	Delayed international support and official notification through the IHR (2005)	Prompt multiagency international support leveraging experience and resources available from existing collaborations
5	Lack of existing trained One Health workforce and delayed mobilization of external capacity surge	Mobilization of existing trained One Health workforce

Key issues

- Proximity of human habitat to animal reservoirs/hosts
- Nonexistent or weak One Health surveillance
- Poor EVD risk perception and minimal community-based risk communication
- High-risk cultural practices that drove human-to-human transmission
- Lack of preparedness and under-resourced health systems
- Delayed international coordination and notification through the IHR

Interventions

- One health surveillance and diagnostics for early detection
- Community-based risk communication, IPC measures, and education
- Case isolation and contact tracing
- Clinical and psychosocial case management
- Social rehabilitation and mobilization

The 2014 West Africa Ebola virus disease (EVD) epidemic—one outbreak in two settings *(cont'd)*

Lessons learned

- *Close human–animal interface* and *weak One Health surveillance* limited detection of zoonotic spillover
- *Minimal diagnostic capacities* challenged reporting systems, particularly from remote areas to central level, resulting in delayed confirmation of etiology and late risk assessment/management which led to continual community-based transmission
- *Poor disease risk perception* and *high-risk cultural and traditional practices* allowed local and international spread
- *Poorly resourced health systems* were unable to ensure containment
- *Delayed coordination and support* with the IHR National Focal Points resulted in challenges with fulfilling mandatory functions as described in IHR (2005)

Recommendations

- Multilevel One Health approach to prevent, detect and respond to infectious disease epidemics must be supported through:
  - Workforce development
  - Readily accessible material resources for rapid mobilization and capacity surge to ensure access to basic healthcare
  - Expanded research to understand biology, transmission dynamics and epidemiological risk factors in the face of cross-border mobility and air travel
  - Expanded vaccine development and research capacity
  - Engagement of local leaders and communities in IPC education and risk communication
  - Integrated multiagency, multisectoral Incident Management Systems
  - Continued global efforts that foster and strengthen networks to support early warning surveillance systems for emerging and reemerging infectious diseases at all levels—global, regional, national, and subnational—with support for LMICs at the highest risk of zoonotic spillover

Vision for the future

One Health is a cutting-edge vision of 21st century global health[83] and represents an important strategy to reduce the risk of global health catastrophes even worse than COVID-19. At the highest political levels, including the leaders of the Group of Seven (G7), modernizing global health security will require "strengthening a One Health approach across all aspects of pandemic prevention and preparedness, recognizing the critical links among human and animal health and the environment."[84] The Quadripartite partners have a broad, yet clear mandate to shape the health research agenda and stimulate the generation, translation, and dissemination of knowledge, with the ultimate goal of attaining the highest level of health by all peoples. One cannot overlook the significance of animal and environmental health sectors on the path to this goal, both as risks and potential solutions. Countries should utilize a shared governance system that prioritizes formal coordination across multiple sectors informed by a One Health approach.

Many countries have made significant progress toward institutionalizing a One Health approach, such as adopting evidence-based frameworks within their governance structures to increase their capacity to prevent, detect, and respond to zoonoses and other infectious disease threats. For example, in Rwanda, this has manifested as a One Health Strategic Plan[85] and the ongoing development of a One Health workforce.[86] Uganda, similarly a hotspot for EID epidemics, has formalized its One Health action plan through its National One Health Platform that includes Memoranda of Understanding among sectors and university networks to strengthen training programs, joint resources pool and management.[87]

Communities, particularly those in rural settings and other hotspots for EID events, are essential for informing One Health policy and implementing One Health approaches. Governments should engage community leaders as part of their zoonotic disease risk-reduction strategies, as they are critical to obtaining buy-in for surveillance activities and communicating One Health threats. Rwanda has mobilized community health workers (CHWs), community animal health workers, NGOs, health clinics, park rangers, and farmers for surveillance of zoonotic diseases, with the goal of moving toward a reimagined "One Health CHW" that uses mobile phone technology to rapidly identify and report unusual events affecting humans, animals, plants, or ecosystems.[88] These are merely a few examples that highlight a global movement toward collaborative governance that creates a sustainable One Health infrastructure for the future.

Actions

On the path to full implementation of One Health approaches at all levels of governance, there will be necessary stopgaps (i.e., short-term solutions) as well as long-term capacity strengthening against zoonotic diseases and other priority challenges. Solutions will not come from one sector alone, nor will they come from research advances, enhanced public health surveillance, or policy implementation independently. Rather it is a necessary continuum, with different groups and strategies informing the others for effective knowledge transfer, implementation, and assessment (Fig. 6.2).

Research has been instrumental in building our understanding of viral zoonosis emergence and has been used to expedite the development of pharmaceutical (e.g., vaccines, monoclonal therapies, drugs) and nonpharmaceutical countermeasures (e.g., avoidance of bats during high virus-shedding periods). Research must be continued and capacity strengthened, particularly in low-resource/high-risk areas. Social sciences, such as anthropology and psychology, are critical components of One Health operationalization and bring necessary research and implementation tools, such as science communication, ethnography, and behavior change theory, to understand and motivate decision-making.

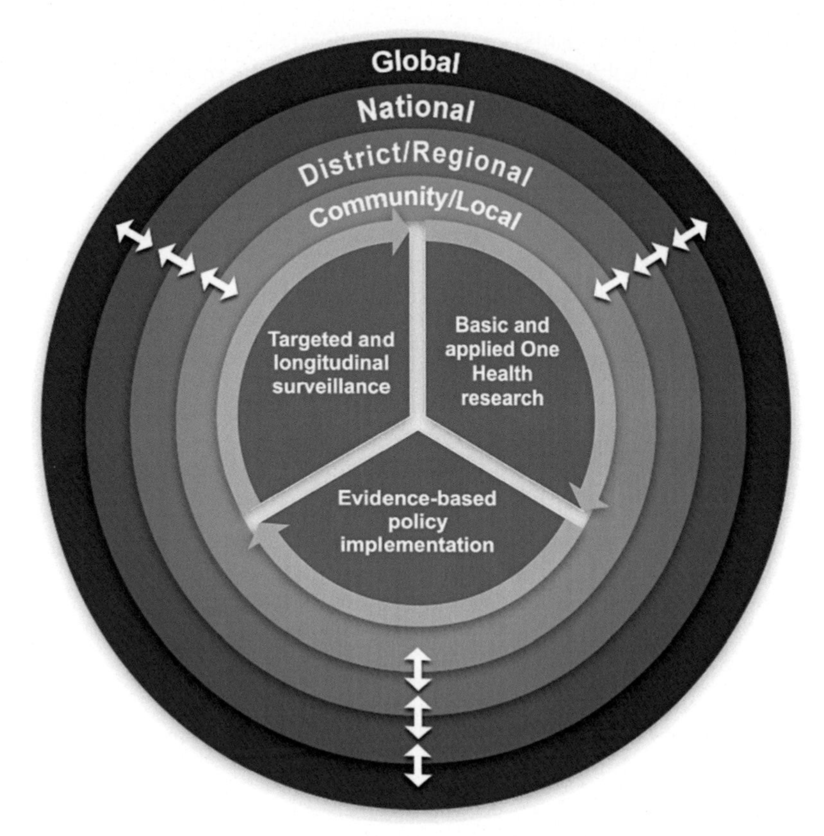

FIGURE 6.2 One Health implementation cycle.

One Health networks must be operationalized for longitudinal public health surveillance and integration of disease surveillance systems, targeting high-risk interfaces and conditions to reduce the probability and frequency of spillover events. Emerging infectious diseases, including "Disease X" (a currently unknown pathogen with the potential to cause a serious international epidemic), must continue to be a priority for the GHSA, with sufficient funds allocated for focused public health surveillance, research, and preparedness. The development of a One Health workforce must be prioritized, including the identification of specific core competencies and cross-cutting skills necessary to translate and communicate risks across disciplines and stakeholders. In addition, public health workforces that are not specifically identified as One Health workforces should be knowledgeable of One Health competencies. Currently, a specific intergovernmental agency dedicated to wildlife health does not exist; therefore, significant gaps can be expected in the implementation of wild animal surveillance programs (risk assessment, reporting, investigation, and management decisions). One Health efforts should include strengthening wildlife health capacity to monitor and mitigate known and novel zoonotic disease risks.

Additionally, communities must be engaged to understand risks, motivations for high-risk behaviors, and potential feasible alternatives and solutions. We must acknowledge that globally, many economic systems and community livelihoods depend on wild animals. As these practices pose known and significant risks to global health security, collectively we must decide how to regulate or otherwise modify these practices to reduce risk. Policymaking must incorporate the knowledge and expertise of professionals within fields that traditionally do not participate in public health policy-making processes but whose expertise may help shed light on emerging zoonoses and their control.

Ultimately, humankind must realign our relationship with the natural world. Given the threats posed by the disruption of natural systems for novel pathogen emergence, a modern global health security agenda must look even further upstream to reduce drivers of spillover. At all levels of government, a One Health approach should provide the framework to mitigate emergence when possible and respond efficiently when necessary.

References

1. Guan Y, Vijaykrishna D, Bahl J, Zhu H, Wang J, Smith GJD. The emergence of pandemic influenza viruses. *Protein Cell.* 2010;1(1):9–13.
2. Beach B, Clay K, Saavedra MH. *The 1918 Influenza Pandemic and its Lessons for COVID-19.* 2020.
3. Yang W, Petkova E, Shaman J. The 1918 influenza pandemic in New York City: age-specific timing, mortality, and transmission dynamics. *Influenza Other Respir Viruses.* 2014;8(2):177–188.
4. Snowden FM. Emerging and reemerging diseases: a historical perspective. *Immunol Rev.* 2008;225(1):9–26.
5. FACT SHEET. https://www.unaids.org/sites/default/files/media_asset/UNAIDS_FactSheet_en.pdf; 2021.
6. Coulibaly I. The impact of HIV/AIDS on the labour force in Sub-Saharan Africa: a preliminary assessment. *Int Labour Organ Res Policy Anal.* 2005;(3).
7. Forsythe S. *How Does HIV/AIDS Affect African Businesses. State of the Art: AIDS and Economics POLICY Project.* Washington, DC: Merck & Company; 2002:30–37.
8. Dixon S, McDonald S, Roberts J. The impact of HIV and AIDS on Africa's economic development. *BMJ.* 2002;324(7331):232–234.
9. Morens DM, Fauci AS. Emerging pandemic diseases: how we got to COVID-19. *Cell.* 2020;182.
10. King N. Security, disease, commerce: ideologies of postcolonial global health. *Soc Stud Sci.* 2002;32:763–789.
11. Noah D, Fidas G. *The Global Infectious Disease Threat and its Implications for the United States.* Washington DC: National Intelligence Council; 2000.
12. FAO-OIE-WHO. *Sharing Responsibilities and Coordinating Global Activities to Address Health Risks at the Animal-Human-Ecosystems Interfaces: A Tripartite Concept Note.* 2010.
13. Global Health Security Agenda (GHSA). https://ghsagenda.org; 2014.
14. Organization WH. *International Health Regulations (2005).* World Health Organization; 2008.
15. *Transforming Our World: The 2030 Agenda for Sustainable Development*; 2015. https://www.un.org/ga/search/view_doc.asp?symbol=A/RES/70/1&Lang=E.
16. FAO. *FAO Honoured to Take the Rotating Chairmanship of the Tripartite*; 2021. https://www.fao.org/news/story/pt./item/1376126/icode/%20(accessed%20on%2010%20November%202021).
17. WHO. *Strengthening WHO Preparedness for and Response to Health Emergencies, Strengthening Collaboration on One Health*; 2022. https://apps.who.int/gb/ebwha/pdf_files/WHA75/A75_19-en.pdf.
18. Smith KF, Goldberg M, Rosenthal S, et al. Global rise in human infectious disease outbreaks. *J R Soc Interface.* 2014;11(101), 20140950.
19. Jones KE, Patel NG, Levy MA, et al. Global trends in emerging infectious diseases. *Nature.* 2008;451(7181):990–993.
20. Rohr JR, Barrett CB, Civitello DJ, et al. Emerging human infectious diseases and the links to global food production. *Nat Sustain.* 2019;2(6):445–456.

21. Redding DW, Atkinson PM, Cunningham AA, et al. Impacts of environmental and socio-economic factors on emergence and epidemic potential of Ebola in Africa. *Nat Commun*. 2019;10(1):4531.
22. Plowright RK, Parrish CR, McCallum H, et al. Pathways to zoonotic spillover. *Nat Rev Microbiol*. 2017;15(8):502–510.
23. Childs JE, Krebs JW, Real LA, Gordon ER. Animal-based national surveillance for zoonotic disease: quality, limitations, and implications of a model system for monitoring rabies. *Prev Vet Med*. 2007;78(3–4):246–261.
24. Haydon DT, Cleaveland S, Taylor LH, Laurenson MK. Identifying reservoirs of infection: a conceptual and practical challenge. *Emerg Infect Dis*. 2002;8(12):1468–1473.
25. Plowright RK, Reaser JK, Locke H, et al. Land use-induced spillover: a call to action to safeguard environmental, animal, and human health. *Lancet Planet Health*. 2021;5(4):e237–e245.
26. Burgner D, Jamieson SE, Blackwell JM. Genetic susceptibility to infectious diseases: big is beautiful, but will bigger be even better? *Lancet Infect Dis*. 2006;6(10):653–663.
27. Morse SS, Mazet JA, Woolhouse M, et al. Prediction and prevention of the next pandemic zoonosis. *Lancet*. 2012;380(9857):1956–1965.
28. Daniels PW, Halpin K, Hyatt A, Middleton D. *Infection and Disease in Reservoir and Spillover Hosts: Determinants of Pathogen Emergence*. New York: Springer; 2007.
29. Montecino-Latorre D, Goldstein T, Gilardi K, et al. Reproduction of East-African bats may guide risk mitigation for coronavirus spillover. *One Health Outlook*. 2020;2(1):2.
30. Vlasova AN, Diaz A, Damtie D, et al. *Novel Canine Coronavirus Isolated from a Hospitalized Pneumonia Patient, East Malaysia*. Clinical Infectious Diseases; 2021.
31. Machalaba C, Uhart M, Ryser-Degiorgis M-P, Karesh WB. Gaps in health security related to wildlife and environment affecting pandemic prevention and preparedness, 2007–2020. *Bull World Health Organ*. 2021;99(5):342–350B.
32. WHO-OIE. *Operational Framework for Good Governance at the Human-Animal Interface: Bridging WHO and OIE Tools for the Assessment of National Capacities*; 2014. https://www.who.int/publications/i/item/who-oie-operational-framework-for-good-governance-at-the-human-animal-interface.
33. Belot G, Caya F, Errecaborde KM, et al. IHR-PVS National Bridging Workshops, a tool to operationalize the collaboration between human and animal health while advancing sector-specific goals in countries. *PLoS One*. 2021;16(6), e0245312.
34. Pinto J, Ben Jebara K, Chaisemartin D, De La Rocque S, Abela B. *The FAO/OIE/WHO Global Early Warning System*. 2011.
35. Wildlife Conservation Society, Wild Health Net. https://oneworldonehealth.wcs.org/Initiatives/WildHealthNet.aspx#2.
36. Wildlife Conservation Society, Smart for Health. https://oneworldonehealth.wcs.org/Initiatives/SMARTforhealth.aspx.
37. Denstedt E, Porco A, Hwang J, et al. Detection of African swine fever virus in free-ranging wild boar in Southeast Asia. *Transbound Emerg Dis*. 2020;68:2669–2675.
38. PREDICT Consortium. *Advancing Global Health Security at the Frontiers of Disease Emergence*. Vol 596. Davis: One Health Institute, University of California; December 2020:2021.
39. Ending Pandemics and Salzburg Global Seminar. *The Salzburg Statement on Metrics for One Health Surveillance*; 2020. https://www.salzburgglobal.org/fileadmin/user_upload/Documents/2010-2019/2019/Session_641/SalzburgGlobal_Statement_641_One_Health.pdf.
40. Keusch GT, Amuasi JH, Anderson DE, et al. Pandemic origins and a One Health approach to preparedness and prevention: solutions based on SARS-CoV-2 and other RNA viruses. *Proc Natl Acad Sci USA*. 2022;119(42).
41. Ito T, Kawaoka Y. Host-range barrier of influenza A viruses. *Vet Microbiol*. 2000;74(1–2):71–75.
42. Antia R, Regoes RR, Koella JC, Bergstrom CT. The role of evolution in the emergence of infectious diseases. *Nature*. 2003;426(6967):658–661.
43. Fan W, Su Z, Bin Y, et al. A new coronavirus associated with human respiratory disease in China. *Nature*. 2020;579(7798):265–269.
44. Damas J, Hughes GM, Keough KC, et al. Broad host range of SARS-CoV-2 predicted by comparative and structural analysis of ACE2 in vertebrates. *Proc Natl Acad Sci USA*. 2020;117(36):22311–22322.
45. Andersen KG, Rambaut A, Lipkin WI, Holmes EC, Garry RF. The proximal origin of SARS-CoV-2. *Nat Med*. 2020;26(4):450–452.

46. Beerens N, Heutink R, Harders F, Bossers A, Koch G, Peeters B. Emergence and selection of a highly pathogenic avian influenza H7N3 virus. *J Virol*. 2020;94(8):e01818–e01819.
47. Peacock TP, Goldhill DH, Zhou J, et al. The furin cleavage site in the SARS-CoV-2 spike protein is required for transmission in ferrets. *Nat Microbiol*. 2021:1–11.
48. Johnson BA, Xie X, Bailey AL, et al. Loss of furin cleavage site attenuates SARS-CoV-2 pathogenesis. *Nature*. 2021;591(7849):293–299.
49. Korber B, Fischer WM, Gnanakaran S, et al. Tracking changes in SARS-CoV-2 Spike: evidence that D614G increases infectivity of the COVID-19 virus. *Cell*. 2020;182(4):812–827.
50. Yurkovetskiy L, Wang X, Pascal KE, et al. Structural and functional analysis of the D614G SARS-CoV-2 spike protein variant. *Cell*. 2020;183(3):739–751.
51. Li Q, Nie J, Wu J, et al. SARS-CoV-2 501Y. V2 variants lack higher infectivity but do have immune escape. *Cell*. 2021;184(9):2362–2371.
52. Karim SSA. Vaccines and SARS-CoV-2 variants: the urgent need for a correlate of protection. *Lancet*. 2021;397(10281):1263–1264.
53. Emary KR, Golubchik T, Aley PK, et al. Efficacy of ChAdOx1 nCoV-19 (AZD1222) vaccine against SARS-CoV-2 variant of concern 202012/01 (B. 1.1. 7): an exploratory analysis of a randomised controlled trial. *Lancet*. 2021;397(10282):1351–1362.
54. Lustig Y, Zuckerman N, Nemet I, et al. Neutralising capacity against Delta (B.1.617.2) and other variants of concern following Comirnaty (BNT162b2, BioNTech/Pfizer) vaccination in health care workers, Israel. *Euro Surveill*. 2021;26(26).
55. COVID-19 Genomics UK (COG-UK) consortiumcontact@cogconsortium.uk. An integrated national scale SARS-CoV-2 genomic surveillance network. *Lancet Microbe*. 2020 Jul;1(3):e99–e100. https://doi.org/10.1016/S2666-5247(20)30054-9. Epub 2020 Jun 2. PMID: 32835336; PMCID: PMC7266609.
56. Rogers TF, Zhao F, Huang D, et al. Isolation of potent SARS-CoV-2 neutralizing antibodies and protection from disease in a small animal model. *Science*. 2020;369(6506):956–963.
57. Hammer AS, Quaade ML, Rasmussen TB, et al. SARS-CoV-2 transmission between mink (Neovison vison) and humans, Denmark. *Emerg Infect Dis*. 2021;27(2):547.
58. Freuling CM, Breithaupt A, Müller T, et al. *Susceptibility of Raccoon Dogs for Experimental SARS-CoV-2 Infection*. BioRxiv; 2020.
59. Shi J, Wen Z, Zhong G, et al. Susceptibility of ferrets, cats, dogs, and other domesticated animals to SARS–coronavirus 2. *Science*. 2020;368(6494):1016–1020.
60. CDC. SARS-CoV-2 Variant Classifications and Definitions. https://www.cdc.gov/coronavirus/2019-ncov/variants/variant-info.html#Consequence.
61. Corbett KS, Edwards DK, Leist SR, et al. SARS-CoV-2 mRNA vaccine design enabled by prototype pathogen preparedness. *Nature*. 2020;586(7830):567–571.
62. Grange ZL, Goldstein T, Johnson CK, et al. Ranking the risk of animal-to-human spillover for newly discovered viruses. *Proc Natl Acad Sci USA*. 2021;118(15).
63. Wille M, Geoghegan JL, Holmes EC. How accurately can we assess zoonotic risk? *PLoS Biol*. 2021;19(4).
64. Anthony SJ, Epstein JH, Murray KA, et al. A strategy to estimate unknown viral diversity in mammals. *mBio*. 2013;4(5).
65. Carlson CJ, Zipfel CM, Garnier R, Bansal S. Global estimates of mammalian viral diversity accounting for host sharing. *Nat Ecol Evol*. 2019;3(7):1070–1075.
66. Machalaba CC, Loh EH, Daszak P, Karesh WB. *Emerging Diseases from Animals*. State of the World; 2015:105–116.
67. Organization WH. *WHO-Convened Global Study of Origins of SARS-CoV-2: China Part*. 2021.
68. Xia W, Hughes J, Robertson D, Jiang X. *How One Pandemic Led to Another: ASFV, the Disruption Contributing to SARS-CoV-2 Emergence in Wuhan*. 2021.
69. Jones BA, Grace D, Kock R, et al. Zoonosis emergence linked to agricultural intensification and environmental change. *PNAS*. 2013;110(21):8399–8404.
70. Wilcox BA, Ellis B. Forests and emerging infectious diseases of humans. *UNASYLVA-FAO*. 2006;57(2):11.
71. Worobey M, Gemmel M, Teuwen DE, et al. Direct evidence of extensive diversity of HIV-1 in Kinshasa by 1960. *Nature*. 2008;455(7213):661–664.
72. Fallah MP, Skrip LA, Enders J. Preventing rural to urban spread of Ebola: lessons from Liberia. *Lancet*. 2018;392(10144):279–280.

73. Cleaveland S, Haydon DT, Taylor L. Overviews of pathogen emergence: which pathogens emerge, when and why? In: Childs JE, Mackenzie JS, Richt JA, eds. *Wildlife and Emerging Zoonotic Diseases: The Biology, Circumstances and Consequences of Cross-Species Transmission*. Berlin, Heidelberg: Springer; 2007:85–111.
74. Shanks S, van Schalkwyk MCI, Cunningham AA. A call to prioritise prevention: action is needed to reduce the risk of zoonotic disease emergence. *Lancet Glob Health*. 2022;23.
75. Aguirre AA. Changing patterns of emerging zoonotic diseases in wildlife, domestic animals, and humans linked to biodiversity loss and globalization. *ILAR J*. 2017;58(3):315–318.
76. Fukushima CS, Tricorache P, Toomes A, et al. Challenges and perspectives on tackling illegal or unsustainable wildlife trade. *Biol Conserv*. 2021;263.
77. Sirleaf EJ, Clark H. Report of the independent panel for pandemic preparedness and response: making COVID-19 the last pandemic. *Lancet*. 2021;398:101–103.
78. Ohuabunwo C, Ameh C, Oduyebo O, et al. Clinical profile and containment of the Ebola virus disease outbreak in two large West African cities, Nigeria, July–September 2014. *Int J Infect Dis*. 2016;53:23–29.
79. Fawole OI, Dalhat MM, Park M, Hall CD, Nguku PM, Adewuyi PA. Contact tracing following outbreak of Ebola virus disease in urban settings in Nigeria. *Pan Afr Med J*. 2017;27(suppl 1), 8-8.
80. Bell BP. Overview, control strategies, and lessons learned in the CDC response to the 2014–2016 Ebola epidemic. *MMWR Suppl*. 2016;65.
81. Kaner J, Schaack S. Understanding Ebola: the 2014 epidemic. *Glob Health*. 2016;12(1):1–7.
82. Hanna L. *A Wake-Up Call: Lessons from Ebola for the World's Health Systems*. Save the Children; 2015.
83. Tobias J. *What if Scientists Already Know How to Prevent the Next Pandemic?*; 2020. https://www.thenation.com/article/world/one-health-pandemic/.
84. House TW. *Carbis Bay G7 Summit Communiqué*; 2021. https://www.whitehouse.gov/briefing-room/statements-releases/2021/06/13/carbis-bay-g7-summit-communique/?mc_cid=0a36c74d2b&mc_eid=UNIQID.
85. Rwanda Ro. *One Health Strategic Plan (2014–2018)*. 2013.
86. Henley P, Igihozo G, Wotton L. One Health approaches require community engagement, education, and international collaborations—a lesson from Rwanda. *Nat Med*. 2021;27:947–948.
87. Esther B, Edwinah A, Peninah N, et al. Operationalizing the one health approach in Uganda: challenges and opportunities. *J Epidemiol Glob Health*. 2020;10(4):250–257.
88. Nyatanyi T, Wilkes M, McDermott H, et al. Implementing One Health as an integrated approach to health in Rwanda. *BMJ Glob Health*. 2017;2(1).

CHAPTER

7

Emergence and dissemination of antimicrobial resistance at the interface of humans, animals, and the environment

Shu-Hua Wang[1,2,a], *Senait Kebede*[3,a], *Ebba Abate*[4], *Afreenish Amir*[5], *Ericka Calderon*[6], *Armando E. Hoet*[7], *Aamer Ikram*[5], *Jeffrey T. LeJeune*[8,b], *Zelalem Mekuria*[1,7], *Satoru Suzuki*[9], *Susan Vaughn Grooters*[7], *Getnet Yimer*[4,10] *and Wondwossen A. Gebreyes*[1,7]

[1]Global One Health initiative (GOHi), The Ohio State University, Columbus, OH, United States; [2]Division of Infectious Diseases, Department of Internal Medicine, The Ohio State University, Columbus, OH, United States; [3]Hubert Department of Global Health, Rollins School of Public Health, Emory University, Atlanta, GA, United States; [4]Ohio State-Global One Health, Addis Ababa, Ethiopia; [5]National Institutes of Health, Islamabad, Pakistan; [6]Inter-American Institute for Technical Cooperation (IICA), San Jose, Costa Rica; [7]Department of Veterinary Preventive Medicine, College of Veterinary Medicine, The Ohio State University, Columbus, OH, United States; [8]The Food and Agriculture Organization of United Nations (FAO), Rome, Italy; [9]Center for Marine Environmental Studies, Ehime University, Matsuyama, Ehime, Japan; [10]Department of Genetics, Perelman School of Medicine, and Penn Center for Global Genomics & Health Equity, Perelman School of Medicine, University of Pennsylvania, Philadelphia, PA, United States

[a]Co-first authors.

[b]The views expressed in this publication are those of the author(s) and do not necessarily reflect the views or policies of the Food and Agriculture Organization of the United Nations.

Modernizing Global Health Security to Prevent, Detect, and Respond
https://doi.org/10.1016/B978-0-323-90945-7.00021-X

Introduction

Global impact

Antimicrobial resistance (AMR) remains a significant concern for people around the world—impacting both lives and livelihoods. More than 5 years since the launch of the World Health Organization (WHO) Global Action Plan on AMR, many of its goals and objectives are still not addressed.[1,2] To achieve its goals a stronger focus and greater commitment are needed.[3] This lack of progress has been further delayed by the Coronavirus Disease 2019 (COVID-19) global pandemic, which has severely impacted health systems and services.[4,5] This disruption in turn has negatively affected AMR prevention and control activities. In addition, the COVID-19 pandemic has clearly exposed the limitations of national health systems as well as collaborations and networking among countries and the global networks.[6]

While the pandemic impact of COVID-19 has rightfully garnered a large amount of attention and resources, AMR is a "silent pandemic" that if not addressed will continue to have very significant consequences over a long period of time. In 2019 alone, an estimated 1.27 million (95% UI, 0.911–1.71) deaths were attributable to antibiotic-resistant infections and 4.95 million (95% UI, 3.62–6.57) AMR-associated deaths.[7] Globally, AMR is responsible for more deaths than HIV and malaria combined.[8,9] Although, AMR affects every age group, economic class, and geographic region, Africa and Asia face the highest burden. If the present trend continues, it is estimated that over 10 million lives per year could be lost due to AMR by 2050, surpassing cancer, the current leading cause of death.[10] Moreover, with increasing AMR, the potential for spreading of highly pathogenic organisms with the potential to cause pandemics and epidemics threatens the GHS.[11]

Risk factor for AMR

The development and spread of resistant infections is multifactorial.[12] AMR results from inheriting or acquiring traits that allow the organism to survive, spread, or evolve in the presence of an antimicrobial at concentrations that would otherwise kill or inhibit their multiplication.[13] While AMR can naturally occur over time, external anthropogenic factors such as transmission through animal source food, environmental contamination in community settings, and poor adherence to infection control measures in healthcare settings increase the spread.[14] Moreover, insufficient healthcare workforce, limited laboratory diagnostic capacity, and inadequate infection prevention and control (IPC) practices amplify the risk for transmission of multidrug-resistant organisms (MDROs) in healthcare settings. Injudicious use of antimicrobials in humans, food animals, and horticulture has accelerated the emergence, persistence, and spread of resistance.[12] The lack of sound policies and pharmaceutical regulations to control substandard and/or counterfeit antimicrobials and the contamination of water sources with residual products also contribute to the rise in AMR.[15]

Impact of AMR

The risk factors in community settings at the interface of animals, humans, and the environment (including plants) lead to consequences in hospitals and other healthcare settings. At the healthcare level, AMR limits treatment options for several communicable diseases such as

HIV, malaria, tuberculosis, and gonorrhea. AMR also compromises the successful management of healthcare-associated infections (HAIs) contributing to higher morbidity and mortality. There is ample evidence that there is an excessive financial and social impact of AMR on the healthcare system.[3,16] Factors that contribute to the rise in healthcare costs include: increased number of hospital admissions; extended inpatient stays; higher antimicrobial costs for newer therapeutics and combination therapy; and the additional cost for isolation and extensive care, including additional trained healthcare workers, personal protective equipment (PPE), and waste management for MDROs. If the current trend continues unchecked, it is projected that AMR cost could be $300 billion to more than $1 trillion USD annually by 2050.[15]

AMR also threatens and complicates food safety and security.[17] Among zoonotic, foodborne, and animal pathogens, it threatens to reduce animal productivity and thus livelihood of populations, with disproportionate impact in low- and middle-income countries (LMICs). One study showed it has a potential for as much as 11% decrease in production by 2050.[18] The ineffectiveness of commonly used antimicrobials will limit the treatment options of animal diseases, increasing morbidity and mortality of animals, making them unavailable as food products. Less livestock production as well as food production from animal origin, impacted by diverse factors including partly antimicrobial use pattern, in face of an increasing demand to feed a growing population will lead to insufficient access to nutritious food, impeding the goal of food security. Likewise, contamination of food with antimicrobial residues and/or MDROs can make food products unsafe for consumption by humans.[19]

Mitigating the impact of AMR

Mitigation of AMR is complex and requires multifaceted strategies addressing the human, animal, plants (mainly horticulture), and environmental factors.[3] The first step in addressing AMR is updating policies and regulations to reflect the necessary measures to reduce the risks for AMR. Important avenues include International Health Regulations (IHR) and the Global Health Security Agenda (GHSA) targeting AMR and disease spread. IHR is a legally binding agreement ratified by 194 World Health Organization (WHO) Member States.[20] It aims to provide a legal framework stipulating countries' obligations and rights in managing public health emergencies with the potential to cross borders while building local capacity to prevent, detect, and respond to disease outbreaks. Key measures to be included in comprehensive policies and regulations include strengthening systematic and integrated AMR surveillance in humans, animals, and the environment; building and enhancing laboratory diagnostic capacity; improving IPC measures; implementing antimicrobial stewardship practices; ensuring access to affordable and effective treatment; and investing in new antimicrobial and vaccine research and development.

Literature review and gap analysis

AMR has a complex ecological and epidemiological cycle with a spectrum of resistant patterns for antimicrobials.[21,22] Typically, an organism is pan-susceptible (susceptible to all antibiotics); after acquiring resistance determinants (genes, plasmids, integrons, etc.) it develops resistance to one or more classes of antimicrobials. Increased antimicrobial use can produce

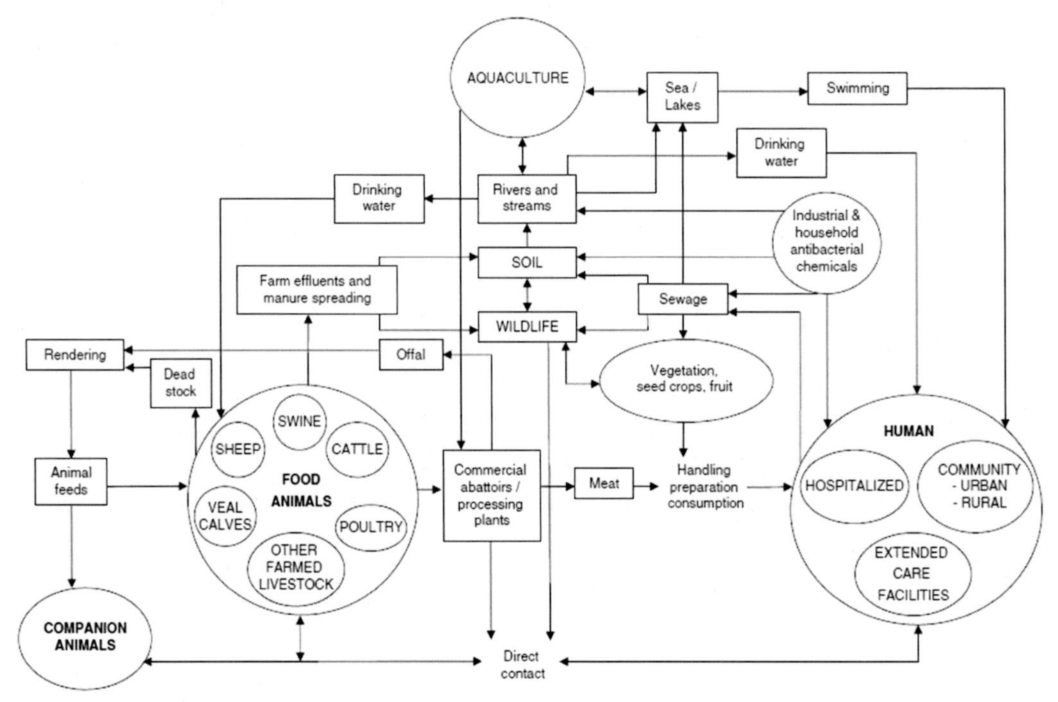

FIGURE 7.1 The ecology of the spread of antimicrobial resistance and of resistance genes. A schematic representation of resistant bacteria and antimicrobial resistance genes transmission routes across the multiple ecological compartments. *This figure is further development (Irwin et al., 2008) of an original one by Linton, 1977 and reproduced with permission. Davies J, Davies D. Origins and evolution of antibiotic resistance. Microbiology and molecular biology reviews. 2010;74(3):417–433.*

selective advantage for resistant mutations and facilitates the acquisition of additional resistant genes, subsequently an organism becomes multidrug resistant and in rare cases, pan-drug-resistant strains can emerge and persist. These MDROs fall in the pathogenic category that emerged and then circulated endemically among humans, animals, and the environment.[23,24] Fig. 7.1 shows a simplified version of the ecology of AMR and transmission of resistance genes across humans, animals, and the environment.

AMR in human health

Humans can acquire AMR microorganisms from healthcare facilities or at the community level. HAIs are acquired after admission to the hospital or at a healthcare facility whereas community-acquired infections are contracted outside of the healthcare setting.[25] Examples of HAIs include central line–associated bloodstream infections, catheter-associated urinary tract infections (UTIs), ventilator-associated pneumonia, and surgical site infections.[25] In a 2015 survey conducted in the United States, 3% of hospitalized patients had one or more HAIs. The WHO estimates that at any given time, the prevalence of HAIs in LMICs varies between 5.7% and 19.1%.[26]

Healthcare-associated MDROs can spill over into the community: 37% of extended-spectrum beta-lactamase (ESBL)-producing *Escherichia coli* infections such as UTI originate in hospitals and transmit in the community through close contacts, food supply, and through exposures at common locations such as sporting facilities, daycare centers, and schools.[27,28] Case study 1 illustrates how ESBL-producing Enterobacterales, such as *E. coli*, can be transmitted between humans and animals.

Community-associated (CA)-MDR pathogens can spread in the community with fatal outcomes.[29] Methicillin-resistant *Staphylococcus aureus* (MRSA) is often identified in individuals with healthcare-associated risk factors such as history of hospitalization, surgery, indwelling devices such as central line, hemodialysis catheter, or urinary catheters, and/or residence in long-term care facilities. Alarmingly, in the 1990s, CA-MRSA was reported in children without any predisposing risk factors.[30] Initially, HA-MRSA and CA-MRSA were genetically different strains USA100 strain type versus USA300, respectively, but currently, the MRSA strains are endemic in both healthcare- and community-associated infections in individuals.[31]

As these infections are transmissible beyond their local settings, a coordinated response to the AMR crisis is needed at least at a national level, but preferably at regional and global scales as indicated earlier. An example of this in practice is the US Department of Health and Human Services' five-year National Action Plan to Prevent Health Care Associated Infections and the Centers of Disease Control and Prevention (CDC)'s National Healthcare Safety Network that tracks HAIs and provides data at a national, state, region, and facility level.[32,33]

AMR in animal health and zoonoses

In veterinary medicine, zoonotic transmission of MDROs involves livestock, poultry, aquaculture, companion animals, recreational animals, and wildlife and indeed humans through zoonotic transmission.[34] The number of microorganisms acquiring AMR in animal populations and the spectrum of resistance in each organism are increasing rapidly. This trend is associated with a wide use of some classes of antimicrobials in food animals not only to treat/prevent infections but also as growth promoters and metaphylaxis (treating animals with subclinical disease). A recent report identified over 20 different AMR pathogens impacting animal health, with a potential for zoonotic disease including ESBL *E. coli* and MRSA.[35] **Case 2** describes livestock-associated (LA)-MRSA infection due to occupational exposure of farm workers. The improper or excessive use of antimicrobials can lead to additional antimicrobial selective pressure resulting in an increased burden of drug-resistant pathogens, affecting animal health, welfare, and production. In addition, the normal flora in these animals, and their food products, can become potential reservoirs of AMR genes that can be transferred to human pathogens. The surveillance and regulation of the nontherapeutic use of antimicrobials in livestock and other animal populations is a critical step for mitigation of AMR globally.[36]

In 2019, it was estimated that households globally own 471 million pet dogs and 373 million pet cats—one in two households in the United States and one in four households in China own a dog or a cat.[37] The rise in pet ownership has been associated with increased transmission of MDROs between humans and animals. Commonly reported outbreaks among household-associated pets include: MDR *Campylobacter* with puppies,[38] MDR

Salmonella with reptile pets such as snakes and turtles,[39] ESBL *E. coli* in dogs,[40] and MRSA in dogs and cats.[41] These animal-origin AMR infections have been associated with primary clinical infections in human. Studies have established identical bacterial clones and similar microbiota between pets and their owners. The transmission link between humans and their companion animals has become more prominent due to the detection of a transmissible colistin-resistance gene (*mrc-1*, a gene encoding a third-line antibiotic used as a last resort for MDROs) within the flora of household dogs.[42] However, due to the lack of a systematic global surveillance system, the degree of risk and impact of companion animal populations on human health needs to be explored further. It is important for public health surveillance systems to include AMR information about companion and recreational animals to improve early detection of emerging resistant pathogens and implement interventions.

AMR in the environment

Unlike AMR in humans or animals, the impact of AMR on the environment has received less attention. The release of antimicrobials into the environment (land and water surface) where they interact with natural bacterial communities has the potential to drive bacterial evolution and to lead to the emergence of more resistant strains.

Water and wastewater ecosystems

Water environments link AMR genes (ARGs) among terrestrial origin bacteria (from humans and animals) with environmental bacteria. Some wastewater treatment plants (WWTPs) with flawed filtering processes allow resistant microorganisms to travel to the water surface.[43] Therefore, WWTPs can be the primary locations for ARGs.[44] In some developing countries, animal and human wastewater and effluent from highly contaminated facilities (health facilities, pharmaceutical industries, etc.) are sometimes released directly to the environment without advanced sanitation or regulatory oversight. Through sewage and wastewater systems, MDROs can be released from healthcare facilities, pharmaceutical waste, and households into water sources such as rivers, lakes, and coastal waters, leading to potential contamination.[45–47] The water sources can eventually be introduced to the food chain through crops, aquaculture, and food animals. These water sources can contain resistant organisms as well as antimicrobial drug residues. Domesticated animals and wildlife that come into contact with discharge from wastewater or livestock farms can also be colonized with resistant strains. One study in the Netherlands detected ESBL *E. coli* in 6% of raw vegetables,[46] while in the United Kingdom, ESBL *E. coli* was detected in 2% of beef and pork and 65% of the chicken meat.[48] Similarly, animals can be exposed to AMR organisms, either through contaminated water and feed sources (i.e., contaminated pastures or grains) or through direct contact with wildlife, insects, and pests, and subsequently these organisms can be transmitted to farmers and handlers, resulting in colonization or infection. Alarmingly, MDROs can also directly enter the drinking water supply; a study in Sweden found that 27.5% of surface water samples were positive for transferable ESBL *E. coli* strains and related MDR genes.[49] More environmental surveillance studies need to be conducted in order to understand the major gaps and impacts of wastewater on AMR emergence and spread.

AMR in aquatic/marine ecosystems

Evidence on prevalence and dynamics of AMR in the marine environment has increased. As noted in the previous section, contaminated wastewater can lead to the release of ARGs into the marine environment. The release of additional contaminants into marine environments can increase the resistance levels of bacteria present. Recent observations show that certain metals and elements, such as arsenic and vanadium, are linked with antibiotic resistance in the marine environment.[50,51] Disinfectants can also contribute to horizontal ARG transfer.[52] Complex contamination and horizontal gene transfer are factors that contribute toward increasing the development and transmission of ARGs in the marine environment.[53]

The spread of AMR bacteria and ARGs is also reported from natural waters contaminated with antibiotic compounds from feed and fish feces.[54] Fluoroquinolones, sulfonamides and tetracyclines, macrolides, and penicillin are used in aquaculture.[55] Studies have indicated that tetracycline and sulfonamide resistance genes persist for a prolonged time in sediment beneath the aquaculture fishing ground long after antibiotic administration has been stopped.[56] There is evidence that marine bacterial communities possess novel marine origin ARGs with potential spread to human environment via aquaculture products or workers. Suspected exposure route of antibiotic-resistant bacteria and antibiotic resistance genes with focus on aquatic environments, thus attention to the gene flow route from the sea to humans and animals is needed to strategically reduce the risks in each sector as indicated in Fig. 7.2.[57]

Other environmental factors and AMR

In addition to contamination of wastewater and aquatic/marine ecosystems with resistant microorganisms, the broader environment, including ground and surface waters, soils, the

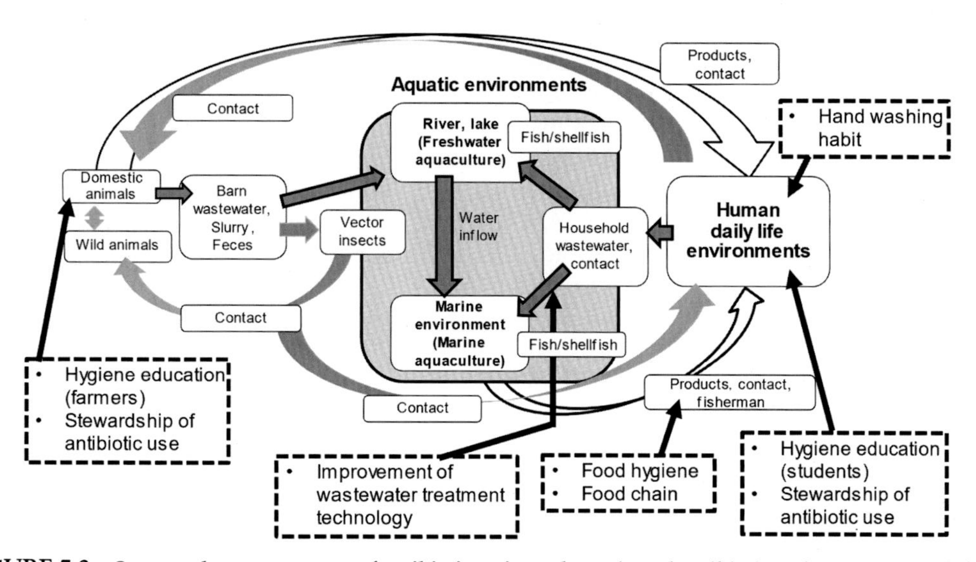

FIGURE 7.2 **Suspected exposure route of antibiotic-resistant bacteria and antibiotic-resistant genes** with focus on aquatic environments. Framed gray arrow, inflow; nonframed gray arrow, contact; and white arrow, intake/contact. Dotted line frame with black arrow, proposed strategy for reducing risk. *Reproduced with permission from Gebreyes et al., 2017.*

built environment,[58–62] and other inanimate objects (e.g., medical and food processing equipment),[63] can be reservoirs for AMR organisms and antimicrobial residues that can serve as contaminants to humans, animals, crops, and other foods. Antimicrobial chemicals applied to crops and aquaculture, used in human medicine and given in the feed of terrestrial food animals, and from pharmaceutical manufacturing enter the environment in vast quantities.[64,65] These pollution sources are primary drivers of AMR development in the environmental selective pressure, contributing to the persistence of AMR in the environment.[66]

Notably, effluent from farms in intensive food animal production systems and the manufacturing of antimicrobials is a source of pollution. Addressing pharmaceutical manufacturing is now a primary target for intervention efforts put forward by the Group of 7 (G7) ministries.[67,68] The G7 recommended that pollution controls be included in good manufacturing practices put forward by the WHO, which focus primarily on quality control of medicines. Additionally, the G7 notes with concern that there are currently no international standards for the safe concentrations of antimicrobials released into the environment from pharmaceutical manufacturing, healthcare facility effluent, agriculture, and aquaculture. Collaborative efforts are needed to establish pollution standards, similar to maximum residue levels in animal products (meat and meat) as they relate to these environmental polluting sectors.

Current efforts to prevent and control AMR

Strategies and action plan

Threats of infectious diseases and increased risk of food safety and security caused by AMR necessitate a globally, coordinated, and intersectoral response (e.g., One Health). The Food and Agriculture Organization (FAO), the World Organization for Animal Health (WOAH) formerly (OIE), and the WHO, collectively known as the Tripartite, have been working together with the United Nations Environment Program (UNEP) on One Health issues, with increasing attention to AMR. International attention and political awareness on AMR have continued to grow. In May 2015, The World Health Assembly adopted the WHO Global Action Plan (GAP) on AMR, which was developed in collaboration with FAO and WOAH, outlining a plan to address AMR. Same year, the Conference organized by FAO endorsed Resolution 4/2015 recognizing the need and urging members to address the problem of AMR in food, agriculture, and the environment. The WOAH has long advocated for prudent antimicrobial use and, in 2016, released The OIE Strategy on AMR and the Prudent Use of Antimicrobials.[69] In 2016, United Nations General Assembly (UNGA) unanimously adopted the High-Level Political Declaration on AMR, acknowledging the global severity of the problem and calling for all nations to implement National Action Plans.[70] The five strategic objectives include improving awareness, strengthening evidence and surveillance, reducing incidence of infection, appropriate antimicrobial use, and sustainable investment. In addition, this Resolution called for the creation of an interagency consultative group (IACG) to provide advice on sustained action by the Tripartite to combat AMR. The IACG released their report, No Time to Wait, in 2019 with many valuable recommendations, including suggestions on how to improve AMR governance at the international level, notable increased collaboration.[71] Of relevance to this book, the GHSA that includes 11 complex global priorities emphasizes prevention and control of AMR as one of the major health security needs across the world. The Action Package on AMR seeks to combat AMR by focusing on the One Health aspects addressing human, animal health, food production, and the environment (ghsagenda.org).

In Africa, in response to the WHO Global Action Plan on AMR, Africa Centers for Disease Control and Prevention developed a Framework for AMR, 2018–23.[72] Concrete actions that have been fulfilled in this regard include the creation of a One Health Global Leadership Group to maintain political momentum and public visibility of the issue and the establishment of an Independent Panel on Evidence for Action to provide UN Member States with information on the science surrounding the evidence and risks associated with AMR.

Global surveillance

Due to inadequate systems of integrated surveillance among the global organizations mentioned in the previous section, the magnitude of emergence and dissemination of AMR in healthcare and community settings is not fully understood.[73] Gaps in the systems include a lack of strong national surveillance system linked with regional and global networks, absence of well-equipped laboratory services, and insufficient credible quality assured data. Fortunately, with increased recognition of the urgent need to address AMR globally, there has been an increase in collaboration across national, regional, and global institutions in promoting and coordinating surveillance, capacity development, research, and the development of operational and monitoring tools (WHO GAP). AMR surveillance monitors dynamics in microbial pathogens, enhances early detection of pathogens of public health importance, and ensures early notification and investigation of outbreaks.[74]

Description of AGISAR and milestones

The WHO Advisory Group on Integrated Surveillance of Antimicrobial Resistance (AGISAR) has been working since 2009 engaging veterinarians, public health, and food safety experts from regulatory and research sectors across the world including every continent.[75] AGISAR's efforts also followed the One Health principle while focusing on drug-resistant foodborne pathogens. The integrations of the surveillance system included antimicrobial usage as well as antimicrobial resistance in animal, food, and human settings.

Another broad initiative of WHO GAP on AMR is the establishment of the Global Antimicrobial Resistance and Use Surveillance System (GLASS) in 2015.[76] Through GLASS, data are collected for selected bacterial pathogens causing common human infections across the partnering nations around the world. The targeted bacterial organisms include *Acinetobacter* spp., *E. coli*, *Klebsiella pneumoniae*, *Neisseria gonorrhoeae*, *Salmonella enterica* (different serovars), *Shigella* spp., *Staphylococcus aureus*, and *Streptococcus pneumoniae* (Table 7.1). GLASS encourages the establishment of national surveillance systems with data coordination centers that incorporate the use of antimicrobials and AMR in the environment and food chain. It also provides a standardized approach to data collection, analysis, and interpretation. This facilitates data-driven decision-making, monitoring of trends and patterns of resistance, and determination of AMR disease burden. Further, expansion to GLASS fungi and GLASS antimicrobial consumption (AMC) has also been initiated with a potential to expand for more pathogens. Apart from isolate-based surveillance, sample-based surveillance is encouraged.

TABLE 7.1 Priority antimicrobial-resistant pathogens[a].

Critical Threats
• Carbapenem-resistant *Acinetobacter* • Carbapenem-resistant Enterobacteriaceae • Carbapenem-resistant *Pseudomonas aeruginosa* • Extended-spectrum beta-lactamases—producing Enterobacteriaceae
Serious Threats
• Vancomycin-resistant *Enterococci* • Methicillin-resistant *Staphylococcus aureus* • Multidrug-resistant *Pseudomonas aeruginosa* • Multidrug-resistant nontyphoidal *Salmonella* • Fluoroquinolone-resistant *Salmonella* serotype Typhi • Drug-resistant *Mycobacterium tuberculosis* (TB) • Fluoroquinolone-resistant and third-generation cephalosporin-resistant *Neisseria gonorrhoeae*
Moderate Threats
• Penicillin-resistant *Streptococcus pneumoniae* • Ampicillin-resistant *Haemophilus* • Fluoroquinolone-resistant *Campylobacter* • Fluoroquinolone-resistant *Shigella* • Erythromycin-resistant group A *Streptococcus* • Clindamycin-resistant group B *Streptococcus* • *Clostridium difficile* • *Candida auris* and drug-resistant *Candida* spp.

[a]Modified from the World Health Organization[113] and The Centers for Disease Control and Prevention[114] Antibiotic resistance threats in the United States.

Though GLASS is critical to fill the knowledge gaps, income-based representation depicts that more of high-income countries (HICs) are implementing the strategy as compared to LMICs. This could be due to the differences in the capacities of countries,[76] scarce availability of resources, and inadequate commitments toward AMR.[15]

The WOAH collects annual surveys about usage of antimicrobials in terrestrial and aquatic animals among members. In the fourth annual report in 2018, many members reported that a lack of regulatory frameworks, human resource constraints, and lack of information technology (IT) tools hinder their ability to provide comprehensive and complete data.[77]

Data on AMR from animals and food sources are mostly from HICs and lacking in most LMICs. FAO is assisting countries to develop surveillance programs for AMR in food and agriculture. Once implemented, the Tripartite has a shared goal of collating the data collected to create an electronic One Health repository for AMR, the Tripartite Integrated Surveillance Systems for AMR (TISSA). Various programs that contribute to AMR policy and control internationally are listed in Table 7.2.

Antimicrobial stewardship programs

Antibiotic stewardship is the effort to promote the appropriate use of antimicrobials to improve patient outcomes, reduce microbial resistance, and decrease the spread of infections

TABLE 7.2 Programs contributing to antimicrobial resistance policy and guidance.

Program	Implementing body	Objectives	Contribution to the AMR field
Advisory group on integrated surveillance of antimicrobial resistance (AGISAR)	WHO (2009)	Advise WHO on ensuring global containment of foodborne antimicrobial resistance (AMR)	It strengthens integrated surveillance using the One Health principle
Global action Plan AMR	WHO (2015)	• To improve awareness and understanding of AMR through effective communication, education, and training • To strengthen the knowledge and evidence base through surveillance and research • To reduce the incidence of infection through effective sanitation, hygiene, and infection prevention measure • To optimize the use of antimicrobial medicines in human and animal health • To develop the economic case for sustainable investment that takes account of the needs of all countries and to increase investment in new medicines, diagnostic tools, vaccines, and other interventions	It ensured the treatment and prevention of infectious diseases with quality-assured, safe, and effective medicines is achievable across all member countries
Global antimicrobial resistance and use surveillance system (GLASS)	WHO (2015)	GLASS provides a standardized approach to the collection, analysis, interpretation, and sharing of data by countries and seeks to actively support capacity building and monitor the status of existing and new national surveillance systems	It provided a platform for AMR data sharing and visualization in all participating countries
Interagency Coordination group on AMR (IACG)	FAO, OIE, WHO	To provide practical guidance for approaches needed to ensure sustained effective global action to address AMR, including options to improve coordination, taking into account the global action plan on AMR	It provided platform for effective coordination between the relevant stakeholders
GLASS-FUNGI		GLASS-FUNGI focuses on the surveillance of invasive fungal bloodstream infections caused by *Candida* spp.	It generated the attention toward the neglected fungal diseases and their surveillance
GLASS-AMC		GLASS-AMC provides a common and standardized set of methods for measuring and reporting antimicrobial consumption (AMC) at country, regional, and global level	It supported the healthcare facilities to make use of the AMC data through point prevalence surveys

FAO, The Food and Agriculture Organization of the United Nations; *OIE*, The World Organisation for Animal Health; *WHO*, World Health Organization.

caused by drug-resistant organisms. Effective antimicrobial stewardship programs (ASPs) must include IPC, diagnostic stewardship, and a strong clinical laboratory. Positively, the implementation of ASPs has been increasing in hospitals in HICs and, more recently, in LMICs. To support the establishment of ASPs the WHO and the US CDC have issued detailed guidelines regarding ASPs and protocols for surveillance, monitoring, and evaluation. The ASP committee or team typically consists of infectious disease physician(s), clinical nursing staff, infection preventionists, microbiologists, and infectious disease pharmacists. According to the US CDC, more than half of antibiotics prescribed in the hospitals were not prescribed consistent with recommended practice guidelines.[78] Antibiotic prescription was not supported in 79% of the patients with community-acquired pneumonia and 77% of patients with UTI.[79] During the COVID-19 pandemic, patients in the ICU with severe disease were more likely to have a co-infection or later develop nosocomial infections. The use of broad-spectrum antibiotics for the management of COVID-19 patients resulted in increased risk for HAIs with MDROs.[80] Rapid diagnostic tests, optimizing pathogen-based antibiotic selection, are for the shortest effective duration which will lead to a decrease in AMR.[77]

The principle of ASP is the same for human and veterinary medicine and focuses on promoting and monitoring judicious antimicrobial use. It is based on basic principles of: a) establishment of committed leadership to set proper coordination and guidance, b) enhancement of infection control, c) development of comprehensive guidance on antimicrobial use including the evaluation of prescription practices, and d) advocacy and education of veterinary clinical personnel, including staff and students.[81] These principles are necessary as colonized/infected animals with AMR bacteria can transmit infection to other animals, contaminate the environment, and expose veterinary personnel.[82] Thus, in food animal veterinary settings where unregulated, unmonitored antimicrobial use at subtherapeutic levels and for prophylactic purposes occur and also in companion animals where therapeutic use still imposes selective pressure (similar to human medicine), the risk is higher for further spread of AMR bacteria into the community of animals and humans. Currently, there is a lack of global momentum for international and national ASP regulatory guidelines, regulatory framework, workforce and surveillance capacity, and infrastructure, for ASP both in human medicine and particularly in veterinary settings.[83,84]

Infection prevention and control

Another vital component of AMR mitigation strategy is pathogen reduction approaches. The general principle is simple and straight forward, if we reduce infection or pathogen emergence and transmission in human (clinical or community) or food and companion animal settings, we will reduce the need for antimicrobial use and thus it will lead to the reduction of AMR. Proper execution of IPC protocols at national and local levels depends on several factors including the type of facility, availability of new evidence and guidance, responsible personnel to implement the guidance, capacity for early detection and management of infection, effectiveness of existing IPC precautions, and availability of supplies and equipment.[85] Hence, IPC recommendations may vary for different settings such as healthcare settings (for both humans and animals), animal production systems, households, and other service providing and educational facilities.[86] Several studies have demonstrated that rigorous implementation of comprehensive IPC measures significantly reduces the rate of infections.[81,86–88] Current global IPC practices including adherence to an updated standard precaution, optimal use of PPE, proper hand hygiene, and appropriate cleaning and disinfecting remain indispensable globally to prevent infections.[87,89,90]

While the principles of IPC are equally important and should be implemented optimally during the presence or absence of a pandemic, executing this guidance across all regions and countries has been difficult. There is a need for LMICs to build their capacity for IPC implementation, strengthen ASP, and increase diagnostic capacity for HAIs (from point of care test to genomic surveillance), interventions that are critical for the prevention and control of HAIs, AMR, and future outbreaks.[91]

The pathogen reduction approach in food animals follows the animal source food value chain starting from on-farm (preharvest) to slaughterhouse (harvest) and retail (postharvest). The most effective pathogen reduction intervention phase is at the slaughterhouses during carcass processing. The US Department of Agriculture Food Safety Inspection Services and its counterparts in other nations use the Hazard Analysis Critical Control Point that aims to reduce pathogens that may originate from farm or acquired at slaughterhouses. Food animal carcasses are treated with various chemical or physical interventions at certain processing steps (typically post-evisceration) and during carcass chilling after dressing is completed. These steps are known as critical control points.[92]

Global gaps in AMR prevention, detection, and response activities

To address global AMR challenges, a comprehensive response against AMR needs to be in place in a holistic manner to contain and mitigate consequences associated with AMR. AMR is a risk to global health safety, and this can be countered by increasing public health awareness and through implementing a coordinated comprehensive global action.

There are multiple gaps in prevention and control of AMR pathogens. First, global surveillance is limited, leading to insufficient progress in assessing the risk, determining the attribution, and mitigating the emergence of MDR pathogens and its impact on public health outcomes in humans, animals, and the environment.[2] Second, a major gap in understanding the magnitude and scope of AMR and its impact in LMICs since most studies are conducted in HICs; communicable diseases remain a major burden in LMICs where antimicrobials are poorly monitored and regulated and can be prescribed without prescription.[93,94] Third, the lack of a One Health approach to manage and monitor AMR across human, animal, and environmental settings systematically fails to address the multifaceted threats and challenges posed from AMR: A coordinated action is required across human medicine, veterinary practice, the agriculture, and the environmental sectors at local, national, continental, and global levels.[3] Without a One Health approach, an integrated system for surveillance and intervention, even in HICs, is impossible due to requirement of a concerted effort and robust allocation of financial resources to strengthen the infrastructure and establish multisectoral networks and interdisciplinary engagement. Fourth, there is an insufficient trained and skilled workforce in the human, animal, and environmental sectors to successfully address the needs of AMR surveillance systems. For example, the WHO has recommended a minimum of 2.3 skilled healthcare staff (physician and nurses/midwives) per 1000 patients.

However, in 44% of its Member States, there is less than 1 skilled physician per 1000 patients.[95] Currently, there is no data on the skilled workforce needs for veterinary and environmental health. Under these circumstances, there is a dire need to enhance One Health workforce development for AMR strategy and implementation of planned activities.

A fifth gap is the current lack of a global emergency response plan on how to identify and respond to major emerging MDR pathogens or related genetic determinants. While countries and intergovernmental agencies have early warning systems to detect and respond to infectious diseases threats, and "clear" triggers that activate global responses to manage outbreaks of infectious agents—a global response plan does not exist. Presently, there are no uniform guidelines and/or recommendations on how the countries are responding to or prevent the MDR pathogens, the steps necessary to take when the new pathogen is circulating in the country, nor the process of implementing timely response to newly identified or emerging MDR strains and understand transmission mechanisms or conduct source-tracking before they become endemic. An example demonstrating this gap was the emergence of resistance to colistin, an antibiotic used as last resort for MDR gram-negative infections, where a global response was mostly in place but was not systematic, fully organized, with clear guidance and recommendations, and all inclusive (using a One Health approach), and, in many cases, left the countries to figure out their response.[96]

As described above, uncontrolled/unmonitored antimicrobial use in humans as well as companion animals in the community (including veterinary) settings is also a major gap not covered within the IHR or the GHS Agenda. Currently, most governments, especially in LMICs, do not monitor or regulate antimicrobial use and AMR in these animal populations. Therefore, the scope of the problem and the potential role in the emergence of epidemic MDR pathogens remain undetermined. It is difficult to implement interventions and control measures to curtail the spread of these microorganisms as only limited national agencies and institutions are currently in charge to lead such a response.

Finally, financial resources are often limited to implement AMR intervention strategies especially in LMICs. To partially address this problem, the Tripartite, in 2019, established a trust fund to invite partnerships and financing to support the implementation of One Health National Action Plans.[97]

Vision

GHS envisages resilient health systems encompassing essential One Health components across the human, animal, and environmental health to support the prevention, detection, and response to emerging health threats.[98] Situation is being complicated by increasing drug resistance in infections threatening our ability to treat even common infections. Developing countries are faced with even tougher challenges where communicable diseases claim an estimated 12 million people every year.[99] A resilient health system in the context of GHS requires multiple elements including strong commitment, political will, sound policies and regulations, ample resources, skilled manpower capacity such as healthcare staff including in veterinary settings, and continuous surveillance and monitoring in integrated ways.[100,101]

Recommendations

- Strengthen integrated global surveillance efforts within IHR and focus on One Health approach to control the spread of AMR infectious diseases in line with the objectives of the GHSA.
- Sustain advocacy and foster collaboration to address the gaps in equitable access to quality essential healthcare services, diagnostics, safe, effective, and quality medicines to prevent AMR and improve IPC, Pathogen Reduction, and ASP best practices in both human and animal health.
- Engage communities and all stakeholders in continuing education and awareness to change behavior in misuse and overuse of antibiotics in human, animal, and agriculture sectors.
- Address the use of antibiotics for growth promotion through sound policies at the highest level.
- Advocacy and support for promoting policies restricting nonprescription use of antibiotics including growth promotion in food animals, unregulated use in humans, and horticulture use of antibiotics (often labeled as pesticides).
- Strengthen applied research and development that will lead to informing policies and strategies targeted for customized interventions considering differences in geographic, epidemiologic, and health system capacity to control AMR.
- Bridge the gap in AMR research areas including development of newer drugs, alternatives to antimicrobials, and diagnostics that can be used for aquaculture, companion, and food animals.
- Develop a mechanism for nations and regions to establish an interoperable data exchange system to track AMR.
- Build capacity to enhance regulation, sustain skilled workforce, strengthen laboratory capacity, and implement prudent antimicrobial use for best practices in human, companion animals, livestock, and agriculture sector including aquaculture.
- Use of available data to design strategies for increased investment and sustainable financing applying economic analysis, projections, and investment case for AMR at local, regional, and global levels.

CASE STUDY 1

Increased exposure to multiple courses of antibiotics leads to multidrug resistance in patients with chronic infections

Chronic urinary tract infection with increased drug resistance

Background

Urinary tract infection (UTI) is one of the common conditions where increasing empiric antibiotic treatment is often offered at an outpatient level.[102,103] In many countries, routine antimicrobial treatments used include beta-lactams, trimethoprim, nitrofurantoin, and quinolones. However, there is increasing report of resistance to these drugs.[104] In the United States alone, AMR among patients with UTIs accounts for over 40% of all HAIs.[105] Enterobacteriaceae in the human, animal, and environment can be contaminants, acquire resistant genes, and serve as a reservoir. The food chain animals colonized with these pathogens can transmit multidrug-resistant organisms (MDROs) such as extended-spectrum beta-lactamase (ESBL)-producing organism to human. Because of limited options for treatment, the ESBL pathogens are a growing threat to human health.[106]

A 65-year-old female with history of chronic UTI was admitted to hospital after failing to respond to multiple courses of oral antibiotics as an outpatient. Laboratory confirmed an ESBL UTI resistant to trimethoprim sulfamethoxazole, nitrofurantoin, ciprofloxacin, and all cephalosporin. She was treated with intravenous (IV) antibiotic ertapenem (a carbapenem) due to the drug resistance profile and limited oral antimicrobial options. Patient was also placed in contact isolation due to drug resistance to multiple classes of antibiotics.

Observations

- Limited treatment options for oral outpatient antibiotics often may require inpatient admission (with significant increase cost) and use of IV antibiotics that may be expensive and have adverse drug reactions.
- Antimicrobial stewardship program (ASP) in the inpatient and outpatient setting can help guide antimicrobial therapy to decrease development of drug resistance.
- Infection prevention and control (IPC) teams needed to place the patient in contact isolation to decrease spread to healthcare workers and prevent transmission of antimicrobial resistance (AMR) organism to other patients.

Key issues

Initial treatment with empirical antibiotic therapy may be responsible for treatment delay due to prescription of incorrect antimicrobials that does not adequately treat multidrug-resistant pathogens. It is important to know the patient's history of prior drug-resistant pathogens, antimicrobial history, and local antibiogram. Rapid laboratory identification of resistant organisms is critical to help guide clinical decision regarding choice of appropriate antibiotics.

Interventions

- The patient was successfully treated with IV antibiotic based on the laboratory determination of resistance and identification of the correct antimicrobial.

CASE STUDY 1 (cont'd)

- Inpatient and outpatient ASP guidelines helped with antibiotic selection, dose, and duration.
- IPC protocol was implemented as per the national guideline.

Lessons learned

- What works:
  - Understanding the growing challenge of AMR in UTI is critical to guide practice.
  - Diagnostic stewardship with improved and rapid diagnostic test assists in de-escalation of antimicrobials.
  - IPC is critical to contain MDROs and prevent transmission and spread to healthcare workers and patients.
- What does not work:
  - Empiric recurrent outpatient treatment of chronic UTIs is a major contributing factor to the development of MDR UTIs.

Recommendations

- Antibiotic stewardship is a critical step in prevention of MDR UTIs.
- National policy and local implementation guides are essential for identification and management of multidrug-resistant organisms and prevention of transmission and spread to healthcare workers and patients.
- ASP and IPC programs are important for surveillance, management, and treatment of patients with AMR and to prevent the spread of AMR pathogens.
- Continuous surveillance is critical to monitor trends and outbreaks.

CASE STUDY 2

***One Health approach is needed to combat antimicrobial resistance**

Methicillin-resistant *Staphylococcus aureus* in livestock with impact on human health

Background

Livestock-associated methicillin-resistant *Staphylococcus aureus* (LA-MRSA) sequence type (ST) 398 has been associated to swine, cattle, and other food chain, companion, and wildlife animals as well as environmental water source.[107,108] First reported in Denmark in farm workers and veterinarians, LA-MRSA ST398 has expanded geographically to Europe, Asia, and Americas and to individuals with no contact with animals.[107–110] Phenotypically, LA-MRSA ST398 is multidrug resistant and genotypically does not harbor common virulent genes associated with human disease, thus causing mostly mild kin and soft tissue infections. However, more recently, reports of septicemia with increase morbidity and mortality as well as genotypic acquisition of virulent genes have been reported.[111]

CASE STUDY 2 *(cont'd)*

A 56-year-old male, with history of skin and soft tissue infection, is admitted to the hospital with fevers, chills, and shortness of breath. He is started on empiric intravenous (IV) antibiotics with vancomycin and piperacillin/tazobactam. Blood culture obtained is positive for MRSA. Drug susceptibility testing showed resistance to tetracycline. The patient works on a pig farm and additional genotyping performed identified it as LA-MRSA ST398.

Observations

- Detection of MRSA in animals in the community—National survey of market hogs in Holland—showed 39% positivity with nasal swab.
- LA-MRSA ST398 clones, potential for zoonotic transmission, have been linked to human disease. All the isolates were nontypeable by standard pulsed-field gel electrophoresis, multilocus sequence type (ST) 398 and three sap types.
- LA-MRSA ST398 strains have been identified in not only farm workers but also others in the community without contact with livestock.
- LA-MRSA with low virulence usually causes skin and soft tissue infections, abscess, or wound but can cause invasive disease with increased morbidity and mortality.
- Nosocomial transmission of LA-MRSA has been reported in hospital outbreaks.
- MRSA was detected in swine veterinarians, market hogs, and retail pork products in the United States has more diversity in *spa* types than in Europe.
- The antimicrobials used without indication in food animals leads to increased resistance. Isolates usually resistant to tetracycline—the use of tetracycline in animal feed may be selective for MRSA resistance.

Key issues

- LA-MRSA ST398 were initially identified in pigs in Denmark with zoonotic potential causing mild disease in human, now have expanded globally with increased morbidity and mortality in both individuals with and without animal association.

Interventions

- Surveillance screening of animal and environment to help determine prevalence in the community.
- Awareness of individual with occupational risks with history of livestock association for isolation, screening, and proper treatment and management.

Lessons learned

- What works:
  - MRSA is a major concern in humans. Detecting MRSA infection and colonization in animal, livestock, and environment identifies potential reservoirs for MRSA in the community.
  - Pigs, cattle, and poultry are colonized with MRSA and the potential for zoonotic transmission to human via direct animal contact, environmental contamination, or meats.
  - Understanding the community prevalence of MRSA in animals can improve the health of the community.
  - LA-MRSA may have increased resistance to certain antimicrobials; knowledge of the drug resistance can assist with treatment regimen.

CASE STUDY 2 *(cont'd)*

- Infection prevention and control is critical to contain multidrug-resistant organisms (MDROs) and prevent transmission and spread to healthcare workers and patients.
- What does not work
  - Public health and veterinary health combating AMR independently rather than using a One Health approach and working together.

Recommendations

- A One Health approach is needed to control, prevent, and contain zoonotic diseases such as LA-MRSA.
- A national policy and regional implementation are important for surveillance identification. A continuous surveillance in human, animal, and environment is critical to monitor trends and outbreaks of zoonotic pathogens.
- Knowledge of prevalence of drug resistance can help in the management of MDROs and prevention of transmission in the community and in the healthcare setting.

Acknowledgement

We would like to thank Michael Kozuch for his help with the chapter.

References

1. Weldon I, Hoffman SJ. Bridging the commitment-compliance gap in global health politics: lessons from international relations for the global action plan on antimicrobial resistance. *Global Publ Health*. 2021;16(1):60–74.
2. Iwu CD, Patrick SM. An insight into the implementation of the global action plan on antimicrobial resistance in the WHO African region: a roadmap for action. *Int J Antimicrob Agents*. 2021;58(4), 106411.
3. Ghebreyesus TA. Making AMR history: a call to action. *Glob Health Action*. 2019;12(suppl 1), 1638144.
4. Nieuwlaat R, Mbuagbaw L, Mertz D, et al. Coronavirus disease 2019 and antimicrobial resistance: parallel and interacting health emergencies. *Clin Infect Dis*. 2021;72(9):1657–1659.
5. Seneghini M, Rüfenacht S, Babouee-Flury B, et al. It is complicated: potential short-and long-term impact of coronavirus disease 2019 (COVID-19) on antimicrobial resistance—an expert review. *Antimicrob Stewardship Health Epidemiol*. 2022;2(1).
6. Lal A, Ashworth HC, Dada S, Hoemeke L, Tambo E. Optimizing pandemic preparedness and response through health information systems: lessons learned from Ebola to COVID-19. *Disaster Med Public Health Prep*. 2022;16(1):333–340.
7. Murray CJL, Ikuta KS, Sharara F, et al. Global burden of bacterial antimicrobial resistance in 2019: a systematic analysis. *Lancet*. 2022;399(10325):629–655.
8. Thompson T. The staggering death toll of drug-resistant bacteria. *Nature*. 2022. https://doi.org/10.1038/d41586-022-00228-x. Epub ahead of print. PMID: 35102288.
9. WHO. *The Global Health Observatory, Explore a World Health Data*; 2022. https://www.who.int/data/gho/data/themes/malaria.

10. Lancet T. Antimicrobial resistance: time to repurpose the Global Fund, https://www.thelancet.com/journals/lancet/article/PIIS0140-6736(22)00091-5/fulltext; 335. Available at. Accessed 10322, 399.
11. Archive CW. *Global Health Security Agenda: GHSA Antimicrobial Resistance Action Package (GHSA Action Package Prevent-1)*; 2014. https://www.cdc.gov/globalhealth/security/actionpackages/antimicrobial_resistance.htm.
12. Holmes AH, Moore LS, Sundsfjord A, et al. Understanding the mechanisms and drivers of antimicrobial resistance. *Lancet*. 2016;387(10014):176–187.
13. Council NR. Committee on drug use in food animals. In: *The Use of Drugs in Food Animals: Benefits and Risks*. 1999.
14. Mellon MG, Benbrook C, Benbrook KL. *Hogging it: Estimates of Antimicrobial Abuse in Livestock*. Union of Concerned Scientists; 2001.
15. O'Neill J. *Tackling Drug-Resistant Infections Globally: Final Report and Recommendations*. 2016.
16. Tompson AC, Chandler CI. *Addressing Antibiotic Use: Insights from Social Science Around the World*. 2021.
17. Boussard J, Trouvé A, Choplin G, Strickner A, Trouvé A. *World Food Summit Plan of Action, World Food Summit, 13-17 Novembre*. Roma: FAO; 2009. agriregionieuropa.30 http://www.fao.org/wfs/index_en.htm. The State of Food Insecurity in the World. Economic crises-impacts and lessons learned.
18. Jonas OBI, Irwin A, Berthe FCJ, Le Gall FG, Marquez PV. *Drug-resistant Infections : A Threat to Our Economic Future*. Vol 2. World Bank Group; 2017. final report (English). HNP/Agriculture Global Antimicrobial Resistance Initiative Washington, D.C. http://documents.worldbank.org/curated/en/323311493396993758/final-report
19. George A. Antimicrobial resistance (AMR) in the food chain: trade, one health and codex. *Trop Med Infect Dis*. 2019;4(1):54.
20. WHO. *International Health Regulations (2005)*. World Health Organization; 2008.
21. Boerlin P, Reid-Smith RJ. Antimicrobial resistance: its emergence and transmission. *Anim Health Res Rev*. 2008;9(2):115–126.
22. EPoB H, Koutsoumanis K, Allende A, et al. Role played by the environment in the emergence and spread of antimicrobial resistance (AMR) through the food chain. *EFSA J*. 2021;19(6), e06651.
23. Bengtsson-Palme J, Kristiansson E, Larsson DJ. Environmental factors influencing the development and spread of antibiotic resistance. *FEMS Microbiol Rev*. 2018;42(1):fux053.
24. Baker S, Thomson N, Weill F-X, Holt KE. Genomic insights into the emergence and spread of antimicrobial-resistant bacterial pathogens. *Science*. 2018;360(6390):733–738.
25. CDC. *Types of Healthcare-Associated Infections*; 2014. https://www.cdc.gov/hai/infectiontypes.html.
26. WHO. Health care-associated infections fact sheet 2011. https://apps.who.int/iris/bitstream/handle/10665/80135/9789241501507_eng.pdf.
27. Day MJ, Hopkins KL, Wareham DW, et al. Extended-spectrum β-lactamase-producing *Escherichia coli* in human-derived and foodchain-derived samples from England, Wales, and Scotland: an epidemiological surveillance and typing study. *Lancet Infect Dis*. 2019;19(12):1325–1335.
28. Naziri Z, Derakhshandeh A, Soltani Borchaloee A, Poormaleknia M, Azimzadeh N. Treatment failure in urinary tract infections: a warning witness for virulent multi-drug resistant ESBL- producing *Escherichia coli*. *Infect Drug Resist*. 2020;13:1839–1850.
29. van Duin D, Paterson DL. Multidrug-resistant bacteria in the community: trends and lessons learned. *Infect Dis Clin*. 2016;30(2):377–390.
30. Herold BC, Immergluck LC, Maranan MC, et al. Community-acquired methicillin-resistant *Staphylococcus aureus* in children with no identified predisposing risk. *JAMA*. 1998;279(8):593–598.
31. Dukic VM, Lauderdale DS, Wilder J, Daum RS, David MZ. Epidemics of community-associated methicillin-resistant *Staphylococcus aureus* in the United States: a meta-analysis. *PLoS One*. 2013;8(1), e52722.
32. HHS. *Office of Infectious Disease and HIV/AIDS Policy (OIDP)*. HAI National Action Plan; 2021. https://www.hhs.gov/oidp/topics/health-care-associated-infections/hai-action-plan/index.html#actionplan_development.
33. CDC. *National Healthcare Safety Network (NHSN)*; 2021. https://www.cdc.gov/nhsn/.
34. Palma E, Tilocca B, Roncada P. Antimicrobial resistance in veterinary medicine: an overview. *Int J Mol Sci*. 2020;21(6):1914.
35. AVMA. *Antimicrobial Resistant Pathogens Affecting Animal Health in the United States, AVMA/committee on Antimicrobial*; 2020. https://www.avma.org/sites/default/files/2020-10/AntimicrobialResistanceFullReport.pdf.
36. Ma Z, Lee S, Jeong KC. *Mitigating Antibiotic Resistance at the Livestock-Environment Interface: A Review*. 2019.

37. Statista. *Number of Dogs and Cats Kept as Pets Worldwide in 2018*; 2022. https://www.statista.com/statistics/1044386/dog-and-cat-pet-population-worldwide/.
38. Montgomery MP, Robertson S, Koski L, et al. Multidrug-resistant Campylobacter jejuni outbreak linked to puppy exposure—United States, 2016–2018. *MMWR (Morb Mortal Wkly Rep)*. 2018;67(37):1032.
39. Marin C, Lorenzo-Rebenaque L, Laso O, Villora-Gonzalez J, Vega S. Pet reptiles: a potential source of transmission of multidrug-resistant Salmonella. *Front Vet Sci*. 2021;7, 613718-613718.
40. Dupouy V, Abdelli M, Moyano G, et al. Prevalence of beta-lactam and quinolone/fluoroquinolone resistance in Enterobacteriaceae from dogs in France and Spain-characterization of ESBL/pAmpC isolates, genes, and conjugative plasmids. *Front Vet Sci*. 2019;6:279.
41. Van Balen JC, Landers T, Nutt E, Dent A, Hoet AE. Molecular epidemiological analysis to assess the influence of pet-ownership in the biodiversity of *Staphylococcus aureus* and MRSA in dog- and non-dog-owning healthy households. *Epidemiol Infect*. 2017;145(6):1135–1147.
42. EurekAlert. *Study Identifies Last-Line Antibiotic Resistance in Humans and Pet Dog*; 2020. https://www.eurekalert.org/news-releases/919789.
43. Segura PA, François M, Gagnon C, Sauvé S. Review of the occurrence of anti-infectives in contaminated wastewaters and natural and drinking waters. *Environ Health Perspect*. 2009;117(5):675–684.
44. Rizzo L, Manaia C, Merlin C, et al. Urban wastewater treatment plants as hotspots for antibiotic resistant bacteria and genes spread into the environment: a review. *Sci Total Environ*. 2013;447:345–360.
45. Mathys DA, Mollenkopf DF, Feicht SM, et al. Carbapenemase-producing Enterobacteriaceae and Aeromonas spp. present in wastewater treatment plant effluent and nearby surface waters in the US. *PLoS One*. 2019;14(6), e0218650.
46. Marathe NP, Berglund F, Razavi M, et al. Sewage effluent from an Indian hospital harbors novel carbapenemases and integron-borne antibiotic resistance genes. *Microbiome*. 2019;7(1):97.
47. Yang CM, Lin MF, Liao PC, et al. Comparison of antimicrobial resistance patterns between clinical and sewage isolates in a regional hospital in Taiwan. *Lett Appl Microbiol*. 2009;48(5):560–565.
48. Randall LP, Lodge MP, Elviss NC, et al. Evaluation of meat, fruit and vegetables from retail stores in five United Kingdom regions as sources of extended-spectrum beta-lactamase (ESBL)-producing and carbapenem-resistant *Escherichia coli*. *Int J Food Microbiol*. 2017;241:283–290.
49. Egervärn M, Englund S, Ljunge M, et al. Unexpected common occurrence of transferable extended spectrum cephalosporinase-producing *Escherichia coli* in Swedish surface waters used for drinking water supply. *Sci Total Environ*. 2017;587–588:466–472.
50. Farias P, Espírito Santo C, Branco R, et al. Natural hot spots for gain of multiple resistances: arsenic and antibiotic resistances in heterotrophic, aerobic bacteria from marine hydrothermal vent fields. *Appl Environ Microbiol*. 2015;81(7):2534–2543.
51. Zhang X-X, Zhang T, Fang HH. Antibiotic resistance genes in water environment. *Appl Microbiol Biotechnol*. 2009;82(3):397–414.
52. Zhang Y, Gu AZ, He M, Li D, Chen J. Subinhibitory concentrations of disinfectants promote the horizontal transfer of multidrug resistance genes within and across genera. *Environ Sci Technol*. 2017;51(1):570–580.
53. Abe K, Nomura N, Suzuki S. Biofilms: hot spots of horizontal gene transfer (HGT) in aquatic environments, with a focus on a new HGT mechanism. *FEMS Microbiol Ecol*. 2020;96(5). fiaa031.
54. Cabello FC, Godfrey HP, Tomova A, et al. Antimicrobial use in aquaculture re-examined: its relevance to antimicrobial resistance and to animal and human health. *Environ Microbiol*. 2013;15(7):1917–1942.
55. Okocha RC, Olatoye IO, Adedeji OB. Food safety impacts of antimicrobial use and their residues in aquaculture. *Publ Health Rev*. 2018;39(1):1–22.
56. Tamminen M, Karkman A, Lohmus A, et al. Tetracycline resistance genes persist at aquaculture farms in the absence of selection pressure. *Environ Sci Technol*. 2011;45(2):386–391.
57. Pruden A, Larsson DJ, Amézquita A, et al. Management options for reducing the release of antibiotics and antibiotic resistance genes to the environment. *Environ Health Perspect*. 2013;121(8):878–885.
58. Andrade L, Kelly M, Hynds P, Weatherill J, Majury A, O'Dwyer J. Groundwater resources as a global reservoir for antimicrobial-resistant bacteria. *Water Res*. 2020;170, 115360.
59. Cho S, Jackson C, Frye J. The prevalence and antimicrobial resistance phenotypes of Salmonella, *Escherichia coli* and Enterococcus sp. in surface water. *Lett Appl Microbiol*. 2020;71(1):3–25.

60. Zhu Y-G, Zhao Y, Zhu D, et al. Soil biota, antimicrobial resistance and planetary health. *Environ Int*. 2019;131, 105059.
61. Jorquera CB, Moreno-Switt AI, Sallaberry-Pincheira N, et al. Antimicrobial resistance in wildlife and in the built environment in a wildlife rehabilitation center. *One Health*. 2021;13, 100298.
62. Phoon HYP, Hussin H, Hussain BM, et al. Distribution, genetic diversity and antimicrobial resistance of clinically important bacteria from the environment of a tertiary hospital in Malaysia. *J Glob Antimicrob Resist*. 2018;14:132–140.
63. Kiros T, Damtie S, Eyayu T, Tiruneh T, Hailemichael W, Workineh L. Bacterial pathogens and their antimicrobial resistance patterns of inanimate surfaces and equipment in Ethiopia: a systematic review and meta-analysis. *BioMed Res Int*. 2021:2021.
64. Oniciuc E-A, Likotrafiti E, Alvarez-Molina A, Prieto M, López M, Alvarez-Ordóñez A. Food processing as a risk factor for antimicrobial resistance spread along the food chain. *Curr Opin Food Sci*. 2019;30:21–26.
65. Booth A, Aga DS, Wester AL. Retrospective analysis of the global antibiotic residues that exceed the predicted no effect concentration for antimicrobial resistance in various environmental matrices. *Environ Int*. 2020;141, 105796.
66. EFSA. Role played by the environment in the emergence and spread of antimicrobial resistance (AMR) through the food chain. *EFSA J*. 2021;19(6), e06651.
67. Kotwani A, Joshi J, Kaloni D. Pharmaceutical effluent: a critical link in the interconnected ecosystem promoting antimicrobial resistance. *Environ Sci Pollut Control Ser*. 2021;28(25):32111–32124.
68. Ministers' GH. *G7 Health Ministers' Declaration*; 2021. https://assets.publishing.service.gov.uk/government/uploads/system/uploads/attachment_data/file/992268/G7-health_ministers-communique-oxford-4-june-2021_5.pdf.
69. OIE. *The OIE Strategy on Antimicrobial Resistance and the Prudent Use of Antimicrobials*; 2016. https://www.oie.int/app/uploads/2021/03/en-oie-amrstrategy.pdf.
70. UN. *Draft Political Declaration of the High-Level Meeting of the General Assembly on Antimicrobial Resistance*. 2016.
71. WHO. No time to wait: securing the future from drug-resistant infection. In: *Report to the Secretary General of the United Nations*; 2019. https://www.who.int/groups/one-health-global-leaders-group-on-antimicrobial-resistance.
72. Perovic O, Ondoa P, Laxminarayan R, et al. Africa centres for disease control and prevention's framework for antimicrobial resistance control in Africa. *Afr J Lab Med*. 2018;7(2):1–4.
73. WHO. *Antimicrobial Resistance: Global Report on Surveillance*. World Health Organization; 2014.
74. Ferreira JP, Battaglia D, García AD, et al. Achieving antimicrobial stewardship on the global scale: challenges and opportunities. *Microorganisms*. 2022;10(8).
75. WHO. *Integrated Surveillance of Antimicrobial Resistance in Foodborne Bacteria: Application of a One Health Approach: Guidance from the WHO Advisory Group on Integrated Surveillanec of Antimicrobial Resistance (AGISAR)*. Geneva: World Health Organization; 2017.
76. WHO. *Global Antimicrobial Resistance Surveillance System (GLASS) Report: Early Implementation 2020*. 2020.
77. OIE. *OIE Annual Report on Antimicrobial Agents Intended for Use in Animals: Better Understanding of the Global Situation*; 2020. https://www.oie.int/app/uploads/2021/03/a-fourth-annual-report-amr.pdf.
78. CDC. *Elements of Hospital Antibiotic Stewardship Programs*. Atlanta, GA: US Department of Health and Human Services; 2019. Available at https://www.cdc.gov/antibiotic-use/core-elements/hospital.html.
79. Donà D, Di Chiara C, Sharland M. Multi-drug-resistant infections in the COVID-19 era: a framework for considering the potential impact. *J Hosp Infect*. 2020;106(1):198–199.
80. Feyes EE, Diaz-Campos D, Mollenkopf DF, et al. Implementation of an antimicrobial stewardship program in a veterinary medical teaching institution. *J Am Vet Med Assoc*. 2021;258(2):170–178.
81. Storr J, Twyman A, Zingg W, et al. Core components for effective infection prevention and control programmes: new WHO evidence-based recommendations. *Antimicrob Resist Infect Control*. 2017;6(1):1–18.
82. Weese J, Giguère S, Guardabassi L, et al. ACVIM consensus statement on therapeutic antimicrobial use in animals and antimicrobial resistance. *J Vet Intern Med*. 2015;29(2):487–498.
83. McEwen SA, Collignon PJ. Antimicrobial resistance: a one health perspective. *Microbiol Spectr*. 2018;6(2):6–10.
84. Feyes EE, Diaz-Campos D, Mollenkopf DF, et al. Implementation of an antimicrobial stewardship program in a veterinary medical teaching institution. *J Am Vet Med Assoc*. 2021;258(2):170–178.
85. Allegranzi B, Kilpatrick C, Storr J, Kelley E, Park BJ, Donaldson L. Global infection prevention and control priorities 2018–22: a call for action. *Lancet Glob Health*. 2017;5(12):e1178–e1180.
86. Bentivegna E, Luciani M, Arcari L, Santino I, Simmaco M, Martelletti P. Reduction of multidrug-resistant (MDR) bacterial infections during the COVID-19 pandemic: a retrospective study. *Int J Environ Res Publ Health*. 2021;18(3):1003.

87. Tartari E, Tomczyk S, Pires D, et al. Implementation of the infection prevention and control core components at the national level: a global situational analysis. *J Hosp Infect*. 2021;108:94–103.
88. WHO. *Guidelines on Core Components of Infection Prevention and Control Programmes at the National and Acute Health Care Facility Level*. World Health Organization; 2016.
89. Adams V, Song J, Shang J, et al. Infection prevention and control practices in the home environment: examining enablers and barriers to adherence among home health care nurses. *Am J Infect Control*. 2021;49(6):721–726.
90. Jara BJ. Infection prevention in the era of COVID-19: 2021 basic procedure review. *J Nucl Med Technol*. 2021;49(2):126–131.
91. Abbas M, Robalo Nunes T, Martischang R, et al. Nosocomial transmission and outbreaks of coronavirus disease 2019: the need to protect both patients and healthcare workers. *Antimicrob Resist Infect Control*. 2021;10(1):7.
92. USDA. Pathogen reduction: Hazard analysis and critical control point (HACCP) systems; final rule. *Fed Regist*. 1996;61:38806–38989.
93. Gebretekle GB, Haile Mariam D, Abebe W, et al. Opportunities and barriers to implementing antibiotic stewardship in low and middle-income countries: lessons from a mixed-methods study in a tertiary care hospital in Ethiopia. *PLoS One*. 2018;13(12), e0208447.
94. Iskandar K, Molinier L, Hallit S, et al. Surveillance of antimicrobial resistance in low-and middle-income countries: a scattered picture. *Antimicrob Resist Infect Control*. 2021;10(1):1–19.
95. WHO. *Global Health Workforce Statistics Database*; 2016. https://www.who.int/data/gho/data/themes/topics/health-workforce.
96. Gharaibeh MH, Shatnawi SQ. An overview of colistin resistance, mobilized colistin resistance genes dissemination, global responses, and the alternatives to colistin: a review. *Vet World*. 2019;12(11):1735.
97. UNDP. *Antimicrobial Resistance Multi-Partner Trust Fund*; 2019. https://mptf.undp.org/factsheet/fund/AMR00.
98. Hill PS. Understanding global health governance as a complex adaptive system. *Global Publ Health*. 2011;6(6):593–605.
99. Boutayeb A. The burden of communicable and non-communicable diseases in developing countries. In: *Handbook of Disease Burdens and Quality of Life Measures*. 2010:531.
100. Davies SE, Kamradt-Scott A, Rushton S. *Disease Diplomacy: International Norms and Global Health Security*. JHU Press; 2015.
101. Ensor T, Weinzierl S. Regulating health care in low-and middle-income countries: broadening the policy response in resource constrained environments. *Soc Sci Med*. 2007;65(2):355–366.
102. Goldstein E, MacFadden DR, Karaca Z, Steiner CA, Viboud C, Lipsitch M. Antimicrobial resistance prevalence, rates of hospitalization with septicemia and rates of mortality with sepsis in adults in different US states. *Int J Antimicrob Agents*. 2019;54(1):23–34.
103. Chin TL, McNulty C, Beck C, MacGowan A. Antimicrobial resistance surveillance in urinary tract infections in primary care. *J Antimicrob Chemother*. 2016;71(10):2723–2728.
104. Ny S, Edquist P, Dumpis U, et al. Antimicrobial resistance of *Escherichia coli* isolates from outpatient urinary tract infections in women in six European countries including Russia. *J Glob Antimicrob Resist*. 2019;17:25–34.
105. US-FDA. *Complicated Urinary Tract Infections: Developing Drugs for Treatment: Guidance for Industry*. US Department of Health and Human Services, Food and Drug Administration; 2018. https://www.fda.gov/media/71313/. Accessed February , 2020.
106. Palmeira JD, Ferreira HMN. Extended-spectrum beta-lactamase (ESBL)-producing Enterobacteriaceae in cattle production—a threat around the world. *Heliyon*. 2020;6(1), e03206.
107. Larsen J, Petersen A, Larsen AR, et al. Emergence of livestock-associated methicillin-resistant *Staphylococcus aureus* bloodstream infections in Denmark. *Clin Infect Dis*. 2017;65(7):1072–1076.
108. He L, Zheng H-X, Wang Y, et al. Detection and analysis of methicillin-resistant human-adapted sequence type 398 allows insight into community-associated methicillin-resistant *Staphylococcus aureus* evolution. *Genome Med*. 2018;10(1):1–14.
109. Nielsen RT, Kemp M, Holm A, et al. Fatal septicemia linked to transmission of MRSA clonal complex 398 in hospital and nursing home, Denmark. *Emerg Infect Dis*. 2016;22(5):900.
110. Sieber RN, Larsen AR, Urth TR, et al. Genome investigations show host adaptation and transmission of LA-MRSA CC398 from pigs into Danish healthcare institutions. *Sci Rep*. 2019;9(1):1–10.
111. Coombs GW, Pang S, Daley DA, Lee YT, Abraham S, Leroi M. Severe disease caused by community-associated MRSA ST398 type V, Australia, 2017. *Emerg Infect Dis*. 2019;25(1):190.

112. Davies J, Davies D. Origins and evolution of antibiotic resistance. *Microbiol Mol Biol Rev*. 2010;74(3):417–433.
113. WHO. *WHO Publishes List of Bacteria for Which New Antibiotics Are Urgently Needed*; 2017. https://www.who.int/news/item/27-02-2017-who-publishes-list-of-bacteria-for-which-new-antibiotics-are-urgently-needed.
114. CDC. *Antibiotic Resistance Threats in the United States, 2019 (2019 AR Threats Report) Is a Publication of the Antibiotic Resistance Coordination and Strategy Unit within the Division of Healthcare Quality Promotion*. National Center for Emerging and Zoonotic Infectious Diseases, Centers for Disease Control and Prevention; 2019. www.cdc.gov/DrugResistance/Biggest-Threats.html.

CHAPTER

8

Toxic and environmentally ubiquitous chemical agents

Michael Bisesi[1] *and Jiyoung Lee*[2]

[1]Global One Health Initiative (GOHi), The Ohio State University College of Public Health, Columbus, OH, United States; [2]College of Food, Agriculture, and Environmental Science, Global One Health Initiative (GOHi), The Ohio State University College of Public Health, Columbus, OH, United States

Introduction

Environmental health science applied to the One Health approach

Environmental health, sometimes referred to as environmental public health, is defined as *the science and practice of preventing human injury and illness and promoting well-being by identifying and evaluating sources and hazardous agents; and, limiting exposures to hazardous physical, chemical, and biological agents in air, water, soils, food, and other environmental media or settings that can adversely affect human health.*[1] Environmental health science is a major discipline, along with other applicable disciplines, to address global health issues using a One Health approach.

One very general and appropriate definition of One Health by the World Health Organization (WHO) is *an approach to designing and implementing programmes, policies, legislation, and research in which multiple sectors communicate and work together to achieve better public health outcomes.*[2] The conventional paradigm for the One Health approach focuses exclusively on hazardous infectious microbiological agents and the associated infectious diseases they cause, especially those that are zoonotic. This is consistent with the historic primary emphasis of global health in general. For example, infectious disease is the major focus of initiatives stated by the Global Health Security Agenda in its past and present framing document,[3] as well as in the International Health Regulations.[4] This emphasis remains justified with a need for continued and expanded attention and resources for prevention and mitigation of infectious diseases worldwide. An expanded and more contemporary paradigm, however, now also includes hazardous toxic chemical agents and the associated noninfectious diseases.

Modernizing Global Health Security to Prevent, Detect, and Respond
https://doi.org/10.1016/B978-0-323-90945-7.00025-7

The shift of the historic paradigm for One Health to now include toxic chemical agents and noninfectious diseases continues to evolve. There has never been a better time than the present for this expanded focus due to growing populations and population densities worldwide. The ongoing and projected population growth and density has corresponding measurable expansion of urbanization, industrialization, agriculture, and transportation. With these expanding sectors are increased manufacturing and use of myriad chemical products and fuels and associated increased generation and intentional and unintentional release of solid, liquid, and gaseous toxic chemical waste contaminants into air and water and onto soil. Contamination with toxic chemical agents, and more intensely, pollution of these major environmental media, in turn, results in higher contact and exposure hazard potentials for humans, animals, and plants and, in turn, higher risks of environmentally related noninfectious diseases. Note that "pollution" occurs when the levels or concentrations of the "contaminants" accumulate and exceed the environmental capacity to dilute, decompose/degrade, detoxify, or disinfect within an acceptable time and/or without anthropogenic intervention.

The One Health approach, which involves multidisciplinary teams addressing issues at the interface of human health, animal health, plant health, and environmental health, is extremely applicable to and beneficial for less siloed integrated preventive and mitigative activities focused on both infectious microbiological agents as well as toxic chemical agents.

The profession and practice of environmental health science addresses a broad spectrum of hazardous agents generated from and detected in all environmental media (i.e., air, water, soil, food) and both indoor and outdoor environmental settings and has a major role as part of the integrated, multidisciplinary teams needed to effectively engage in activities using a One Health approach. More specifically, environmental health scientists and practitioners address sources of hazardous agents, contaminated environmental media, and pathways for human exposures, and the related diseases. More specifically, this specialization of "environmental science" is involved in sampling, analyzing, and characterizing air, water, soil, food, and other media for sources, types, levels, and fate of microbial and chemical contaminants in various environmental settings; however, the practice expands to include determining:

- Pathways of hazardous agent exposures to humans
- Modes and routes of exposure
- Types of hazards, levels of exposures, and overall exposure hazard potential
- Related risks for associated health effects to humans

Although the profession of environmental health science focuses on pathways for and modes of exposure and the adverse impacts to humans, the practice can be applied to the broader spectrum of the One Health approach and include the interfaces between and among humans, animals, plants, and the environment (Figs. 8.1 and 8.2). Exposure to hazardous agents, including infectious microorganisms, toxic chemicals, as well as physical agents including nonionizing radiation, ionizing radiation, and noise, is best framed within the context of both direct and indirect exposures. These categories of exposure are presented later in this chapter.

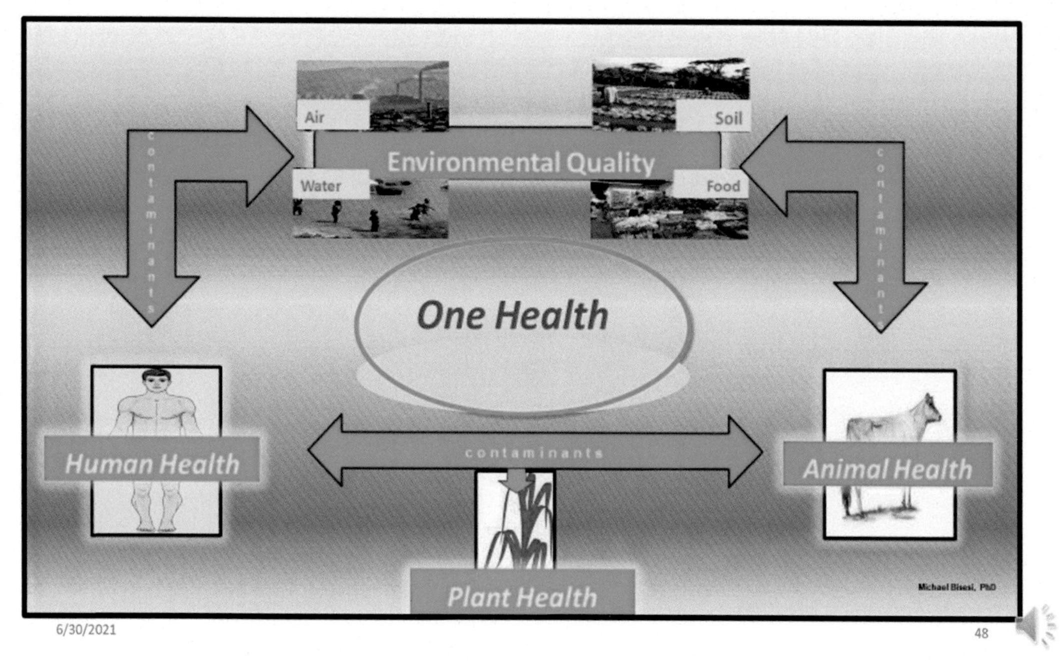

FIGURE 8.1 Expanded One Health paradigm showing the interface between and among the environment, humans, animals, and plants.

Literature review and gap analysis

Some factors associated with the global environmental burden of disease

According to reported WHO data, 24% of all diseases worldwide are attributable to the environment.[5] There are a variety of factors that directly or indirectly influence the environmental burden of disease.[6,7] Historically, two foundational contributing factors are population size and related population density. Both population size and population density have increased in most countries with the highest observed growth presently occurring in the lower- and middle-income countries. According to a United Nations report, population growth is projected to continue to increase despite an overall global decline in fertilization rate.[8]

In relation to the expansive and continued growth of the population and related population density worldwide, there has been a growing number of population-dense urban or metropolitan areas. Nonetheless, rural areas with lower population densities than urban areas remain both sources and receptors of environmental contaminants as well.[9]

Population growth and related increase in population density have shown corresponding increased demand for products and services and increased use and consumption of natural and anthropogenic resources. Accordingly, coinciding with increased population growth has been the increase in anthropogenic sources of environmental contaminants. Major fixed area or stationary sources of environmental contaminants include agriculture, manufacturing, and power-producing industries, plus residential areas. Moving or mobile

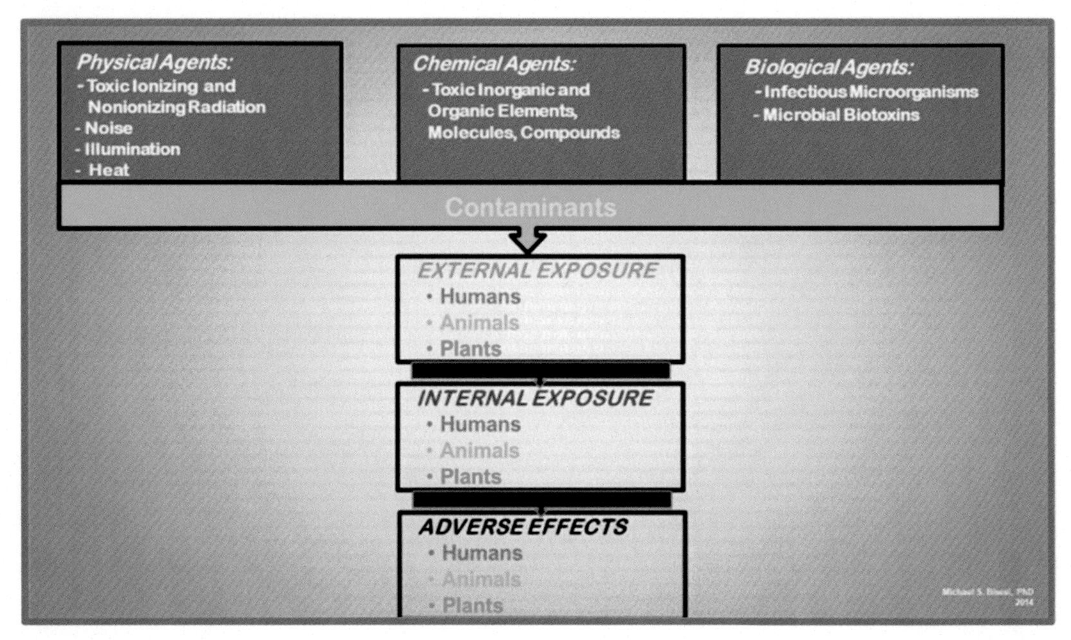

FIGURE 8.2 Categories of human, animal, and/or plant exposures to hazardous physical, chemical, and biological agents.

sources of contaminants and pollutants include combustion engine vehicles (i.e., cars, motorbikes, trucks, buses), airplanes, trains, boats, and ships. Both stationary and mobile categories of sources substantially contribute to increased types and levels of toxic chemical contaminants via controlled and uncontrolled releases into the environment.[10]

The increase of toxic chemical contaminants in most countries worldwide is a result of increased synthesis of anthropogenic chemical substances for a variety of purposes. This is aligned, for example, with increased industrialization and transportation worldwide as two major factors that have contributed to the increased need to synthesize and produce chemicals as raw materials and ingredients for manufacturing products, but also for use as refined fossil fuels (e.g., diesel, gasoline) for various types of land, air, and water transportation sources. Demands for and use of fossil fuels such as coal, oil, and natural gas also increased to produce electricity and provide fuels for stationary sources such as residential, commercial, industrial, and other built environment settings. In addition, manufacturing of myriad pharmaceutical and commercial products, such as medications, paints, cleaners, plastics, and other materials, is a dominant output from manufacturing. Industrialization and transportation were also correlated with increases in population growth and resulting needs and wants as well as population density associated with expanded urbanization.[11] High-income countries experienced the initial and most rapid and robust growth and expansion of industrialization and transportation. During recent decades to the present, lower- and middle-income countries are observing the expanded population growth, industrialization, and transportation as well. Indeed, this corresponds to higher economic growth and gross domestic product (GDP) as indicators.[12]

Major sources of toxic chemical contaminants released into the environment

Beyond some natural sources of toxic chemical contaminants released into the environment, there are numerous, if not more, anthropogenic sectors and contributors involved as sources as well. Major anthropogenic sources, but certainly not all, include the agriculture sector, manufacturing and energy-producing industrial sectors, transportation sector, and the residential including general consumer sector.

There has been a dramatic increase of the agriculture sector to address the ongoing increased needs and demands for food security associated with increased population growth of humans.[13] Accordingly, there has been a concomitant increase in population growth and population density of agricultural livestock, especially cattle, swine, and poultry, among other food animals essential to provide the growing human protein demand. Various vegetable, grain, and fruit crops are grown and processed in increasing amounts too, as food for humans and to raise livestock. Management of crops and animal livestock often includes increased use of various types of toxic chemicals including growth hormones, antibiotics, surface biocides, pesticides, herbicides, and fertilizers. There is also increased generation of vegetative green wastes and animal manures.[14]

Industrial production continues in the higher- and middle-income countries and is expanding rapidly in many lower- and middle-income countries to address and meet need and demand for various products, energy, and services. Although there are reports with evidence that toxic chemical pollution was occurring prior to the original industrial revolution,[15] there was an increase associated with the expansive growth of industrialization worldwide during the 1800s.[16] Increased industrialization results in increased use of chemical raw materials and ingredients for manufacturing and power production as well as various services which, in turn, results in increased generation of process by-products and wastes generated into air, water, and soil media.

Transportation accounts for all mobile sources of toxic chemical contaminants released into the environment. The magnitude of transportation expansion and increase worldwide has been described as an observed revolution similar to what was observed with the Industrial Revolution.[17] Based on the data for the period 1975 to 2005, the expansion of transportation and associated energy use has increased the most in high-income countries and continues to increase as well in low-income countries.[18] Air is the medium most affected by the contaminants, in the form of exhaust emissions, released from transportation sources using gasoline- or diesel-fueled combustion engines.

Residential and consumer sources also contribute substantially to the release of toxic chemicals into the environment. Combustion sources such as open fires and cook stoves, as well as contemporary stoves and furnaces, generate toxic chemical emissions as gases and particulate matter. These toxic chemical air contaminants affect humans and animals both indoors and outdoors.[19] Use of consumer, commercial, and personal care products also continue to increase.

Waste generation

Corresponding with the historic and ongoing increases and expansions of industrialization, transportation, and agriculture are the increased energy and other natural and anthropogenic resource needs and demands, use of vast land areas, as well as generation of wastes discharged into air and water and onto land. Wastes are referred to as used materials or by-products from materials or processes that are intentionally or unintentionally released (emitted; discharged; discarded) typically without concern for return or fate. The levels of waste management and associated regulations vary between and among high-, low-, and middle-income countries. It is the waste products, including the hazardous infectious microbial agents and toxic chemical agents, that contaminate the various environmental media of air, water, and soil. Contaminated and polluted air, water, and soil contribute to contamination of food. The contaminated environmental media, in turn, are the major sources for exposures to humans, animals, and even plants that correspond with adverse impacts in the form of infectious and noninfectious diseases.

Wastes are generated in many forms and include urinary and fecal matter (biological waste), food matter ("garbage"; biological waste), paper-, plastic-, glass-, and metal-based matter ("trash"; physical wastes), and solvents, metals, residual cleaners, biocides, paints, pharmaceutical matter, etc. (chemical wastes). In relation, wastes are generated in many ways including release of gaseous and particulate chemical emissions from stationary and mobile sources into the air, some of which settle down to soil and surface waters, as well as, direct dumping into surface waters and onto soils.

Major categories and types of toxic chemicals

Whereas there are various types, categories, characteristics, and sources of hazardous infectious microorganisms (and microbial toxins), there are also various toxic chemicals of both natural and, more so, anthropogenic origin. Toxic chemicals, and in fact chemical agents in general, are divided into inorganic chemicals and organic chemicals. Each category includes subcategories of specific toxic chemicals.

Inorganic chemicals include toxic metals and various salts and other compounds. These inorganics are used as raw materials and catalysts for manufacturing other chemicals and chemical-based products and are also associated with waste discharges and emissions. Asbestos and metals are naturally occurring compounds and elements, respectively. Asbestos is not used in manufacturing much anymore, anywhere; however, asbestos-containing materials remain prevalent in many building materials and structures. Toxic metals, such as cadmium and chromium, are present in various materials including various pigments used for colorants and paints and dyes. Many metals, originating from various natural and anthropogenic sources, exhibit adverse toxic effects to humans following excessive exposure.[20] Metals have also been shown to adversely affect exposed animals too and can be incorporated into the food chain.[21] Plants are a major component of the food chain and are also susceptible to toxic effects from metals as well.[22]

Organic, carbon-containing, chemicals are also prevalent everywhere. Most of all consumer goods, manufacturing chemicals, and fossil fuels are organic. Overall, common categories of toxic organic chemicals include solvents, paints, plastics, insecticides, rodenticides, fungicides, herbicides, most pharmaceuticals, and cleaners and biocidal disinfectants. Organic solvents are foundational ingredients in many chemical mixtures and commercial products. Many if not most organic solvents are volatile and evaporate to form gaseous vapors which contaminate and diffuse through the air medium. Environmental exposures are associated with numerous adverse impacts to humans.[23]

Solvents and chemical residues can also accumulate in animals, including livestock, and cause adverse health effects.[24] Chemical insecticides, fungicides, rodenticides, and herbicides are most frequently organic chemicals and are used heavily for commercial and agricultural applications, as well as for use in other settings including residential. In addition to posing disease risks to both humans and animals, these and other organic chemicals can be assimilated by plants via root uptake and cause adverse effects and bioaccumulation.[25] This is a concern relative to chemical pharmaceuticals for humans and animals where by-products from production and use end up in waste streams contaminating the environment.[26]

Antimicrobial resistance and the direct and indirect adverse impacts relative to treating human, animal, and plant infectious diseases are most commonly attributable to excessive use and misuse of organic antibiotic pharmaceuticals. There are reports of pharmaceutical chemical antibiotics being detected in wastewater and animal manures, and the soils where applied, and entering the food chain.[27] However, in addition to pharmaceutical antibiotic drugs, antimicrobial resistance is also associated with excessive use and misuse of disinfectant and sanitizing chemicals in numerous commercial and consumer products for their biocidal properties.[28]

The myriad pesticides and related biocidal chemicals are also known to be used extensively and excessively to control unwanted infestations and growths. Humans, animals (aquatic and terrestrial), and plants are susceptible to exposure and related adverse impacts associated with these toxic chemicals.[29] In relation, exposure to heavy metals in the food animal systems has been shown to be strongly associated with increase in antimicrobial resistance.[30] Similar results have also been shown between use of biocidal disinfectants in human hospitals and environment with that of increase in multidrug resistance among bacteria.[31]

Both inorganic and organic chemicals can move through various environmental media either unchanged or altered forms. For example, inorganic chemicals such as metals in elemental and salt forms, as well as certain categories and types of organic chemicals, are relatively resistant to rapid degradation. Accordingly, many chemicals can persist in the environment due to the recalcitrance to decay. Many toxic organic chemicals, including polycyclic aromatic hydrocarbons, polychlorinated (PCBs) and polybrominated biphenyls, dioxins, and per- and polyfluoroalkyl substances (PFASs), among others, are referred to persistent organic pollutants and have been detected in various environmental matrices.[32] Since these chemicals are recalcitrant to biotransformation and decay, many of these same chemicals are associated with bioaccumulation and biomagnification within the food chain. For example, recalcitrant PFAS chemical compounds have been detected in water, soil, plants, animals, and humans and PCBs.[33,34]

Major media and fate of toxic chemicals

Toxic chemical contamination and pollution of the environment is ubiquitous worldwide as a result of the ongoing prolific use of chemical raw materials, fuels, and products and the coinciding intentional and unintentional release of the corresponding chemical wastes into the environment. The three major, but not the only, environmental media adversely affected are air, water, and soil. Each environmental medium is susceptible to direct contamination by toxic chemicals and in turn may serve as a pathway to human, animal, and plant exposure to toxic chemicals. Animal- and plant-based food is an additional medium that is susceptible to environmental contamination but typically becomes contaminated as a result of contact with contaminated air, water, and/or soil.

There are characteristics and capacities of our environment and its various complementary ecosystems. Environment consists of various dynamic ecosystems. It is essential to have biodiverse ecosystems with sustained equilibrium for high level of environmental quality and healthy environments. In relation, there are essential environmental factors for propagation and sustainment of life within ecosystems. Unfortunately, release of various types of contaminants and pollutants into the environment is associated with alterations of factors that disrupt the equilibrium of an ecosystem, including the intra- and interecosystem food chains and dependences.

There are related consequences that coincide with population growth/population density, including increased utilization of natural and anthropogenic resources (e.g., food, water, energy, manufactured products, etc.), as well as resource depletion; expansive growth of agriculture and industrialization to meet needs, wants, and demand; plus expansive growth of modes and numbers of transportation vehicles.

The fate of toxic chemicals is associated with and influenced by many factors. Fundamentally, the physical state and physical properties of the released chemicals can influence the fate in various environmental media and settings. Toxic chemicals are nonliving elements, molecules, and molecules as compounds that are in the physical state or form of solids, liquids, and gases. In addition, toxic chemicals also have physical characteristics or properties, based on the chemical composition and characteristics, that influence whether they are volatile and will evaporate from liquid and/or solid state to a gaseous form, polar and soluble in water, or nonpolar and insoluble in water but fat soluble instead.

Toxic chemicals released into the environment may also undergo changes from the original or parent chemical structure to new biotransformed molecules. The biotransformed "metabolites" will have modified chemical composition and structure which in turn influences whether they have altered physical properties and/or toxicity. For example, the solvent benzene in its parent form is nonpolar, therefore lipophilic and more fat soluble. As a result, benzene can more readily be absorbed across cell membranes and cause systemic effects in humans and animals. Biotransformation can alter the chemistry of benzene and change its solubility to more water-soluble forms. In humans, benzene is a procarcinogenic chemical meaning it first requires biotransformation to active carcinogenic metabolites.[35] Benzene and other toxic chemicals can undergo biotransformation in environmental media, such as water and soil, via microbial catalytic activity.[36] Indeed, benzene is one of countless examples of biotransformation of organic parent compounds into various altered forms as products. Summaries of examples of the fate of discharged toxic chemicals relative to physical properties and physical characteristics are shown in Tables 8.1 and 8.2.

The air medium is especially susceptible to contamination with various types of toxic chemicals as a result of emissions from both stationary and mobile sources. Polydisperse particulate matter consisting of both nonrespirable and respirable fractions is a major component

TABLE 8.1 Physical states of toxic chemicals and the generic fate when released into the environment.

Physical state	Released into environment
Solids	• Discharged/deposited on and into soil • Discharged/deposited on and into surface waters and leach into ground waters • Discharged into air (dusts, fumes, smoke) • Deposited on surfaces
Liquids	• Discharged/deposited on and into soil • Discharged/deposited on and into surface waters and leaches into ground waters • Discharged into air (mists, sprays)
Gases (including vapors)	• Deposited on surfaces • Diffuse into and through soil • Diffuse and dissolve into surface waters and ground waters • Diffuse and discharged into air

TABLE 8.2 Physical properties of toxic chemicals and the generic fate when released into the environment.

Physical property	Released into environment
Volatile (high vapor pressure)	• Evaporate from solid or liquid into a gaseous state referred to as airborne vapors
Hydrophilic (polar)	• Soluble in water
Lipophilic (nonpolar)	• Insoluble in water

of air pollutants. Inhaled particulate matter is associated with various adverse health effects in humans as well as animals.[37,38] In addition, air pollution has been shown to adversely affect plants, including reducing agriculture crop production.[39]

Air pollution occurs when the air contains gaseous or particulate contaminants in harmful amounts, relative to the health or comfort of humans and animals or which could cause damage to plants and materials. The substances that cause air pollution are called air contaminants and air pollutants. Pollutants that are pumped into our atmosphere and directly pollute the air are called primary pollutants. Primary pollutant examples include carbon monoxide from combustion engine vehicle exhaust and sulfur dioxide from the combustion of coal. Further pollution can arise if primary pollutants in the atmosphere undergo chemical reactions to form compounds called secondary pollutants. Photochemical smog and acid rain are examples of these.

Whereas air contaminants are associated with both direct and indirect exposures to humans, animals, and plants, select air contaminants also contribute to climate change, including the greenhouse effect and global warming.[40] Indeed, the WHO has identified climate change and air pollution as top environmental threats.[41]

Both surface and ground water are susceptible depots for toxic chemical contamination too.[42] Irrigation water from sources such as wastewater treatment process may be contaminated with toxic chemicals which in turn contaminate plants, soil, and underlying ground water when used to irrigate crops.[43] There are other reports demonstrating environmental distribution and biological uptake of trace toxic chemicals associated with pharmaceutical and personal care products.[44]

In general, soil is a depot for contamination, including the persistent chemicals that are more recalcitrant to biotransformation and decomposition. Soil can be contaminated directly due to discharging and/or dumping wastes onto the soil surface or injecting into the soil matrix. However, deposition of settled air contaminants in the form of particulate matter can contribute to the contamination of soil as well as contamination of surface bodies of water. In addition, contaminated soil can move as runoff into surface bodies of water too as another mode of indirect contamination. Accordingly, it is clear that the initial contamination of one environmental medium can contribute to subsequent contamination of the other environmental media too.

Major modes, routes, and types of exposure

Exposure begins with a contaminated environmental medium, such as air, water, soil, or food. Depending on whether direct or indirect contact with the contaminated medium occurs determines if and how exposure occurs. The mode of exposure refers to how initial contact (exposure) occurs, and each mode is associated with a distinct route or portal of contaminant contact and entry into human and animal systems (Table 8.3). For example, if outdoor or indoor air is contaminated with a toxic chemical there is the likely possibility that exposure will occur as a result of inhalation of the contaminant-containing air. This is how humans and animals can be exposed. Plants can be exposed to air contaminants too via transpiration. Bioavailability of toxic chemicals is related to the ability of humans, animals, and plants to uptake various toxic elements, molecules, or compounds via absorption into their tissues and systems following initial contact with a contaminated source.

Ingestion is another major mode of exposure. This is most commonly associated with human and animal consumption of contaminated food and/or water. Dermal contact is another common mode of exposure that may result in manifestation of effects locally at the point of contact or following absorption systemically.

TABLE 8.3 Modes and routes of toxic chemical contaminant contact/entry for humans and animals.

Modes ("how")	Routes/Portals ("where")
Inhalation	Respiratory system
Ingestion	Gastrointestinal system
Contact/Injection	Integumentary system (dermal)

Direct vs. indirect exposure to toxic chemicals

Direct and indirect contact and resulting exposures to toxic chemicals may contribute to the overall environmental burden of diseases. Exposures to toxic chemicals based on several factors, including the levels of exposure and duration, can individually and/or collectively adversely affect humans, animals, and plants.

Direct exposure involves the initial contact with a primary source or contaminated matrix. For example, if a human or animal is present in an outdoor or indoor area where the air is contaminated with toxic chemical vapor, inhalation of the toxic vapor–containing air results in direct exposure to the toxic chemical in a gaseous state. Plants may have similar types of contact and exposures with contaminated air, water, and/or soil. Examples of other media are shown in Fig. 8.3.

Indirect exposure involves an initial direct exposure to a plant or animal receptor, for example, and subsequent contact and exposure to the plant or animal, for example, a human ingesting chemically contaminated plants. There are numerous examples of this indirect exposure resulting from bioaccumulation of chemical toxicants and passage through the food chain to humans. For example, reclaimed wastewater contaminated with pharmaceutical compounds was used to irrigate produce. In turn, the soil and produce were contaminated, and the human consumers showed biomarkers indicating contamination as well.[45] Examples of other media are shown in Fig. 8.4.

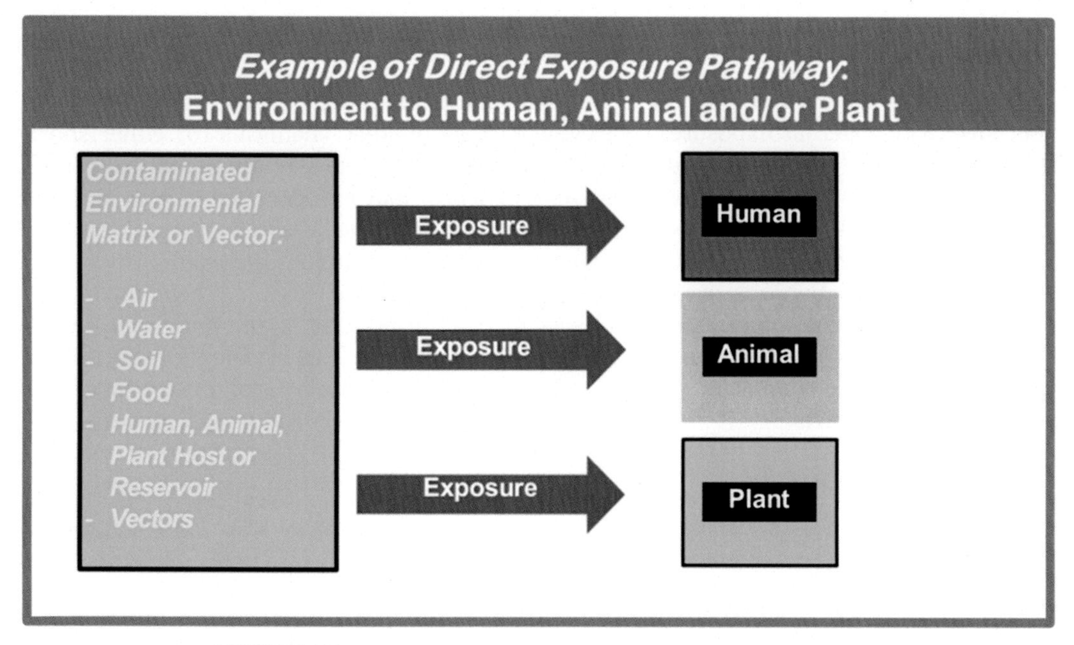

FIGURE 8.3 Direct contact and exposure to a contaminated matrix.

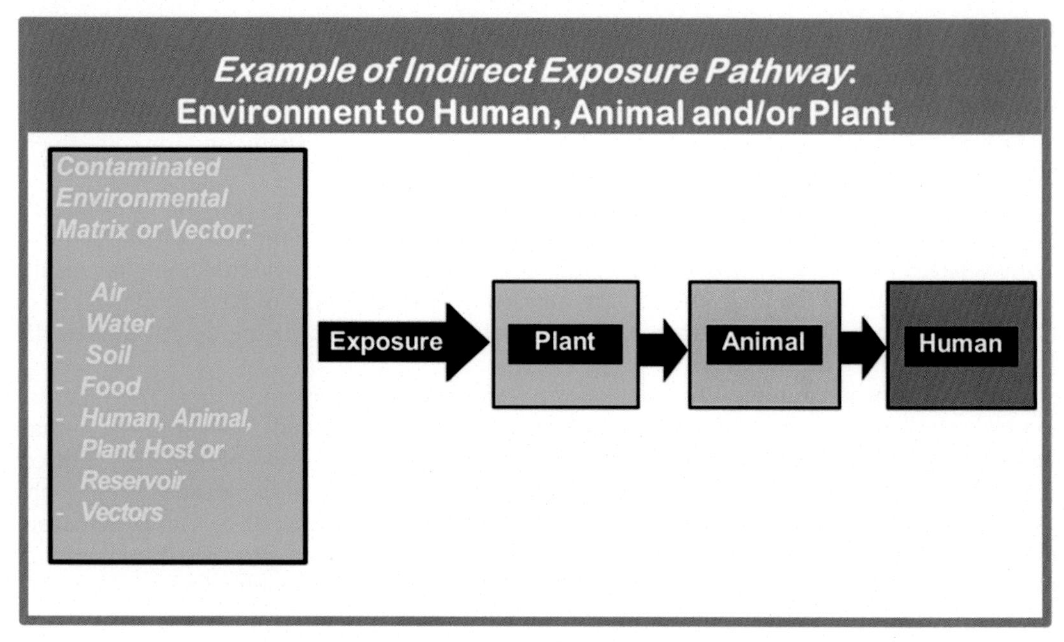

FIGURE 8.4 Indirect contact and exposure to a contaminated matrix.

Categories of noninfectious diseases associated with toxic chemical exposure

There are numerous acute and chronic noninfectious diseases affecting humans that are associated with external exposures to elevated levels of toxic chemicals for either short-, but typically, long-term durations. There are numerous factors environmental (e.g., type, level, duration) as well as biological (e.g., age, genetics) that influence the risk of initiating the diseases associated with exposures to toxic chemicals. The diseases are categorized in various ways but typically based on human organ systems including the respiratory system ("pneumotoxicity"), nervous system ("neurotoxicity"), hepatic system ("hepatotoxicity"), hematic system ("hematotoxicity"), renal system ("nephrotoxicity"), and endocrine system ("endocrine toxicity"; "endocrine disruption") as examples. Another categorization of the disease-causing toxic chemicals is designation of effects as toxic effects, mutagenic effects, and carcinogenic effects.

Recommendations to fill gaps

Most of the approaches to address infectious microbial agents and toxic chemical agents are fragmented or siloed. For example, environmental health and other applicable scientists or practitioners who specialize in infectious agents may only characterize air, water, soil, and/or food for microbial agents of concern. Others specializing in toxic chemical agents may ignore infectious microbial agents and only characterize air, water, soil, or food for chemical agents of concern. In reality, there are many examples when a source is associated with both categories of hazardous agents. Accordingly, characterization of environmental media as well as potential corresponding exposures to humans, animals, and plants should involve more multidisciplinary planning for and implementation of integrated sampling, analysis, and interpretation. There is a similar need as well for scientists, clinicians, and other practitioners who focus on the adverse impacts separately to humans, animals, or plants associated with environmental exposures to expand communications and interactions.

A major gap from a more general global health perspective is the low level of integrative, multidisciplinary approaches to characterize environmental media for contamination with various hazardous biological and chemical agents. As stated, historically to present, most emphasis is on infectious microbes. The encouraging news is that the increased awareness of the need to allocate resources and efforts to prevent and mitigate environmental contamination with toxic chemicals has never been higher. Indeed, the United Nations has recognized toxic chemicals as a global threat to humans, animals, and ecosystems in general.[46]

An overall consensus is that environmental contamination and related pollution control is attainable but will require the combination of science and policy. This will require international cooperation. For example, as a good step, the USEPA works cooperatively with the UN Environmental Protection (UNEP) program, as stated in a Memorandum of Understanding,[47] focusing on the "One Planet" initiatives. It is framed as a deliberate One Health approach with stated goals and surrounding initiatives clearly aligned with a focus beyond human health alone. Having said this, the gap can be narrowed more if there is formal adoption of an operational integrative One Health approach explicitly incorporated into the programs and implemented relative to science and policy.

The One Health approach is a major answer to promote the more integrated and therefore less siloed preventive and mitigative activities; however, adoption and implementation remain a work in progress. Indeed, there are numerous reports of both challenges[48] and impactful successes[49] associated with the One Health approach.

CASE STUDY

Toxic chemical mercury in the food chain.[50]

Background

- An industrial manufacturing facility located on a major river in the state of Maine operated a chloralkali process for approximately 33 years.
- The chloralkali process manufactures the common industrial materials chlorine and sodium hydroxide as the primary products.
- There are variations of the chloralkali process. It was reported that the mercury cell process was used and was once one of the common past approaches as part of the chlorine and sodium hydroxide manufacturing.

Observations

- Waste brine sludge containing the inorganic toxic metal mercury was discharged directly from the sewers from the industrial manufacturing facility into the river for years.
- The inorganic metal mercury–containing wastes were considered a major source of mercury contamination of the river.

Key issues

- Following discharge into the environment, in this case, surface water, the inorganic form of the metal mercury can

Continued

CASE STUDY *(cont'd)*

be biotransformed via specific bacteria to form organic methylmercury which is a more bioavailable form of the metal. This organic form of mercury, methylmercury, is more bioavailable since it has a physical property of being relatively nonpolar and lipophilic. Accordingly, it can more readily and passively cross cell membranes of animals and humans.

- The U.S. Environmental Protection Agency and the Maine Department of Environmental Protection sampling and analytical data revealed that elevated levels of inorganic mercury and organic methylmercury were detected and measured downstream of the industrial manufacturing facility, considered a major source. The toxic metal was found both in the water and the bottom sediment.
- In addition, analysis of tissues from various downstream aquatic organisms confirmed that the bioavailable methylmercury was being absorbed. There was corresponding evidence that the methylmercury was being bioaccumulated and biomagnified moving up the aquatic food chain.
- Higher level animals and humans consuming contaminated fish also absorb the methylmercury. Various health effects may result including neurotoxicity, renal toxicity, and reproductive toxicity. Wildlife studies suggested disruptions to reproduction of bird and wildlife in the area.

Interventions

- Regulatory and legal actions to intervene.
- A risk assessment based on toxic effects observed elsewhere indicated that there needed to be limits to human consumption of fish, shellfish, and eels from the river due to the biomagnified levels of the highly absorbable and very toxic methylmercury in their tissues.
- Pregnant women were told not to consume any fish, shellfish, and eels due to higher risk to their developing embryo or fetus.

Lessons learned

- Proactive controls and elimination of indiscriminate waste discharge of toxic chemicals onto soil and/or into air and water, including the metal mercury presented in this case, reduce the level of environmental contamination and, in turn, reduce the exposure hazard potential for humans (plus animals and plants) and the associated disease risks.

Recommendations

- Continue to learn from the past manufacturing and the associated indiscriminate waste management practices.
- Increase the specificity of applicable federal, state, and local regulations, and increase the number of regulatory "consultants" with One Health education and training.
- Enhance and expand surveillance activities that apply the multidisciplinary and integrated One Health approach. For example, relative to this case and myriad others that are very similar, an ideal approach would engage and involve integrated teams consisting of human medical practitioners, animal medical practitioners, plant health practitioners, and environmental health practitioners.

CASE STUDY (cont'd)

Well-planned and coordinated integrated sample, specimen, and other data collection with corresponding team-based data interpretation would add both increased efficiency to the surveillance activities as well as effectiveness of proactive and reactive decision-making and action.

Vision for the future

A foundational vision for the future is elimination of the siloed approach to prevention and mitigation of environmental hazards and related contamination and adverse impacts. Inherent to this is the need for increased worldwide awareness, adoption, and implementation of a well-coordinated, professionally multidisciplinary, operationally integrated One Health approach with the corresponding capacity of the applicable people, places, and things. Environmental health scientists and practitioners are the essential nonmedical (i.e., human; animal) members of a successful multidisciplinary One Health team. Adopting and incorporating more environmental health practice into the One Health approach will contribute to enhanced collaborative and systematic activities that address both infectious microbiological agents and toxic chemical agents, and increased prevention and mitigation of the respective diseases caused.

Actions

The applicable governmental and nongovernmental agencies, professional organizations, and academic institutions within and between countries worldwide must begin establishing and adopting the guiding principles to convene and catalyze the changes necessary, followed by the development and implementation of relevant policies, procedures, and practices for robust global One Health competency and practice. This requires expanded capacity building of the major clinical and scientific professions prepared to become part of the One Health workforce.

Inherent to this need is ongoing expanded focus with increased emphasis on including toxic chemical agents and noninfectious diseases as part of the expanding paradigm that goes beyond, but is in addition to, infectious microbial agents and infectious diseases. Indeed, the frameworks for global health, including Global Health Security (GHS) and International Health Regulations (IHR), among others, should more explicitly incorporate the integrated multidisciplinary One Health approach plus the broader focus and related initiatives associated with prevention and mitigation of exposures to and diseases associated with both infectious microbiological agents and toxic chemical agents.

References

1. National Environmental Health Association Workgroup. New Perspectives on Environmental Health: The Approval of New Definitions. *NEHA J Env Health*. 2013:72–73. October.
2. World Health Organization. *What is One Health?*; 2017. https://www.who.int/news-room/q-a-detail/one-health.
3. GHSA. *Global Health Security Agenda 2024 Framework*; 2018. https://ghsagenda.org/wp-content/uploads/2020/06/ghsa2024-framework.pdf.
4. World Health Organization (WHO. *International Health Regulations*. 3rd ed.; 2008. https://www.who.int/publications/i/item/9789241580496".
5. WHO. *Preventing Disease through Healthy Environments: A Global Assessment of the Burden of Disease Environmental Risks*; 2016. https://www.who.int/news/item/14-03-2016-preventing-disease-through-healthy-environments-a-global-assessment-of-the-burden-of-disease-from-environmental-risks.
6. Pimentel D, Cooperstein S, Randell H, et al. Ecology of increasing diseases: population growth and environmental degradation. *Hum Ecol: An Interdiscip J*. 2007;35(6):653–668. https://doi.org/10.1007/s10745-007-9128-3.
7. Prüss-Ustün A, Wolf J, Corvalán C, Bos R, Neira M. *Preventing Disease through Healthy Environments: A Global Assessment of the Burden of Disease from Environmental Risks*. 2016.
8. United Nations. *World Population Prospectus*; 2019. ST/ESA/SER.A/423 https://population.un.org/wpp/Publications/Files/WPP2019_Highlights.pdf.
9. Rosner M, Ritchie H, Ortiz-Ospina E. *World Population Growth*; 2013. https://ourworldindata.org/world-population-growth.
10. Fayiga AO, Ipinmoroti MO, Chirenje T. Environmental pollution in Africa. *Environ Dev Sustain*. 2018;20(1):41–73. A Multidisciplinary Approach to the Theory and Practice of Sustainable Development.
11. Sinha KC. Sustainability and urban public transportation. *J Transport Eng*. 2003;129(4):331–341.
12. Ndiaya N, Lv K. Role of industrialization on economic growth: the experience of Senegal (1960-2017). *Am J Ind Bus Manag*. 2018;8:2072–2085. https://doi.org/10.4236/ajibm.2018.810137.
13. World Bank. *Ending Poverty and Hunger by 2030: An Agenda for the Global Food System*. 2nd ed. 2015 (Washington, DC).
14. Novotny V. Diffuse pollution from agriculture - a worldwide outlook. *Water Sci Technol*. 1999;39(3):1–13.
15. Uglietti C, Gabrielli P, Cooke C, Vallelonga P, Thompson L. Widespread pollution of the South American atmosphere predates the industrial revolution by 240 y. *Proc Natl Acad Sci U S A*. 2015;112(8):2349–2354.
16. Haupert M. A brief history of cliometrics and the evolving view of the industrial revolution. *Eur J Hist Econ Thought*. 2019;26(4):738–774.
17. Connors J, Gwartney JD, Montesinos HM. Transportation and communication revolution: 50 years of dramatic change in economic development. *Cato J*. 2020;40(1):153–198.
18. Poumanyvong P, Kaneko S, Dhakal S. Impacts of urbanization on national transport and road energy use: evidence from low, middle, and high income countries. *Energy Pol*. 2012;46:268–277.
19. Lai A, Deng ASM, Carter E, Yang X, Baumgartner J, Schauer J. Differences in chemical composition of PM2.5 emissions from traditional versus advanced combustion (semi-gasifier) solid fuel stoves. *Chemosphere*. 2019;233:852–861.
20. Tchounwou PB, Yedjou CG, Patlolla AK, Sutton DJ. Heavy metal toxicity and the environment. *Experientia Suppl*. 2012;101:133–164. https://doi.org/10.1007/978-3-7643-8340-4_6.
21. Okereafor U, Makhatha M, Mekuto L, Uche-Okereafor N, Sebola T, Mavumengwana V. Toxic metal implications on agricultural soils, plants, animals, aquatic life and human health. *Int J Environ Res Publ Health*. 2020;17(7):2204. https://doi.org/10.3390/ijerph17072204.
22. Pichhode M, Nikhil K. Effect of heavy metals on plants: an overview. *Int J Appl Innov Eng Manage*. 2016;5(3):56–66.
23. Joshi DR, Adhikari N. An overview of common organic solvents and their toxicity. *J Pharmaceut Res*. 2019;28(3):1–18.
24. Shull LR, Cheeke PR. Effects of synthetic and natural toxicants on livestock. *J Anim Sci*. 1983;57(suppl_2):330–354.
25. Collins C, Fryer M, Grosso A. Plant uptake of non-ionic organic chemicals. *Environ Sci Technol*. 2006;40:45–52.
26. Gaffney V, Almeida C, Rodrigues A, Ferreira E, Benoliel MJ, Cardoso VV. *Occurrence of Pharmaceuticals in a Water Supply System and Related Human Health Risk Assessment*. Vol 72. 2015:199–208.

27. Pan M, Chu LM. Transfer of antibiotics from wastewater of animal manure to soil and edible crops. *Environ Pollut*. 2017;231:829–836.
28. Gnanadhas DP, Marathe SA, Chakravortty D. Biocides – resistance, cross-resistance mechanisms and assessment. *Expet Opin Invest Drugs*. 2013;22(2):191–206.
29. Lushchak VI, Matviishyn TM, Husak VV, Storey JM, Storey KB. Pesticide toxicity: a mechanistic approach. *EXCLI J*. 2018;17:1101–1136.
30. Medardus JJ, Molla BZ, Nicol M, et al. In-feed use of heavy metal micronutrientds in U.S. swine production systems and its role in persistence of multi-drug resistant salmonellae. *Appl Environ Microbiol*. 2014;80(7):2317–2325.
31. Srinivasan VB, Rajamohan G, Gebreyes WA. Role of AbeS, a novel efflux pump of the SMR family of transporters, in resistance to antimicrobial agents in*Acinetobacter baumannii*. *Antimicrob Agents Chemother*. 2009;53(12):5312–5316.
32. Avino P, Russo MV. A comprehensive review of analytical methods for determining persistent organic pollutants in air, soil, water, and waste. *Curr Org Chem*. 2018;22(10):939–953.
33. Nakayama SF, Yoshikane M, Onoda Y, et al. Worldwide trends in tracing ply- and perfluoroalkyl substances (PFAs) in the environment. *Trends Anal Chem*. 2019;121:1–20.
34. Kassa H, Bisesi M. Levels of polychlorinated biphenyls (PCBs) in fish: the influence on local decision-making about fish consumption. *J Environ Health*. 2001;63(8):29–35.
35. Yardley-Jones A, Anderson D, Parke DV. The toxicity of benzene and its metabolism and molecular pathology in human risk assessment. *Br J Ind Med*. July 1991;48(7):437–444.
36. Agency for Toxic Substances and Disease Registry. *Toxicologic Profile for Benzene*. U.S. Department of Health and Human Services; 2007.
37. Hime NJ, Marks GB, Cowie CT. A comparison of health effects of ambient particulate matter from five emission sources. *Int J Environ Res Publ Health*. 2018;15(6):1206.
38. Catarina L, Perillo A. Particulate matter air pollution and respiratory impact on humans and animals. *Environ Sci Pollut Res*. 2018;25:33901–33910.
39. Wang Z, Wei W. Effects of modifying industrial plant configuration on reducing air pollution-induced agricultural loss. *J Clean Prod*. 2020;277, 124046.
40. Ramanathan V, Feng Y. Air pollution, greenhouse gases and climate change: global and regional perspectives. *Atmos Environ*. 2009;43(1):37–50.
41. Campbell-Lendrum D, Prüss-Ustün A. Climate change, air pollution and noncommunicable diseases. *Bull World Health Organ*. 2019;97(2):160–161.
42. Ritter L, Solomon K, Sibley P, et al. Sources, pathways, and relative risks of contaminants in surface water and groundwater. *J Toxicol Environ Health*. 2002;65(1):1–142.
43. Elgallal M, Fletcher L, Evans B. Assessment of potential risks associated with chemicals in wastewater used for irrigation in arid and semiarid zones: a review. *Agric Water Manag*. 2016;177:419–431.
44. Colon B, Toor GS. A review of uptake and translocation of pharmaceuticals and personal care products by food crops irrigated with treated wastewater (ch. 3). In: Sparks DL, ed. *Advances in Agronomy*. Vol 140. Academic Press; 2016:75–100.
45. Paltiel O, Fedorova F, Tadmor G, Kleinstern G, Maor Y, Chefetz B. Human exposure to wastewater-derived pharmaceuticals in fresh produce: a randomized controlled trial focusing on carbamazepine. *Environ Sci Technol*. 2016;50(8):4476–4482.
46. United Nations Environment Programme (UNEP. *Harmful Substances and Hazardous Wastes*. UNEP; 2010. https://www.unep.org/resources/report/harmful-substances-and-hazardous-waste-united-nations-environment-programme.
47. USEPA. *EPA's Role in the United Nations Environment Programme (UNEP) Memorandum of Understanding*; 2021. https://www.epa.gov/system/files/documents/2021-09/unep-epa-mou-sept-8-2021_finaltext.pdf.
48. Johnson I, Hansen A. The challenges of implementing an integrated One Health surveillance system in Australia. *Zoonoses Public Health*. 2018;65:e229–e236.
49. Rostal MK, Ross N, Machalaba C, Cordel C, Paweska JT, Karesh WB. Benefits of a one health approach: an example using Rift Valley Fever. *One Health*. 2018;5:34–36.
50. *Maine People's Alliance and Natural Resources Defense Council, Plaintiffs, Appellees,v. Mallinckrodt, Inc., Defendant, Appellant. No. 05-233*. United States Court of Appeals, First Circuit; December 22, 2006. https://scholar.google.com/scholar_case?case=5082666576500876781&hl=en&as_sdt=6,36.

Further reading

1. WHO. https://www.who.int/quantifying_ehimpacts/publications/preventing-disease/en/.
2. United States Environmental protection Agency. https://www.epa.gov/pfas/basic-information-pfas.

CHAPTER

9

Global climate change impacts on vector ecology and vector-borne diseases

Rafael F.C. Vieira[1,2], *Sebastián Muñoz-Leal*[3], *Grace Faulkner*[2], *Tatiana Sulesco*[4], *Marcos R. André*[5] *and Risa Pesapane*[6,7]

[1]Department of Public Health Sciences, The University of North Carolina at Charlotte, Charlotte, NC, United States; [2]Center for Computational Intelligence to Predict Health and Environmental Risks (CIPHER), The University of North Carolina at Charlotte, Charlotte, NC, United States; [3]Departamento de Ciencia Animal, Facultad de Ciencias Veterinarias, Universidad de Concepción, Chillán, Chile; [4]Laboratory of Entomology, Institute of Zoology, Chisinau, Moldova; [5]Laboratório de Imunoparasitologia, Departamento de Patologia, Reprodução e Saúde Única, Faculdade de Ciências Agrárias e Veterinárias, Universidade Estadual Paulista (FCAV/UNESP), Jaboticabal, Brazil; [6]Department of Veterinary Preventive Medicine, College of Veterinary Medicine, The Ohio State University, Columbus, OH, United States; [7]School of Environment and Natural Resources, College of Food Agricultural and Environmental Science, The Ohio State University, Columbus, OH, United States

Introduction

Arthropod vectors and vector-borne diseases (VBDs) remain a major concern among the global array of diseases that affect humans, animals, and plants, particularly in low-income countries.[1] The recent emergence and reemergence of Zika virus, dengue virus, and chikungunya virus[2–4] in novel ecosystems due to globalization is a stark reminder of the interdependence between our social and ecological systems. Despite concerted efforts to control VBDs, over 80% of the global population lives in areas at risk from at least one VBD, largely due to anthropogenic factors such as climate change, globalization, and deforestation.[5] The rise in

Modernizing Global Health Security to Prevent, Detect, and Respond
https://doi.org/10.1016/B978-0-323-90945-7.00026-9

global temperatures associated with climate change may significantly alter the suitable habitat for some arthropod vectors, potentially expanding the geographic regions at risk for VBDs.[6]

The rapid geographic expansion of ticks in North America accompanied by an exponential increase in the incidence and distribution of tick-borne diseases is just one example of the impacts of climate change. The global trade of animals, wood, and tires coupled with increased global travel may also facilitate the spread of exotic vectors. Specifically, the trade of used tires has contributed to the global expansion of *Aedes albopictus* Skuse, 1895, in Europe,[7] the Americas, Africa, and Australia[8] and the trade of livestock is likely responsible for the spread of *Haemaphysalis longicornis* Neumann, 1901, from East Asia to Australia, New Zealand, the Pacific Islands, and the United States.[9,10]

International travelers, and their pets, visiting areas where VBDs are endemic are at risk of contracting and spreading vector-borne pathogens when returning to their home countries or providing rapid air transportation for hitch-hiking vectors.[9,11] Habitat fragmentation as a result of deforestation is the primary cause of the decline of wild vertebrates,[12] alters the composition of ecological communities, shifts the distribution of reservoir hosts, and can increase the geographic overlap of hosts and vectors, all of which may influence infectious disease dynamics. These drivers, in conjunction with the lack of multisectoral collaboration among professionals in entomology, acarology, public health, and veterinary medicine at regional, national, and global scales, substantially increase the odds of VBD outbreaks.[13]

The World Health Organization (WHO) and the Pan-American Health Organization (PAHO) both emphasize that sustained resources dedicated to public health surveillance (PHS) and control of arthropod vectors and VBDs are imperative for reducing the burden of VBD in the long term.[5] Maintaining these efforts is the most critical, and simultaneously the most challenging, in the midst of crisis. The protracted conflict has been associated with an increase in communicable diseases due to poor water quality, precarious sanitation, hygiene and health infrastructure conditions, and overcrowding, mainly in the less developed regions of the world.[14,15]

Currently, the COVID-19 pandemic has presented a different type of crisis. Surveillance and control of vectors, as well as VBDs, eroded as public health organizations were forced to redirect a majority of their attention, workforce, and resources to quell the pandemic.[16] Consequently, vectors and VBDs flourished.[17,18] Restrictions on human movement during the COVID-19 pandemic affected routine vector control measures, resulting in mosquito population growth in residential areas, ultimately leading to an increased risk of VBDs.[19]

For example, some Latin American and Caribbean countries struggled to control an unprecedented dengue epidemic in 2020.[20] The COVID-19 crisis also altered human behavior and human–animal–environmental relationships, increasing the exposure to vectors.[21] Increased outdoor recreation facilitates contact between vectors, humans, and companion animals raising the risk of VBDs both directly through increased exposure and indirectly through an accumulation of municipal waste discarded during recreational activities that provide artificial breeding habitat, resulting in a greater abundance of vectors.[22] Ownership of companion animals also reportedly increased in many countries.[23] Although human–animal interactions may improve social relationships and well-being,[23] companion animals may also increase exposure to vectors and VBDs through shared human–animal environments or activities.[24] Together, these examples along with the aforementioned anthropogenic-induced factors driving an increased risk in VBD globally highlight the importance of implementation of a One Health approach against VBDs.

Literature review and gap analysis

Vector-borne diseases are parasitic, viral, or bacterial illnesses transmitted to humans often through the infectious bites of insect and/or arthropod vectors, such as mosquitoes, ticks, flies, fleas, and lice. The WHO estimates that VBD accounts for more than 17% of all infectious disease and cause over 700,000 deaths globally every year.[5]

Malaria is considered to be one of the most life-threatening VBD, with most deaths occurring in children <5. Vectored by *Anopheles* mosquitoes, this febrile illness is caused by *Plasmodium* parasites and disproportionately affects African countries, which accounted for 95% of all cases and 96% of all deaths in 2020. This disease is endemic in 85 countries worldwide.[25]

Another VBD of interest, which is considered a priority neglected tropical disease is Leishmaniasis. This disease, caused by 20 different species of protozoan *Leishmania* and vectored by 90 different *Phlebotomine* sandfly species, is estimated to have one million infections annually. This disease is endemic in 98 countries, mainly affecting Africa, Asia, and Latin America. Often associated with poverty and displacement, infections can be observed in the more common cutaneous form, which is characterized by skin lesions and sores, and the more severe mucosal-cutaneous form, which causes damage to the nose, mouth, and throat, and the visceral form, characterized by the enlargement of the spleen and liver and can be fatal if left untreated.[26]

Environmental impacts

Changes in climate and land use patterns are major drivers of global environmental change resulting in biodiversity loss, altered microclimates, and shifts in the burden and distribution of infectious diseases.[27–29] Climate change is predicted to disturb seasonal and long-term trends in rainfall and temperature, ultimately increasing climatic variability and the frequency and severity of extreme weather events.[30] The risk of VBD is influenced by multiple environmental conditions including the direct effects of temperature and rainfall on vector mortality rates, rates of development, host-seeking activity, available breeding habitat, and the duration of the extrinsic incubation period of vector-borne pathogens.[31,32]

Indirectly, changes in climate may affect vector abundance and vector–host interactions as a function of habitat quality, host availability, and the composition of ecological communities of hosts for vector blood meals.[32] The following primary consequences of climate warming in temperate regions can be highlighted:

1. increased risk from endemic VBDs due to long-term changes in temperature and rainfall patterns;
2. geographic expansion of VBDs and arthropod vectors into novel regions as a result of local climatic suitability;
3. introduction and establishment of exotic tropical or subtropical vectors and VBDs in nonendemic countries as a byproduct of climate-mediated human migration and globalized trade.[33]

For example, the increasing density of the tick *Ixodes ricinus* Linnaeus 1758 and its expansion to the northern regions of Europe, linked with increased human cases of tick-borne encephalitis in Sweden and Lyme borreliosis in northern Europe, are correlated with the effects of climate change on growing seasons, host populations, and tick development, survival and seasonal activity.[34] Moreover, abnormally high temperatures in summer have been linked with West Nile virus (WNV) outbreaks in south-eastern Europe and western Asia in 2010,[35] and western and northern Europe in 2020.[36,37] The increase in human cases of WNV in Italy in 2018 has been linked to higher temperatures in early spring, which accelerated virus amplification in the avian and mosquito populations.[38]

According to the US Centers for Disease Prevention and Control (CDC), since 1999, WNV became the most common virus transmitted to humans by mosquitoes in the United States[39] and it keeps on moving throughout Europe.[37] Climate change affects not only the presence, abundance, and species diversity of arthropod vectors, but also the reservoir hosts (animals and humans) which maintain the pathogen in nature and may transport them to unaffected areas. Recent studies demonstrate that changes in temperature cause a phenological shift in bird migrations.[40]

Predictions of future VBD threats both over- and underestimate the effects of climate change. Underestimations of the effect of climate change result from a narrow focus on the proximal impacts of climate on the ecology of infectious diseases, while omitting the distal effects on society's capacity to control and prevent VBDs.[27,41] Overestimations fail to account for the full range of vector physiology by considering only thermal minima and not thermal maxima. Regions that warm beyond arthropod thermal optima may actually experience decreased transmission.[42] Robust public health policies and programs must recognize this interplay of vectors, hosts, and humans in a dynamic environment to effectively reduce the impact of VBDs.

Globalization, human activity, and zoonotic spillover

The global spread of exotic invasive mosquitoes *Ae. albopictus* and *Ae. aegypti* Linnaeus 1762 due to globalization and increasing international trade and travel[8] is another example of the change in VBDs. Subsequently, *Ae. albopictus* became an important vector of a tropical disease caused by the chikungunya virus in Europe with several outbreaks in Italy and France.[2,43] This mosquito vector was also involved in the local circulation of dengue virus[3] and Zika virus[44] in Europe.

Dengue virus has experienced a rapid expansion into new areas and countries worldwide and a dramatic increase in disease incidence over the past. The WHO estimates that there are as many as 50 million new cases every year. However, this number could be as high as 100–200 million infections annually.[45,46] Zika, which became a worldwide VBD of concern in the 2015 outbreak in Brazil, quickly spread into Central America, the Caribbean, and the United States and was declared a Public Health Emergency of International Concern due to the increase in neurological disorders and neonatal malformations.[47]

Deforestation, as an example of human activity, is linked to an increase in the abundance and spatial distribution of vectors, changes in vector-feeding behavior, and altered epidemiology of VBDs.[48] For example, deforestation in the Amazon rainforest increased the prevalence of malaria by modifying the breeding behavior of malaria vector *Anopheles darlingi* Root 1926, as a result of the changes in breeding site availability, microclimate, vegetation, hydrology, and increased contact with humans.[49] The WHO has estimated 229 million cases of malaria with over 400,000 deaths worldwide at the end of 2019. Although this is an overall decrease in disease incidence, malaria prevention progress has plateaued since 2015.[25] About 60 other examples of changes in malaria vector ecology and malaria incidence associated with deforestation and agricultural development have been reported.[50] Recent outbreaks of leishmaniasis in Panama, Central America, have been linked with the deforestation of high-humidity forest ecosystems that previously restricted the geographical distribution of sand flies.[48] Another example is the taiga tick *Ixodes persulcatus* Schulze 1930, an effective vector of Lyme borreliosis spirochetes and tick-borne encephalitis virus in the Palaearctic region. Deforestation of the taiga forest led to the increase of *I. persulcatus* population density as a result of fragmentation and mosaic character of the landscape which is more favorable for the development of hard ticks.[51]

Urbanization is associated with an increase in VBDs due to increased densities of vectors and other pests, causing a greater burden to vector and pest control programs. In the past, tropical diseases were primarily associated with rural poverty. However, urbanization, as a result of population growth and human rural-to-urban migration is closely related to the "ruralization" of urban areas, such as raising livestock and patterns of water use and storage. The United Nations estimates that 68% of the global population will live in urban areas by 2050, with the largest moves projected in Africa and Asia.[52] Consequently, the same diseases have now become more closely associated with densely packed urban areas.[53]

Further, some vectors are highly suited to urban areas and prefer to live around human dwellings. *Ae. aegypti*, which vectors yellow fever, dengue, chikungunya, and Zika virus, is a domesticated, invasive species that has seen expansion around the globe. Following World War II, *Ae. aegypti* has been implicated in the spread of the dengue virus in Southeast Asia and the Americas due to high vector-suitability in these urban settings.[54,55]

Many arboviruses, such as WNV and tick-borne encephalitis viruses, infect people accidently via direct spillover from enzootic cycles, when a vector first bites a viremic enzootic host (wildlife) and then transmits disease during a subsequent feeding on a human. Viruses like Japanese encephalitis and Rift Valley are able to undergo secondary amplification in domestic animals to increase levels of circulation and human spillover infections. Some arboviruses, such as Zika virus, dengue virus, and yellow fever virus can undergo direct human amplification and vertical transmission, thereby bypassing enzootic hosts, having the potential to infect more people and to spread more rapidly and widely via infected travelers.[56]

Yellow fever, endemic in parts of Africa and South America, caused 30,000 deaths in 2018.[57] Despite vaccination coverage, Yellow fever experienced its largest outbreak in the past 20 years in Angola, Africa, in 2015. This led to the importation of disease from infected migrant workers to China in March of 2016.[58] In addition to evidenced travel-related importation of the disease, current vaccine shortages have raised future concerns of disease introduction into naïve human populations, specifically to southern Asia.

Moreover, the possibility of vector-adaptive mutations by arboviruses facilitates virus replication, a genetic adaptation that contributes to global epidemics. For instance, the chikungunya virus is mainly transmitted by *Ae. aegypti* mosquitoes in tropical and subtropical regions. The chikungunya virus adaptation to *Ae. albopictus* through genetic mutation, facilitated its transmission in the temperate climate of Europe, which lacks the typical vector *Ae. aegypti* but *Ae. albopictus* is established and distributed in the region.[59] Several local outbreaks of human chikungunya infection in Europe occurred in Italy[43,60] and France[9,61] sustained by *Ae. albopictus.* According to the European Centre for Disease Prevention and Control (ECDC), approximately 800 suspected autochthonous human cases were reported in Italy.[62]

Diagnostics

Historically, epidemiologic studies have used serological assays, but the quality of antigens and frequency of serological cross-reactions hampered an accurate identification of the VBD pathogen involved. Recombinant DNA technologies and proteomic analyses of antigens have enabled the selection of molecules that improve the specificity and sensitivity of ELISA-based serological tests.[63]

Additionally, these superior antigens can be used in immunochromatographic rapid tests, which are more cost-effective and thus can be used to increase capacity in low-resource settings. Polymerase chain reaction (PCR) followed by sequencing and phylogenetic analyses has offered novel opportunities for investigating the diversity of vector-borne pathogens. However, along with great advances in the identification of specific pathogens, these molecular techniques brought new challenges. The choice of the target gene and its phylogenetic signal has a huge influence on the accurate identification of a certain etiological agent, continuing the need for vector-borne pathogens cell-culture isolation for newly discovered organisms.[64,65]

With more affordable prices, Whole Genome Sequencing platforms have become more popular, allowing rapid and deep descriptions of vector-borne pathogens at a molecular level in concert with phylogenomic assessment.[66,67] Real-time quantitative PCR (qPCR) and digital droplet (dd) PCR assays allow the detection of vector-borne pathogens in a multiplex platform, with high sensitivity and specificity.[68,69] The use of loop-mediated amplification of DNA (LAMP) may be a good alternative for molecular diagnosis of VBDs in low-income countries due to the isothermal-based DNA amplification and the development of the results by colorimetric or turbidity methods.[69,70] These techniques enable highly sensitive detection of VBDs without the need for expensive equipment and, in several cases, without the need for DNA extraction procedures.

Technological advances have similarly aided the identification of arthropod vectors themselves, which historically relied on morphological characteristics requiring the use of entomological keys, an understanding of the taxonomic literature and revisions over time, and specialized training in entomology. PCR-based methods followed by the phylogeny of nuclear and mitochondrial genes,[71,72] as well as methods such as matrix-assisted laser desorption/ionization coupled with time-of-flight mass spectrometry (MALDI-TOF) have gained widespread popularity because they enable more accurate and accessible identification of ticks,[73] mosquitoes,[74] fleas,[75] sand flies,[76] and lice.[77] Likewise, the molecular detection of vector-borne pathogens in arthropod vectors by PCR-based methods has taken advantage of proteomic[78] and microbiome approaches,[79] improving our understanding of the complexity of vector–pathogen relationships in a changing world. If we are to understand the true identities of both vectors and pathogens, an integrative approach combining morphological and molecular data is to be implemented.

Capacity building and public health surveillance

Local capacity building and technical and scientific support to prepare future qualified entomologists and acarologists, particularly in low-income countries, with interdisciplinary expertise that are able to minimize VBDs in a sustainable manner by assisting with the prevention, diagnosis, and control VBDs and arthropod vectors is critically lacking.[80] For instance, most public health organizations, even those with active mosquito surveillance, have inadequate training or capacity to conduct tick surveillance and identification. The limited access of entomologists and acarologists to research facilities and expertise of reference laboratories for more effective PHS of arthropod vectors and the pathogens they transmit (e.g., molecular screening of field-collected vectors for pathogens), associated with the lack of multisectoral collaboration among professionals in entomology, acarology, public health, molecular methods, veterinary medicine, ecology, and zoonotic infectious disease at regional, national, and global scales, represents an important gap in our capacity to reduce VBDs.

For example, *Borrelia* spirochetes responsible for tick-borne relapsing fever (TBRF) in South America have been neglected partly because the symptoms of TBRF overlap with Chagas disease, malaria, and other arboviruses (see Case study). Additionally, databases of the distribution of vectors and VBDs that are maintained by a network of vector professionals (such as the VectorNet project used in European countries) are currently lacking in South American, Asian, and African countries.[81] The availability of such databases is an essential tool for stakeholders. For example, the VectorNet project, which was established in 2014, has been used in European countries to track and share data on the distribution of arthropod vectors of both medical and veterinary pathogens.[81]

In general, many vector-borne pathogens are extremely under surveilled. Ticks, fleas, lice, and mites are typically prioritized below mosquitoes by vector control programs, due to the real or perceived expectations of lower public health burden and often limited resources. As a result, many VBD outbreaks may be unreported, and the distributions and burdens of these diseases are likely underestimated, undermapped, or entirely unknown.[82] For example, the louse-borne disease epidemic typhus (*Rickettsia prowazekii* (da Rocha-Lima 1916)), relapsing fever (*Borrelia recurrentis* (Lebert 1874)), and trench fever (*Bartonella quintana* (Schmincke 1917)) still affect populations particularly living in poor-hygiene conditions, such as homeless individuals, prisons, and refugee camps, where body lice infestations are prevalent. However, routine surveillance of these populations and the implementation of efficient delousing strategies, which would prevent major epidemics of known and/or other potentially emerging louse-borne pathogens, is lacking.[83]

Regarding flea-borne diseases, while murine typhus caused by *Rickettsia typhi* (Wolbach and Todd, 1920) is most prevalent in coastal areas (especially port cities), where rats are the primary mammalian host and rat fleas are the vector,[84] there is an increasing incidence and geographic distribution of the disease in the United States[85] in a cycle of transmission involving opossums and cat fleas (*Ctenocephalides felis* (Bouche, 1835)).[84] This highlights the need for further investigations, particularly in areas where these peridomestic animals cohabitate. Cat fleas are also involved in the transmission of flea-borne spotted fever, caused by *Rickettsia felis*.

Both flea-borne diseases are undifferentiated febrile illnesses that often go unrecognized by physicians.[86] Similarly, mite-borne scrub typhus (*Orientia tsutsugamushi* (Hayashi 1920)), which is transmitted by several species of Trombiculid mites, is quickly emerging as one of the most important yet neglected tropical rickettsioses.[87] Despite over one million cases of scrub typhus reported annually, and up to 10% of those being fatal, very little is known about the enzootic cycle, variations in epidemiology and pathogenesis, and drivers of the emergence of scrub typhus in new regions or increasing incidence in endemic regions.[87]

Lastly, the separation of human health from animal health is a construct of human social and political systems and support for PHS and control of vectors or VBDs is historically episodic, essentially as a reaction in response to a new threat. There are several global disease PHS systems for either human or animal disease, such as the WHO Global Outbreak and Response Network or the Food and Agriculture Organization of the United Nations Emergency Prevention System for Transboundary Animal and Plant Pests and diseases, which have identified zoonotic disease outbreaks.[88] However, these PHS systems have not prevented transmission of disease from animals to humans, and often human cases are detected first. There have been efforts to integrate these systems, such as the US CDC establishment of the National Center for Zoonoses, Vector-borne, and Enteric Diseases and the implementation of ArboNet, a US national PHS system for arboviruses.[89] Recent efforts to develop guidelines specific to vectors and VBD include the WHO Global Vector Control Response 2017–30 to help strengthen vector control worldwide.[5]

Among the priority activities for 2017–22 was to strengthen and integrate national vector PHS systems with health information systems.[5] However, these programs are subjected to fluctuations in funding. For example, in 2011, the United States proposed a budget cut of $26.7 million to the CDC vector-borne disease program, which could have threatened mosquito surveillance for WNV and other VBDs. It is believed that the lack of funding would have caused disease surveillance to be reactive rather than a preventative health service.[90]

CASE STUDY

Tick-borne relapsing fever in tropical South America

Background

Tick-borne relapsing fever (TBRF) spirochetes (*Borrelia*) occur in sylvatic transmission cycles in tropical and subtropical areas of the world infecting wild vertebrates and mainly soft ticks of genus *Ornithodoros*.[101] Humans bitten (Fig. 9.1A) by infected ticks can present typical recurrent febrile episodes, accompanied by nausea, vomiting, preterm labor, and miscarriage.[101] In tropical areas of the globe, and particularly in South America, undifferentiated febrile syndrome prevails,[102] but the etiologies remain unknown. Currently, relapsing fever spirochetes are neglected as a

(Continued)

CASE STUDY *(cont'd)*

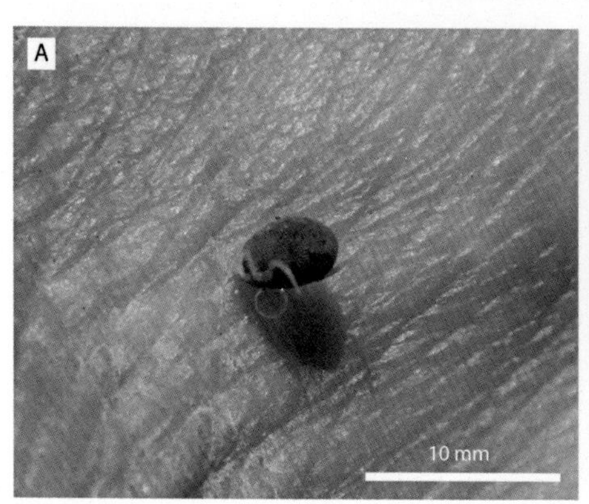

FIGURE 9.1 Soft ticks in which tick-borne relapsing fever spirochetes have been detected in Brazil. (A). *Ornithodoros rietcorreai* feeding on human, Pernambuco State, northeastern Brazil. (B). *Ornithodoros rudis* in between debris of a bird nest, Maranhão State, northeastern Brazil.

causative agent, most likely because the symptoms of TBRF overlap with Chagas disease, malaria, and arboviruses. Thus, the diagnosis is often masked by etiologies with similar manifestations.[103] As the knowledge on TBRF is deficient in many regions where *Ornithodoros* spp. do parasitize humans, the infection with spirochetes might be overlooked.

Observations

South American TBRF is endemic in Colombia and Venezuela. However, human cases were reported only during the first half of the 20th century. The vector of South American TBRF, *Ornithodoros rudis* Karsch, 1880 (Fig. 9.1B), was rediscovered in northeastern Brazil in 2017, and the etiological agent *Borrelia venezuelensis* (Brumpt 1921) was isolated from those ticks.[104] Moreover, at least four soft tick species that often bite humans in Brazil harbor TBRF borreliae of unknown pathogenic roles.[105]

Key issues

- TBRF occurs in South American tropical and subtropical ecosystems in soft ticks that carry the spirochetes, can invade human dwellings,[105] and bite humans.
- TBRF has yet to be recognized in patients, even though in Brazil the risk for humans to contract the disease has been demonstrated.
- To understand the impact on human health, surveillance for TBRF in South America should be mandatory.
- Isolating the etiological agents is imperative to accomplish successful surveillance.
- Serological assays using antigens of local *Borrelia* strains and research proposals to detect where these pathogens emerge.

Lessons learned

- TBRFs are neglected tick-borne diseases globally.[101]
- Medical studies performed in Africa have recurrently demonstrated that TBRF overlaps clinical manifestations with other febrile illnesses.[103]

CASE STUDY *(cont'd)*

- Human case reports are imminent in the region as the African epidemiological scenario regarding the disease mirrors the situation of many rural areas in tropical South America.

Recommendations

- There is a need to educate physicians in South America so they should include the disease as a differential diagnosis regarding undifferentiated febrile syndrome.
- There is a need for studies to identify wild reservoirs of the spirochetes to understand the eco-epidemiology of the disease in the region.
- Public health sector and tick researchers must work together, not only in South America, but also in other regions of the world where the disease prevails.

Vision for the future

Given the inherent interdependence of VBD dynamics and human activities exemplified by the illustrations above, the future of VBD management must involve a fully integrated, adaptive approach to human, animal, and environmental health. If we are to improve our capacity to predict, detect, and respond to future risks of VBD, it is imperative that the public health sector includes the expertise of vector ecologists, biologists, and taxonomists to understand how rapidly changing environmental conditions might shift the distribution, biting habits, abundance, and even the evolution of vectors.

To illustrate, the continued efforts by the WHO and member states in the global campaign dedicated to the control and elimination of malaria estimated 1.7 billion cases and 10.6 million deaths have been averted from 2000 to 2020 due to rapid diagnostic testing, antimalarial therapeutics and vaccines, insecticide-treated nets, indoor residual spraying and vector control, elimination, and research activities.[25] Policy-makers at the national level should also enact legislation that will help minimize vectors and thus VBDs. For instance, recent commitments to environmental protection and restoration could reduce the risks of VBD. At the UN Climate Change Conference, COP26, the Glasgow Climate Pact had 137 countries committed to halting and reversing deforestation by 2030 in the "Glasgow Leader's Declaration on Forest's and Land Use." Twenty-eight countries pledged to protect forests through sustainable development and agricultural trade.[91]

An up-to-date understanding of the distribution of vectors relies on the accurate identification of arthropods and a thorough investigation of the host travel history of humans and animals. Pathologists, clinicians, and even parasitologists often fail to accurately identify arthropods,[92,93] which may hamper differential diagnosis and treatment of VBD. Institutionalizing entomological and acarological expertise within public health is therefore a critical component of effective VBD management, and needs to be complemented by taxonomists working with both a morphological and molecular approach to understand species concepts.

Lastly, the separation of human health from animal health is a construct of human social and political systems.

Vectors may utilize a wide variety of animal hosts, including humans, and these preferences may vary by life stage or ecological community, making surveillance of all animals important. Public health surveillance for novel or expanding vectors and VBDs in nonhuman animals[94] could increase our capacity to detect human health risks at local levels when they first emerge or act as harbingers of spread to new geographic regions, providing precious time to proactively prepare measures for prevention and response. Vector surveillance systems need to reflect this interdependence through integrated databases of information on vectors and VBDs reported from humans, companion animals, livestock, wildlife, and active environmental surveillance via dragging, flagging, and trapping.

This ecosystem approach to the PHS of vectors and VBDs must be supported by sustained investment and capacity-building efforts over generational timescales. Support for PHS and control of vectors or VBDs is historically episodic, essentially as a reaction in response to a new threat.[25,90,95] If we are to be proactive, there must be sufficient, sustained funding of public health to maintain and build capacity in vector surveillance programs in the long term. For example, even the WHO World Report on Malaria, which has experienced continued international funding, had an estimated $3.3 billion USD investment in 2020, despite the estimated $6.8 billion USD needed to reach the Global Technical Strategy for Malaria 2016–30 milestones.[25] Similarly, developed countries fell short in finance goals for climate change.

The Organization for Economic Cooperation and Development estimated that in 2019 $79.6 billion was raised for climate finance, which was about $20 billion short of the promised funds.[96] Thus, the COP26 reaffirmed the commitment of developed countries to mobilize at least $100 billion per year in climate finance to achieve their primary goals: global net zero emissions by the mid-century and adaptions to protect communities natural habitats.[91] This could see rippling effects on VDB due to the intricate relationship between vectors with temperature, the environment, and human VBD risk. A workforce of vector personnel and their associated resources are necessary to bridge sectors, advance knowledge, train future professionals, and mobilize in response to emerging threats.

Capacity building could be accelerated by the creation of international infrastructure projects that improve access to resources and vector-borne communities of practice, especially in resource-limited regions. Resources may include personalized training, standardized protocols, extracts, or strains at little to no cost to the end user. Similarly, centralized repositories for voucher specimens, high-quality images, and molecular data that are fully digitized in a searchable database would allow for comparative studies and facilitate broad geographic modeling of VBD risk. Such models are not only valuable to understanding the current risk, but also for identifying areas at high risk for spillover of vector-borne pathogens or the introduction of exotic vectors. Mapping of current and predicted risk areas can enable targeted PHS efforts and prioritized distribution of resources to areas with the most concerning risk-to-resource ratio.

Resources for PHS are limited and often only available in regions where surveillance is justified based on the known distribution of vectors meaning surveillance for invading vectors is neglected. Predictive tools would allow for the targeted surveillance of areas with a high risk of invasion. Similarly, in the United States, vector PHS and control are conducted at the county level, and resources associated with these activities are distributed accordingly, but some counties contain substantially more habitat for vectors equating to a greater burden for control. Fine-scale vector distribution models would enable quantification of suitable habitat within counties so that resources could be prioritized to those with greater burden. It is highly recommended to collect, connect, correlate, and analyze epidemiological data together with entomological and big environmental data by using the analytical strength of advanced predictive tools and data-driven models to implement the proper VBDs surveillance actions and predict the short and long term entomological and epidemiological risks.

PHS must not be myopic in its focus on only the most abundant VBDs, rather it should be inclusive of the totality of VBD threats, including those that are rare or challenging so that we have a clear understanding of what is endemic or emerging at local scales. As the incidence of historically understudied VBDs increases, like Powassan virus or TBRF, we have a limited understanding of the ecology and epidemiology of these vector-borne pathogens to effectively manage risk, or perhaps how changes in ecology or epidemiology might be driving new risks.

Long-term comprehensive PHS programs can provide baseline information that allows us to quickly adapt to shifting threats and identify broad trends across taxonomic groups of vectors and VBDs. Malaria has been a long-surveilled VBD with annual WHO world reports. These reports document global trends in cases, the status of elimination programs, investment in eradication and research programs, vector research and distribution, and the distribution of malaria prevention, diagnosis, and treatment systems, all of which inform their development goals to reduce the burden of disease.[25]

Just as ecosystems do not conform to geopolitical boundaries, PHS and control of vectors or VBDs must bridge jurisdictional boundaries at local, regional, national, and international scales. Transboundary agreements are needed to coordinate PHS along trade and travel routes that might facilitate the spread of vectors or hosts infected with vector-borne pathogens. The recent outbreaks of Ebola[97] and COVID-19[98] illustrate how uncoordinated approaches across jurisdictions can impede efforts to limit the spread of infectious diseases in a highly connected world. Wherever possible, approaches to vector surveillance and management should be consistent within distinct ecological communities irrespective of political boundaries.

Actions

The following actions are recommended as initial steps for catalyzing this envisioned approach to the management of VBD:

1. Develop transdisciplinary communities of practice for VBDs that include entomologists, acarologists, ecologists, human (including public health) and animal health professionals, data scientists, microbiologists, and mathematicians across academic,

research, and regulatory sectors to enhance research, identify and address gaps in knowledge, and optimize policy approaches to the surveillance and control. National and local governments should direct sustained funding to support these actions.

   a. Provide resources and time allowances for interagency VBD working groups of members with diverse expertise.
   b. Provide collaborative transdisciplinary VBD grant funding mechanisms.

2. Establish or enhance national and international programs with dedicated funding for strategic capacity-building activities and training relating to vectors and VBDs from local to global scales.
   a. All national governments of IHR Member States should update their national preparedness plans with appropriate and relevant skills, logistics, and funding available to cope with future health crises related to VBD.
   b. Developed countries to support COP26 in the mobilization at least $100 billion in climate finance per year, in which some funds can be used to address VBD
3. Create sustainable repositories of data and samples that include information on vector distributions, pathogen detection, and cases of human and animal VBDs along with banked voucher specimens in a fully digital, searchable format.
   a. Expansion of VectorNet into other continents/countries—WHO should create a vector and VBD data management system that is open access, and offers secure technology transfer.
   b. Leverage opportunities for inclusion of civilian vector data collected digitally through mobile applications like iNaturalist, or physically through passive public vector submission programs.
4. Support the development of international infrastructure projects that improve access to research facilities and reference laboratory expertise for the effective PHS of arthropod vectors and vector-borne pathogens (e.g., Infravec2 project).
   a. Establish stronger regional capacities for manufacturing, distribution, and usage of VBD diagnostics, therapeutics, and other essential supplies.
5. Establish and institutionalize formal pathways for multisectoral cooperation and coordination across agencies or ministries at national and international scales. This would be possible through international and national calls for this purpose.
6. Develop international cooperative agreements that aim to improve the surveillance and prevention of VBDs in humans and animals engaged in international travel or trade across geopolitical boundaries, supported by ministries at national and international scales.
   a. Inclusion of VBD in differential diagnosis list for physicians serving transient populations (migrant workers, refugee camps, immigrants)—most VBD are absent in refugee camp health assessments including those used by United States, EU, and WHO.
7. Estimate geographic transmission risk maps and the potential for spillover events through data on wildlife reservoirs and enzootic transmission patterns at the global scale.
   a. Inclusion of VBD modeling during policy assessment, creation, and implementation.
8. Support disease surveillance of rare events under current surveillance practices, or when tick vectors are challenging to sample, such as soft ticks in the genus

Ornithodoros[99] and other questing ticks. National and local capacity building, technical and scientific support to prepare future qualified entomologists and acarologists.

9. Develop standard practices for collecting and reporting on vectors or VBD, specifically for neglected taxa, and the publication of raw materials under a uniform set of criteria (e.g., data harmonization).[100]

10. Integrate climate policies and actions into public health policy and planning, merge expertise on existing climate adaptation policies inside and outside the health sector (e.g., construction, housing, transportation, agriculture, energy sectors), and recognize the relevant control/mitigation practices, and quantify their importance for the well-being and health of citizens/workers, animals, and the environment.

References

1. Thomson MC, Stanberry LR. Climate change and vectorborne diseases. *N Engl J Med*. 2022;387(21):1969–1978. https://doi.org/10.1056/NEJMra2200092.
2. Delisle E, Rousseau C, Broche B, et al. Chikungunya outbreak in Montpellier, France, September to October 2014. *Euro Surveill*. 2015;20(17). https://doi.org/10.2807/1560-7917.es2015.20.17.21108. pii=21108.
3. Succo T, Leparc-Goffart I, Ferré JB, et al. Autochthonous dengue outbreak in Nimes, south of France, July to September 2015. *Euro Surveill*. 2016;21(21). https://doi.org/10.2807/1560-7917.ES.2016.21.21.30240. pii=30240.
4. Sikka V, Chattu VK, Popli RK, et al. The emergence of Zika virus as a global health security threat: a review and a consensus statement of the INDUSEM joint working group (JWG). *J Global Infect Dis*. 2016;8:3–15. https://doi.org/10.4103/0974-777X.176140.
5. *Global Vector Control Response 2017–2030*. Geneva: World Health Organization; 2017. Licence: CC BY-NC-SA 3.0 IGO. ISBN: 9789241512978.
6. Petersen LR, Beard CB, Visser SN. Combatting the increasing threat of vector-borne disease in the United States with a national vector-borne disease prevention and control system. *Am J Trop Med Hyg*. 2019;100(2):242–245. https://doi.org/10.4269/ajtmh.18-0841.
7. Suk JE, Semenza JC. From global to local: vector-borne disease in an interconnected world. *Eur J Public Health*. 2014;24(4):531–532. https://doi.org/10.1093/eurpub/cku041.
8. Kraemer MU, Sinka ME, Duda KA, et al. The global distribution of the arbovirus vectors *Aedes aegypti* and *A. albopictus*. *Elife*. 2015;4, e08347. https://doi.org/10.7554/eLife.08347.
9. Egizi A, Bulaga-Seraphin L, Alt E, et al. First glimpse into the origin and spread of the Asian longhorned tick, *Haemaphysalis longicornis*, in the United States. *Zoonoses Public Health*. 2020;67(6):637–650. https://doi.org/10.1111/zph.12743. Epub July 7, 2020.
10. Hoogstraal H, Roberts FHS, Kohls GM, Tipton VJ. Review of *Haemaphysalis* (*Kaiseriana) longicornis* Neumann (resurrected) of Australia New Zealand New Caledonia Fiji Japan Korea and Northeastern China and USSR and its parthenogenetic and bisexual populations (Ixodoidea Ixodidae). *J Parasitol*. 1968;54:1197–1213.
11. Molaei G, Andreadis TG, Anderson JF, Iii KC. An exotic hitchhiker: a case report of importation into Connecticut from Africa of the human parasitizing tick, *Hyalomma truncatum* (Acari: Ixodidae). *J Parasitol*. 2018;104(3):302–305. https://doi.org/10.1645/18-13.
12. Laurance WF, Goosem M, Laurance SGW. Impacts of roads and linear clearings on tropical forests. *Trends Ecol Evol*. 2009;24:659–669. https://doi.org/10.1016/j.tree.2009.06.009.
13. Dente MG, Riccardo F, Bortel WV, et al. Enhancing preparedness for arbovirus infections with a One Health approach: the development and implementation of multisectoral risk assessment exercise. *BioMed Res Int*. 2020. https://doi.org/10.1155/2020/4832360.
14. Mowafi H. Conflict, displacement and health in the Middle East. *Global Publ Health*. 2011;6:472–487. https://doi.org/10.1080/17441692.2011.570358.
15. Fouad FM, Sparrow A, Tarakji A, et al. Health workers and the weaponisation of health care in Syria: a preliminary inquiry for the Lancet–American University of Beirut Commission on Syria. *Lancet*. 2017;390:2516–2526. https://doi.org/10.1016/S0140-6736(17)30741-9.

16. Gates B. Responding to Covid-19 - a once-in-a-century pandemic? *N Engl J Med*. 2020;382(18):1677–1679. https://doi.org/10.1056/NEJMp2003762.
17. Pan American Health Organization/World Health Organization. *Control of* Aedes aegypti *in the Scenario of Simultaneous Transmission of COVID-19*. Washington, DC: PAHO/WHO; 2020. https://www.paho.org/en/documents/control-aedes-aegypti-scenario-simultaneous-transmission-covid-19.
18. WHO urges countries to ensure the continuity of malaria services in the context of the COVID-19 pandemic. In: *World Health Organization*. Geneva, Switzerland: WHO; March 25, 2020. https://www.who.int/news-room/detail/25-03-2020-who-urges-countries-to-ensure-the-continuity-of-malaria-services-in-the-context-of-the-covid-19-pandemic.
19. Seelig F, Bezerra H, Cameron M, et al. The COVID-19 pandemic should not derail global vector control efforts. *PLoS Neglected Trop Dis*. 2020;14(8), e0008606. https://doi.org/10.1371/journal.pntd.0008606.
20. Pan American Health Organization/World Health Organization. *Epidemiological Update: Dengue and Other Arboviruses*. Washington, DC: PAHO/WHO; 2020. https://www.paho.org/en/documents/epidemiological-update-dengue-and-other-arboviruses-10-june-2020.
21. Aschwanden C. How COVID is changing the study of human behaviour. *Nature*. 2021;593(7859):331–333. https://doi.org/10.1038/d41586-021-01317-z.
22. Krystosik A, Njoroge G, Odhiambo L, Forsyth JE, Mutuku F, LaBeaud AD. Solid wastes provide breeding sites, burrows, and Food for biological disease vectors, and urban zoonotic reservoirs: a call to action for solutions-based research. *Front Public Health*. 2020;7:405. https://doi.org/10.3389/fpubh.2019.00405.
23. Morgan L, Protopopova A, Birkler RID, et al. Human–dog relationships during the COVID-19 pandemic: booming dog adoption during social isolation. *Humanit Soc Sci Commun*. 2020;7:155. https://doi.org/10.1057/s41599-020-00649-x.
24. Overgaauw PAM, Vinke CM, Hagen MAEV, Lipman LJA. A one health perspective on the human-companion animal relationship with emphasis on zoonotic aspects. *Int J Environ Res Publ Health*. 2020;17(11):3789. https://doi.org/10.3390/ijerph17113789. Published May 27, 2020.
25. World Health Organization. *World Malaria Report 2020: 20 Years of Global Progress and Challenges*. Geneva; 2020.
26. Pan American Health Organization. *Plan of Action to Strengthen the Surveillance and Control of Leishmaniasis in the Americas 2017–2022*. IRIS PAHO Home; 2017. https://iris.paho.org/handle/10665.2/34147.
27. Newbold T, Hudson LN, Hill SL, et al. Global effects of land use on local terrestrial biodiversity. *Nature*. 2015;520:45–50. https://doi.org/10.1038/nature14324.
28. Ogden NH. Climate change and vector-borne diseases of public health significance. *FEMS Microbiol Lett*. 2017;364(19). https://doi.org/10.1093/femsle/fnx186.
29. Song X-P, Hansen MC, Stehman SV, et al. Global land change from 1982 to 2016. *Nature*. 2018;560:639–643. https://doi.org/10.1038/s41586-018-0411-9.
30. Intergovernmental Panel on Climate Change (IPCC). Climate change 2013: the physical science basis. In: Stocker TF, Qin D, Plattner G-K, et al., eds. *Contribution of Working Group I to the Fifth Assessment Report of the Intergovernmental Panel on Climate Change*. Cambridge, UK: Cambridge University Press; 2013:1535.
31. Caminade C, Medlock JM, Ducheyne E, et al. Suitability of European climate for the Asian tiger mosquito *Aedes albopictus*: recent trends and future scenarios. *J R Soc Interface*. 2012;9:2708–2717. https://doi.org/10.1098/rsif.2012.0138.
32. Ogden NH, Lindsay LR. Effects of climate and climate change on vectors and vector-borne diseases: ticks are different. *Trends Parasitol*. 2016;32:646–656. https://doi.org/10.1016/j.pt.2016.04.015.
33. Cable J, Barber I, Boag B, et al. Global change, parasite transmission and disease control: lessons from ecology. *Philos T Roy Soc B*. 2017;372, 20160088. https://doi.org/10.1098/rstb.2016.0088.
34. Jaenson TGT, Hjertqvist M, Bergstrom T, Lundkvist A. Why is tick-borne encephalitis increasing? A review of the key factors causing the increasing incidence of human TBE in Sweden. *Parasites Vectors*. 2012;5:184. https://doi.org/10.1186/1756-3305-5-184.
35. Paz S, Semenza JC. Environmental drivers of West Nile fever epidemiology in Europe and western Asia: a review. *Int J Environ Res Publ Health*. 2013;10:3543–3562. https://doi.org/10.3390/ijerph10083543.
36. Pietsch C, Michalski D, Münch J, et al. Autochthonous West Nile virus infection outbreak in humans, Leipzig, Germany, August to September 2020. *Euro Surveill*. 2020;25(46), 2001786. https://doi.org/10.2807/1560-7917.ES.2020.25.46.2001786.
37. Bakonyi T, Haussig Joana M. West Nile virus keeps on moving up in Europe. *Euro Surveill*. 2020;25(46). https://doi.org/10.2807/1560-7917. ES.2020.25.46.2001938. pii=2001938.

38. Marini G, Calzolari M, Angelini P, et al. A quantitative comparison of West Nile virus incidence from 2013 to 2018 in Emilia-Romagna, Italy. *PLoS Neglected Trop Dis*. 2020;14(1), e0007953. https://doi.org/10.1371/journal.pntd.0007953.
39. CDC. *West Nile Virus in the United States: Guidelines for Surveillance, Prevention, and Control*. 4th revision. Fort Collins, CO: US Department of Health and Human Services, CDC; 2013. Available from: http://www.cdc.gov/westnile/resources/pdfs/wnvguidelines.pdf.
40. Horton KG, Sorte FAL, Sheldon D, et al. Phenology of nocturnal avian migration has shifted at the continental scale. *Nat Clim Change*. 2020;10:63–68.
41. Bowles DC, Butler CD, Morisetti N. Climate change, conflict and health. *J R Soc Med*. 2015;108(10):390–395. https://doi.org/10.1177/0141076815603234.
42. Mordecai EA, Caldwell JM, Grossman MK, et al. Thermal biology of mosquito-borne disease. *Ecol Lett*. 2019;22(10):1690–1708. https://doi.org/10.1111/ele.13335.
43. Rezza G, Nicoletti L, Angelini R, et al, CHIKV study group. Infection with chikungunya virus in Italy: an outbreak in a temperate region. *Lancet*. 2007;370:1840–1846. https://doi.org/10.1016/S0140-6736(07)61779-6.
44. Giron S, Franke F, Decoppet A, et al. Vector-borne transmission of Zika virus in Europe, southern France, August 2019. *Euro Surveill*. 2019;24(45), 1900655. https://doi.org/10.2807/1560-7917.ES.2019.24.45.1900655.
45. Nathan MB, Dayal-Drager R, Guzman M. Epidemiology, burden of disease and transmission. In: *Dengue Guidelines for Diagnosis Treatment Prevention and Control: New Edition*. World Health Organization; 2009:3–17.
46. Bhatt S, Gething PW, Brady OJ, et al. The global distribution and burden of dengue. *Nature*. 2013;496(7446):504–507. https://doi.org/10.1038/nature12060.
47. Pan American Health Organization/World Health Organization. *Timeline of the Emergence of Zika Virus in the Americas*. Washington, DC: PAHO/WHO; April 2016. www.paho.org.
48. Valderrama A, Tavares MG, Filho JD. Anthropogenic influence on the distribution, abundance and diversity of sandfly species (Diptera: Phlebotominae: Psychodidae), vectors of cutaneous leishmaniasis in Panama. *Mem Inst Oswaldo Cruz*. 2011;106(8):1024–1031. https://doi.org/10.1590/s0074-02762011000800021.
49. Vittor AY, Pan W, Gilman RH, et al. Linking deforestation to malaria in the Amazon: characterization of the breeding habitat of the principal malaria vector, *Anopheles darlingi*. *Am J Trop Med Hyg*. 2009;81:5–12.
50. Yasuoka J, Levins R. Impact of deforestation and agricultural development on Anopheline ecology and malaria epidemiology. *Am J Trop Med Hyg*. 2007;76:450–460.
51. Bugmyrin SV, Bespyatova LA, Anikanova VS, Ieshko EP. Abundance of larvae and nymphs of the taiga tick *Ixodes persulcatus* Schulze (Acarina, Ixodidae) on small mammals in deforested areas in the middle taiga subzone of Karelia. *Entmol Rev*. 2010;90(1):116–122. https://doi.org/10.1134/S0013873810010094.
52. United Nations. *68% of the World Population Projected to Live in Urban Areas By 2050, Says UN*. UN Department of Economic and Social Affairs. United Nations; 2018. https://www.un.org/development/desa/en/news/population/2018-revision-of-world-urbanization-prospects.html.
53. Knudsen AB, Slooff R. Vector-borne disease problems in rapid urbanization: new approaches to vector control. *Bull World Health Organ*. 1992;70:1–6.
54. Gubler DJ. Dengue, urbanization and globalization: the unholy trinity of the 21(st) century. *Trop Med Health*. 2011;39(4 Suppl):3–11. https://doi.org/10.2149/tmh.2011-S05.
55. Beaty BJ, Black WC, Eisen L, et al. The intensifying storm: domestication of *Aedes aegypti*, urbanization of arboviruses, and emerging insecticide resistance. In: *Global Health Im-pacts of Vector-Borne Diseases: Workshop Summary*. National Academies Press; September 2016. https://www.ncbi.nlm.nih.gov/books/NBK390432/.
56. Weaver SC. Prediction and prevention of urban arbovirus epidemics: a challenge for the global virology community. *Antivir Res*. 2018;156:80–84. https://doi.org/10.1016/j.antiviral.2018.06.009.
57. Essential Programme on Immunization, Immunization, Vaccines and Biologicals. *Yellow Fever: Vaccine Preventable Diseases Surveillance Standards*. World Health Organization; May 2020. Retrieved from https://www.who.int/publications/m/item/vaccine-preventable-diseases-surveillance-standards-yellow-fever.
58. Shearer FM, Longbottom J, Browne AJ, et al. Existing and potential infection risk zones of yellow fever worldwide: a modelling analysis. *Lancet Global Health*. 2018;6(3):e270–e278. https://doi.org/10.1016/S2214-109X(18)30024-Xv.
59. Tsetsarkin KA, Chen R, Leal G, et al. Chikungunya virus emergence is constrained in Asia by lineage-specific adaptive landscapes. *Proc Natl Acad Sci U S A*. 2011;108(19):7872–7877. https://doi.org/10.1073/pnas.1018344108.

60. Venturi G, Luca M, Fortuna C, et al. Detection of a chikungunya outbreak in Central Italy, August to September 2017. *Euro Surveill.* 2017;22(39). pii=17-00646.
61. Grandadam M, Caro V, Plumet S, et al. Chikungunya virus, southeastern France. *Emerg Infect Dis.* 2011;17:910–913. https://doi.org/10.3201/eid1705.101873.
62. European Centre for Disease Prevention and Control (ECDC). *Autochthonous Transmission of Chikungunya Virus in EU/EEA, 2007–2020 Stockholm*; 2020. https://www.ecdc.europa.eu/en/all-topics-z/chikungunya-virus-disease/surveillance-and-diseasedata/autochthonous-transmission. Accessed August 27, 2021.
63. Travi BL, Cordeiro-da-Silva A, Dantas-Torres F, Miró G. Canine visceral leishmaniasis: diagnosis and management of the reservoir living among us. *PLoS Neglected Trop Dis.* 2018;12(1), e0006082. https://doi.org/10.1371/journal.pntd.0006082.
64. Drancourt M, Raoult D. Sequence-based identification of new bacteria: a proposition for creation of an orphan bacterium repository. *J Clin Microbiol.* 2005;43:4311–4315. https://doi.org/10.1128/JCM.43.9.4311-4315.2005.
65. Hanage WP, Fraser C, Spratt BG. Sequences, sequence clusters and bacterial species. *Philos Trans R Soc Lond Ser B Biol Sci.* 2006;361:1917–1927. https://doi.org/10.1098/RSTB.2006.1917.
66. André MR. Diversity of *Anaplasma* and *Ehrlichia/Neoehrlichia* agents in terrestrial wild carnivores worldwide: implications for human and domestic animal health and wildlife conservation. *Front Vet Sci.* 2018;5:293. https://doi.org/10.3389/fvets.2018.00293.
67. André MR, Dumler JS, Herrera HM, et al. Assessment of a quantitative 5′ nuclease real-time polymerase chain reaction using the nicotinamide adenine dinucleotide dehydrogenase gamma subunit (*nuoG*) for *Bartonella* species in domiciled and stray cats in Brazil. *J Feline Med Surg.* 2016;18(10):783–790. https://doi.org/10.1177/1098612X15593787.
68. Maggi RG, Richardson T, Breitschwerdt EB, Miller JC. Development and validation of a droplet digital PCR assay for the detection and quantification of *Bartonella* species within human clinical samples. *J Microbiol Methods.* 2020;176, 106022. https://doi.org/10.1016/j.mimet.2020.106022.
69. Tomita N, Mori Y, Kanda H, Notomi T. Loop-mediated isothermal amplification (LAMP) of gene sequences and simple visual detection of products. *Nat Protoc.* 2008;3(5):877–882. https://doi.org/10.1038/nprot.2008.57.
70. Notomi T, Okayama H, Masubuchi H, et al. Loop-mediated isothermal amplification of DNA. *Nucleic Acids Res.* 2000;28(12):E63. https://doi.org/10.1093/nar/28.12.e63.
71. Sanches GS, Évora PM, Mangold AJ, et al. Molecular, biological, and morphometric comparisons between different geographical populations of *Rhipicephalus sanguineus* sensu lato (Acari: Ixodidae). *Vet Parasitol.* 2016;215:78–87. https://doi.org/10.1016/j.vetpar.2015.11.007.
72. Chandra S, Ma GC, Burleigh A, et al. The brown dog tick *Rhipicephalus sanguineus* sensu Roberts, 1965 across Australia: morphological and molecular identification of *R. sanguineus* s.l. tropical lineage. *Ticks Tick Borne Dis.* 2020;11(1), 101305. https://doi.org/10.1016/j.ttbdis.2019.101.
73. Yssouf A, Flaudrops C, Drali R, et al. Matrix-assisted laser desorption ionization–time of flight mass spectrometry for rapid identification of tick vectors. *J Clin Microbiol.* 2013;51(2):522–528. https://doi.org/10.1128/JCM.02665-12.
74. Yssouf A, Socolovschi C, Flaudrops C, et al. Matrix-assisted laser desorption ionization - time of flight mass spectrometry: an emerging tool for the rapid identification of mosquito vectors. *PLoS One.* 2013;8(8), e72380. https://doi.org/10.1371/journal.pone.0072380.
75. Yssouf A, Socolovschi C, Leulmi H, et al. Identification of flea species using MALDI-TOF/MS. *Comp Immunol Microbiol Infect Dis.* 2014;37(3):153–157. https://doi.org/10.1016/j.cimid.2014.05.002.
76. Mathis A, Depaquit J, Dvořák V, et al. Identification of phlebotomine sand flies using one MALDI-TOF MS reference database and two mass spectrometer systems. *Parasites Vectors.* 2015;8:266. https://doi.org/10.1186/s13071-015-0878-2.
77. Ouarti B, Laroche M, Righi S, et al. Development of MALDI-TOF mass spectrometry for the identification of lice isolated from farm animals. *Parasite.* 2020;27:28. https://doi.org/10.1051/parasite/2020026.
78. El Hamzaoui B, Laroche M, Almeras L, Bérenger JM, Raoult D, Parola P. Detection of *Bartonella* spp. in fleas by MALDI-TOF MS. *PLoS Neglected Trop Dis.* 2018;12(2), e0006189. https://doi.org/10.1371/journal.pntd.0006189.
79. Lejal E, Estrada-Peña A, Marsot M, et al. Taxon appearance from extraction and amplification steps demonstrates the value of multiple controls in tick microbiota analysis. *Front Microbiol.* 2020;11:1093. https://doi.org/10.3389/fmicb.2020.01093.

80. Jourdain F, Picard M, Sulesco T, et al. Identification of mosquitoes (Diptera: Culicidae): an external quality assessment of medical entomology laboratories in the MediLabSecure network. *Parasites Vectors*. 2018;11:553. https://doi.org/10.1186/s13071-018-3127-7.
81. European Centre for Disease Prevention and Control. European network for medical and veterinary entomology (VectorNet). *ECDC*; November 10, 2022. https://www.ecdc.europa.eu/en/about-us/partnerships-and-networks/disease-and-laboratory-networks/vector-net.
82. Lippi CA, Ryan SJ, White AL, Gaff HD, Carlson CJ. Trends and opportunities in tick-borne disease geography. *J Med Entomol*. 2020, tjab086. https://doi.org/10.1093/jme/tjab086.
83. Badiaga S, Brouqui P. Human louse-transmitted infectious diseases. *Clin Microbiol Infect*. 2012;18(4):332–337. https://doi.org/10.1111/j.1469-0691.2012.03778.x.
84. Azad AF. Epidemiology of murine typhus. *Annu Rev Entomol*. 1990;35:553–569. https://doi.org/10.1146/annurev.en.35.010190.003005.
85. Murray KO, Evert N, Mayes B, et al. Typhus group rickettsiosis, Texas, USA, 2003–2013. *Emerg Infect Dis*. 2017;23(4):645–648. https://doi.org/10.3201/eid2304.160958.
86. Caravedo Martinez MA, Ramírez-Hernández A, Blanton LS. Manifestations and management of flea-borne rickettsioses. *Res Rep Trop Med*. 2021;12:1–14. https://doi.org/10.2147/RRTM.S274724.
87. Paris DH, Shelite TR, Day NP, Walker DH. Unresolved problems related to scrub typhus: a seriously neglected life-threatening disease. *Am J Trop Med Hyg*. August 7, 2013;89(2):301–307. https://doi.org/10.4269/ajtmh.13-0064.
88. World Health Organization. *Strengthening Who Preparedness for and Response to Health Emergencies*. Geneva, Switzerland: Seventy-fifth World Health Assembly; May 6, 2022. https://apps.who.int/gb/ebwha/pdf_files/WHA75/A75_19-en.pdf.
89. National Research Council (US). *Committee on Achieving Sustainable Global Capacity for Surveillance and Response to Emerging Diseases of Zoonotic Origin. Achieving an Effective Zoonotic Disease Surveillance System. Sustaining Global Surveillance and Response to Emerging Zoonotic Diseases*; March 18, 2009. https://www.ncbi.nlm.nih.gov/books/NBK215315/.
90. Vazquez-Prokopec GM, Chaves LF, Ritchie SA, Davis J, Kitron U. Unforeseen costs of cutting mosquito surveillance budgets. *PLoS Neglected Trop Dis*. 2010. https://doi.org/10.1371/journal.pntd.0000858.
91. UN Climate Change Conference. *COP26 The Glasgow Climate Pact*; 2021. https://ukcop26.org/wp-content/uploads/2021/11/COP26-Presidency-Outcomes-The-Climate-Pact.pdf.
92. Bush SE, Gustafsson DR, Tkach VV, Clayton DH. A misidentification crisis plagues specimen-based research: a case for guidelines with a recent example (Ali et al., 2020). *J Parasitol*. 2021;107(2):262–266. https://doi.org/10.1645/21-4.
93. Laga AC, Granter SR, Mather TN. Proficiency at tick identification by pathologists and clinicians is poor. *Am J Dermatopathol*. 2021. https://doi.org/10.1097/DAD.0000000000001977.
94. Vieira TSWJ, Collere FCM, Ferrari LDR, et al. Novel Anaplasmataceae agents *Candidatus* Ehrlichia hydrochoerus and *Anaplasma* spp. infecting Capybaras, Brazil. *Emerg Infect Dis*. February 2022;28(2):480–482. https://doi.org/10.3201/eid2802.210705.
95. Hadler JL, Patel D, Nasci RS, et al. Assessment of arbovirus surveillance 13 Years after introduction of West Nile virus, United States. *Emerg Infect Dis*. 2015;21(7):1159–1166. https://doi.org/10.3201/eid2107.140858.
96. COP26. Finance. *UN Climate Change Conference (COP26) at the SEC – Glasgow*; 2021. https://ukcop26.org/cop26-goals/finance/.
97. Ebola outbreak 2021- North Kivu, democratic Republic of Congo. In: *World Health Organization*. Geneva, Switzerland: WHO; October 28, 2021. https://www.who.int/emergencies/situations/ebola-2021-north-kivu.
98. Coronavirus disease (COVID-19). In: *World Health Organization*. Geneva, Switzerland: WHO; October 28, 2021. https://www.who.int/emergencies/diseases/novel-coronavirus-2019.
99. Donaldson TG, Pèrez de León AA, Li AY, et al. Assessment of the geographic distribution of *Ornithodoros turicata* (Argasidae): climate variation and host diversity. *PLoS Neglected Trop Dis*. 2016;10(2), e0004383. https://doi.org/10.1371/journal.pntd.0004383.
100. Estrada-Peña A, Cevidanes A, Sprong H, Millán J. Pitfalls in tick and tick-borne pathogens research, some recommendations and a call for data sharing. *Pathogens*. 2021;10(6):712. https://doi.org/10.3390/pathogens10060712.

101. Talagrand-Reboul E, Boyer PH, Bergström S, Vial L, Boulanger N. Relapsing fevers: neglected tick-borne diseases. *Front Cell Infect Microbiol.* April 4, 2018;8(APR). Available from: http://journal.frontiersin.org/article/10.3389/fcimb.2018.00098/full.
102. Moreira J, Bressan CS, Brasil P, Siqueira AM. Epidemiology of acute febrile illness in Latin America. *Clin Microbiol Infect.* 2018;24(8):827–835. https://doi.org/10.1016/j.cmi.2018.05.001.
103. Nordstrand A, Bunikis I, Larsson C, et al. Tickborne relapsing fever diagnosis obscured by Malaria, Togo. *Emerg Infect Dis.* 2007;13(1):117–123.
104. Muñoz-Leal S, Faccini-Martínez ÁA, Costa FB, et al. Isolation and molecular characterization of a relapsing fever *Borrelia* recovered from *Ornithodoros rudis* in Brazil. *Ticks Tick Borne Dis.* May 2018;9(4):864–871. Available from: http://linkinghub.elsevier.com/retrieve/pii/S1877959X17305733.
105. Muñoz-leal S, Faccini-martínez ÁA, Teixeira BM, et al. Relapsing fever group borreliae in human-biting soft ticks, Brazil. *Emerg Infect Dis.* 2021;27(1):321–324.

CHAPTER

10

Assessment of critical gaps in prevention, control, and response to major bacterial, viral, and protozoal infectious diseases at the human, animal, and environmental interface*

Muhammed S. Muyyarikkandy[1], *Kalmia Kniel*[2], *William A. Bower*[3], *Antonio R. Vieira*[3], *María E. Negrón*[3] *and Siddhartha Thakur*[1]

[1]Department of Population Health and Pathobiology, North Carolina State University, Raleigh, NC, United States; [2]Department of Animal and Food Sciences, University of Delaware, Newark, DE, United States; [3]Bacterial Special Pathogens Branch, Division of High-Consequence Pathogens and Pathology, National Center for Emerging and Zoonotic Infectious Diseases, Centers for Disease Control and Prevention, Atlanta, GA, United States

Introduction

Zoonotic infectious diseases pose a significant threat to global health and economic growth. The top 13 zoonotic infections result in 2.4 billion cases and 2.2. million deaths annually.[1] More than two-thirds of emerging infectious diseases in humans have their links to animal-source food, water contamination, and agricultural production.[2] Hence, a One-Health approach is critical to tackling this growing concern. Current strategies to contain emerging infectious diseases are mainly focused on postemergence outbreak control

* Contribution of authors William A. Bower, Antonio R. Vieira, María E. Negrón and Siddhartha Thakur is subject to public domain.

https://doi.org/10.1016/B978-0-323-90945-7.00024-5

measures. However, the impact of outbreaks varies greatly depending on the socioeconomic status of the affected areas and the causative agents.[3] Hence, detecting the infectious agent early in the outbreak chain is critical. Early detection can be achieved by maintaining an active public health surveillance system, especially in areas prone to outbreaks.[4]

While emerging zoonotic outbreaks gather international attention, endemic zoonotic bacterial infections transmitted by food or water go largely neglected yet inflict a higher day-to-day burden on humans, agriculture, and the economy, particularly in low- and middle-income countries (LMIC). While people from all communities are susceptible to these infections, people associated with livestock and agricultural production are at greater risk.[1] Moreover, the lack of data on impact of endemic zoonotic diseases hinders the efforts to devise proper control strategies. Many current international policies, such as International Health Regulations (IHR) and Global Health Security Agenda (GHSA), mainly focus more on emerging zoonotic infections, which have broader international concerns than endemic zoonotic infections, which are primarily endemic in LMIC.[5]

Three primary endemic bacterial zoonotic infections in LMIC largely neglected by global health initiatives include anthrax, brucellosis, and leptospirosis. Although of high public health and economic impacts, these three diseases were not recognized among the 17 NTDs prioritized by WHO as priority emerging neglected tropical diseases.[6,7]

Anthrax is predominantly a disease of concern in livestock and wildlife. Humans can become secondarily infected by people working in close proximity to animals such as farmers, veterinarians, and butchers who are at greater risk. Conditions in LMIC, such as food insecurity, can increase the number of human cases. Moreover, strategies to control and prevent endemic anthrax in LMIC are of international concern due to its use as a bioweapon.[8] Brucellosis is a neglected disease in endemic LMICs with wide geographical distribution and can cause substantial morbidity in affected human and livestock populations. Livestock species act as a significant reservoir of the pathogen, and the disease is mainly transmitted to humans through direct contact with affected animals or consumption of contaminated milk or milk products.[9] Leptospirosis has emerged as a global threat both in the rural and urban environments, especially in LMIC, where living conditions increase the risk of exposure. In addition, early leptospirosis illness often begins as an undifferentiated acute febrile illness which could be mistaken for other diseases in endemic areas. Due to a lack of adequate medical infrastructure in LMIC, leptospirosis is often misdiagnosed, leading to delays in diagnosis, resulting in high morbidity and mortality.[6,10] Hence, it is imperative to consider human–animal–environmental interaction when devising strategies to control these pathogens.

Small-scale livestock production and agriculture play an essential role in the economy of LMICs, yet they are associated with an increased prevalence of zoonotic infectious diseases.[11] The threat of zoonotic bacterial infections can be mitigated by better understanding livestock rearing practices, disease transmission, and the associated risks. Many livestock species act as a significant reservoir for various zoonotic infectious diseases. Susceptible hosts are then infected by these pathogens either through environmental transmission, direct contact, or food. Studies have shown that contact with livestock animals and poultry increases the risk of diarrhea in children and is positively correlated with increased infections of enteric pathogens, including multiple pathotypes of diarrheagenic *E. coli*.[12–15]

Animal and agricultural production in LMICs is often linked to food- and waterborne infections, which are significant global public health concerns, with varying morbidity and mortality rates among developed and developing countries. Food- and waterborne infections usually present as gastrointestinal upset with nausea, vomiting, diarrhea, and abdominal cramps. These infections are rarely fatal in industrialized countries, unlike in low- and middle-income countries.[16] However, severe illness can occur among the elderly, children, pregnant women, and people with compromised immune systems. Systemic consequences of infection can occur several days or weeks after the initial exposure, such as hemolytic uremic syndrome (HUS) developing secondary to Shiga toxin-producing *Escherichia coli* (STEC).[17] A variety of bacteria, viruses, and parasites transmitted through contaminated food and water can cause diarrheal diseases, and the severity and duration of the illness often vary.

Most studies indicate that the actual pathogen cannot be determined in more than half of foodborne illnesses.[18] Of the identified bacterial pathogens, *Campylobacter* spp., *Clostridium perfringens*, *Escherichia coli*, and nontyphoidal *Salmonella* spp. are consistently the four most commonly associated bacterial pathogens; however, their relative rankings vary between studies. The nontyphoidal *Salmonella* and *Campylobacter* are commonly associated with severe illness resulting in hospitalization. *Salmonella*, *Campylobacter*, and *Listeria* are the organisms predominantly related to mortality due to foodborne diseases in the United States.[16,19] Norovirus accounts for 69%–99% of illnesses attributed to virus and is responsible for 7% of all mortalities due to food- and water-related infections.[20] *Cryptosporidium*, *Cyclospora*, and *Giardia* are the most common foodborne protozoan agents causing gastrointestinal upset, and it is estimated that 48% of all parasitic infections are foodborne.[21,22] The list of major bacterial, viral, and protozoan infectious agents associated with food- and waterborne illnesses are provided in Table 10.1.

The actual burden of food- and waterborne illnesses worldwide is largely unknown; however, the number of suspected mortalities worldwide from foodborne pathogen exposure is staggering. Diarrheal disease is a global burden, a leading cause of child mortality and morbidity globally and commonly results from contaminated food and water sources. Each year worldwide, foodborne infections cause 600 million cases and 42,000 deaths.

TABLE 10.1 List of major food- and water-borne bacterial, viral, and protozoal agents causing diseases in humans.[17]

Bacterial pathogens	Viral pathogens	Protozoal agents
Aeromonas spp.	*Aichivirus*	*Blastocystis hominis*
Bacillus cereus	*Astroviruses*	*Cryptosporidium* spp.
Campylobacter spp.	*Coronavirus*	*Cyclospora cayetanensis*
Clostridium botulinum	*Enteric adenovirus*	*Dientamoeba fragilis*
Clostridium perfringens	*Hepatitis A virus*	*Entameba histolytica*
Enterobacter sakazakii	*Hepatitis E virus*	*Giardia duodenalis*
Escherichia coli (STEC, EPEC, ETEC, EAEC, EIEC, and DAEC)	*Norovirus*	*Isospora belli*
	Parvoviruses	*Microsporidium*[a]
Listeria monocytogenes	*Picobirnaviruses*	*Toxoplasma gondii*
Plesiomonas shigelloides	*Reovirus*	
Salmonella spp. (typhoidal and nontyphoidal)	*Rotavirus*	
Shigella spp.	*Sapro-like virus*	
Staphylococcus aureus	*Torovirus*	
Vibrio spp.		
Yersinia spp.		

[a]*No longer considered as protist: recognized as fungal or fungal-like organism.*

Globally, one out of 10 daily child deaths is due to diarrheal diseases, and 2195 children die daily. Almost 88% of these deaths are due to the lack of clean water and poor hygiene. The world health organization has estimated that the European region has the lowest burden of foodborne illnesses, with more than 23 million illnesses resulting in 5000 deaths. The burden hits harder on LMIC because of the high prevalence and insufficient control measures, poor food safety, and sanitation. It is estimated that Africa and South-East Asia regions are hit hardest. In Africa, 91 million people fall ill due to foodborne diseases, with 157,000 mortalities, and 70% of the cases are manifested as diarrheal diseases.[23]

The International Health Regulations (IHRs) are a legally binding agreement between 196 countries to build the capacity to detect, report, and respond to potential public health emergencies worldwide. The WHO-IHR guidelines serve as an essential tool for prevention, control, and response to international disease spread. The guidelines have not been updated since 2005., and it is high time to strengthen the policies with the additional information raised since 2005. The scope of this chapter is to assess the critical public health policy gaps in prevention, control, and response before and during incidents of common bacterial, viral, and protozoan infections in LMIC and developed nations. Since the scope of the chapter is expansive, it will focus on five bacterial (anthracosis, brucellosis, cholera, colibacillosis, and leptospirosis), two viral (Rotavirus and Norovirus infections), and two protozoan parasites (cryptosporidiosis and cyclosporiasis). Furthermore, this chapter aims to help drive the action plan to prevent the spread and reduce the challenges arising from endemic zoonotic infections.

Literature review and gap analysis

Escherichia coli

There are multiple pathotypes of *E. coli* associated with food- and waterborne transmission resulting in diarrheal diseases.[24] Of these, Enterohemorrhagic *E. coli* (EHEC) mainly affects industrialized countries resulting in often called "bloody diarrhea," and Enterotoxigenic *E. coli* (ETEC) typically impacts developing countries causing acute gastrointestinal disturbances. EHEC is the most common *E. coli* infection resulting in severe foodborne diseases.[25] Contaminated foods such as meat products, milk, and fresh produce serve as a significant infection source in developed countries.[26–29]

EHEC is characterized by the production of Shiga toxins, which is the major virulence factor of the organism. Shiga toxin is encoded by one or more *stx* genes and is therefore referred to as "Shiga toxin-producing *E. coli*."[24] In the United States, EHEC is estimated to cause 100,000 infections annually. EHEC infections are caused by *E. coli* belonging to either O157:H7 or non-O157 serotypes (O26, O111, O103, O121, O45, and O145 occur with some frequency, among others).[30] Approximately 15% of EHEC infections caused by the most common serotype (O157:H7) develop serious hemolytic uremic syndrome.[30,31] EHEC infections are mainly managed by supportive therapy because antibiotic therapy may increase the risk for HUS, especially in children.[32] Enteropathogenic *E. coli* (EPEC) was the first strain of *E. coli* responsible for infantile diarrhea outbreaks in developed countries in 1940 and 1950s; for unknown reasons, they are no longer considered a major concern in developed countries. However, in LMIC, EPEC is major contributor to infantile diarrhea and in some cases, it accounts for up to 30% mortality rate.[33]

Enterotoxigenic *E. coli* is the most common cause of traveler's diarrhea.[34] ETEC are genetically heterogeneous from EHEC and produce heat-labile and or heat-stable enterotoxins that cause secretory diarrhea. ETEC is responsible for millions of infections annually and thousands of mortalities, especially among children in the developing world.[35] Contaminated food and water are the primary sources of infections.[36] Several ETEC infection clusters are associated with travel on cruise ships that visit ports where ETEC is endemic. Domestic outbreaks in developed countries are often related to increased international food trade from areas with high ETEC prevalence.[37,38] Contaminated foods such as seafood and salads were the primary source of infections among these outbreaks.[36]

Vibrio cholerae

Cholera is a public health threat globally, mainly affecting LMICs, and is an indicator of social inequity. Cholera primarily affects people in areas with poor water, sanitation, and hygiene conditions. Cholera infection is manifested as an acute, rapidly dehydrating diarrheal disease and is transmitted through water or contaminated food. It is estimated that there are 1.3 − 4 million cholera cases annually. Even though mortality can be prevented with vaccination, timely rehydration therapy, oral rehydration solution or IV fluid administration, 21,000−143,000 people die due to cholera each year.[39]

WHO estimates that only 5%−10% of Cholera cases are reported globally because many countries lack proper public health surveillance and diagnostic systems, and others may fail to report cases altogether due to fear of trade and travel sanctions.[39,40] Over 200 serotypes of *V. cholerae* are reported, and strains O1 and O139 are major outbreak-causing strains. *V. cholerae* O1 is classified as classical or El Tor biotypes based on phenotypic characteristics or into Inaba and Ogawa based on antigenic determinants of its O antigen.[41,42] *V. cholerae* can survive in hostile environments such as estuaries and salty waters and form biofilm when nutrients supply is depleted and in response to unfavorable environmental conditions.[43]

The pathogen also has acquired and developed antibiotic resistance, through superintegron in its genome, to various classes of antibiotics, including sulfonamides, quinolones, aminoglycoside, chloramphenicol, and azithromycin.[43] With a low infectious dose of 10^3−10^8 cells, cholera infection could develop within an incubation period of less than 24 h to 5 days.[44] Cholera can be an endemic or epidemic depending on the region; therefore, control strategies should be devised based on the affected region.

Cholera often affects regions with limited resources; hence, it is crucial to diagnose the disease as quickly as possible to prevent disease transmission. Stool test remains the gold standard for cholera diagnosis; however, rapid diagnostic tests are more reliable in endemic regions because it enables immediate preventive care. Acute cholera cases can be managed with oral or intravenous rehydration and antibiotic therapy as needed. Antibiotics can be administered in patients with serious dehydration and should be selected based on antibiotic susceptibility patterns; doxycycline is the drug of choice in most countries, and azithromycin can be used in pregnant women and children. Mass cholera vaccine administration can achieve short- and medium-term management in communities with a history of recurrent infections.[45] There are three cholera vaccines, Dukoral (manufactured by SBL Vaccines), Shan-Chol (manufactured by Shantha Biotec in India), and Euvichol-Plus/Euvichol (manufactured by Eubiologics) are prequalified by WHO. Long-term disease control requires serious investment and should focus on proper hygiene, sanitation, access to clean water, and targeted vaccine administration.[46]

Leptospira

Leptospirosis is caused by a gram-negative spirochetal bacterium belonging to the genus *Leptospira*. Leptospirosis is considered one of the most common zoonotic diseases in the world. The prevalence is higher in the tropical and subtropical regions of the world, with the highest incidences in South and Southeast Asia, Oceania, the Caribbean, Latin America, and sub-Saharan Africa. Leptospirosis is estimated to cause greater than 1 million cases each year worldwide, resulting in over 58,000 deaths.[47] Rodents are important reservoirs for leptospirosis; however, essentially all mammals can be infected with *Leptospira* spp., including dogs, livestock, pigs, horses, and wildlife. Many animal species have adapted to become chronically infected and serve as maintenance hosts with chronic kidney infections which leads to persistent shedding in the environment. *Leptospira* are shed in the urine of infected animals and can survive in water or wet soil for weeks to months.[48] Humans and other animals are infected via contact with urine or urine-contaminated water or soil.[49–52] Exposures can also happen by ingestion of contaminated food or sources of drinking water. The bacteria can enter the human body through mucous membranes, eyes, and skin abrasions or wounds. Macerated skin resulting from prolonged water exposure is suspected to be another route of infection. Risk factors for leptospirosis include wading, swimming, swallowing, or bathing in floodwater or contaminated freshwater.

The incubation period of leptospirosis is usually 5–14 days, with a range of 2–30 days. The majority (~90%) of clinically apparent conditions present as a mild, self-limiting, acute febrile illness lasting approximately 7 days.[53,54] Most common symptoms include fever, headache, myalgia (especially calf and lower back), conjunctival suffusion, and gastrointestinal illness (vomiting, diarrhea, abdominal pain). A rash is sometimes noted. These nonspecific symptoms often lead to misdiagnosis as a viral syndrome and underreporting of the true incidence of leptospirosis. In some cases (~10%) of leptospirosis, after a temporary resolution of the mild, nonspecific symptoms will progress to a more severe illness with multi-organ dysfunction characterized by kidney failure and jaundice, aseptic meningitis, or pulmonary hemorrhage. The fatality rate among patients with severe disease is 5%–15%. Rarely patients can develop severe pulmonary hemorrhagic syndrome, which has up to a 60% mortality rate. Doxycycline along with supportive fluid therapy is the line of treatment. For patients who have contraindications or cannot tolerate doxycycline, alternatives are either azithromycin or amoxicillin.[55–57]

Control of leptospirosis in LMICs requires investment in public health infrastructure to provide access to clean water sources and rodent control in and around living and work areas. Early diagnosis and treatment of leptospirosis improve outcomes. Investments in public health surveillance to identify outbreaks would allow for earlier detection following weather events, such as flooding, that can lead to an increase in cases. Use of point-of-care diagnostic tests would also improve the timeliness of diagnosis and treatment.

Anthrax

Anthrax is a zoonotic disease caused by *Bacillus anthracis*, a gram-positive, rod-shaped, and spore-forming bacterium that primarily affects herbivorous livestock and wildlife, usually fatal among these animals. Humans can acquire anthrax via cutaneous, ingestion, inhalation, or injection routes. Most human infections with *B. anthracis* are cutaneous and result from handling spore-infected animals, their carcasses, or their meat, hides, or wool. Foodborne ingestion anthrax occurs following consumption of meat from sick animals or animals who died of anthrax.[58] The disease usually develops typically 1–7 days after eating contaminated meat. Symptoms include fever, chills, nausea, vomiting, diarrhea, severe sore throat, difficulty swallowing, swelling of the neck, and regional lymphadenopathy; airway compromise can occur. Later symptoms can include shortness of breath and altered mental status, with shock and death occurring 2–5 days after disease onset.

Specimens for anthrax diagnosis include blood, skin lesion exudates, pleural or ascitic fluid, cerebrospinal fluid, or stool. Culture and gram stains will likely be negative if specimens are collected after antibiotic therapy has been initiated, regardless of the form of the disease. Antigen and molecular tests are available for anthrax diagnosis, but the likelihood that these tests will be positive decreases with the length of antibiotic treatment before sample collection. All types of anthrax infection can be treated with antibiotics. First-line agents include ciprofloxacin or doxycycline; clindamycin is an alternative, as are penicillin if the isolate is penicillin-susceptible. Another option for treatment is anthrax antitoxins administered in conjunction with antibiotics. Patients with severe cases of anthrax will need to be hospitalized and may require aggressive treatment.[59]

Policies and interventions to prevent anthrax cases in LMICs focus on controlling the disease in livestock and wildlife. This can be achieved via livestock vaccination, promotion of proper carcass disposal methods, and community awareness of anthrax as a threat to their livestock and their health. Early recognition of the disease in livestock can prevent human cases. Government-sponsored livestock vaccination is key in areas where anthrax is endemic.[60–62]

Brucella

Brucellosis is a zoonotic disease caused by a group of gram-negative intracellular bacteria from the genus *Brucella.* Brucellosis is mainly a livestock disease; however, *Brucella*'s hosts are quite extensive (e.g., dogs, swine, rodents, bats, marine mammals, frogs). Brucellosis has a worldwide distribution, and areas around the Mediterranean Basin, South and Central America, Eastern Europe, Asia, Africa, and the Middle East have the highest disease burden in both animals and humans. Estimates suggest that over 500,000 new human cases are diagnosed annually worldwide.[63] The disease in humans is primarily contracted through consumption of unpasteurized dairy products and occupational exposures in persons working closely with infected animals or animal tissues, such as livestock workers, veterinarians, laboratory technicians, and abattoir workers. On average, people develop symptoms 2–4 weeks following infection. However, symptoms can start as early as 5 days or as late as 6 months following infection. Fever is the most common symptom observed, and other symptoms include night sweats, malaise, arthralgia, myalgia, chills, fatigue, headache, and back pain.

Culturing the bacteria is the gold standard for diagnosing the disease. However, culture is not performed regularly due to the risk of laboratory-acquired infections to those performing these tasks in the lab, the difficulty in growing the organism, and the low bacterial load in the collected specimens.[64] Thus, serological tests that detect *Brucella* antibodies are mainly used to diagnose brucellosis. Polymerase chain reaction (PCR) assays to detect *Brucella* sp. DNA are more sensitive, faster, and safer than culture. However, the diagnostic use in medicine is limited. Due to low bacterial load, PCR may fail to identify bacterial DNA. If bacterial DNA is identified by PCR, PCR can detect bacterial DNA up to 2 years following successful treatment.[65] Oral doxycycline and rifampin are the most common choice administered for uncomplicated brucellosis. Streptomycin or gentamycin for 2–3 weeks in addition to doxycycline for 6 weeks is an alternative option that is used often in cases with complications associated with brucellosis.[66] Human brucellosis is primarily prevented by controlling the disease in the animal population through a successful vaccination program and pasteurization of dairy products. Personal protective equipment (PPE) should include adequate protection to minimize direct contact (to skin and mucous membranes) and aerosol exposures depending on the type of work being performed and exposure risk.[66,67] A risk assessment should be performed following any potential exposures to identify the level of risk and subsequently postexposure follow-up.

Norovirus

Viruses account for an estimated 10%–67% of foodborne infections worldwide. Person-to-person and fecal–oral transmission routes are the most common route of infection in developed countries. However, in low- and middle-income countries, contaminated food and water are the predominant routes of disease transmission.[68] Moreover, in industrialized countries, Norovirus (NoV) encompasses the single most significant group of pathogens in terms of numbers of cases and outbreaks.[69] NoV is divided into five different genogroups, and GI and GII are the most common isolates found in humans.[69,70]

The virus has a very low infectious dose of <10 virions, and the most common sources of infection include fish, raw oysters, clams, salad items, and fresh produce. In addition to low infectious dose, other important virulence factors include their ability to survive in harsh environmental conditions and rapid spread from person to person and in water.[71–73] The clinical symptoms of NoV are usually indistinguishable from the symptoms of other food- and waterborne pathogens. The incubation period is longer and the duration of the illness is shorter compared to a bacterial pathogen. However, the number of affected individuals in a single outbreak is significantly higher in a NoV outbreak with a median of 25 than an outbreak caused by a bacterial pathogen with a median of 15.[74]

Rapid detection of the disease is vital in the control of norovirus infection. Kaplan's criteria are useful, especially in regions with limited laboratory facilities. Kaplan's criteria include (a) vomiting in more than half of symptomatic cases, (b) mean (or median) incubation period of 24–48 h, (c) mean (or median) duration of illness of 12–60 h, and (d) no bacterial pathogen isolated in stool culture. Norovirus antigen detection tests and molecular diagnostic tests also could be employed to detect norovirus from various samples.[75]

Effective control measures include strict hygiene, sanitation, and environmental decontamination procedures.[76] Supportive therapy including rehydration and electrolytes is the primary treatment strategy for noroviral infection. Antiemetics and antimotility drugs also could be used in some patients. The virus inflicts a greater burden on children, the elderly, and immunocompromised patients, and nitazoxanide has been demonstrated to be effective in reducing the symptoms in these groups.[77] Furthermore, immunoglobulin supplements and immunosuppressive therapy also could be useful in patients with severe immunocompromised conditions. Moreover, there are no proven prophylactic treatment strategies to control noroviral infections.[75]

Rotavirus

Rotavirus is a significant cause of acute gastroenteritis due to food- and waterborne outbreaks and is the most common cause of severe diarrheal disease in young children globally. Even though the disease is preventable with vaccination, more than 215,000 children die annually.[78] The major routes of transmission include the fecal-to-oral route and through contaminated surfaces. The huge viral load of 10^8 to 10^{11} viral particles per gram feces starts on day one of diarrhea.[79] Contaminated fresh produce, consumption of meat of infected animals, and food contamination after cooking also serve as common sources of infections.[80]

Even though it is not proven in humans, rotavirus can also spread through the respiratory route via aerosols.[81] Rotavirus can be concentrated in shellfish; however, infections have not been linked with seafood consumption.[82,83] Rotaviruses can survive in potable and recreational waters for weeks and at least 4 h on human hands. The viruses are relatively resistant to commonly used disinfectants.[84] Rotavirus is divided into various serological groups based on their antigens, from A to E with 2–3 subgroups and 11 serotypes. Serogroups F and G groups are provisional for the present.[85] Rotavirus serotypes, particularly group A, are considered the primary cause of viral gastroenteritis globally in infants and young children. However, the mortality rate is high in developing countries (20%) than in developed countries.[79]

Enzyme-linked immunosorbent assay is commonly used for the detection of rotavirus in clinical samples such as stool samples; however, it requires 10^4–10^7 virions for successful detection. Another drawback of immunoassay is their inability to detect nongroup A rotavirus. Electron microscopy allows to visualize the pathognomonic wheel-like appearance of the virus. PCR is another powerful technique to detect the virus and has almost 1000 times more sensitivity than immunoassays. Adults shed fewer virus in the stool samples compared to children.[86] Treatment of rotavirus infection is mainly focused on symptomatic relief, including rehydration therapies. Additional measures should be taken to prevent the volume and frequency of diarrhea. In addition, oral administration of human serum immunoglobulins also could be beneficial. Live oral vaccines are available to control the virus; however, the efficacy is lower in low- and middle-income countries because of the higher incidence rate than industrialized nations.[87]

Cryptosporidium

Cryptosporidium is the most common diarrhea-causing protozoan parasite worldwide, resulting in 8.9 million illnesses and 3759 deaths globally.[88] Cryptosporidium is ubiquitous with multiple species infecting domestic animals, wildlife, and humans. The symptoms are usually self-limiting; however, the mortality rate is high among children under 5 and immunocompromised individuals such as HIV and cancer patients.[89] The disease is highly underreported, even in the developed countries such as the, with United States, <2% of the estimated 748,000 cases reported annually.[90,91] Humans are susceptible to nearly 20 species of *Cryptosporidium*, and *C. hominis* and *C. parvum* are the major species affecting humans globally.[69,92] *Cryptosporidium* is mainly transmitted through contaminated water-a situation worsened by climate change. The Global Waterborne Pathogen model for human *Cryptosporidium* emissions predicts that while *Cryptosporidium* emissions in developing countries will decrease by 24% in 2050, emissions to surface water will increase by up to 70% in Africa.[93]

Cryptosporidium oocyst is highly resilient and can survive in harsh environmental conditions for months before being ingested by a new host.[94] The new hosts usually acquire the oocysts through contaminated potable water, recreational waters, and fresh produce irrigated with contaminated water and spread new oocysts to the environment through feces.[93]

Diagnosis of cryptosporidiosis depends on the laboratory techniques since there are no pathognomonic symptoms. Fresh, frozen, or preserved diarrheic feces could be used for diagnosis. Acid-fast or fluorescent staining, immunoassays, or molecular diagnosis commonly utilize fresh or frozen feces, while preserved feces could be used for staining after concentration and immunoassays.[95] A combination of different drugs is implemented in the treatment of cryptosporidiosis. Nitazoxanide is a broad-spectrum antiparasitic drug effective against *Cryptosporidium*; however, it is not effective in AIDS patients.[96] Paromomycin has been effective when combined with a protease inhibitor or recombinant IL-12, including AIDS patients. Since no drug can completely eliminate *Cryptosporidium* from the host, supportive therapy is also recommended in humans and domestic animals.[97]

Cyclospora

Cyclosporiasis is a gastrointestinal disease caused by *Cyclospora* spp., an obligate intracellular parasite with a life cycle that is not fully characterized.[98] Sporulated oocyst is the infective stage of *Cyclospora,* and factors that influence the sporulation of the parasite in poorly known. The infectious dose is very low, which is found to be 10–100.[99] The common routes of *Cyclospora* infection in humans are food or water contaminated with the parasite. The symptoms are usually exhibited as mild to moderate self-limiting diarrhea. However, the disease can result in severe intestinal injury and prolonged diarrhea in individuals with compromised immune system.[100]

In developed countries, modes of transmission and risk factors are well studied and documented. In those countries, the main modes of infection are through the consumption of contaminated fresh produce and travel to regions with endemic cyclosporiasis. However, epidemiological factors are poorly studied in less-developed countries, and disease transmission mainly occurs due to exposure to contaminated potable and recreational waters.[99] *Cyclospora* oocysts can survive in water at 4°C for 2 months and at 37°C for 7 days, and oocysts are resistant to commonly used disinfectants.[101] The prevalence rates of *Cyclospora* are often underestimated because the parasite may be present at subclinical levels, and the sensitivity of conventional detection methods is low.[99] Improving environmental sanitation and health education could be the most practical intervention for controlling cyclosporiasis in low- and middle-income countries. Growing globalization and international trade are critical factors in industrialized countries in preventing and controlling cyclosporiasis.[99]

Oocyst testing from the clinical samples is the commonly employed diagnostic tool for *Cylclosopora,* and multiple fecal samples should be tested to rule out the disease. Staining and molecular detection are also viable laboratory techniques. However, there is no commercially available serological technique for detecting the pathogen, and priority should be given to developing such diagnostic kits for field application.[102] Trimethoprim-sulfamethoxazole (TMP-SMX) is the drug of choice and is effective in immunocompetent and immunocompromised patients to treat cyclosporiasis.[103] Nitazoxanide, a broad-spectrum antiparasitic drug, has been effective for treating cyclosporiasis, especially during mixed parasitic infections.[102]

CASE STUDY

Lessons learned from *Cryptosporidium* outbreaks in developed and developing countries

Background

Globally *Cryptosporidium species* are a common nonbacterial cause of diarrhea in children under 5 years of age and people with compromised immune systems. The disease is mainly transmitted through ingestion of infectious oocysts, which are found in water contaminated with the feces of an infected individual or animal. Transmission can also occur associated with consumption of fresh produce irrigated with the contaminated water. While disease outbreaks in developed countries are most likely well detected and documented, the cases go unreported in low- and middle-income countries.[106,107]

Observations

In Nepal, fecal samples were collected from 253 children aged 5 years or below during the period of April 1998 to March 2004. Out of 253 children, 153 were exhibiting acute diarrhea and 100 were healthy controls. The study found that 35.5% of the children with acute diarrhea had protozoal infections and *Cryptosporidium* was detected in nine cases (5.6%). Positive cases were strongly associated with coinfection with HIV and TB cases, malnutrition, lack of potable water, and increased cryptosporidium emission rates.[108]

The US CDC has estimated that there were 444 cryptosporidiosis outbreaks in the United States during 2009–17, which resulted in 7465 cases and the leading causes of these outbreaks included swallowing contaminated recreational water, contact with infected livestock, and contact with infected persons in childcare settings.[109]

Key issues

The key issues associated with cryptosporidiosis in developing and developed countries vary greatly and are as follows.[106–109]

Developing countries:

- Overcrowded living conditions
- Poor potable drinking water quality
- Sharing the same water resources with livestock and animal contact
- Lack of toilet facilities
- Infected individuals living in the same household
- Increased *Cryptosporidium* emission rate
- Prevalence of comorbidity factors such as tuberculosis and AIDS.

Developed countries:

- Contaminated recreational water
- Travel to regions where this parasite is endemic
- International food trade from endemic regions
- Animal contact through farm visits or petting zoo visits.

Interventions

The key intervention strategy should be focused on three different stages of disease transmission:

- Survival of viable oocysts in the feces of animals.

(Continued)

CASE STUDY *(cont'd)*

- Contamination of water and produce with feces from infected animals.
- Preventive measures including proper hygiene, surveillance, and vaccination in animals.

Lessons learned

- Poor WASH (water, sanitation, and hygiene) conditions are principal contributors to the spread of gastrointestinal infections such as *Cryptosporidium.*
- Animal contact is a major issue in both developed and LMICs; however, livestock animals are major contributors in LMIC while pets and zoo animals play a major role in developed countries.

Recommendations

- Improve sanitation coverage in developing countries for a sustainable solution.
- Animal contact and case contact are relevant in both developed and developing countries; hence, better awareness and hygiene are required.
- Increase capacity building.
- Provide training for laboratory staff.
- Build better facilities.

Vision for the future

Increased consumption of imported and preprocessed food and drinks result in increased food- and waterborne infections in developed countries. However, the key issue among LMIC countries is different from those developed nations since most of their outbreaks are from contaminated water resources. Often, infected livestock and humans share unsanitary common water resources, which leads to epidemics. Here, we are dealing with two entirely different issues; in developed countries, the problem is primarily with the food supply chain and early detection failure. At the same time, the developing countries are affected by the lack of clean water. Therefore, the prevention, control, and response guidelines should approach the issue in developed and developing countries with different strategies.

Globalization has increased the food trade between different countries, which has resulted in the transmission of various pathogens between other countries.[104] Moreover, the quality standards vary among countries depending on the geopolitical situation. The methods and policies that helped developed nations reduce the incidences of various infections should be studied to create guidelines for low- and middle-income countries. However, geopolitical situations should be taken into consideration while forming these guidelines.

The basic infrastructure and knowledge gaps vary widely between developed and LMIC; therefore, "one size fits all" guidelines are not practical. Since the prominent disease agents and comorbidities vary between these countries, appropriate care should be taken while tailoring policies for these LMICs. For example, Norway has a very low prevalence of many common foodborne infections. Also, certain diseases such as parasitic outbreaks are well detected and documented in developed countries, while those diseases are neglected in LMIC. The end goal should be to propose preventive and control measures for LMIC by extracting the information which worked and not worked for other countries.

The gap in technology and bioinformatics capabilities also plays a significant role in the high prevalence of foodborne infections in LMIC. Developed nations routinely employ whole genome sequencing and bioinformatics to monitor and identify food- and waterborne pathogens. However, WGS and bioinformatics tools are limited in academia and research in LMIC. Moreover, the knowledge gap and difficulty in using molecular biology and bioinformatics tools appear to be the biggest hurdle for proper public health surveillance of food- and waterborne infections in LMIC. Collaboration and training in WGS and bioinformatics tools would reduce the gap between LMIC and developed nations and improve disease monitoring globally.[105] Table 10.2 lists the key issues, challenges and recommendations to prevent food- and water-borne infections in LMIC.

TABLE 10.2 Key issues and challenges involved in infections transmitted by food and water in low- and middle-income countries (LMICs).

Issue/parameter	LMIC challenge	Recommendation
Potable water	Lack of clean potable water for humans and animals.	Protect the main water source from contamination from animal excreta and runoff water from the surrounding environment. Household treatment and safe storage of the water.
Food production	Direct interaction and contact between animals, humans, and agricultural activities.	Proper cleaning of hands and fresh produce before and after handling livestock.
Contamination	Shared common water resources between animals and humans. Lack of proper sanitation facilities.	Avoid allowing livestock direct access to primary water sources such as ponds or streams; instead, provide water in separate containers. Proper management of both human and animal excreta.
Food processing	Several broad issues occur across food processing situations, including poor storage conditions, temperature abuse, cross-contamination, inadequate cooking and heating, and lack of proper implementation of HACCP and other regulations like good handling/ agricultural practices.	Maintain cold chain and perform routine inspections. Boil or filter water used in food processing if possible.
Comorbidities	Comorbidities such as AIDS and tuberculosis.	Improved nutrition, along with appropriate health care and treatment for comorbidities.

(*Continued*)

TABLE 10.2 Key issues and challenges involved in infections transmitted by food and water in low- and middle-income countries (LMICs).—cont'd

Issue/parameter	LMIC challenge	Recommendation
Testing	Lack of adequate facilities equipment and technically trained staff.	Capacity building and sharing of laboratory tools and equipment. Increase the availability of point-of-care tests.
Case reporting	Underreporting of illness and cases corresponding to outbreaks.	Implementation of an adequate public health surveillance system building up to the development of an epidemiological framework for disease detection and education dissemination.
WGS and bioinformatics	Lack of facilities and a knowledge gap in bioinformatics	Capacity building, training, and collaboration within each nation and developed nations. Develop a system where neighboring countries can communicate, provide training, and share knowledge and resources between developed and LMIC.

Actions

- Use One Health approach to achieve optimal health outcomes by recognizing the interconnection between people, animals, plants, and their shared environment.
- Enforce water, sanitation, and hygiene controls for livestock animals.
- Integrate public health surveillance of diseases globally by implementing a sentinel public health surveillance system in areas where the disease is not prevalent, and actively monitor outbreaks and the emergence of new infections.
- As much as possible, provide livestock access to clean water sources. Practically, this may be a challenge, as in LMIC, animals commonly use ponds or streams; instead, provide water in separate containers.
- Implement proper cleaning and hygiene practices before and after handling livestock.
- Maintain cold chain during food processing, transport, and storage.
- As much as possible, make proper veterinary care and treatment as well as nutrition available to support livestock.
- Capacity building for LMIC and sharing resources between LMIC and developed nations to build critical resources.
- Build training platforms and systems mainly within and between LMIC.
- Redefine sanitation; current definitions of sanitation aim to separate humans from their excreta but not from animal feces.
- Pay more attention to sanitation to reduce contact with animal feces.
- Dig the latrines deep enough so that the oocyst will not contaminate the soil or water.
- Avoid exposure to food and water potentially contaminated with zoonotic pathogens, such as animals that have suddenly died or water in rat-infested areas.

- Boiling or chemically treating potentially contaminated drinking water is advised to prevent waterborne infections.
- Wear protective clothing in high-risk areas for potential exposure to zoonotic infectious diseases, including additional precautions such as footwear and covering cuts and abrasions with occlusive dressings.
- Avoid direct and indirect contact with animal carcasses and do not eat meat from animals butchered after being found dead or ill. Even cooking contaminated meat does not completely eliminate the risk of contracting ingestion anthrax.
- Inspect meat to prevent anthrax in endemic regions. The risk of contracting anthrax is minimal to none in countries and areas where meat inspectors routinely inspect animals at abattoirs before, during, and after slaughter, or regions where anthrax in livestock is uncommon.
- Monitor for illness and use of postexposure prophylaxis after ingestion of meat from an animal suspected to be sick or to have died from anthrax, regardless of method of preparation.
- Develop rapid field diagnostic kits because low- and middle-income countries are often less equipped to diagnose endemic diseases.
- Develop easy-to-use and accessible "Point-of-Care Tests" (PCTs) because low-income countries are faced with human and financial constraints to successfully diagnose various diseases.
- Develop ready-to-use rapid diagnostic tests (RDTs) for areas with limited laboratory facilities.
- Follow basic principles of Codex Alimentarius while preparing guidelines for monitoring and public health surveillance programs (Box 10.1).
- Use the F-diagram to identify the key interception points and interception methods in disease transmission of food- and water-borne infections (Fig. 10.1).

BOX 10.1

Basic principles of codex alimentarius, CSG 94 –2021: Guidelines on integrated monitoring and public health surveillance of foodborne antimicrobial resistance

These guidelines are intended to provide government agencies with the assistance needed in the design and implementation of integrated monitoring and public health surveillance programs. The guidelines are tailored with a specific focus on foodborne antimicrobial resistance in the food chain and food production environment. The guidelines are created on nine basic principles. These principles should be followed whenever an AMR guideline is being developed.[110]

Principle 1: A One Health approach should be applied whenever possible and applicable when establishing monitoring and public health surveillance programs for foodborne AMR, contributing to the food safety component of such an approach.

Continued

BOX 10.1 *(cont'd)*

Principle 2: Monitoring and public health surveillance programs are an important part of national strategy(ies) to minimize and contain the risk of foodborne AMR.

Principle 3: Risk analysis should guide the design, implementation, and evaluation of monitoring and public health surveillance programs.

Principle 4: Monitoring and public health surveillance programs should be designed to generate data on AMR and AMU, in relevant sectors to inform risk analysis.

Principle 5: Monitoring and public health surveillance programs should be tailored to national priorities and should be designed and implemented to allow continuous improvement as resources permit.

Principle 6: Priority for implementation of monitoring and public health surveillance programs should be given to the most relevant foodborne AMR and/or AMR food safety issues (which are the defined combinations of the food commodity, the AMR microorganism and determinants, and the antimicrobial agent(s) to which resistance is expressed as described in CXG 77–2011) from a public health perspective, taking into account national priorities.

Principle 7: Monitoring and public health surveillance programs should incorporate, to the extent practicable, the identification of new and emerging foodborne AMR or trends and should be designed to inform the epidemiological investigation.

Principle 8: Laboratories involved in monitoring and public health surveillance should have effective quality assurance/ management systems in place.

Principle 9: Monitoring and public health surveillance programs should aim to harmonize laboratory methodology, data collection, analysis, and reporting across sectors according to national priorities and resources as part of an integrated approach. Use of internationally recognized, standardized, and validated methods and harmonized interpretative criteria, where available, contributes to the comparability of data, facilitates the multisectoral exchange and analysis of data, and enhances an integrated approach to data management, analysis, and interpretation.

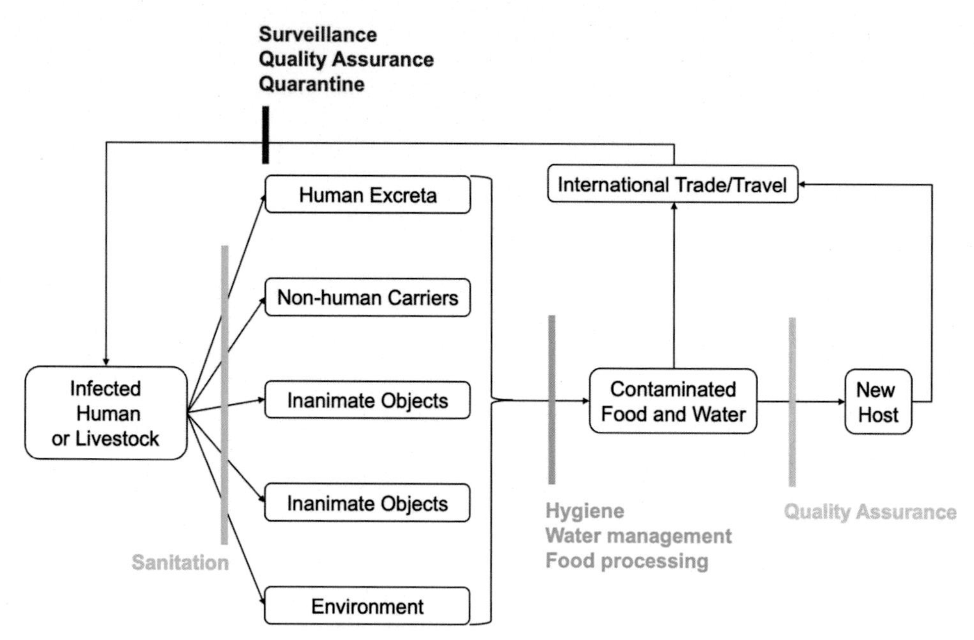

FIGURE 10.1 F-diagram of key interception points in disease transmission of food- and water-borne pathogens.

References

1. Grace D, Mutua F, Ochungo P, et al. *Mapping of Poverty and Likely Zoonoses Hotspots*. 2012.
2. Heymann DL, Chen L, Takemi K, et al. Global health security: the wider lessons from the west African ebola virus disease epidemic. *Lancet*. 2015;385(9980):1884–1901.
3. Phua K, Lee LK. Meeting the challenge of epidemic infectious disease outbreaks: an agenda for research. *J Publ Health Pol*. 2005;26(1):122–132.
4. Allen T, Murray KA, Zambrana-Torrelio C, et al. Global hotspots and correlates of emerging zoonotic diseases. *Nat Commun*. 2017;8(1):1–10. https://doi.org/10.1038/s41467-017-00923-8. https://www.nature.com/articles/s41467-017-00923-8. Accessed August 17, 2021.
5. Cleaveland S, Sharp J, Abela-Ridder B, et al. One health contributions towards more effective and equitable approaches to health in low- and middle-income countries. *Phil Trans Biol Sci*. 2017;372(1725):20160168. https://doi.org/10.1098/rstb.2016.0168. https://royalsocietypublishing.org/doi/full/10.1098/rstb.2016.0168. Accessed August 17, 2021.
6. WHO. Control of neglected tropical diseases. https://www.who.int/teams/control-of-neglected-tropical-diseases/overview.
7. Mableson HE, Okello A, Picozzi K, Welburn SC. Neglected zoonotic diseases—the long and winding road to advocacy. *PLoS Neglected Trop Dis*. 2014;8(6):e2800.
8. Fasanella A, Galante D, Garofolo G, Jones MH. Anthrax undervalued zoonosis. *Vet Microbiol*. 2010;140(3):318–331. https://doi.org/10.1016/j.vetmic.2009.08.016. https://www.sciencedirect.com/science/article/pii/S0378113509003769. Accessed August 17, 2021.
9. Rubach MP, Halliday JEB, Cleaveland S, Crump JA. Brucellosis in low-income and middle-income countries. *Curr Opin Infect Dis*. 2013;26(5):404–412. https://doi.org/10.1097/QCO.0b013e3283638104. https://www.ncbi.nlm.nih.gov/pmc/articles/PMC3888775/. Accessed August 17, 2021.
10. Hartskeerl RA, Collares-Pereira M, Ellis WA. Emergence, control and re-emerging leptospirosis: dynamics of infection in the changing world. *Clin Microbiol Infection*. 2011;17(4):494–501.
11. Lowenstein C, Waters WF, Roess A, Leibler JH, Graham JP. Animal husbandry practices and perceptions of zoonotic infectious disease risks among livestock keepers in a rural parish of Quito, Ecuador. *Am J Trop Med Hyg*.

2016;95(6):1450–1458. https://doi.org/10.4269/ajtmh.16-0485. https://www.ncbi.nlm.nih.gov/pubmed/27928092.

12. Zambrano LD, Levy K, Menezes NP, Freeman MC. Human diarrhea infections associated with domestic animal husbandry: a systematic review and meta-analysis. *Trans R Soc Trop Med Hyg*. 2014;108(6):313–325.
13. Al-Ghamdi MS, El-Morsy F, Al-Mustafa ZH, Al-Ramadhan M, Hanif M. Antibiotic resistance of *Escherichia coli* isolated from poultry workers, patients and chicken in the eastern province of Saudi Arabia. *Trop Med Int Health*. 1999;4(4):278–283.
14. Locking ME, O'BRIEN SJ, Reilly WJ, et al. Risk factors for sporadic cases of *Escherichia coli* O157 infection: the importance of contact with animal excreta. *Epidemiol Infect*. 2001;127(2):215–220.
15. Belongia EA, Chyou P, Greenlee RT, Perez-Perez G, Bibb WF, DeVries EO. Diarrhea incidence and farm-related risk factors for *Escherichia coli* O157: H7 and campylobacter jejuni antibodies among rural children. *J Infect Dis*. 2003;187(9):1460–1468.
16. Newell DG, Koopmans M, Verhoef L, et al. Food-borne diseases — the challenges of 20years ago still persist while new ones continue to emerge. *Int J Food Microbiol*. 2010;139:S3–S15. https://doi.org/10.1016/j.ijfoodmicro.2010.01.021. https://www.sciencedirect.com/science/article/pii/S0168160510000383. Accessed March 30, 2021.
17. Acheson DWK. Food and waterborne illnesses. *Encyclopedia Microbiol*; 2009:365–381. https://doi.org/10.1016/B978-012373944-5.00183-8. https://www.ncbi.nlm.nih.gov/pmc/articles/PMC7173519/. Accessed March 31, 2021.
18. OUTBREAK, what IS A FOODBORNE ILLNESS. 13. Introduction to Foodborne Illness Outbreak Investigations CopyRight.
19. Scallan E, Hoekstra RM, Angulo FJ, et al. Foodborne illness acquired in the United States—major pathogens. *Emerg Infect Dis*. 2011;17(1):7.
20. Fleckenstein JM, Bartels SR, Drevets PD, Bronze MS, Drevets DA. Infectious agents of food-and water-borne illnesses. *Am J Med Sci*. 2010;340(3):238–246.
21. Torgerson PR, Devleesschauwer B, Praet N, et al. World health organization estimates of the global and regional disease burden of 11 foodborne parasitic diseases, 2010: a data synthesis. *PLoS Med*. 2015;12(12). https://doi.org/10.1371/journal.pmed.1001920. https://www.ncbi.nlm.nih.gov/pmc/articles/PMC4668834/. Accessed April 2, 2021.
22. Ortega YR, Kváč M. Foodborne protozoa. In: *Guide to Foodborne Pathogens*. John Wiley and Sons, Ltd; 2013:303–316. https://onlinelibrary.wiley.com/doi/abs/10.1002/9781118684856.ch19. Accessed April 2, 2021.
23. *WHO's First Ever Global Estimates of Foodborne Diseases Find Children under 5 Account for Almost One Third of Deaths*; 2015. https://www.who.int/news/item/03-12-2015-who-s-first-ever-global-estimates-of-foodborne-diseases-find-children-under-5-account-for-almost-one-third-of-deaths. Accessed February 4, 2021.
24. Croxen MA, Finlay BB. Molecular mechanisms of *Escherichia coli* pathogenicity. *Nat Rev Microbiol*. 2010;8(1):26–38.
25. Tarr PI, Gordon CA, Chandler WL. Shiga-toxin-producing *Escherichia coli* and haemolytic uraemic syndrome. *Lancet*. 2005;365(9464):1073–1086.
26. Bell BP, Goldoft M, Griffin PM, et al. A multistate outbreak of *Escherichia coli* o157: H7—associated bloody diarrhea and hemolytic uremic syndrome from hamburgers: the Washington experience. *JAMA*. 1994;272(17):1349–1353.
27. Wendel AM, Johnson DH, Sharapov U, et al. Multistate outbreak of *Escherichia coli* O157: H7 infection associated with consumption of packaged spinach, August–September 2006: the Wisconsin investigation. *Clin Infect Dis*. 2009;48(8):1079–1086.
28. Cody SH, Glynn MK, Farrar JA, et al. An outbreak of *Escherichia coli* O157: H7 infection from unpasteurized commercial apple juice. *Ann Intern Med*. 1999;130(3):202–209.
29. Breuer T, Benkel DH, Shapiro RL, et al. A multistate outbreak of *Escherichia coli* O157: H7 infections linked to alfalfa sprouts grown from contaminated seeds. *Emerg Infect Dis*. 2001;7(6):977.
30. Griffin PM, Ostroff SM, Tauxe RV, et al. Illnesses associated with *Escherichia coli* 0157: H7 infections: a broad clinical spectrum. *Ann Intern Med*. 1988;109(9):705–712.
31. Brooks JT, Sowers EG, Wells JG, et al. Non-O157 Shiga toxin–producing *Escherichia coli* infections in the United States, 1983–2002. *J Infect Dis*. 2005;192(8):1422–1429.

32. Wong CS, Jelacic S, Habeeb RL, Watkins SL, Tarr PI. The risk of the hemolytic–uremic syndrome after antibiotic treatment of *Escherichia coli* O157: H7 infections. *N Engl J Med*. 2000;342(26):1930–1936.
33. Deborah Chen H, Frankel G. Enteropathogenic *Escherichia coli*: unravelling pathogenesis. *FEMS Microbiol Rev*. 2005;29(1):83–98.
34. Shah N, DuPont HL, Ramsey DJ. Global etiology of travelers' diarrhea: systematic review from 1973 to the present. *Am J Trop Med Hyg*. 2009;80(4):609–614.
35. Fleckenstein JM, Hardwidge PR, Munson GP, Rasko DA, Sommerfelt H, Steinsland H. Molecular mechanisms of enterotoxigenic *Escherichia coli* infection. *Microb Infect*. 2010;12(2):89–98.
36. Beatty ME, Adcock PM, Smith SW, et al. Epidemic diarrhea due to enterotoxigenic *Escherichia coli*. *Clin Infect Dis*. 2006;42(3):329–334.
37. Daniels NA, Neimann J, Karpati A, et al. Traveler's diarrhea at sea: three outbreaks of waterborne enterotoxigenic *Escherichia coli* on cruise ships. *J Infect Dis*. 2000;181(4):1491–1495.
38. Hall JA, Goulding JS, Bean NH, Tauxe RV, Hedberg CW. Epidemiologic profiling: evaluating foodborne outbreaks for which no pathogen was isolated by routine laboratory testing: United States, 1982–9. *Epidemiol Infect*. 2001;127(3):381–387.
39. World Health Organization. *Cholera*; 2021. https://www.who.int/en/news-room/fact-sheets/detail/cholera. Accessed March 4, 2021.
40. Ali M, Nelson AR, Lopez AL, Sack DA. Updated global burden of cholera in endemic countries. *PLoS Neglected Trop Dis*. 2015;9(6):e0003832.
41. Ajayi A, Smith SI. Recurrent cholera epidemics in Africa: which way forward? A literature review. *Infection*. 2019;47(3):341–349.
42. López AL, You YA, Kim YE, Sah B, Maskery B, Clemens J. The global burden of cholera. *Bull World Health Organ*. 2012;90(3).
43. Clemens JD, Nair GB, Ahmed T, Qadri F, Holmgren J. Cholera. *Lancet*. 2017;390(10101):1539. https://doi.org/10.1016/S0140-6736(17)30559-7. https://www.sciencedirect.com/science/article/pii/S0140673617305597.
44. Schmid-Hempel P, Frank SA. Pathogenesis, virulence, and infective dose. *PLoS Pathog*. 2007;3(10):e147. https://doi.org/10.1371/journal.ppat.0030147. https://journals.plos.org/plospathogens/article?id=10.1371/journal.ppat.0030147. Accessed April 3, 2021.
45. mondiale de la Santé O, World Health Organization. Cholera vaccines: WHO position paper–August 2017–Vaccins anticholériques: note de synthèse de l'OMS–août 2017. *Weekly Epidemiol Rec = Relevé épidémiologique hebdomadaire*. 2017;92(34):477–498.
46. Hsiao A, Hall AH, Mogasale V, Quentin W. Vaccine. *Vaccine*. 1983;36(30):4404–4424. https://www.sciencedirect.com/science/article/pii/S0264410X18307990.
47. Costa F, Hagan JE, Calcagno J, et al. Global morbidity and mortality of leptospirosis: a systematic review. *PLoS Neglected Trop Dis*. 2015;9(9):e0003898.
48. Ko AI, Goarant C, Picardeau M. Leptospira: the dawn of the molecular genetics era for an emerging zoonotic pathogen. *Nat Rev Microbiol*. 2009;7(10):736–747.
49. Reis RB, Ribeiro GS, Felzemburgh RD, et al. Impact of environment and social gradient on leptospira infection in urban slums. *PLoS Neglected Trop Dis*. 2008;2(4):e228.
50. Khalil H, Santana R, de Oliveira D, et al. Poverty, sanitation, and leptospira transmission pathways in residents from four Brazilian slums. *PLoS Neglected Trop Dis*. 2021;15(3):e0009256.
51. Jesus MS, Silva LA, Lima KMda S, Fernandes OCC. Cases distribution of leptospirosis in city of Manaus, state of Amazonas, Brazil, 2000-2010. *Rev Soc Bras Med Trop*. 2012;45:713–716.
52. Stern EJ, Galloway R, Shadomy SV, et al. Outbreak of leptospirosis among adventure race participants in Florida, 2005. *Clin Infect Dis*. 2010;50(6):843–849.
53. Katz AR, Ansdell VE, Effler PV, Middleton CR, Sasaki DM. Assessment of the clinical presentation and treatment of 353 cases of laboratory-confirmed leptospirosis in Hawaii, 1974–1998. *Clin Infect Dis*. 2001;33(11):1834–1841.
54. Vanasco NB, Schmeling MF, Lottersberger J, Costa F, Ko AI, Tarabla HD. Clinical characteristics and risk factors of human leptospirosis in Argentina (1999–2005). *Acta Trop*. 2008;107(3):255–258.
55. Brett-Major DM, Coldren R. Antibiotics for leptospirosis. *Cochrane Database Syst Rev*. 2012;(2).
56. Suputtamongkol Y, Niwattayakul K, Suttinont C, et al. An open, randomized, controlled trial of penicillin, doxycycline, and cefotaxime for patients with severe leptospirosis. *Clin Infect Dis*. 2004;39(10):1417–1424.

57. Panaphut T, Domrongkitchaiporn S, Vibhagool A, Thinkamrop B, Susaengrat W. Ceftriaxone compared with sodium penicillin G for treatment of severe leptospirosis. *Clin Infect Dis*. 2003;36(12):1507–1513.
58. World Health Organization. *International Health Regulations (2005)*. World Health Organization; 2008.
59. CDC. Anthrax. https://www.cdc.gov/anthrax/.
60. Shadomy SV, Smith TL. Anthrax: zoonosis update. *JAVMA*. 2008;233(1):63–72.
61. Fasanella A. *Anthrax Undervalued Zoonosis Antonio Fasanella*, Domenico Galante, Giuliano Garofolo and Martin Hugh Jones B 2*. 2009.
62. Vieira AR, Salzer JS, Traxler RM, et al. Enhancing surveillance and diagnostics in anthrax-endemic countries. *Emerg Infect Dis*. 2017;23(Suppl 1):S147.
63. Pappas G, Papadimitriou P, Akritidis N, Christou L, Tsianos EV. The new global map of human brucellosis. *Lancet Infect Dis*. 2006;6(2):91–99.
64. Yagupsky P, Morata P, Colmenero JD. Laboratory diagnosis of human brucellosis. *Clin Microbiol Rev*. 2019;33(1):73.
65. Vrioni G, Pappas G, Priavali E, Gartzonika C, Levidiotou S. An eternal microbe: Brucella DNA load persists for years after clinical cure. *Clin Infect Dis*. 2008;46(12):e131–e136.
66. CDC. Brucellosis Reference Guide: Exposures,Testing, and Prevention. https://www.cdc.gov/brucellosis/pdf/brucellosi-reference-guide.pdf.
67. WHO. Brucellosis in Humans and Animals. https://www.who.int/publications/i/item/9789241547130.
68. Bennett JE, Dolin R, Blaser MJ. *Mandell, Douglas, and Bennett's Principles and Practice of Infectious Diseases: 2-volume Set*. Vol 2. Elsevier Health Sciences; 2014.
69. Patel MM, Hall AJ, Vinjé J, Parashar UD. Noroviruses: a comprehensive review. *J Clin Virol*. 2009;44(1):1–8.
70. Hutson AM, Atmar RL, Estes MK. Norovirus disease: changing epidemiology and host susceptibility factors. *Trends Microbiol*. 2004;12(6):279–287.
71. Teunis PF, Moe CL, Liu P, et al. Norwalk virus: how infectious is it? *J Med Virol*. 2008;80(8):1468–1476.
72. Ozawa K, Oka T, Takeda N, Hansman GS. Norovirus infections in symptomatic and asymptomatic food handlers in Japan. *J Clin Microbiol*. 2007;45(12):3996–4005.
73. Martinez A, Dominguez A, Torner N, et al. Epidemiology of foodborne norovirus outbreaks in Catalonia, Spain. *BMC Infect Dis*. 2008;8(1):1–7.
74. Widdowson M, Sulka A, Bulens SN, et al. Norovirus and foodborne disease, United States, 1991–2000. *Emerg Infect Dis*. 2005;11(1):95.
75. Robilotti E, Deresinski S, Pinsky BA. Norovirus. *Clin Microbiol Rev*. 2015;28(1):134–164.
76. Dolin R, Treanor JJ. Noroviruses and other caliciviruses. In: *Mandell, Douglas, and Bennett's Principles and Practice of Infectious Diseases*. 7th ed. Philadelphia: Churchill Livingstone; 2010:2399–2406.
77. Rossignol J, EL-GOHARY YM. Nitazoxanide in the treatment of viral gastroenteritis: a randomized double-blind placebo-controlled clinical trial. *Aliment Pharmacol Ther*. 2006;24(10):1423–1430.
78. Svraka S, Duizer E, Vennema H, et al. Etiological role of viruses in outbreaks of acute gastroenteritis in the Netherlands from 1994 through 2005. *J Clin Microbiol*. 2007;45(5):1389–1394.
79. Bajolet O, Chippaux-Hyppolite C. Rotavirus and other viruses of diarrhea. *Bull Soc Pathol Exot*. 1998;91(5 Pt 1–2):432–437.
80. Richards GP. Enteric virus contamination of foods through industrial practices: a primer on intervention strategies. *J Ind Microbiol Biotechnol*. 2001;27(2):117–125.
81. Vesikari T, Matson DO, Dennehy P, et al. Safety and efficacy of a pentavalent human–bovine (WC3) reassortant rotavirus vaccine. *N Engl J Med*. 2006;354(1):23–33.
82. Cook N, Bridger J, Kendall K, Gomara MI, El-Attar L, Gray J. The zoonotic potential of rotavirus. *J Infect*. 2004;48(4):289–302.
83. Lees D. Viruses and bivalve shellfish. *Int J Food Microbiol*. 2000;59(1–2):81–116.
84. Ansari SA, Springthorpe VS, Sattar SA. Survival and vehicular spread of human rotaviruses: possible relation to seasonality of outbreaks. *Rev Infect Dis*. 1991;13(3):448–461.
85. Acha PN, Szyfres B. *Zoonoses and Communicable Diseases Common to Man and Animals*. Vol 580. Pan American Health Org; 2003.
86. Anderson EJ, Weber SG. Rotavirus infection in adults. *Lancet Infect Dis*. 2004;4(2):91–99.

87. Tate JE, Burton AH, Boschi-Pinto C, Steele AD, Duque J, Parashar UD. 2008 estimate of worldwide rotavirus-associated mortality in children younger than 5 years before the introduction of universal rotavirus vaccination programmes: a systematic review and meta-analysis. *Lancet Infect Dis*. 2012;12(2):136–141.
88. Ryan U, Hijjawi N, Xiao L. Foodborne cryptosporidiosis. *Int J Parasitol*. 2018;48(1):1–12.
89. Chalmers RM, Davies AP. Minireview: clinical cryptosporidiosis. *Exp Parasitol*. 2010;124(1):138–146.
90. Centers for Disease Control and Prevention, (CDC). *Cryptosporidiosis Summary Report —National Notifiable Diseases Surveillance System, United States, 2017*; 2019. https://www.cdc.gov/healthywater/surveillance/pdf/2017-Cryptosporidiosis-NNDSS-Report-508.pdf. Accessed January 3, 2022.
91. Painter JE, Hlavsa MC, Collier SA, Xiao L, Yoder JS. Cryptosporidiosis surveillance—United States, 2011–2012. *Morb Mortal Wkly Rep - Surveillance Summ*. 2015;64(3):1–14.
92. Ryan UM, Feng Y, Fayer R, Xiao L. Taxonomy and molecular epidemiology of cryptosporidium and giardia–a 50 year perspective (1971–2021). *Int J Parasitol*. 2021;51(13–14):1099–1119.
93. Hofstra N, Bouwman AF, Beusen AHW, Medema GJ. Theœ science of the total environment. *Sci Total Environ*. 2013;442:10–19. https://www.sciencedirect.com/science/article/pii/S0048969712012958.
94. Medema G, Teunis P, Blokker M, et al. *Risk Assessment of Cryptosporidium in Drinking Water*. World Health Organization; 2009:143.
95. Chalmers RM, Katzer F. Looking for cryptosporidium: the application of advances in detection and diagnosis. *Trends Parasitol*. 2013;29(5):237–251.
96. Amadi B, Mwiya M, Sianongo S, et al. High dose prolonged treatment with nitazoxanide is not effective for cryptosporidiosis in HIV positive zambian children: a randomised controlled trial. *BMC Infect Dis*. 2009;9(1):1–7.
97. Rossle NF, Latif B. Cryptosporidiosis as threatening health problem: a review. *Asian Pac J Trop Biomed*. 2013;3(11):916–924. https://doi.org/10.1016/S2221-1691(13)60179-3. https://www.sciencedirect.com/science/article/pii/S2221169113601793. Accessed April 4, 2021.
98. Sun T, Ilardi CF, Asnis D, et al. Light and electron microscopic identification of cyclospora species in the small intestine: evidence of the presence of asexual life cycle in human host. *Am J Clin Pathol*. 1996;105(2):216–220.
99. Chacín-Bonilla L. Epidemiology of cyclospora cayetanensis: a review focusing in endemic areas. *Acta Trop*. 2010;115(3):181–193. https://doi.org/10.1016/j.actatropica.2010.04.001.
100. Shields JM, Olson BH. Cyclospora cayetanensis: a review of an emerging parasitic coccidian. *Int J Parasitol*. 2003;33(4):371–391.
101. Smith HV, Paton CA, Mitambo MM, Girdwood RW. Sporulation of cyclospora sp. oocysts. *Appl Environ Microbiol*. 1997;63(4):1631–1632.
102. Almeria S, Cinar HN, Dubey JP. Cyclospora cayetanensis and cyclosporiasis: an update. *Microorganisms*. 2019;7(9):317.
103. Mathison BA, Pritt BS. Cyclosporiasis—updates on clinical presentation, pathology, clinical diagnosis, and treatment. *Microorganisms*. 2021;9(9):1863.
104. Jurabaevich SN, Bulturbayevich MB. Directions for food security in the context of globalization. *Innova Technol: Method Res J*. 2021;2(01):9–16.
105. Apruzzese I, Song E, Bonah E, et al. Investing in food safety for developing countries: opportunities and challenges in applying whole-genome sequencing for food safety management. *Foodborne Pathog Dis*. 2019;16(7):463–473. https://doi.org/10.1089/fpd.2018.2599. https://www.liebertpub.com/doi/abs/10.1089/fpd.2018.2599.
106. Pumipuntu N, Piratae S. Cryptosporidiosis: a zoonotic disease concern. *Vet World*. 2018;11(5):681.
107. Innes EA, Chalmers RM, Wells B, Pawlowic MC. A one health approach to tackle cryptosporidiosis. *Trends Parasitol*. 2020;36(3):290–303.
108. Mukhopadhyay C, Wilson G, Pradhan D, Shivananda P. Intestinal protozoal infestation profile in persistent diarrhea in children below age 5 years in western Nepal. *Southeast Asian J Trop Med Publ Health*. 2007;38(1):13–19. Accessed April 5, 2021.
109. Gharpure R. Cryptosporidiosis outbreaks — United States, 2009–2017. *MMWR Morb Mortal Wkly Rep*. 2019;68. https://doi.org/10.15585/mmwr.mm6825a3. https://www.cdc.gov/mmwr/volumes/68/wr/mm6825a3.htm. Accessed April 5, 2021.
110. FAO and WHO. Codex Alimentarius. https://www.oie.int/fileadmin/Home/eng/Media_Center/docs/EN_TripartiteZoonosesGuide_webversion.pdf. Accessed May 26, 2022.

CHAPTER

11

Urbanization, human societies, and pandemic preparedness and mitigation

Gonzalo M. Vazquez-Prokopec[1], *Laura E. Binkley*[2,3], *Hector Gomez Dantes*[4], *Amanda M. Berrian*[2,3], *Valerie A. Paz Soldan*[5], *Pablo C. Manrique-Saide*[6] *and Thomas R. Gillespie*[1]

[1]Department of Environmental Sciences, Emory University, Atlanta, GA, United States; [2]Department of Veterinary Preventive Medicine, College of Veterinary Medicine, The Ohio State University, Columbus, OH, United States; [3]Global One Health initiative (GOHi), The Ohio State University, Columbus, OH, United States; [4]Center for Health System Research, National Institute of Public Health, Cuernavaca, Mexico; [5]Department of Tropical Medicine, Tulane School of Public Health and Tropical Medicine, New Orleans, LA, United States; [6]Department of Zoology, School of Biological Sciences, Autonomous University of Yucatan, Merida, Yucatan, Mexico

Introduction

In the history of humans on earth, the number, size, social, technological, and environmental impacts of cities are unprecedented. As of 2007, the world passed a major demographic milestone, with more than half of people living in cities (30% increase from 1950[1]) (Fig. 11.1). The populations in all countries of the world are becoming increasingly urbanized, with current projections from the United Nations (UN) Population Division predicting that by 2050, 68% of the world's population (totaling 6.7 billion people) will live in an urban area.[1] According to these estimates, most human population growth over the next century will be accounted for by the growing number of city dwellers in low- and middle-income countries (LMICs) in Africa and Southeast Asia[1] (Fig. 11.1). There is also much heterogeneity in what constitutes an

Modernizing Global Health Security to Prevent, Detect, and Respond
https://doi.org/10.1016/B978-0-323-90945-7.00014-2

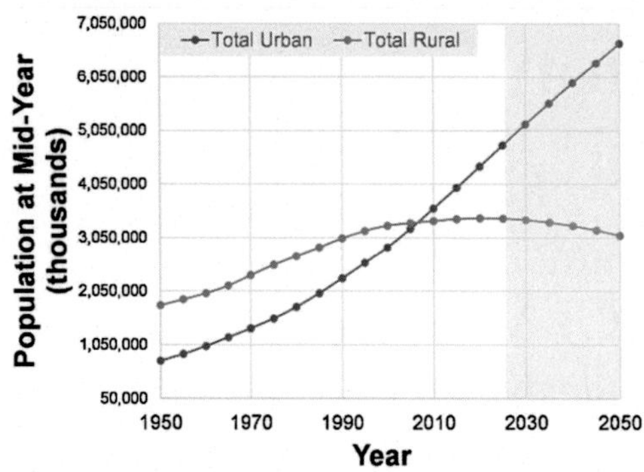

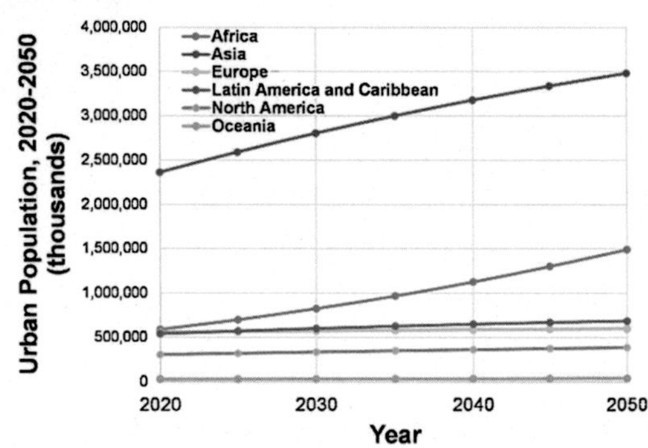

FIGURE 11.1 **Global urbanization trends.** Cities are unique landscapes where social, economic, and environmental factors shape the location and type of land-use (top panel). Such anatomies of urban environments, combined with the unique opportunities provided by urban living, have fueled rapid urban population growth compared to rural settings (bottom left panel). Projections of the global population size show increasing urban populations across all continents, with Africa experiencing the most rapid urban expansion (bottom right panel).

urban area or city, and the types of growth within each: while mega-cities (metropolitan areas with more than 10 million inhabitants) have increased from 14 in 1995 to 34 in 2017, most of the future growth is expected to occur in cities of less than 500,000 inhabitants.[1] The drivers fueling this major demographic trend primarily involve a surplus of births over deaths, migration from rural to urban areas with the expectation of better jobs, healthcare, education, higher living standards, and rapid urbanization of formerly rural areas (Fig. 11.1).[1–5]

Despite this rapid urbanization trend, most cities in LMICs have grown with minimal planning and limited resources and infrastructure to adapt to their growing population and needs. A direct consequence of such unplanned growth has been the proliferation of urban slums (defined as "a group of individuals that live under the same roof that lack one or more of the following conditions: access to improved water and sanitation, sufficient living space, durability of housing, and secure tenure"[6]). While it may be difficult to identify general urban trends that apply to all urban areas (e.g., across levels of economic development and regardless of their culture, political systems, social structure, geography, and built environment), there is universal global consensus about the pervasive occurrence and negative economic, social, and public health impacts of inequality (including the growth of slums in

LMICs) in urban areas.[6–8] Moreover, urban slum dwellers and those living in underserved communities tend to have less favorable health outcomes for both infectious and noncommunicable diseases (NCDs), compared to those who are better-off.[7,9,10]

Urban areas are also remarkably heterogeneous with regard to the distribution of infectious diseases and NCDs, with a higher burden of disease concentrated in some areas or segments of the population.[9–12] The impact of emerging infectious diseases, such as COVID-19, both in high- and low-income countries, has been particularly profound in impoverished urban areas.[8,13,14] For COVID-19, high-density living conditions (e.g., multiple families in one home, crowded neighborhoods), reliance on overcrowded public transport, informal work situations (e.g., day-to-day wage earners), and lack of access to healthcare or information technology (e.g., digital literacy) have made it difficult to implement public health social measures, also known as nonpharmaceutical interventions (NPIs), and to follow recommended measures such as social distancing, self-isolation, or work from home.[15] Similarly, the recent Zika virus pandemic had a more prominent impact on those populations lacking access to proper mosquito prevention tools and reproductive health options (see case study below).[16] Monkeypox, the latest emerging pathogen declared a Public Health Emergency of International Concern (PHEIC) by the World Health Organization (WHO), is also disproportionately impacting stigmatized urban populations that, in LMICs, do not have the same access to effective vaccines as in developed nations.[17] Therefore, addressing human health and epidemic preparedness in cities requires an explicit consideration of the spatial heterogeneity in the fabric of social, environmental, and economic determinants of health in which urban dwellers live.[18]

Despite its many challenges, urbanization is also considered a catalyst for innovation and global sustainable development.[19,20] The UN Sustainable Development Goals (SDGs) 2030 explicitly acknowledge that "fostering sustainable cities and communities" are key mechanisms to reduce poverty and address the most pressing global challenges, from climate change to pandemics.[8] Although urbanized areas occupy only 2% of the world's landmass, these account for about 60% of all residential water use, 75% of energy use, 80% of wood used for industrial purposes, 80% of human greenhouse gas emissions, and over 90% of all COVID-19 cases worldwide.[21] Emphasis on public health surveillance and infectious disease mitigation efforts in urban areas is appropriate, particularly in LMICs, and has implications at local, regional, and global levels. However, the 2005 International Health Regulations (IHR)[22] do not explicitly provide guidance for public health surveillance and containment of health threats in urban areas, revealing both a prior lack of consideration of their key epidemiological role, and an opportunity for future change. Similarly, while the GHS Agenda (GHSA) has helped build capacity in Africa and Asia on immunization and One Health approaches, urban areas are not outlined as critical to strengthen public health surveillance and pandemic preparedness.[23]

Reaching the goal of sustainable and healthy cities will require a systems-thinking approach—an approach with a larger view of the multiple components of the system and how these interact with one another—with explicit consideration of the complexity inherent to urban environments and societies and the role they have in fueling pathogen transmission from local to global scales. This chapter focuses on identifying major determinants of infectious disease transmission and pandemic spread in urban areas including: (a) the complexity that dominates urban processes can influence the rate of transmission and extent of public health response to pandemics; (b) human behavior and social interactions as enablers of infectious disease transmission; (c) the human–animal interface in cities playing a key role in urban zoonotic pathogen spillover and transmission; (d) the complex political and administrative structures in cities that often stall planning against and response to infectious disease threats; and (e) ports of entry in cities playing a vital role in pathogen introduction and circulation. In the following sections, we provide an overview of the global health relevance of such factors and identify key gaps in policies that are relevant for future revisions of the IHR.

Literature review and gap analysis

While an extensive body of literature on the impact of urbanization on infectious diseases exists,[7,9,11,18,24–26] there is ample consensus that multiple drivers influence specific disease systems differently. This section outlines the major gaps in knowledge, as well as in the IHR and other policies (e.g.,[23]), that pertain to the transmission and prevention of infectious diseases in urban areas.

Complexity dominates urban processes

Research over the past 50 years, led by urban scientists and ecologists, has shown that human influence on urban ecosystem functioning and response is the result of coupled human (social) and ecological processes that occur within a changing economic and urban landscape.[4,20,27–31] This complexity was captured by urban scientists, who generated a conceptual framework known as the rural-to-urban gradient[4,32] (Fig. 11.2) to depict the natural transition from rural areas to peri-urban, middle-class housing, industrial terrain, and to the central business district (or core administrative area). Throughout this "urban development gradient," there are cross-cutting and interconnected heterogeneous elements that are constantly evolving including: ecological (e.g., biodiversity, tree cover, air pollution), social (e.g., human mobility and interactions), and public health (e.g., healthcare infrastructure, access to clean water, exposure to infectious diseases, exposure to vector-borne or zoonotic diseases) processes[4,28,32] (Fig. 11.2). What makes this framework useful is the concept that multiple conditions (e.g., poverty, environmental degradation, infectious disease transmission) may be concentrated in some areas. If multiple drivers of poor health are addressed in these target areas, overall health and welfare metrics could be improved in the entire city.[33] Understanding of only one element of the system is insufficient for the planning and management of health services.[4,28,32]

As the Nobel-Prize winning economist Paul Romer has argued[34]:

> The urban environment that humans are so busily creating is many things: a biological environment, a social environment, a built environment, a market environment, a business environment, and a political environment. It includes not only the versions of these environments that exist inside a single city, but also those that are emerging from the interaction between cities.

A good example of complexity in urban processes can be exemplified by the emergence of urban segregation and inequality. For instance, the spatial configuration of Sao Paulo, Brazil,

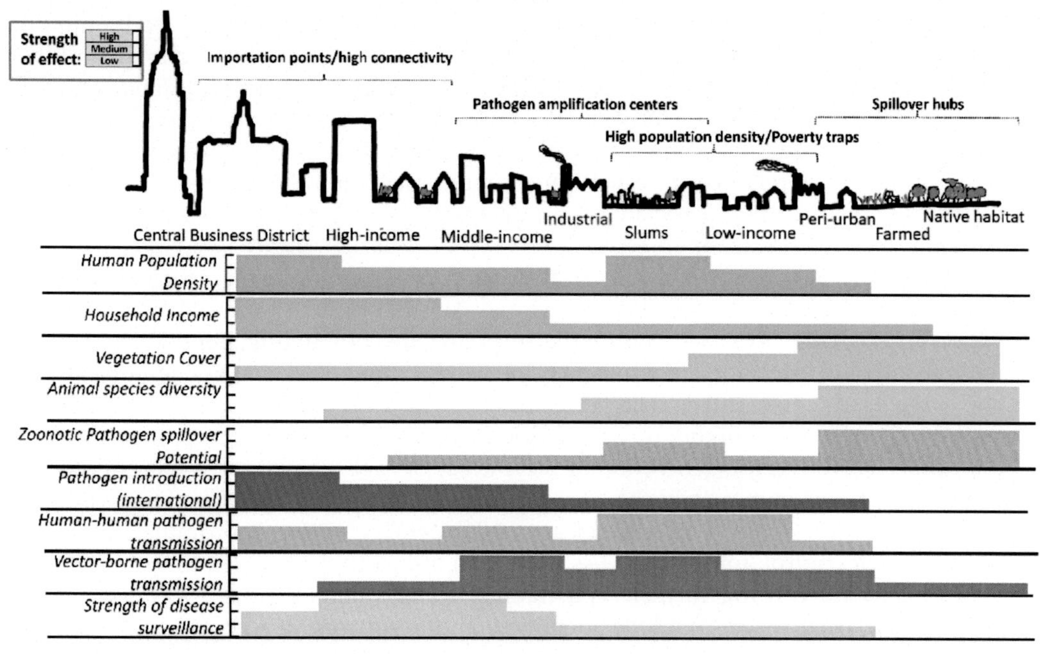

FIGURE 11.2 **The rural to urban gradient.** Described by urban ecologists and used to characterize the complexities of the urban ecosystem, this gradient provides a framework to map infectious disease transmission, their drivers, and potential areas of interest for development and public health.

was shaped by political, social, and economic changes occurring in the 1980s.[35,36] The urban poor were unable to afford living in the central neighborhoods or to build houses in irregular settlements, leading to the proliferation of *favelas* (i.e., slums) in abandoned central areas (often too environmentally risky for formal housing developments) found close to wealthy neighborhoods. This new growth pattern modified the prevailing notion of environmental (landscape) and social homogeneity (where closer areas share similar socio-economic background) into a complex pattern where social-economic-environmental drivers, such as land value, race and gender differences, economic disparities, land-use, and human mobility behavior combine (Fig. 11.2), resulting in a complex patchwork of urban settlement.

The complex environmental, social, and political structures within cities imply a need for integrated solutions to address the root causes of infectious diseases and for strategic investments that strengthen health resiliency to infectious disease threats.[37] A landmark example of a radical solution to a complex urban health problem—integrating multiple sectors and actors—occurred in Victorian London where typhoid fever infected thousands and cholera epidemics were large and common.[38] It took a civil engineer, Sir Joseph Bazalgette, to address the unsanitary sewer discharges into streets and their runoff into the Thames River by building London's first sewer network.[39]

A modern example of integration across epidemiological, social, and environmental dimensions was developed for emerging urban arboviruses such as Zika, dengue, and chikungunya, all transmitted by the highly urbanized and invasive *Aedes aegypti* mosquito. In urban areas, transmission of such viruses cluster in hotspot areas that are characterized by multiple conditions (e.g., high population density, poor access to water, high crime) and that contribute up to 50% of all reported cases in a city.[40] The Pan-American Health Organization (PAHO) has taken such findings and turned them into a new framework that utilizes spatial analysis and public health information to stratify urban areas based on arbovirus transmission risk,[41] a shift from reliance on responding to reported symptomatic cases (i.e., syndromic surveillance), which is often associated with delays and ineffective containment of disease outbreaks.[42]

This framework involves using historical dengue case reports to map their distribution within a city as counts of standardized cases per census block per year over 5 years and then applying spatial analyses to identify the areas that are consistently identified as hotspots of high transmission intensity.[40] For instance, the PAHO framework allowed identification of dengue hotspots in nine highly endemic Mexican cities and even confirmed that such dengue hotspots overlapped with chikungunya and Zika hotspots. Current efforts are identifying the main factors involved in the generation of these areas of high transmission intensity to help guide integrated vector management approaches (e.g., targeted indoor residual spraying, control of productive larval habitats, release of *Wolbachia*-carrying mosquitoes, or similar innovative vector control approaches). As ministries of health gain access to tools for stratifying urban areas according to their vulnerability to infectious disease transmission, root causes for their occurrence could be identified and resources (e.g., personnel, supplies) more efficiently used.[40]

Human behavior and social interactions disproportionately influence infectious disease transmission in urban areas

Density-dependent, directly transmitted pathogens depend on the occurrence of epidemiologically relevant contacts between infectious and susceptible individuals.[43] Urban areas experience significantly higher opportunities for pathogen transmission compared to rural areas due to the synergy between their high and concentrated population size and the abundant and diverse mobility patterns of people as they go about their daily lives.[44–46] Human mobility behavior within cities shape pathogen transmission networks and has been a key driver of the local and global transmission of recent pandemics (e.g., severe acute respiratory syndrome (SARS),[47] the 2009 A/H1N1pdm influenza,[48] Zika,[49] and COVID-19[50]), as well as endemic infectious diseases such as dengue fever.[51]

Thanks to technological advancements in global positioning systems (GPS) and expansion of mobile phone usage, research is unveiling how human mobility and social contacts within cities shape pathogen transmission networks.[44,52] Heterogeneous contacts emerging from variable human mobility patterns lead to infectious disease super-spreaders (i.e., a small fraction of individuals who are responsible for a disproportionately large number of transmission events), considered to be catalysts for epidemic propagation.[47,53] While super-spreaders have been described throughout history (e.g., Typhoid Mary), more formal analytical descriptions have emerged from the study of "core groups" in sexually transmitted diseases[54] and from contact tracing studies after epidemics such as SARS in 2002,[47] A/H1N1pdm influenza in 2009[48] and, more recently, COVID-19.[50,55]

Control strategies that identify and target super-spreaders will show increased efficiency and effectiveness compared to population-based measures.[47] Unfortunately, and with some exceptions for sexually transmitted diseases,[54] super-spreaders of most infectious diseases are not characterized by a specific trait that can be used a priori to identify and target them.[56] For instance, pathogen transmission by asymptomatic carriers and latency of some infections are important super-spreading drivers.[47,57] A "silent majority" of infectious individuals are unaware of being infected and hence, of their potential infectiousness to others. Consequently, they often go undetected by passive public health surveillance systems despite being responsible for a disproportionally large number of pathogen transmission events. For dengue fever, asymptomatic individuals contribute ~84% to the force of viral infection.[57] Similarly, 44% of the 2009 A/H1N1pdm influenza virus transmission was estimated to be due to asymptomatic carriers.[58] Asymptomatic carriers, given their disease-unaffected behavior and social interactions, have also been identified as super-spreaders for COVID-19 in multiple settings.[59] Population-level enforced mobility reductions due to COVID-19, all of which impact both symptomatic and asymptomatic carriers (e.g., social distancing, curfews, and travel restrictions), have evidenced the relevance of contacts and super-spreading in the progression of disease outbreaks and a path for slowing transmission during surges.[60]

The social fabric of urban dwellers is equally as important as their mobility behavior in shaping infectious disease transmission networks.[18,61] Strong and informal ties characterize social networks in rural areas, compared to weaker and more formal ties in urban settings.[62] This weakening of social connections, which may be influenced by culture and race, is reflected in how individuals treat, trust, and depend on each other, treat their shared spaces, and respond to infectious disease threats.[63] For instance, new migrants to urban slums may take time to develop a sense of belonging in their new space: they may not know their neighbors, perceive their new living quarters as transitory, may be exposed to increased violence and drug and alcohol abuse in some neighborhoods, and sense a lack of collective respect for the local environment.[64] The absence of connection with one's surroundings transfers responsibility to others (e.g., government, neighbor, women) when what is truly needed is to increase community participation in developing solutions to local problems. This absence of commitment may make the difference in the strength of social messaging and adoption of public health measures in certain demographic and social groups.

Within urban settings, there are populations that are more vulnerable than others (e.g., children, women, elderly, disabled, and other marginalized individuals). As an example related to gender, a recent review of the impact of COVID-19 on mental health identified multiple drivers such as stress due to parenting roles during lockdowns, increased partner violence in confined quarters for long periods of time, and depression, disproportionately impacting women.[65] Similarly, Zika virus transmission is an example of complex social-behavioral dynamics with a strong gender bias, as evidenced by the influence of infection on woman's mental and reproductive health (see Case Study).[66] The 2022 monkeypox outbreak is currently propagating predominantly among gay and bisexual men (primarily intimate and skin to skin exposure), evidencing acute social and gender differences in access to vaccines and therapeutics.[17] Clearly, in urban areas, a better understanding and accurate quantification of the role of social determinants of health, such as gender,[67] is a focal point for developing improved public health interventions that are both effective and equitable.

Connection among humans, animals, and the environment are relevant for emerging infectious disease emergence, spillover, and spread in urban areas

Human activities and needs impact ~75% of the planet's land surface.[68] Urban dwellers' need for agricultural products dominates this land surface use, with crop and livestock production now covering over one-third of the globe with projections to cover half of the globe if current trends continue unabated.[68] Hence, what urban dwellers choose to eat and how their needs are supplied will largely shape the global food and land-use systems. To ensure that these choices will not further degrade natural systems, heighten food insecurity, and exacerbate poverty, there is a push to rethink our agriculture and food system at all levels and within all sectors to promote healthy diets, substantially reduce food waste, and promote regenerative agricultural practices.[68] In relation, commercial agriculture is expanding rapidly in regions of tropical forests. This process increases the risk of pathogen spillover (Chapter 6) by: (1) intensifying the interface between wildlife, arthropod vectors, and people and (2) promoting novel behaviors by wild animals often related to seeking out new food sources or habitats as their longstanding natural sources disappear or become less dependable (e.g., wildlife becoming unwelcome guests/pests in homes, livestock enclosures, and/or crop fields).

Cities themselves also play a focal role in zoonotic pathogen spillover and spread. Urban wet markets, in which live animals of domestic and wild origin are sold and slaughtered, provide a backdoor for pathogen spillover.[69] Crowded conditions, poor food safety practices, and the mixing of humans (e.g., retailers, customers, tourists) with animals provide optimal conditions for infectious disease emergence.[69] The first SARS epidemic, caused by a coronavirus which is known to be of pandemic concern, most likely emerged from those individuals in contact with the live-game trade and subsequently disseminated through wet markets in the Guandgong province in late 2002.[70,71] One plausible source of SARS-CoV-2 (later known as COVID-19) emergence was in Wuhan City wet markets where, in the years prior to 2019, animals from 38 terrestrial wild species, many with high zoonotic potential, were harvested and sold both from legal and illegal sources.[72]

However, zoonotic spillover is not limited to wet markets. New ecological landscapes that have resulted from urbanization not only bring human and animal populations into closer contact with one another, as discussed in Chapter 6, but are often exploited by species that adapt favorably to them (e.g.,[73]). New vector species invade urban habitats and increase the risk of vector-borne and zoonotic diseases transmission (Chapters 6 and 9). Human-adapted wildlife species that can flourish in these urban landscapes, such as rodents, birds, smaller carnivores, and bats, among others, also tend to be efficient transmitters of zoonotic pathogens such as rabies virus, *Sin Nombre orthohantavirus* (hantavirus pulmonary syndrome), coronaviruses, and *Yersinia pestis* (plague).[25,73] Clumping of resources in urban areas can also lead to increased birth rates among urban-adapted species which, in turn, could provide further opportunities for pathogen transmission by increasing the abundance of susceptible juvenile hosts with lower immunity.[24] Urban environments also introduce stressors on many species through factors such as noise and light pollution, the heat island effect, and chemical exposure, that can result in lowered immunity and increased pathogen shedding within these populations.[24]

Transport infrastructure via air, land, and sea—and related transportation hubs—concentrated in cities can spread zoonotic pathogens rapidly over long distances, leading to regional and global spread within hours.[74] The 2009 A/H1N1pdm influenza pandemic likely emerged in rural poultry farms.[75] First detection of the virus occurred in cities such as San Diego (USA) and Oaxaca (Mexico), where it likely propagated further through road and air traffic networks.[75] Factors that may exacerbate transmission also include human behaviors related to livestock-keeping practices (e.g., increased crop and livestock raiding), production systems, any movements of livestock and animal products (e.g., wet markets),[25] diversion of water sources, and clumping of other resources (e.g., waste, slaughter plants). Much of this activity takes place at the edges of cities which often serve as transitionary zones where biophysical factors, biological activity, and ecological evolutionary processes are both concentrated and intensified.[25] Therefore, these edges can represent important local-scale interfaces within which zoonotic agents can circulate and infect both wild and domestic animals, as well as humans.[25] Similarly, the reurbanization of yellow fever in the Americas and Africa has become a plausible concern due to the transmission beyond forests and into peri-urban areas on the edge of major urban centers.[76]

Urbanization can also result in significant changes to the structure of wildlife communities, often resulting in reduced biodiversity with a subsequent increase in abundance of certain generalist species.[4] Loss of biodiversity can have implications for pathogen transmission risk within a given environment, either increasing or decreasing it.[77] Mathematical models and laboratory experiments have shown that a "dilution effect," in general described as a decrease in pathogen transmission risk with increasing animal diversity, can occur under a wide range of conditions.[78] When it occurs, the effect depends on the lower availability of competent hosts for a pathogen with regard to the other (less competent) hosts.[77,79] The urban environment is a key determinant of which species will remain and, thus, whether or not a dilution effect will be observed. For example, in the case of Lyme disease, reduced host biodiversity (e.g., absence of predators) in forest fragments surrounding urban and suburban areas resulted in an increase in the abundance of the most competent reservoirs for *B. burgdorferi*, the white-footed mouse.[79] Consequently, this increased prevalence of infection in host-seeking ticks and overall disease incidence.[80]

Urban areas often face political/administrative challenges to deal with pandemic threats

Given the population size, rapid growth, and role of urban areas as centers of human activity, city governments are a powerful bridge between local issues and national policies. Urban governance, involving a range of actors and institutions, plays a critical role in shaping the physical and social structure of cities, influences the quantity and quality of local services, and determines the share of costs and the distribution of resources throughout the population.[81] While powerful urban governance occurs in developed urban centers, very often what affects the life of the urban poor lies outside the control of city administrations. Other actors, such as private businesses, nongovernmental organizations (NGOs), agencies of the central government, or the civil society, become more relevant for the lives of these urban dwellers.[81] In fact, the private sector is often not included in pandemic decisions yet, it represents an important social, economic, and political engine of cities.[82] Such a patchwork of governance has enabled disorganized urban growth and exacerbated inequities in access to resources and goods—from access to green space to transportation options for mobility within the city. For instance, access to basic services, such as reliable and safe piped water and sanitation, is a major challenge for the urban poor and a top SDG milestone the world has failed to achieve.[8]

While the public health response to diseases in urban areas is the direct responsibility of local ministries of health, root causes that contribute to diminished urban health such as sanitation, poverty, and violence, are generally the responsibility of separate government units with their own priorities, budgets, and goals. The rapid turnover of high-level authorities in leading government institutions—the same institutions expected to respond quickly—results in fragmentation and lack of multisectoral integration.

Responses to infectious disease threats often encounter the same disorganized patchwork of responsibilities and policies within a complex system in which units do not necessarily engage with one another and often fail to understand the problem from multiple angles, contributing to delays in epidemic and pandemic response. When the Zika virus invaded the continental United States, uncoordinated responses to transmission by local governments led to failures to contain the virus.[83] Similarly, COVID-19 has shown that while cities may have strong economic power, decisions often become politicized between city, state, and federal governments.[37,84] Therefore, more local power and resources—centered within specific nonpolitical institutions or some type of "One Health board," for the purpose of long-term planning and implementation by a multisectoral group to detect and rapidly contain emerging pathogens with pandemic potential within cities is crucial. Current efforts emerging from the GHSA have trained large number of professionals who are engaged in multisectoral initiatives with government, NGOs and the private sector of the global south with the ultimate aim of strengthening public health surveillance and response to emerging disease threats.[23]

Ports of entry and points of amplification

Epidemic and pandemic preparedness and response must focus on ports of entry and points of amplification, and, to the extent possible, efforts must focus on finding evidence-based approaches or interventions. While the IHR emphasizes "ports of entry" as focal detection points, previous pandemics have shown that detection should also occur in the points of amplification such as large and complex urban environments. As seen from the spread of SARS, Zika, 2009 A/H1N1pdm influenza, and COVID-19, pandemics follow major flight travel patterns connecting these large and complex urban environments, both domestically and internationally. Investments in epidemic preparedness at the city level would counterbalance the large local, regional, and global costs of pandemics.[37,85] Ensuring that urban settings are prepared to respond to infectious disease threats is crucial. The financial cost for global preparedness (~US$4.5 billion/year) would represent a fraction of the estimated (pre-COVID-19) global pandemic costs (ranging from ~US$60 to US$570 billion USD/year).[86,87] The International Monetary Fund (IMF) has estimated that the COVID-19 pandemic has cost the global economy at least US$11 trillion,[23] with further costs on lives, poverty, and international security not yet estimated but expected to at least double the direct costs of the pandemic. Investing in better public health surveillance and response plans has become an ultimate need, particularly in urban areas with strong regional and international connections.

Vision for the future

While the previous sections show that there is no single factor that can be linked to infectious disease transmission and pandemic spread in urban areas, they also emphasize the need for integrated approaches that embrace the role of a myriad of actors while acknowledging the complexity of improving human, animal, and environmental health and pandemic preparedness in urban areas. The economic and social shock waves generated by recent pandemics can no longer be ignored and urban governance structures, in partnership with the private sector, should be strengthened to improve local and global health security.[23,87] History has shown this is both a feasible and worthy investment. During the 1800s in the southern United States, urban yellow fever outbreaks obliterated cities, trade, and slave trade routes (i.e., and were a major source of mortality and urban decay).[88] It took a joint effort by the commerce sector to galvanize local and federal authorities into forming public health boards that still exist today.[88] Taking this a step further to establish multidisciplinary, intersectoral, One Health task forces and/or governing bodies at the local level may help overcome many of the barriers we face today. Updating the IHR and GHSA to explicitly acknowledge the pivotal role of cities for epidemic preparedness is a logical first step.

As the urban poor are not benefiting from the majority of city living advantages, the value of improving cities and the lives of all its citizens (listed as the SDG-11)[8] must become mainstream.[86] In 2019, one in four urban residents (around one billion people) globally reported living in a slum.[8] The World Bank estimates show that, as of October 2021, COVID-19 has increased the number of people living in poverty (i.e., with less than $1.9 a day) by 97 million,[89] particularly in less-developed countries, emphasizing the social and economic impact of pandemics on global development. We must ask ourselves how vibrant cities in LMICs can continue their rapid growth trend while also generating a more equitable and balanced environment for all its citizens.[37] Global health security can be improved by recognizing that urban areas are complex and heterogeneous environments in which the health sector must be integrated with other key sectors (e.g., urban development, housing, environmental protection, animal health, agriculture, business) into a framework aimed at addressing the root causes of poor health such as poverty, inequality, and food insecurity, and at increasing the detection and response to infectious disease threats.

Actions

Significantly increase funding for epidemic preparedness within cities. Properly funded and well-developed plans for pathogen detection and epidemic response within cities can have benefits that transcend the local domain into regional and global scales. Global health security can be increased by explicitly addressing, both in the IHR, GHSA, and other instruments, the pivotal role of urban areas as foci for introduction, amplification, and pandemic propagation of infectious and emerging disease threats. Strengthening of public health surveillance efforts, including a medical workforce better trained to diagnose emerging infectious disease threats, should be part of this effort. Increased funding will also reflect a political commitment to reduce the chances that the next pandemic becomes as costly to life and economic resources than COVID-19 has been.[90]

Establish One Health approaches within urban areas to strengthen public health surveillance and mitigate the root causes of their emergence and propagation within cities. The zoonotic origin of most recent pandemics, which is catalyzed in key urban locations such as wet markets or urban slums, emphasizes the need for a more holistic approach to detect and contain them within cities. Enhanced surveillance and research to identify key reservoir and sentinel species for priority pathogens in urban settings to inform targeted surveillance will be critical. Further, the establishment of interagency One Health task forces and/or governing bodies may help catalyze connection and information sharing among environmental, public health, veterinary, academic, business, and governmental sectors. For instance, the GHSA is building a One Health workforce in Asia and Africa by training thousands of current health professionals on concepts and approaches integrating humans, wildlife, and environment.[23] COVID-19 has led to updates to the IHR pertaining to identification and response to the pandemic.[91] Future updates should go beyond COVID-19 updates to more broadly incorporate One Health as a framework for increased preparedness against future pandemics.

Addressing and reducing urban inequality will increase urban resiliency and preparedness against pandemic threats. Urban environments vary in their contribution to infectious disease burden and transmission (Fig. 11.2), with urban slums and marginalized areas contributing a disproportionately higher risk for both pathogen spillover and rapid propagation. Policies aimed at reducing poverty and urban inequality could carry the added benefit of buffering societies against pandemics. Funding such initiatives to reduce inequality should be a priority and could be offset by the dramatic costs in pandemic mitigation emphasizing that both the government and private sectors should partner toward the common goal of increased resilience against infectious disease threats.

CASE STUDY

Zika Infection: From Tropical Medicine to Global Public Health

Background

The emergence of the Zika virus across the Latin America and Caribbean (LAC) region was declared a Public Health Emergency of International Concern by WHO in 2016,[92] due to rapid dissemination and transmission in a region with a fully susceptible population infested with the highly competent and urban-adapted *Aedes aegypti* mosquito.[93] The Zika epidemic was first detected in urban centers where conditions for *Aedes aegypti* development and high infestation rates were enhanced by poor urban housing conditions, crowding, irregular provision of public services such as running water and recycling containers to reduce mosquito breeding sites, and complex social and high-risk behaviors.

CASE STUDY (cont'd)

Observations

The emergence of severe Zika congenital syndrome (ZCS) and neurological complications (e.g., microcephaly, Guillain–Barre syndrome) demanded different responses from national health ministries in LAC,[94,95] where primary healthcare units, emergency rooms, obstetric, and pediatric wards, along with specialized and intensive care units, needed to coordinate an emergency response for which the healthcare system was not prepared. The lack of diagnostic capabilities (e.g., PCR, serology, ultrasound) in primary health care units, coupled with the incorrect use of tapes and standard tables used to measure head circumference, led to misdiagnoses and under-reporting of Zika congenital syndrome (ZCS).[96] Once microcephaly was identified, treatment of neurological complications (e.g., seizures), and early stimulation therapies were restricted—if available—to highly specialized hospitals in capital cities. The implementation of various mosquito control methods did not translate into measurable reductions in Zika burden, primarily due to the limited evidence available regarding their effectiveness to prevent outbreaks.[97] Finally, maternal and childcare services were most affected by holding women responsible for preventing infection during pregnancy and, when infection occurred, were often blamed and stigmatized. On the other hand, the use of condoms by men to prevent sexual transmission was not seen as a priority.[98]

What did not work?

- The lack of a coordinated effort to respond to a vector-borne disease with neonatal consequences evidenced the challenges that Zika imposed to the healthcare sector.
- Weak disease surveillance structures detected microcephaly late and long after Zika transmission was confirmed.
- Poor interaction between entomology and epidemiology structures within Ministries of Health led to a fragmented response and messaging.
- Communication strategies exacerbated gender biases by feminizing the pandemic and the responsibility for preventing microcephaly.
- Primary healthcare, antenatal, birth delivery, specialized medical care, and psychological and social support were not in place nor strengthened to face these challenges.

What worked?

- Technical support from international agencies (PAHO, USAID, CDC, etc.) was deployed in the region; however, efforts fell into disintegrated initiatives.
- Vector control programs were strengthened by the epidemic which helped in the control of dengue transmission as well.
- Community participation schemes were developed and supported in response to the epidemic due to its visible impact on family health.

Recommendations

- Improve epidemiological surveillance of arboviral diseases with better diagnostic infrastructure at the local level.
- Improve training of health personnel in the diagnosis and management of Zika

(Continued)

CASE STUDY (cont'd)

complications for opportune interventions.

- Create social and community resources to support the families affected by Zika complications.
- Enhance vector control programs at the local level to prevent arbovirus transmission.

Interventions

- Promote healthy urban environments: housing, provision of public services, recycling, community-based interventions.
- Deploy preventive vector control interventions in high-risk areas and monitor their effectiveness.
- Target enhanced surveillance: case detection and contact tracing in early pregnancy and reproductive age groups.
- Create clinical, psychological, and community support teams in primary healthcare units.

References

1. United Nations Department of Economic and Social Affairs PD. *World Urbanization Prospects: The 2018 Revision*. ST/ESA/SER.A/420. New York: United Nations; 2019.
2. Ledent J. The factors of urban population growth: net immigration versus natural increase. *Int Reg Sci Rev*. 1982;7(2):99–125.
3. Lucas REB. Internal migration in developing economies: an overview of recent evidence. *Geopolit Hist Int Relat*. 2016;8(2):159–191.
4. Forman RTT. *Urban Ecology: Science of Cities*. Cambridge University Press; 2014.
5. Schneider A, Woodcock CE. Compact, dispersed, fragmented, extensive? A comparison of urban growth in twenty-five global cities using remotely sensed data, pattern metrics and census information. *Urban Stud*. 2008;45(3):659–692.
6. United Nations–Habitat. *The Challenge of Slums. Global Report on Human Settlements*. London, UK: Earthscan Publications Ltd.; 2003.
7. Ezeh A, Oyebode O, Satterthwaite D, et al. The history, geography, and sociology of slums and the health problems of people who live in slums. *Lancet*. 2017;389(10068):547–558.
8. United Nations. *The Sustainable Development Goals Report 2021*. New York: United Nations; 2021.
9. McMichael AJ. The urban environment and health in a world of increasing globalization: issues for developing countries. *Bull World Health Organ*. 2000;78(9):1117–1126.
10. Van de Poel E, O'Donnell O, Van Doorslaer E. Are urban children really healthier? Evidence from 47 developing countries. *Soc Sci Med*. 2007;65(10):1986–2003.
11. Reyes R, Ahn R, Thurber K, et al. Urbanization and infectious diseases: general principles, historical perspectives, and contemporary challenges. In: Fong IW, ed. *Challenges in Infectious Diseases*. New York, NY: Springer New York; 2013:123–146.
12. Rosen G. *A History of Public Health*. Revised Expanded Edition. Baltimore, MD: Johns Hopkins University Press; 2015.
13. Zhang CH, Schwartz GG. Spatial disparities in coronavirus incidence and mortality in the United States: an ecological analysis as of May 2020. *J Rural Health*. 2020;36(3):433–445.

14. Malani A, Shah D, Kang G, et al. Seroprevalence of SARS-CoV-2 in slums versus non-slums in Mumbai, India. *Lancet Global Health*. 2021;9(2):e110–e111.
15. Saltiel F. Who can work from home in developing countries? *COVID Econ*. 2020;7:104–118.
16. United States Agency for International Development. Lessons for USAID on the effects of the Zika outbreak on MCH services: learning from the past to prepare for the future. In: *Case Studies in Five Latin American and Caribbean Countries*; 2020. Washington, DC https://www.harpnet.org/wp-content/uploads/2020/10/ENGLISH-version-Cross-Cutting-Report-for-Dissemination-FINAL.pdf.
17. Gonsalves GS, Mayer K, Beyrer C. Deja vu all over again? emergent monkeypox, delayed responses, and stigmatized populations. *J Urban Health*. 2022;99(4):603–606.
18. World Health Organization. Our cities, our health, our future: acting on social determinants for health equity in urban settings. In: *Report to the WHO Commission on Social Determinants of Health from the Knowledge Network on Urban Settings*. Geneva, Switzerland: World Health Organization; 2008.
19. Raudsepp-Hearne C, Peterson GD, Tengö M, et al. Untangling the environmentalist's paradox: why is human well-being increasing as ecosystem services degrade? *Bioscience*. 2010;60(8):576–589.
20. Wu J. Urban ecology and sustainability: the state-of-the-science and future directions. *Landsc Urban Plann*. 2014;125:209–221.
21. United Nations. *Policy Brief: COVID-19 in an Urban World*. New York: United Nations; 2020.
22. World Health Organization. *International Health Regulations*. 2nd ed. 2005. Geneva, Switzerland.
23. The Global Health Security Agenda. *Strengthening Health Security across the Globe: Progress and Impact of U.S. Government Investments in the Global Health Security Agenda*. 2020.
24. Bradley CA, Altizer S. Urbanization and the ecology of wildlife diseases. *Trends Ecol Evol*. 2007;22(2):95–102.
25. Hassell JM, Begon M, Ward MJ, et al. Urbanization and disease emergence: dynamics at the wildlife-livestock-human interface. *Trends Ecol Evol*. 2017;32(1):55–67.
26. Alirol E, Getaz L, Stoll B, et al. Urbanisation and infectious diseases in a globalised world. *Lancet Infect Dis*. 2011;11(2):131–141.
27. Alberti M, McPhearson T, Gonzalez A. *Embracing Urban Complexity*. 2018:45–67.
28. Alberti M. *Urban patterns and ecosystem function. Advances in Urban Ecology*. New York: Springer; 2008.
29. McDonnell MJ, MacGregor-Fors I. The ecological future of cities. *Science*. 2016;352(6288):936–938.
30. Ahern J. Urban landscape sustainability and resilience: the promise and challenges of integrating ecology with urban planning and design. *Landsc Ecol*. 2013;28(6):1203–1212.
31. Des Roches S, Brans KI, Lambert MR, et al. Socio-eco-evolutionary dynamics in cities. *Evol Appl*. 2021;14(1):248–267.
32. Alberti M, Marzluff JM, Shulenberger E, et al. Integrating humans into ecology: opportunities and challenges for studying urban ecosystems. *Bioscience*. 2003;53(12):1169–1179.
33. Lilford RJ, Oyebode O, Satterthwaite D, et al. Improving the health and welfare of people who live in slums. *Lancet*. 2017;389(10068):559–570.
34. Romer P. In: Romer P, ed. *The City as Unit of Analysis*; 2013. https://paulromernet/the-city-as-unit-of-analysis/.
35. Torres H, Marques E, Ferreira MP, et al. *Poverty and Space: Patterns of Segregation in São Paulo*. 2002. Austin, TX.
36. Caldeira T. *City of Walls: Crime, Segregation, and Citizenship in São Paulo*. Berkeley, CA: UC Press; 2001.
37. Daszak P, Keusch GT, Phelan AL, et al. Infectious disease threats: a rebound to resilience. *Health Aff*. 2021;40(2):204–211.
38. Tulchinsky TH. *John Snow, cholera, the broad street pump; waterborne diseases then and now. Case Studies in Public Health*. 2018:77–99.
39. Brewer T, Pringle Y. Beyond Bazalgette: 150 years of sanitation. *Lancet*. 2015;386(9989):128–129.
40. Dzul-Manzanilla F, Correa-Morales F, Che-Mendoza A, et al. Identifying urban hotspots of dengue, chikungunya, and Zika transmission in Mexico to support risk stratification efforts: a spatial analysis. *Lancet Planet Health*. 2021;5(5):e277–e285.
41. Pan American Health Organization. *Technical Document for the Implementation of Interventions Based on Generic Operational Scenarios for* Aedes aegypti *Control*. Washington, D.C.: PAHO; 2019.
42. Cavany SM, Espana G, Lloyd AL, et al. Optimizing the deployment of ultra-low volume and targeted indoor residual spraying for dengue outbreak response. *PLoS Comput Biol*. 2020;16(4).
43. Anderson RM, May R. *Infectious Diseases of Humans: Dynamics and Control*. 1992. Oxford, UK.

44. Vazquez-Prokopec GM, Bisanzio D, Stoddard ST, et al. Using GPS technology to quantify human mobility, dynamic contacts and infectious disease dynamics in a resource-poor urban environment. *PLoS One*. 2013;8(4).
45. Bharti N. Linking human behaviors and infectious diseases. *Proc Natl Acad Sci U S A*. 2021;118(11).
46. Stoddard ST, Morrison AC, Vazquez-Prokopec GM, et al. The role of human movement in the transmission of vector-borne pathogens. *PLoS Neglected Trop Dis*. 2009;3(7).
47. Lloyd-Smith JO, Schreiber SJ, Kopp PE, et al. Superspreading and the effect of individual variation on disease emergence. *Nature*. 2005;438(7066):355–359.
48. Halloran ME, Ferguson NM, Eubank S, et al. Modeling targeted layered containment of an influenza pandemic in the United States. *Proc Natl Acad Sci U S A*. 2008;105(12):4639–4644.
49. Zhang Q, Sun K, Chinazzi M, et al. Spread of Zika virus in the Americas. *Proc Natl Acad Sci U S A*. 2017;114(22):E4334–e4343.
50. Chang S, Pierson E, Koh PW, et al. Mobility network models of COVID-19 explain inequities and inform reopening. *Nature*. 2021;589(7840):82–87.
51. Stoddard ST, Forshey BM, Morrison AC, et al. House-to-house human movement drives dengue virus transmission. *Proc Natl Acad Sci U S A*. 2013;110(3):994–999.
52. Wesolowski A, Buckee CO, Engo-Monsen K, et al. Connecting mobility to infectious diseases: the promise and limits of mobile phone data. *J Infect Dis*. 2016;214(suppl_4):S414–S420.
53. Woolhouse ME, Dye C, Etard JF, et al. Heterogeneities in the transmission of infectious agents: implications for the design of control programs. *Proc Natl Acad Sci U S A*. 1997;94(1):338–342.
54. Koopman JS, Simon CP, Riolo CP. When to control endemic infections by focusing on high-risk groups. *Epidemiology*. 2005;16(5):621–627.
55. Lau MSY, Grenfell B, Thomas M, et al. Characterizing superspreading events and age-specific infectiousness of SARS-CoV-2 transmission in Georgia, USA. *Proc Natl Acad Sci U S A*. 2020;117(36):22430–22435.
56. Vazquez-Prokopec GM, Perkins TA, Waller LA, et al. Coupled heterogeneities and their impact on parasite transmission and control. *Trends Parasitol*. 2016;32(5):356–367.
57. Ten Bosch QA, Clapham HE, Lambrechts L, et al. Contributions from the silent majority dominate dengue virus transmission. *PLoS Pathog*. 2018;14(5).
58. Van Kerckhove K, Hens N, Edmunds WJ, et al. The impact of illness on social networks: implications for transmission and control of influenza. *Am J Epidemiol*. 2013;178(11):1655–1662.
59. You Y, Yang X, Hung D, et al. Asymptomatic COVID-19 infection: diagnosis, transmission, population characteristics. *BMJ Support Palliat Care*. 2021. https://doi.org/10.1136/bmjspcare-2020-002813.
60. Xylogiannopoulos KF, Karampelas P, Alhajj R. COVID-19 pandemic spread against countries' non-pharmaceutical interventions responses: a data-mining driven comparative study. *BMC Publ Health*. 2021;21(1):1607.
61. Graham H, White PC. Social determinants and lifestyles: integrating environmental and public health perspectives. *Publ Health*. 2016;141:270–278.
62. Mair CA, Thivierge-Rikard RV. The strength of strong ties for older rural adults: regional distinctions in the relationship between social interaction and subjective well-being. *Int J Aging Hum Dev*. 2010;70(2):119–143.
63. Boessen A, Hipp JR, Smith EJ, et al. Networks, space, and residents' perception of cohesion. *Am J Community Psychol*. 2014;53(3–4):447–461.
64. Chandola T. Spatial and social determinants of urban health in low-, middle- and high-income countries. *Publ Health*. 2012;126(3):259–261.
65. Almeida M, Shrestha AD, Stojanac D, et al. The impact of the COVID-19 pandemic on women's mental health. *Arch Womens Ment Health*. 2020;23(6):741–748.
66. Arenas-Monreal M, Pina-Pozas M, Gomez H. Challenges and inputs of the gender perspective to the study of vector borne diseases. *Salud Publica Mex*. 2015;57:66–75.
67. Phillips SP. Defining and measuring gender: a social determinant of health whose time has come. *Int J Equity Health*. 2005;4.
68. Díaz S, Settele J, Brondízio ES, et al. Pervasive human-driven decline of life on Earth points to the need for transformative change. *Science*. 2019;366(6471).
69. Naguib MM, Li R, Ling J, et al. Live and wet markets: food access versus the risk of disease emergence. *Trends Microbiol*. 2021;29(7):573–581.

70. da Costa VG, Moreli ML, Saivish MV. The emergence of SARS, MERS and novel SARS-2 coronaviruses in the 21st century. *Arch Virol*. 2020;165(7):1517–1526.
71. Xu R-H, He J-F, Evans MR, et al. Epidemiologic clues to SARS origin in China. *Emerg Infect Dis*. 2004;10(6):1030–1037.
72. Xiao X, Newman C, Buesching CD, et al. Animal sales from Wuhan wet markets immediately prior to the COVID-19 pandemic. *Sci Rep*. 2021;11(1).
73. Bingham J. Canine rabies ecology in southern Africa. *Emerg Infect Dis*. 2005;11(9):1337–1342.
74. Centers for Disease Control and Prevention. Why It Matters: The Pandemic Threat.
75. Swerdlow D, Finelli L, Bridges C. The 2009 H1N1 influenza pandemic: field and epidemiologic investigations. *Clin Infect Dis*. 2011;52(suppl l_1).
76. Possas C, Martins RM, Oliveira RL, et al. Urgent call for action: avoiding spread and re-urbanisation of yellow fever in Brazil. *Mem Inst Oswaldo Cruz*. 2018;113(1):1–2.
77. Ostfeld RS, Keesing F. Effects of host diversity on infectious disease. *Annu Rev Ecol Evol Syst*. 2012;43(1):157–182.
78. Keesing F, Holt RD, Ostfeld RS. Effects of species diversity on disease risk. *Ecol Lett*. 2006;9(4):485–498.
79. Keesing F, Belden LK, Daszak P, et al. Impacts of biodiversity on the emergence and transmission of infectious diseases. *Nature*. 2010;468(7324):647–652.
80. Turney S, Gonzalez A, Millien V. The negative relationship between mammal host diversity and Lyme disease incidence strengthens through time. *Ecology*. 2014;95.
81. Devas N. *Urban Governance Voice and Poverty in the Developing World*. Taylor & Francis; 2012.
82. Acuto M, Larcom S, Keil R, et al. Seeing COVID-19 through an urban lens. *Nat Sustain*. 2020;3(12):977–978.
83. Gridley-Smith C. *NACCHO Report: Vector Control Assessment in Zika Virus Priority Jurisdictions*. 2017.
84. Frumkin H. COVID-19, the built environment, and health. *Environ Health Perspect*. 2021;129(7).
85. Lee VJ, Ho M, Kai CW, et al. Epidemic preparedness in urban settings: new challenges and opportunities. *Lancet Infect Dis*. 2020;20(5):527–529.
86. World Economic Forum. *Outbreak Readiness and Business Impact: Protecting Lives and Livelihoods across the Global Economy*. 2019. Geneva.
87. National Academies of Sciences. Commission on a Global Health Risk Framework for the Future. Washington, DC: National Academies Press (US).
88. Humphreys M. *Yellow Fever and the South*. Johns Hopkins University Press; 1999.
89. World Bank. COVID-19 Leaves a Legacy of Rising Poverty and Widening Inequality.
90. The Independent Panel for Pandemic Preparedness and Response. *COVID-19: Make it the Last Pandemic*; 2021. https://theindependentpanelorg/wp-content/uploads/2021/05/COVID-19-Make-it-the-Last-Pandemic_finalpdf.
91. World Health Organization. Statement on the tenth meeting of the international health Regulations. In: *Emergency Committee Regarding the Coronavirus Disease (COVID-19) Pandemic*. 2005.
92. Heymann DL, Hodgson A, Sall AA, et al. Zika virus and microcephaly: why is this situation a PHEIC? *Lancet*. 2016;387(10020):719–721.
93. Mustafa MS, Ramasethu R. Zika: an enormous public health challenge for a miniscule virus. *Med J Armed Forces India*. 2018;74(1):61–64.
94. Brady OJ, Osgood-Zimmerman A, Kassebaum NJ, et al. The association between Zika virus infection and microcephaly in Brazil 2015–2017: an observational analysis of over 4 million births. *PLoS Med*. 2019;16(3).
95. Pan American Health Organization. *Regional Zika Epidemiological Update (Americas)*. August 25, 2017.
96. Cardenas VM, Paternina-Caicedo AJ, Salvatierra EB. Underreporting of fatal congenital Zika syndrome, Mexico, 2016–2017. *Emerg Infect Dis*. 2019;25(8):1560–1562.
97. Bowman LR, Donegan S, McCall PJ. Is dengue vector control deficient in effectiveness or evidence?: systematic review and meta-analysis. *PLoS Neglected Trop Dis*. 2016;10(3).
98. Brito MB, Fraser IS. Zika virus outbreak and the poor Brazilian family planning program. *Rev Bras Ginecol Obstet*. 2016;38(12):583–584.

SECTION III

Movement of people and things: The challenge of pandemic spread

Ann Marie Kimball[1,2] *and Wondimagegnehu Alemu*[3,4]

[1]Department of Epidemiology, University of Washington, Bainbridge Island, WA, United States; [2]Chatham House, London, United Kingdom; [3]International Health Consultancy, LLC, Decatur, GA, United States; [4]Rollins School of Public Health, Emory University, Decatur, GA, United States

The dynamic interactions between humans (the host), infectious agents (microbes), and the environment sometimes cause huge pandemics resulting in a high burden of death and illness across the world. In addition, pandemics such as COVID-19 have resulted in negative consequences, unprecedented in modern times, due to the social, economic, and political disruptions. While many levers were "pulled" by authorities—lock downs, travel blocks, masking, social distancing, contact tracing, vaccinations, and more—the pandemic remains unrivalled in modern times for the level of damage, sickness, and death. What did we miss? This section will describe the most critical pathway of transmission for a global pandemic: the movement of people and goods across the expanse of the globe.

While many microbial pathogens have motility, none are capable of swimming oceans and hopping continents. Time and again, from the sailing of the first ships and overland caravans, people and cargo have carried microbes from their point of origin to new destinations with vulnerable populations. As the people of the world and their goods have become more and more mobile and continuously in motion within a very short span of time in huge quantities, science has struggled to keep up with necessary speed to institute timely measures to prevent spread and mitigate impact of infectious diseases. Epidemiology's core metrics of person, place, and time rely on fixed coordinates to carry out analytics—just as the early study of microbes relied on fixed and stained slides under a light microscope. Importantly, microbial change through mutation can be fast given microbes' prolific strategy of reproduction in hosts. While most may end up harmless with no consequence to human health and well-being, rarely some pathogens end up causing pandemics with unprecedented and grave consequences. This is especially true for RNA viruses (such as coronavirus) which do not have genetic correction mechanisms. So, key questions are (1) How can communities best insure against the exportation of pathogens from their populations? How should they armor themselves against the incursion of infection through importation of goods, other people, or animals? (2) What aspects of transportation systems for humans and goods enable transmission during the voyage? What could be done about it and by whom? (3) Do distance, processing or environmental conditions, and speed of transport affect the natural history of microbial evolution? Are there conditions which accelerate change in the nature of virii or bacteria as they are transported by air, sea, or overland? (4) How can innovations in vaccines, therapies, and other protective strategies be generalized to support protection across the globe with urgency and equitability?

Balancing the risk of disease transmission and importation with the economic and social benefit of trade and travel has figured as a primary priority for nations coming together for centuries. As nations began to trade overland, and by sea plagues came carried by traders or in the stuffs being traded. Trading is likely as old as humankind. In the 14th century, the first quarantine laws were passed by the city states of the Mediterranean to assure against the delivery of plague by ships and their human and rodent passengers. The International Health Regulations (IHR) of today (most recently revised in 2005) descend from this long history. Following the post-World War II incorporation of the World Health Organization in 1948, the passage of the International Sanitary Regulations in 1951 was one of the first orders for the new global organization. In fact, the charge is embedded within the constitution of the organization:

Under Articles 21(a) and 22, the Constitution of WHO confers upon the World Health Assembly the authority to adopt regulations "designed to prevent the international spread of disease" which, after adoption by the Health Assembly, enter into force for all WHO Member States that do not affirmatively opt out of them within a specified time period.[a]

Indeed, the volumes and characteristics of the business of travel and trade are critical to the health, social order, and economic well-being of nation states and their populations. This fact cannot be overstated as this section of this book will explore.

The body of IHR 2005 focusses on these issues. The reader is referred specifically to Articles 18–44 for recommendations and regulations most

[a]IHR 2007 revision published in 2016, World Health Organization.

pertinent to this section. Despite attempts to foresee to the spread of infection via transit of people and goods, examples of previously unanticipated vulnerabilities are legion. Consider the reintroduction of cholera in Peru in 1991. First thought to have been discharged in ballast from ships into intensely settled coastal communities, further investigation has clarified that story to be over simplified.[b] Newer genetic analysis suggests multiple sources of introduction, but also presents the possibility of microbial evolution occurring in transmission leading to the transnational outbreak. In another example, transmission of SARS 1 has been demonstrated aboard aircraft.[1] This threat remains on top of mind with coronavirus, and air circulation on aircraft is a key engineering concern. Mosquitoes hitchhiking in aircraft have been recognized as a threat since the inception of air travel, and yet airport malaria continues to occur.[2] In sum, the IHR cannot embrace all specific threats in the space of mobility. The venture of transporting people and goods in high volume and across wide geographic spaces is inherently risky.

The IHR recognized the broad array of cooperating partners necessary for successful engagement in this domain.[3] During the revision process in the early 2000s, WHO elected to include the World Trade Organization (WTO) and the Food and Agricultural Organization (FAO) as observers, reserving the ownership of the regulations to WHO. Consequently, no comprehensive attempt was made to harmonize regimes across these critical sectors. At a global level, such harmonization is a herculean task which would have delayed the revision by years.

Operational cooperation has not been optimal across these various entities in the face of pandemics. Countries have little guidance at national level with how to do so within their existing legal and policy frameworks. The exception to this generality is the area of "One Health" linking animal health, human health, and the environment described earlier in this volume.

Over the past 30 years, additional regional organizations focused on trade have incorporated emerging infections and pandemic threat into their policy portfolios. These include the European Union, Asia Pacific Economic Cooperation, ASEAN, MercoSur, ECOWAS, and COMESA (Africa). This nonexhaustive list[c] suggests that broad partnerships and active intersectoral collaboration will be increasingly required to assure the global safety net is fit for the purpose of global health security. The compliance of State parties to act collaboratively and in partnership within the allowable provisions of international conventions and regulations is critical to safeguard health and well-being of populations wherever they are and whoever they are.

The following chapters seek to inform the reader about the "global mobile express" of peoples and goods in trade and travel in contemporary times. The current state of global traffic will be described. Risk assessment, a

[b]https://wwwnc.cdc.gov/eid/article/16/7/10-0131_article

[c]http://culturalrelations.org/the-10-major-regional-trading-blocs-in-the-world-economy/

cornerstone of Global Health Security, is set forth and explained. Examples of economic crises resulting from pandemics will be examined, and possible mitigation strategies discussed. Finally, measures taken at ports of entry will be explored.

While the broader scientific community has not accepted that COVID itself could be imported in goods, the People's Republic of China has published a number of studies[d] which have detected the pathogen in products of packaging in frozen imported goods. Indeed, that member state has put stringent handling procedures in place to prevent onward transmission from imported frozen goods. This remains an active area of discussion, as the creation of microfilms on packaging through refrigeration and freezing could represent a vulnerability of note.

In this time of global pandemic, the reader is invited into this complex ecosystem of mobility and the forces that shape it. In view of a potential new "pandemic treaty," how can partners from the sectors of trade and travel best be included in the development and governance of such an accord? Can the regimes be harmonized providing clear guidance to member states? In view of the evidence, leaving national governments to find their way alone during this and future emergencies poses risk. The IHR are a "living document." Successful protection from transnational pandemic activity remains elusive.

References

1. Olsen SJ, Chang HL, Cheung TYY, et al, Transmission of the severe acute respiratory syndrome on aircraft. N Engl J Med 2003 Dec 18;349(25):2416–2422. https://doi.org/10.1056/NEJMoa031349.
2. Isaäcson M. Airport malaria: a review. Bull World Health Organ 1989;67(6):737–742. PMCID: PMC2491318 PMID: 2699278.
3. IBID endnote I page 3 paragraph 4.

[d]https://www.sciencedirect.com/science/article/pii/S0048969723000037.

CHAPTER

12

The interconnected world of trade, travel, and transportation networks

Mohamed Moussif[1], *Marissa Morales*[2] *and Ryan Rego*[3]

[1]Casablanca International Airport, Ministry of Health, Casablanca, Morocco; [2]Hubert Department of Global Health, Rollins School of Public Health, Emory University, Atlanta, GA, United States; [3]Center for Global Health Equity, University of Michigan, Ann Arbor, MI, United States

Introduction

Travel and transportation networks have directly facilitated the spread of respiratory infectious diseases, as evidenced during the SARS outbreak in 2003, the H1N1 pandemic in 2009, and the recent COVID-19 pandemic.[1]

International trade, transport, and travel are the backbone of the world's economy. Seamless movement of goods and people has been the number one contributing factor to economic and societal development in the past century and contributes significantly to improved health and well-being. However, when new infections emerge that pose risks to life and health, the transboundary movement of people and animals can facilitate the spread of new pathogens with serious consequences on the world economy.[2,3] Left unchecked, this can lead to the global proliferation of onboard conveyances carried-emerging pathogens and, ultimately, triggering pandemics and traffic losses.

Indeed, global passenger traffic recovered modestly in 2021 from 2020. The number of human passengers worldwide was 2.3 billion or 49% below prepandemic (2019) levels, up from the 60% drop seen in 2020.[4]

Similarly, during health crises, unregulated trade of essential goods (medical supplies and food) may result in lopsided stockpiling of goods, either through manufacturing nations not exporting goods or wealthier countries importing more than they need.[5] In a pandemic, it is essential that the movement of people, animals, and goods is regulated and conducted safely and sustainably.

In response to the COVID-19 pandemic, many nations introduced draconian policies on the movement of people, often closing borders nearly completely.[6] This policy was shown

Modernizing Global Health Security to Prevent, Detect, and Respond
https://doi.org/10.1016/B978-0-323-90945-7.00022-1

to generally not work in preventing COVID-19 from crossing borders, but it decimated various sectors of the economy – including hospitality, tourism, and international trade.[7] In some parts of the world, this adverse economic impact has far-reaching consequences on human life and health (it is well known that the state of the economy and individual finances are closely tied with health). The inability for people to travel also prevented the movement of essential personnel to combat the pandemic, such as public health practitioners, healthcare providers, and humanitarian staff.[6]

Similarly, many nations affected by waves of the pandemic introduced export bans on essential supplies, such as food or medical products, conserving these materials for their own populations while preventing those supplies from going where they were most needed. Where nations allowed exports, the goods were often sent to countries able to bid the highest price – again, very often, not where they were most needed. The 2005 iteration of the International Health Regulations (IHR) contained articles attempting to prevent this from happening, primarily the IHR article 43,[8] but their implementation was widely unsuccessful.

Additionally, various international organizations such as the International Civil Aviation Organization (ICAO), the International Maritime Organization (IMO), and the World Trade Organization (WTO) have their own policies on trade and transport during pandemics.

In this chapter, we describe a historically unprecedented major disruption to global trade and the movement of people during the COVID-19 pandemic. We also discuss what regulations and systems were introduced in an effort to prevent this disruption and methods for bringing these regulations and systems into place.

Literature review and gap analysis

During pandemics, public health recommendations related to travel and trade are dealt with in the IHR (2005). Article 2 of the IHR states, "The purpose and scope of these regulations are to prevent, protect against, control and provide a public health response to the international spread of disease in ways that are commensurate with and restricted to public health risks, and which avoid unnecessary interference with international traffic and trade".[8] The IHR explicitly states the need for avoidance of unnecessary interference with international traffic and trade; throughout the COVID-19 pandemic, we have seen a disregard for these regulations by many Member States. Article 43 of the IHR adds that when Member States determine whether to implement health measures for pandemic control, their decisions must be based on scientific principles, available scientific evidence, and specific guidance or advice from the World Health Organization (WHO).[8] If health measures that interfere with international traffic are implemented, the public health rationale and scientific information for such health measures must be reported to the WHO. However, this is seldom the case due to in part to the unenforceable nature of the IHR.

Disruptions in the movement of goods and people

The movement of people and goods can quickly become restricted during pandemics due to travel restrictions, border closures, and lockdowns. Supply chain issues persist with an inability to move people and goods as usual. During the COVID-19 pandemic, major supply chain challenges persevered as cargo was backlogged, sailings were canceled by ocean carriers, and truck drivers necessary to pick up containers were in short supply due to travel restrictions in many areas.[9]

These same logistic issues occurred in China at the beginning of the COVID-19 pandemic until the virus was contained and the government adapted trucking policies to be more favorable for the movement of goods through a waiver of national highway tolls and the removal of quarantine requirements for trucks carrying essential goods.[9] Other countries and companies have also sought opportunities to facilitate the movement of people and goods, including investing in alternate modes of transportation such as repurposing passenger aircraft for cargo while passengers' aircraft were grounded.[10]

Environmental measures for pandemic control

Preventing the transmission of infection among people on aircraft and vessels is paramount. Article 24 of the IHR describes the responsibilities for conveyance operators, including the responsibility to apply measures to keep conveyances under their responsibility free from sources of infection or contamination.[8]

On-board interventions

High-efficiency particulate air (HEPA) filters are effective in removing airborne pathogens and mitigating against the spread of disease on aircraft.[11] A policy of some airlines has been to require face masks on flights to mitigate the spread of respiratory infectious diseases.[12] Others have avoided providing food to passengers unless the flights are longer than 5 h.[13] Some studies have found evidence of reducing transmission risk with particular flight-related interventions.[14] One study concluded that the risk of transmission is reduced when flights are 8 h and 40 min in duration.[15] Several other studies noted that distanced seating could minimize transmission risk, and a two-row buffer between rows has been recommended to reduce transmission risk further.[16]

Contactless measures

Contactless measures should be enforced whenever possible. Physical distancing should be encouraged in areas prone to the congregation, such as check-in, baggage drop, and security check areas in airports. There are opportunities for check-in processes and health screenings to be completed on personal devices, which allow more travelers to avoid crowded queues and in-person interaction with airline agents. However, in-person services should still be considered to maintain travel access for those with limited technology literacy or access to personal technology. More opportunities can be made for contactless payment services for those who choose to make purchases at the airport. Bathrooms can be equipped with contactless water and soap dispensers and contactless hand dryers.[17]

Travel-related hygiene

Masks are one form of equipment that people wear to protect themselves. During respiratory disease pandemics, some people opt to wear masks voluntarily during travel, and others wear them to follow policies set forth by the airport or transportation companies they are traveling with.[13] In some cases, hygiene kits have been provided to aircraft passengers. One model confirms that transport services should be providing hygiene kits to travelers, particularly during pandemics.[18] These kits can contain anything from alcohol-based disinfectant rubs, masks, and gloves, to biohazard bags and seals for infected waste. When in airports, on transportation conveyances, or in other areas with high population densities, travelers should practice hygiene measures such as handwashing, cough and sneeze etiquette, and hygienic waste disposal.[19] Not only is this proper hygienic practice, but it also reduces the free flow of respiratory droplets in the air and virus particles on surfaces. Transport services have an opportunity to facilitate this practice by making handwashing stations visible, readily available, and well-maintained.

Commuting/staggering

Reductions in typical short-term travel, such as commuting to and from work or school, can contribute to a reduction in cases. When people commute individually rather than through modes of public transportation, there is less exposure to others, and this may reduce the number of new infections on public transportation. However, the risk of infection is still present when people arrive at their destination; in response, some institutions have implemented staggering interventions to reduce the amount of direct, in-person interaction people have with one another.[20]

Travel deferment

Individuals who are unable or do not wish to partake in staggering can choose to defer all travel, though this may pose health, economic, and social challenges, among others. Travel deferment can be a personal and an institutional decision.[21] People may choose to avoid travel or public transportation when they are able, and institutions may implement protocols to continue operations remotely.

Communication/information exchange

Communication is of utmost importance during any emergency to keep all parties informed about the status of the hazard and any prevention and control measures that can be taken to mitigate or lessen risk. In the case of respiratory infectious diseases, clear communication is necessary about the usefulness of face masks, particularly targeted to those most likely to benefit and in activities in which the impact may be more significant for both the travelers and the points of entry staff.[22] In fact, the goal of risk communication targeting travelers and relevant stakeholders at ports, airports, and ground crossings is to[23]:

- Establish, build, and maintain trust through ongoing two-way communication that regularly addresses misunderstandings, misinformation, rumors, and frequently asked questions;

- Encourage travelers and borders personnel to adopt protective behaviors;
- Manage expectations and communicate uncertainties;
- Coordinate and encourage collaboration among response partners;
- Provide information and guidance.

Furthermore, announcements, health-related signs, and pamphlet distribution should be placed at strategic locations to better sensitize and communicate with travelers. As the context of the crisis changes, active engagement with the relevant national coordinating bodies and government agencies is required.

Indeed, in meeting the enormous challenges of the COVID-19 crisis, governments face diverging and competing requests from different ministries and other authorities. Consequently, the immediate response prioritizes public health measures and those aimed at limiting the overall impact on the economy, rather than the recovery of specific sectors such as aviation and tourism. Travel restrictions are a clear demonstration of the efforts being made to curb the virus notwithstanding the potential negative impacts on aviation and its contribution to the economic recovery.[24]

However, as countries prepare for recovery from the pandemic, it is important that authorities adopt a risk-based approach to ensure safe and stepwise ease of travel and trade restrictions.

Personal protective equipment

Personal protective equipment has been recommended widely to prevent the spread and contraction of disease during pandemics. Mask use has been recommended for individuals to prevent the contraction and spread of respiratory infectious diseases.[25] Both trade and travel-related hygiene have also been recommended as methods for disease prevention.[26] This includes regularly disinfecting high-contact surfaces, promoting handwashing, and the use of alcohol-based disinfectant rubs, cough and sneeze etiquette, and hygienic waste disposal.[26] Trade and transportation networks have an opportunity to promote these interventions by combining engineering, administrative, and personal protective control levels to create environmental changes that will make personal precautions a default choice for people directly involved in these networks.

Health measures on arrival and departure

During the COVID-19 pandemic, border closures occurred in 130 countries.[27] These border closures have been shown ineffective in preventing the spread of COVID-19,[28] and rather hampered the COVID-19 response through a reduction in trade (as many goods are carried on passenger aircraft) and essential personnel.[29] While the WHO spoke against these border closures, the WHO and IHR regulations had little power in preventing them. For the most part, these measures were mostly based on public opinion and theatrics rather than science.[1] While less draconian but still present, border screening measures introduced were also ineffective in preventing the spread of disease while at a high cost and allowing for a false sense of security.[50]

Entry and exit screening

Border screening is a less restrictive measure that many countries have employed to prevent respiratory infectious diseases during pandemics. In some cases, border screening may occur for travelers upon exiting a region or country and upon entry to another. Border screening measures have not been found to prevent the spread of disease entirely.[30] However, they may help control the spread of these diseases when implemented as prescribed.[31] As vessels operate with varying procedures for health screenings, there is only a limited opportunity for infected individual passengers to be detected.[31]

Exit and entry screening includes measures like checking for signs and symptoms of disease and interviewing passengers about respiratory infection symptoms and any exposure to high-risk contacts, which can contribute to active case finding among sick travelers. Symptomatic travelers and identified contacts should be guided to further medical examination, followed by testing. Confirmed cases for infectious diseases such as COVID-19 should be isolated and offered treatment as required. Temperature screening alone, at exit or entry, is likely to be only partially effective in identifying infected individuals since infected individuals may be in the incubation period, may not express apparent symptoms early in the course of the disease, or could even dissimulate fever through the use of antipyretic medications. Where resources are limited, entry screening is advisable and should be prioritized for passengers arriving on direct flights from areas with community transmission.[32]

In addition, passengers may complete a form informing health authorities about their possible exposure to cases within the last 2 weeks (contact with patients among healthcare workers, visits to hospitals, sharing accommodation with an infected person). The form should include relevant contact details of passengers who may need to be reached after travel when, for instance, they are identified as a possible contact of a case. It is recommended that such a form be filled out during the flight to avoid crowds at the arrival. Authorities may also require arriving passengers to download and utilize mobile applications to facilitate contact tracing, as occurred during the COVID-19 pandemic.[32]

Crowd control should be put in place to prevent transmission in areas where travelers gather, such as areas for interviews. A more unified approach is needed if border screening measures are to be implemented.

Travel advisories

One article discussed how individual countries should discourage their residents from traveling to affected areas during pandemics if they travel for nonemergency purposes.[33] Public health models have shown that travel advisories can cause significant reductions in travel but do not impact the global spread of disease as they are often implemented too late and are not always followed by residents.[34]

General advice for travelers includes personal and hand hygiene, respiratory etiquette, maintaining physical distance of at least 1 meter from others, and the use of a mask as appropriate. Sick travelers and persons at risk, including elderly travelers and people with serious chronic diseases or underlying health conditions, should postpone travel internationally to and from areas with community transmission.[32]

Travel restrictions/bans

Travel advisory compliance and screening effectiveness demonstrate challenges in implementing such a strict policy as travel restrictions. Travel restrictions at points of entry do not prevent the spread of disease but rather slow the exportation of cases.[35] Fear and lack of scientific understanding may drive governing bodies to ignore IHR recommendations and implement travel restrictions when not scientifically warranted.[36] Additionally, a probable effect of the travel restrictions on the spread of the pandemic might be noticed only during the early stages of the pandemic.[37]

Quarantine

The quarantine of persons is the restriction of activities of or the separation of persons who are not ill but who may have been exposed to an infectious agent or disease to monitor their symptoms and ensure the early detection of cases.[38]

Quarantine is included within the legal framework of the IHR,[8] specifically:

- Article 30 – Travelers under public health observation;
- Article 31 – Health measures relating to entry of travelers;
- Article 32 – Treatment of travelers.

Quarantine may not be recommended even after passengers have been exposed to border screening measures and on-board interventions; instead, rapid detection and disembarkation may prevent further contamination.[39] There are some recommendations for travelers who have been in close contact with these symptomatic individuals to quarantine.[40] Some countries even recommend that all travelers coming from another country should quarantine prophylactically.[41] Even with all these recommendations in place, quarantine is ineffective and impractical in most cases.[42]

If quarantine of international travelers is implemented in the arrival country, ensure that a risk-based approach is used in decision-making, and the dignity, human rights, and fundamental freedoms of travelers are respected, and any discomfort or distress minimized, as per the provisions of the IHR.[8]

Repatriation

Repatriation is the return of a resident to their home country, whether voluntarily or forcibly. This type of intervention can be quite complex when policymakers begin to consider its logistics. Considerations of the country's travel restrictions and access to flights back to the home country must be made for repatriation to work. Needs and risk assessments can be conducted to evaluate the potential success of these measures.[43]

Lockdowns

During pandemics, particularly the COVID-19 pandemic, lockdowns have been a public health measure implemented to slow the spread of contagious disease. Many countries assume that human mobility will decrease with the implementation of lockdown measures and that this has the potential to reduce the number of new cases.[44] The complexity of lockdowns due to essential social interactions makes them a challenging policy to implement and enforce.[45]

Trade measures for pandemic control

The IHR calls for public health response measures not to inhibit trade more than necessary, particularly trade of necessary resources.[8] However, during the COVID-19 pandemic, trade suffered due to export bans and price wars, compounded by resource scarcity. For example, during the initial outbreak, Chinese exports of PPE to the United States fell by 19% due to export bans on PPE[46–48]; high-income countries out-bid low-income countries for PPE, vaccines, and other needed medical supplies; and the United States and European Union put in place export bans that were never before seen—even stockpiling supplies rather than exporting them to where needed.[46–48] This inequitable allocation of resources would cause significant humanitarian implications, particularly in low- and middle-income countries and countries that heavily rely on importing these crucial medical supplies.

The WTO Agreement on Trade-Related Aspects of Intellectual Property Rights (TRIPS) is an agreement on the intellectual property rights of WTO members. It has protected disease-related information from being shared widely with the international community during pandemics.[49] During pandemics, this Agreement is a source of ethical dispute as countries attempt to create and produce vaccines and diagnostic tests to achieve domestic policy objectives.

WHO[51] recommends conducting a risk-assessment approach during the implementation of public health risk mitigation measures for international travel.

The risk assessment should take into consideration the following factors:

- Epidemiological situation in departure and destination countries
- The volume of travelers and existing bilateral or multilateral agreements between countries to facilitate free movement, like the Public Health Corridor principle promulgated by ICAO[52]
- The capacity and performance of public health and health services in involved countries
- The public health and social measures implemented in departure and destination countries.

Vision for the future

An interdisciplinary, collaborative approach is needed to prevent trade, travel, and transport disruption during future pandemics. When used in the context of pandemics, the public health measures identified in Table 12.1 highlight the importance of including but not limited to all the involved stakeholders: public health departments, aviation sector, maritime sector, tourism sector, the scientific community, healthcare providers, engineers, occupational safety experts, policymakers, business owners, and civilians.

The critical lesson for the future is to implement the culture and the mechanisms avoiding that the involved sectors in transport and trade do not work in silos. Instead, bridges should

TABLE 12.1 Themes identified for pandemic control.

Public health measures	Border measures	Trade measures

• Contactless measures
• On-board interventions
• Commuting/staggering
• Travel deferment
• Personal protective equipment
• Travel-related hygiene
• Communication/information exchange
	• Border screening (both entry and exit)
	  ◦ Entry screening
	  ◦ Exit screening
	• Lockdowns
	• Quarantine
	• Repatriation
	• Travel advisories
	• Travel restrictions/bans
		• Trade restrictions/bans
		• Trade regulations
		• Global supply chain
		  ◦ Personnel (cargo handlers) + capacity (less flights) + infrastructure

be built between them in terms of information sharing and joint mutually accepted procedures.

The aviation sector and the maritime sector are highly regulated by actors such as the ICAO, IMO, and International Air Transport Association (IATA). To have public health requirements incorporated within transportation rules, all concerned actors need to harmonize the whole set of procedures.

The COVID-19 pandemic demonstrated how travel and trade could be devastated due to disease outbreaks. Therefore, it is crucial to find the right balance between implementing efficient mitigation measures to protect health while avoiding unnecessary interference with travel and trade. Additionally, it is essential to recall that according to the IHR article 43, restrictions on travel and trade should be based on scientific principles and underpinned by evidence-based data.

Furthermore, on December 1, 2021, the 194 members of WHO reached a consensus to kick-start the process to draft and negotiate a pandemic treaty to strengthen pandemic prevention, preparedness, and response. It also includes access to countermeasures such as vaccines, therapeutics, and diagnostics. Accordingly, such an international, legally binding treaty would undoubtedly upgrade surveillance and response mechanisms and provide better access to the end-to-end global supply chain, reducing the harmful effects of any potential pandemic.

Actions

The first COVID-19 case was recorded on December 9, 2019, in Wuhan, China, and quickly spread internationally due to increasing globalization. The indirect spread of respiratory infectious diseases occurs through trade networks as trade, travel, and transportation are highly interconnected. With travel, reduction comes a reduction in trade as many products are transported with the movement of people to protect their residents from infectious respiratory diseases. It is essential to consider trade, travel, and transportation networks in pandemic preparedness and response efforts.

In order to respond to future respiratory infectious disease pandemics, recommendations by international regulatory organizations such as WHO, IMO, and ICAO should be at the forefront of trade, travel, and transportation decisions made by individual countries, particularly when these decisions have the potential to impact other countries. Those recommendations need to be proactively harmonized to allow Member States to operationalize them smoothly during a crisis such as the COVID-19 pandemic. IHR requirements should also be utilized during pandemic preparedness and response efforts. Recommendations from these sources seek to benefit the global community and are evidence-based where possible. When countries make decisions that contradict these recommendations, they should report to international authorities such as the WTO and WHO their evidence-based rationale for why the decision was made to act against international recommendations.

Models are useful tools that can be employed when creating pandemic preparedness plans. Decision-makers should make assumptions for the models based on the characteristics of the area in which the decision is to be made while also considering available data from previous pandemics. It should be noted that models often do not create scenarios exactly as they will play out in a real-life pandemic, so they should be used with caution and should be used alongside other evidence-based recommendations.

Passenger aircraft should be equipped with HEPA filters that promote filtration and circulation of cabin air. Air transportation industries have the ability to require and regulate this technology to ensure they are up to necessary standards.

Trade, travel, and transportation industries can promote personal protection by creating policies that would require people to wear masks and physically distance themselves when using specific facilities. These policies should be created as a collaborative effort among all both the public and private sectors, including civilian activities. To be more effective, these policies should have some level of enforcement, which may include repercussions for those who do not comply. Before putting those policies in place, these industries should train their staff members on how to effectively follow the new policies and how to communicate them to those they serve in the industry. The industries should also provide their staff and customers with digestible information about the policies in place and the rationale behind them.

When creating preparedness response plans, technology, available tools, human resources, relevant training, and the timing of these interventions should be considered. Trade, travel, and transportation industries have opportunities to improve pandemic response efforts and mitigate the spread of respiratory infectious diseases. Furthermore, particular attention should be granted to the design of public health emergency contingency plans at the national level and at the "points of entry" level to mitigate public health risks. These emergency plans should also be tested through simulation exercises to assess the level of preparedness and the mechanisms of coordination and collaboration between the stakeholders.

CASE STUDY

Issues in the global PPE supply chain

Background

- Due to failures in the global supply chain, the beginning of the COVID-19 pandemic was punctuated by shortages in much needed personal protective equipment (PPE). Between March and May 2020 in the United States, 87% of nurses reported needing to reuse single-use, disposable N95 masks, and 27% of nurses reported requiring to treat patients without adequate PPE (10.1016/j.ypmed.2020.106263). This story was repeated worldwide and was a major contributing factor in the close to 200,000 COVID-related healthcare worker deaths seen globally during the pandemic (WHO).

Observations and key challenges

- Several key challenges are cited for failures in the global PPE supply chain, including:
  - A lack of cargo transport capacity, as a majority of cargo travels on scheduled passenger flights which were not running due to travel restrictions and decreased demand;
  - Export bans and high tariffs on PPE and other medical goods, preventing import into countries without their own manufacturing capabilities;
  - Prioritization of export from countries with manufacturing capabilities in order to turn profits while neglecting their internal market; and
  - Not including supply chain workers on key worker lists. These were all a result of inadequate global regulations and enforcement of regulations to prevent supply chain issues in such a crisis.

Interventions

- Several interventions were deployed in the immediate phase to combat supply chain issues, such as:
  - Using empty passenger planes to transport PPE;
  - Reducing qualification needs for the manufacture of PPE, resulting in increased supply and production capacity, particularly in import-dependent countries;
  - Global coordination of PPE using a just-in-time approach to make sure PPE is used where and when appropriate;
  - Placing supply chain workers on essential workers lists and allowing cargo-only flights to travel into countries with otherwise bans on travel.

Lessons learned

- What Worked
  - Overall, there was a scarcity of policies which worked in preventing issues in the global PPE supply chain. However, some supply chain interventions, such as the use of empty passenger planes and nontraditional means of transportation, reducing the qualification requirements for PPEs, using a globally coordinated just-in-time approach, and easing travel restrictions on supply chain workers, reduced the severity of the supply chain issues.

CASE STUDY (cont'd)

- What Didn't Work
  - The overarching regulatory system for the supply chain failed and did not work in ensuring a stable PPE supply chain. The policies failed to negate bans on travel, did not account for disruption in scheduled conveyances and border closings, and resulted early on in supply chain workers being unable to travel. These key issues, along with others, contributed to the global PPE supply chain crisis.

Overall recommendations

- Dependencies on global supply chains, while economically advantageous, can lead to issues during a worldwide surge in demand. To avoid this, well thought out, enforceable plans should be in place to ensure global supply chain contingency. These include plans to continue cargo flights if passenger flights are not running, international laws against excessive tariffs and export bans, and consideration of supply chain workers as key workers. If possible, national manufacturing capabilities should also be planned for and stockpiles of key goods established.

References

1. Lee K, Worsnop CZ, Grépin KA, Kamradt-Scott A. Global coordination on cross-border travel and trade measures crucial to COVID-19 response. *Lancet*. May 23, 2020;395(10237):1593–1595.
2. Burns A, Van der Mensbrugghe D, Timmer H. *Evaluating the Economic Consequences of Avian Influenza(1)*. January 1, 2008.
3. Fan VY, Jamison DT, Summers LH. *The Inclusive Cost of Pandemic Influenza Risk*. National Bureau of Economic Research; March 2016 (Working Paper Series). Report No.: 22137. Available from: https://www.nber.org/papers/w22137.
4. *The Impact of COVID-19 on Global Air Passenger Traffic in 2021*. Uniting Aviation; 2022. Available from: https://unitingaviation.com/news/economic-development/the-impact-of-covid-19-on-global-air-passenger-traffic-in-2021/.
5. Kelland K. Rich nations stockpiling a billion more COVID-19 shots than needed -report. *Reuters*; February 19, 2021. https://www.reuters.com/business/healthcare-pharmaceuticals/rich-nations-stockpiling-billion-more-covid-19-shots-than-needed-report-2021-02-19/.
6. Devi S. Travel restrictions hampering COVID-19 response | EndNote Click. Available from: https://click.endnote.com/viewer?doi=10.1016%2Fs0140-6736%2820%2930967-3&token=WzM2MDc4NjksIjEwLjEwMTYvczAxNDAtNjczNigyMCkzMDk2Ny0zIl0.ijljQEx6LhEhk_70yD7eFngooSI.
7. Hall JV, Koks EE, Verschuur J. Observed impacts of the COVID-19 pandemic on global trade | EndNote Click. Available from: https://click.endnote.com/viewer?doi=10.1038%2Fs41562-021-01060-5&token=WzM2MDc4NjksIjEwLjEwMzgvczQxNTYyLTAyMS0wMTA2MC01Il0.sB_udXzewm5R3T9jiQ7j1O9jSzQ.

8. *International Health Regulations*. 3rd ed.; 2005. Available from: https://www.who.int/publications-detail-redirect/9789241580496.
9. The Impact of COVID-19 on Logistics. Available from: https://www.ifc.org/wps/wcm/connect/Industry_EXT_Content/IFC_External_Corporate_Site/Infrastructure/Resources/The+Impact+of+COVID-19+on+Logistics.
10. Repurposing Aircraft Passenger Cabins for Transport of Cargo. Available from: https://www.icao.int/safety/OPS/OPS-Normal/Pages/Airworthiness%20TCPC.aspx.
11. Cabin Air & Low Risk of On-Board Transmission. Available from: https://www.iata.org/en/youandiata/travelers/health/low-risk-transmission/.
12. Why are Masks Mandatory when Traveling by Air. Available from: https://www.iata.org/en/youandiata/travelers/health/masks/.
13. EASA ECDC COVID-19 Aviation Health Safety Protocol. EASA. Available from: https://www.easa.europa.eu/document-library/general-publications/covid-19-aviation-health-safety-protocol.
14. van Doremalen N, Bushmaker T, Morris DH, et al. Aerosol and surface stability of SARS-CoV-2 as compared with SARS-CoV-1. *N Engl J Med*. April 16, 2020;382(16):1564–1567.
15. Hoehl S, Karaca O, Kohmer N, et al. Assessment of SARS-CoV-2 transmission on an international flight and among a tourist group. *JAMA Netw Open*. August 18, 2020;3(8):e2018044.
16. Cotfas LA, Delcea C, Milne RJ, Salari M. Evaluating classical airplane boarding methods considering COVID-19 flying restrictions. *Symmetry*. July 2020;12(7):1087.
17. ACI World updates guidance on restart and recovery for airports - ACI World. Available from: https://aci.aero/2021/03/16/aci-world-updates-guidance-on-restart-and-recovery-for-airports/.
18. Suspected Communicable Disease Universal Precaution Kit - Recherche Google. Available from: https://www.google.com/search?q=Suspected+Communicable+Disease+Universal+Precaution+Kit&sxsrf=APq-WBt_qE9Rq9w5JSOrpARrL-yeJZ75yg%3A1645573098990&ei=6nMVYpbmO4rqkwXSnJ_QBQ&ved=0ahUKEwiWg5HAvZT2AhUK9aQKHVLOB1oQ4dUDCA4&uact=5&oq=Suspected+Communicable+Disease+Universal+Precaution+Kit&gs_lcp=Cgdnd3Mtd2l6EAMyBQghEKABSgQIQRgASgQIRhgAUABYAGDzDmgAcAF4AIABogGIAaIBkgEDMC4xmAEAoAECoAEBwAEB&sclient=gws-wiz.
19. Aircraft Module - Hazardous Waste. Available from: https://www.icao.int/covid/cart/Pages/Aircraft-Module—-Hazardous-Waste.aspx.
20. He L, Li J, Sun J. How to promote sustainable travel behavior in the post COVID-19 period: a perspective from customized bus services. *Int J Transp Sci Technol*. 2021;12(1):19–33. Available from: https://www.sciencedirect.com/science/article/pii/S2046043021000836.
21. Elena PR. The impact of covid-19 pandemic on global tourism industry. *Rev Econ*. 2021;73(3):162–170.
22. Malecki KMC, Keating JA, Safdar N. Crisis communication and public perception of COVID-19 risk in the era of social media. *Clin Infect Dis*. February 15, 2021;72(4):697–702.
23. World Health Organization. *Risk Communication and Community Engagement Readiness and Response to Coronavirus Disease (COVID-19): Interim Guidance*, 2020. World Health Organization; March 19, 2020. Report No.: WHO/2019-nCoV/RCCE/2020.2. Available from: https://apps.who.int/iris/handle/10665/331513.
24. Council Aviation Recovery Taskforce (CART). Available from: https://www.icao.int/covid/cart/Pages/default.aspx.
25. World Health Organization. *Rational Use of Personal Protective Equipment (PPE) for Coronavirus Disease (COVID-19): Interim Guidance*, 2020. World Health Organization; March 19, 2020. Report No.: WHO/2019-nCoV/IPC PPE_use/2020.2. Available from: https://apps.who.int/iris/handle/10665/331498.
26. Management of ill travellers at points of entry – international airports, seaports and ground crossings – in the context of COVID-19 outbreak. Available from: https://www.who.int/publications-detail-redirect/10665-331512.
27. Devi S. Travel restrictions hampering COVID-19 response. *Lancet Lond Engl*. 2020;395(10233):1331–1332.
28. Hoffman SJ, August 27 PFO published on PO, 2020. Why pandemic-era border closures are about symbolism, not science – Policy Options. Available from: https://policyoptions.irpp.org/magazines/august-2020/why-pandemic-era-border-closures-are-about-symbolism-not-science/.
29. Rising Up to the Challenge: COVID-19 Guidelines Amidst Border Closures and a Pandemic - World. ReliefWeb. Available from: https://reliefweb.int/report/world/rising-challenge-covid-19-guidelines-amidst-border-closures-and-pandemic.
30. Quilty BJ, Clifford S, Flasche S, Eggo RM. Effectiveness of airport screening at detecting travellers infected with novel coronavirus (2019-nCoV). *Euro Surveill*. February 6, 2020;25(5):2000080.

31. van Seventer JM, Hochberg NS. Principles of infectious diseases: transmission, diagnosis, prevention, and control. *Int Encycl Public Health.* 2017:22–39.
32. Public health considerations while resuming international travel. Available from: https://www.who.int/news-room/articles-detail/public-health-considerations-while-resuming-international-travel.
33. Beirman D. Government travel advisories. In: Wilks J, Pendergast D, Leggat PA, Morgan D, eds. *Tourist Health, Safety and Wellbeing in the New Normal.* Singapore: Springer; 2021:445–466. Available from: https://doi.org/10.1007/978-981-16-5415-2_18.
34. Bielecki M, Patel D, Hinkelbein J, et al. Reprint of: air travel and COVID-19 prevention in the pandemic and peri-pandemic period: a narrative review. *Trav Med Infect Dis.* November 1, 2020;38:101939.
35. Hollingsworth TD, Ferguson NM, Anderson RM. Will travel restrictions control the international spread of pandemic influenza? *Nat Med.* May 2006;12(5):497–499.
36. Madhav N, Oppenheim B, Gallivan M, Mulembakani P, Rubin E, Wolfe N. Pandemics: risks, impacts, and mitigation. In: Jamison DT, Gelband H, Horton S, et al., eds. *Disease Control Priorities: Improving Health and Reducing Poverty.* 3rd ed. Washington (DC): The International Bank for Reconstruction and Development/The World Bank; 2017. Available from: http://www.ncbi.nlm.nih.gov/books/NBK525302/.
37. Chu AMY, Tiwari A, Chan JNL, So MKP. Are travel restrictions helpful to control the global COVID-19 outbreak? *Trav Med Infect Dis.* 2021;41:102021.
38. World Health Organization. *Considerations for Quarantine of Individuals in the Context of Containment for Coronavirus Disease (COVID-19): Interim Guidance*, 2020. World Health Organization; March 19, 2020. Report No.: WHO/2019-nCoV/IHR_Quarantine/2020.2. Available from: https://apps.who.int/iris/handle/10665/331497.
39. Mouchtouri VA, Dirksen-Fischer M, Hadjichristodoulou C. Health measures to travellers and cruise ships in response to COVID-19. *J Trav Med.* May 18, 2020;27(3): taaa043.
40. Abdulrahman A, AlSabbagh M, AlAwadhi A, et al. Quarantining arriving travelers in the era of COVID-19: balancing the risk and benefits a learning experience from Bahrain. *Trop Dis Travel Med Vaccines.* January 12, 2021;7(1):1.
41. León A. Study of the effectiveness of partial quarantines applied to control the spread of the Covid-19 virus. *medRxiv*; 2021. https://doi.org/10.1101/2021.04.03.21254727. Available from: https://www.medrxiv.org/content/10.1101/2021.04.03.21254727v1. Available from:.
42. Alqahtani D. *Public Health Policy: An Ethical Analysis of Quarantine. Electron Theses Diss*; May 10, 2019. Available from https://dsc.duq.edu/etd/1756.
43. Kawuki J, Papabathini SS, Obore N, Ghimire U. Evacuation and repatriation amidst COVID-19 pandemic. *SciMedicine J.* June 22, 2021;3(0):50–54.
44. Atalan A. Is the lockdown important to prevent the COVID-19 pandemic? Effects on psychology, environment and economy-perspective. *Ann Med Surg.* August 1, 2020;56:38–42.
45. Roy S. COVID-19 pandemic: impact of lockdown, contact and non-contact transmissions on infection dynamics. *medRxiv*; 2020. https://doi.org/10.1101/2020.04.04.20050328. Available from: https://www.medrxiv.org/content/10.1101/2020.04.04.20050328v1.
46. Legal and Policy Implications of COVID-19-related Export Restrictions. International Economics. Available from: https://www.tradeeconomics.com/iec_publication/legal-and-policy-implications-of-covid-19-related-export-restrictions/.
47. Export restrictions do not help fight COVID-19 | CNUCED. Available from: https://unctad.org/fr/node/33253.
48. WTO | COVID-19: Measures affecting trade in goods. Available from: https://www.wto.org/english/tratop_e/covid19_e/trade_related_goods_measure_e.htm.
49. WTO | Intellectual property (TRIPS) - gateway. Available from: https://www.wto.org/english/tratop_e/trips_e/trips_e.htm.
50. Meier BM, Bueno de Mesquita J, Burci GL, Chirwa D, et al. Travel restrictions and variants of concern: global health laws need to reflect evidence. *Bull World Health Organ.* 2022;100(3):178. https://doi.org/10.2471/BLT.21.287735.
51. https://www.who.int/news-room/articles-detail/policy-and-technical-considerations-for-implementing-a-risk-based-approach-to-international-travel-in-the-context-of-covid-19.
52. https://www.icao.int/safety/CAPSCA/Pages/Public-Health-Corridor-(PHC)-Implementation-.aspx.

CHAPTER

13

Mitigating negative economic impacts of pandemics

Jarjieh Fang[1], *Erin Holsted*[2], *Krishna Patel*[2], *Ryan Rego*[3] *and Thomas Kingsley*[4]

[1]**ICAP at Columbia University, New York City, NY, United States;** [2]**Rollins School of Public Health, Emory University, Atlanta, GA, United States;** [3]**Center for Global Health Equity, University of Michigan, Ann Arbor, MI, United States;** [4]**American Action Forum, Washington, DC, United States**

Introduction

Recent outbreaks of infectious diseases have had dramatic impacts on global economic growth and development. Most notably, the COVID-19 pandemic has caused national gross domestic products (GDPs) to contract by up to 20%.[2] Similar stories can also be seen in other outbreaks, including the HIV outbreak which was associated with an up to 15% GDP decrease in the most afflicted countries.[3] When containment of the disease is not possible, these negative economic impacts are due in large part to failures in economic responses to pandemics, including failures by local and national governments, as well as failures by international organizations such as the World Health Organization (WHO) or World Trade Organization (WTO).

Failures in economic responses to disease outbreaks take place through all phases of the outbreak. Recognizing that disease outbreaks are cyclical rather than finite, the WHO has defined four pandemic phases: Interpandemic, when there is no current outbreak; alert, when there is a newly emerged public health threat; pandemic, when the newly emerged threat is spreading globally; and transition, when risk is reduced, either through disease elimination or endemicity (with elimination being incredibly rare). For COVID-19, the interpandemic phase is the period before December 2019, the alert phase was December 2019 (when the first case was recorded), and the pandemic phase likely began between February and March of 2020, which is still ongoing at the time of writing.[4] Numerous failings can be seen throughout all phases of the COVID-19 response, to name just a few: a lack of robust

Modernizing Global Health Security to Prevent, Detect, and Respond
https://doi.org/10.1016/B978-0-323-90945-7.00010-5

systems and enforceable regulations through the International Health Regulations created during the interpandemic phase; a lack of information sharing and improvement of response plans during the alert phase; and poor economic stimulus and the imposition of travel and trade restrictions not based on science during the pandemic phase. It is critical that these failings be avoided for the next pandemic.[5]

In this chapter, we first examined the impacts that pandemics have had on economies in the past. We then explained what solutions may be used in the future to improve responses, in particular the creation of frameworks and regulations, and how these solutions may be developed and deployed.

Literature review and gap analysis

Overview

We conducted a literature review of academic and gray literature to summarize the adverse economic impacts of infectious disease outbreaks. Focusing first on GDP, the review traced the economic impacts of notable infectious disease outbreaks, including severe acute respiratory syndrome (SARS), severe acute respiratory syndrome coronavirus 2 (SARS-CoV-2), HIV/AIDS, foot and mouth disease, and H1N1 on GDP. Of these, SARS and SARS-CoV-2 are unique in triggering monetary and fiscal interventions designed to buoy impacted industries and workers and impart stability throughout regional and national economies. The review examined SARS and SARS-CoV-2-related national fiscal and monetary policies, identifying gaps in policy responses that limit their effectiveness or lead to unintended knock-on effects. Finally, the review used data from the ongoing COVID-19 pandemic to assess the relationship between the severity, duration, and distribution of an outbreak and economies.

H1N1 (1918 pandemic)

Outbreaks impact economies. Even prior to the advent of economic indicators used today, contemporaries grappling with their own pandemics observed that economic activity was a victim of diseases. For example, during the fall of 1918, as pandemic flu swept the country; newspaper articles from cities across the United States documented influenza's economic impacts.

An article in the Arkansas Gazette titled "How Influenza Affects Business" claimed that merchants' business declined 40 percent and retail grocery business reduced by one-third.[6] Other common themes that appeared in the contemporary news and magazine articles were churches, schools, and theaters closings. The 1918 influenza pandemic or great influenza pandemic would go on to result in the deaths of an estimated 24.7 and 39.3 million people around the world. Cases would stretch from the United States to New Zealand, spreading in three waves between 1918 and 1920.[7]

While contemporary economists paid little attention to the pandemic that raged around them, modern economists estimate that the pandemic reduced real per capita GDP by 6.2% for the typical country.[8,9] Sweden's robust and detailed administrative data provide additional insight into the pandemic's impact. Analysis that attempts to disaggregate the impact of World War I and the pandemic find widespread economic shock from 1918 through 1920 that worsens through 1930. From 1918 to 20, capital income per capita declined and the proportion of the population living in public poorhouses, which provide shelter and economic support to those in need, increased. In the postpandemic period from 1921 to 30, these effects deepened sharply.[10]

Human immunodeficiency virus

Just 60 years later, the first recognized case of another novel viral infection would jump-start the next global epidemic, and, as recognition of HIV/AIDS spread, so did our understanding of its economic impacts. A wide range of outcomes related to GDP have been reported in the analysis of the economic consequences of HIV/AIDS. Generally, as the national prevalence of HIV/AIDS increases, the economy contracts. For example, an analysis measuring GDP and AIDS cases from 1980 to 92 in 128 countries reported that each additional AIDS case per 1000 persons per year was associated with 0.86 percentage point reduction in the average annual rate of per capita income growth.[11] (65) More recent retrospective analysis of sub-Saharan African countries between 1997 and 2005, GDP contraction is a function of HIV/AIDS prevalence. For HIV/AIDS prevalence of 0.05, GDP fall rate/capita is −0.024, and for HIV/AIDS prevalence of 0.3, GDP fall rate/capita is −0.04.

Countries not yet approaching HIV epidemic control experience ongoing negative impacts to national economic health. Results from a modeling study from 1997 to 2005 showed that HIV/AIDS depressed GDP in Trinidad and Tobago by −4.2% and Jamaica by −6.4%.[12] In Tanzania, in 2010, there was a real GDP percent change of −15%.[3] The 2010 of Malawi GDP was expected to be reduced from a real GDP of 5.03 billion Kwacha without AIDS to 4.814.77 billion Kwacha.[13] Using data from Honduras, the impact of a mature HIV/AIDS epidemic would be approximately 0.012% of GDP and 0.027% GDP in a persistently high prevalence context.[14]

Foot and mouth disease

Foot and mouth disease is an epizootic disease which causes significant morbidity and mortality to animals with cloven hoofs (but not humans), including pigs, sheep, and cows. In the 2001, the United Kingdom experienced an outbreak of foot and mouth disease, which led to 2026 infections and the killing of six million exposed animals to prevent spread.[15] In addition to the huge impact on animal life and supply of food products, foot and mouth disease had a significant effect on the UK's economy. The UK's economy shrunk by 2.3 billion pounds in 2001, 1.5 billion pounds in 2002, and 0.5 billion pounds in both 2003 and 2004 before resolving in 2005. While one may believe this contraction was due to a lack of production of animal products given, it was in mostly fact due to a reduction in tourism (with around 80% of the contraction of GDP between 2001 and 2004 attributable to reductions in tourism income).[16] The UK suffered a reduction in tourism during this time frame for two main reasons. Firstly, to curb the spread of foot and mouth disease, the UK's government closed many rural country footpaths and attractions, a major draw for tourists; and many zoos and other animal dependent attractions also closed. Secondly, despite foot and mouth disease having no major risk to human health, tourists were reluctant to travel to the UK.

Indeed, as part of this study, it was estimated that 30% of visitors to the UK changed their plans directly due to the foot and mouth disease. As a result of this, it was recommended that the UK government and agricultural sector take into account tourism in future disease control strategies and other agricultural policies.[17]

Severe acute respiratory syndrome

In November 2002, SARS emerged as the first novel infectious disease in the Guangdong Province of China. SARS proved both virulent and highly infectious leading to outbreaks throughout Guangdong led to more than 300 cases by February 2003. As case counts grew, communication gaps and restrictions on public information related to SARS impeded critical outbreak information.[18] This lack of transparency led to underreactions by public health professionals throughout the region and around the world as outbreaks in 29 countries emerged. China, Hong Kong, Taiwan, Canada, and Singapore eventually reported the highest cumulative case counts.[19,20]

At the start of Hong Kong's SARS outbreak in February 2003, Hong Kong's economy was still recovering from the Asian financial crisis. Unemployment had surged to record levels by the end of 2002, and many households struggled with mortgage debts.

In most economies, the acute impact of the SARS outbreak was borne by tourism, transportation, and entertainment industries left without tourists and domestic consumers as most sheltered at home.[21] In response, most relief packages focused specifically on buying impacted sectors time.

In China, a $3.5 billion USD relief package brought temporary tax relief and subsidies for affected industries. Malaysia and Taiwan provided grants to tourism-related industries, and Taiwan went further by providing partial reimbursement of SARS-related business losses.[22]

The government of Hong Kong announced a $1.5 billion USD package that rebated salaries tax (income tax) and waived water and sewage charges. The package reduced commercial rents in public buildings and reduced property taxes. Particularly hard-hit sectors, including tourism and entertainment, received access to special low-interest loans designed to support payroll retention.[22]

Singapore followed a similar approach providing property tax rebates to impacted commercial properties, especially shops, restaurants, and hotels, reducing fees for tourist hotels, and extending low-interest bridge loans to tourism-related small and medium-sized businesses. The transportation sector received rebates and reductions to fuel taxes, landing fees, and port dues.[22]

As the SARS outbreak came under control, the overall economic impact of SARS proved acute and short-lived, thanks notably to relatively low case counts and virulence of SARS. For the period following the outbreak, consumer expenditure exceeded pre-SARS levels, pointing to at least some consumers spending down savings accrued during the pandemic, fueling a rapid, V-shaped recovery.

SARS-CoV-2

In December 2019, a cluster of patients with pneumonia of unknown cause were identified in Wuhan, China. By January 2020, a novel coronavirus was identified as the causative pathogen. The coronavirus was named SARS-CoV-2 in February 2020. As of January 2022, SARS-CoV-2 has caused more than 300 million confirmed COVID-19 cases and 5.5 million confirmed deaths, reaching every continent and country. The pandemic also caused an estimated 3.2% contraction in global GDP in 2020.[23]

While the purpose of this chapter is not to opine directly on the success of individual macro-economic policies (nor would it be responsible to suggest a direct link that can be quantified or demonstrated between a country's macroeconomic policies and that country's economic health) that a strong correlation exists is a comfortable assumption. Comparing countries that have invested heavily in their COVID macroeconomic policy response and countries that have not and using quarterly GDP fluctuation as a proxy for economic health reveal trends and gaps that policymakers can consider when tailoring their economic policy response.

The following figure shows the quarterly GDP movement of six Organization for Economic Cooperation and Development (OECD) countries; the three countries that spent the most on their COVID policy response as a percentage of that country's GDP and the three that spent the least.

		2020				2021		
	% GDP	**Q1**	**Q2**	**Q3**	**Q4**	**Q1**	**Q2**	**Standard deviation**
United States	25.5	−1.3	−8.9	7.5	1.1	1.5	1.6	5.3
New Zealand	19.3	−1.4	−9.9	13.9	−1	1.4	2.8	7.7
United Kingdom	19.3	−2.8	−19.5	16.9	1.3	−1.6	4.8	11.8
Denmark	3.4	−0.7	−6.4	6.1	0.9	−0.9	2.3	4.1
Saudi Arabia	2.6	−1.6	−5.5	2.1	1.8	−0.5	0.6	2.8
Mexico	0.7	−0.9	−17.3	12.7	3.3	1.1	1.5	9.8

Data source: Congressional Research Services[23]

The following figure considers the fiscal and monetary policy responses of those same six countries.

		United States	New Zealand	United Kingdom	Denmark	Saudi Arabia	Mexico
Fiscal	Additional funding for health sector	X	X	X	X	X	X
	Allocation of funding for covid testing and vaccines	X	X	X			
	Direct payments to individuals	X					
	Small business grants	X	X	X			X
	Sector specific grants	X	X	X	X		
	Grants specifically targeting vulnerable populations	X	X	X			
	Tax deferral, reduction, or freezes	X	X	X	X	X	X
	Tax increases on foreign goods	X				X	
	Subsidies	X				X	
	Wage support or replacement	X	X	X	X	X	
	Enhanced unemployment benefits	X		X			
	Trade credit insurance	X		X			
	Skills training schemes	X	X	X			
	Public infrastructure spending	X	X	X			
	Housing programs	X	X				X
	Immigration and expat worker restrictions eased	X				X	
	Student loan repayments frozen	X					
Monetary and macrofinancial	Quantitative easing	X	X	X		X	
	Interest rate manipulation	X	X	X	X	X	X
	Extension of government overdraft facility	X		X			
	Extraordinary lending facilities	X	X	X	X	X	X
	Loan guarantee schemes	X	X	X			
	Enhancement of swap line arrangements	X	X	X	X		X
	Freezing or reducing bank capital requirements	X	X	X	X		X
	Mortgage guarantee schemes	X		X			

—cont'd

	United States	New Zealand	United Kingdom	Denmark	Saudi Arabia	Mexico
LTV suspension		X				
Suspension of dividends, buybacks, and bonuses	X	X	X	X		X
Personal loan repayment freezes			X			
Mortgage moratoria	X	X	X			
Waiving central bank fees					X	

Data source: IMF[24]

To reiterate, while a low standard deviation is demonstrative of economic health, an analysis that considers only the economic policy responses is fundamentally limited view that does not consider the strength of the pre-existing economy or central bank system (or public health system, for that matter).

With that in mind, these results nonetheless suggest avenues of approach for policymakers in considering their policy response to mitigate the financial impact of a pandemic.

Significant emergency public spending is not a necessary requirement for success

Denmark stands out as having spent relatively little as a percentage of GDP on its COVID policy response but having suffered comparatively little economic shock. While an argument can be made that Denmark's policy response was therefore the most efficient possible and would make the fiscal and monetary COVID response policies employed by Denmark worthy of particular study, this suggestion is potentially undercut by the relative maturity of the social safety net in place in Denmark prior to the onset of the pandemic. A robust unemployment benefit already in place, for example, required Denmark to change its fiscal and monetary policies little to cope with the novel demands of the pandemic. This would seem to lend credence to the idea that while significant *emergency* spending is not a necessary requirement for success, this is typically true only of countries with a mature and well-financed public health framework already in place.

True cost of COVID policies not yet known

Despite COVID-specific policies representing a small fraction of Denmark's GDP, these policies are not "free"—the cost was simply paid at another time, in the ordinary course of Danish government.

Similarly, policymakers are encouraged to consider the macroeconomic implications of emergency pandemic-related spending. While a flat GDP quarterly variance is a positive indicator that the economy is not undergoing a financial shock, that measure is relatively short term and might not be worth the long-term implications including the potential impact of vastly swelling country deficits.

Successful implementation of economic relief policies as important as their content

The US government allocated $46.5 bn toward a rental aid program that, a year into the pandemic, had only disbursed 11% to renters. The potential causes of this are varied and deserve consideration by policymakers; these might include a significant burden on under-equipped local authorities is one of the perils of a decentralized policy response, or complex terms for the aid may confuse the intended recipients and prevent uptake. In any case, significant funds were tied up in a lengthy and bureaucratic process that resulted in the worst possible scenario—funds expended to no economic benefit.

Some policies would appear to be universal

Most countries with a central bank apparatus appear to have employed interest rate manipulation and, to a lesser extent, quantitative easing as part of their economic policy response to COVID. Wage support or wage replacement proved equally common.[24] Knowing that these tools will likely be employed again in the event of similar circumstances suggests that policymakers should review the effectiveness of these tools as applied to COVID and lessons for future use.

Summary of key gaps

Going back beyond even the 1918 flu pandemic, policymakers and the populace alike have recognized the heavy burden of infectious disease outbreaks on economies. The emergence of contemporary novel infectious diseases has furthered this awareness, spurring countries to attempt measures to mitigate negative economic consequences. However, the relatively short history of economic mitigation measures, first with SARS and now with SARS-CoV-2, highlights several key gaps in the global approach to containing economic fallout.

Evaluations of economic interventions are needed to guide planning and future action

Prior to the current COVID-19 pandemic, SARS represented the only other novel infectious disease outbreak that triggered widespread economic interventions. Despite billions of dollars in relief, there is a dearth of literature rigorously estimating the impact of different measures. The economic impacts and interventions triggered by the COVID-19 pandemic are being scrutinized at sub-national, national, and regional levels, generating over 22000 volumes of COVID-19 related economic literature since 2020. Synthesizing, collating, and interpreting these findings will be critical to informing policymakers of both short- and long-term impact of policy responses.

Many countries were forced to develop and implement policies in real time

The wide-ranging economic consequences of the COVID-19 pandemic caught most countries off-guard. As macroeconomic instability and weakness emerged, policymakers scrambled to assess and respond in parallel, resulting in disparate outcomes depending on the selection and implementation of the policy. This reactive approach weakened countries' abilities to mitigate pandemic-related economic impacts.

CASE STUDY

Problem and background

In November 2002, SARS was first reported in Asia. SARS is a viral respiratory, airborne, illness caused by a coronavirus called SARS-CoV.[25] SARS can spread through small droplets of saliva in a similar way to the common cold and influenza.[25] Most SARS cases were diagnosed in Asia, but the spread was worldwide. The virus spread to more than two dozen countries globally before containment in July 2003.

Observations and key issues

- Infectious disease outbreaks put a heavy burden on economics.
  - The most significant economic impacts of SARS were related to a country's overall decline in GDP.
  - There were estimated losses of 12.3–28.4 billion US dollars within Asian countries, and an estimated contract of GDP by 1% in China and 0.5% in Southeast Asia.[26]
  - Toronto, Ontario, Canada also experienced a decline in economic activity.
  - Due to SARS's mode of transmission, sectors that relied heavily upon people's movement and interactions were most severely affected economically during the outbreak, such as tourism, hotels, and restaurants.

Interventions

- To mitigate negative economic impact, governments passed billions of US dollars' worth of fiscal programs to stimulate the economy and target affected industries. Some governments also passed spending packages that provided help for medical and public health sectors to combat the spread of SARS. A few stimulus packages in response to SARS are documented below.[22,27]
  - China ($3.5 billion)
    - Tax reliefs and subsidies for affected industries
    - Free medical treatment for the poor
    - Some price controls on SARS-related drugs and good
  - Hong Kong ($1.5 Billion)
    - Tax reliefs for affected industries
    - Job creation
    - Loan guarantee schemes
  - Malaysia ($1.92 billion)
    - Loan programs
    - Support for tourism-related industries
    - Job training
  - Singapore ($132 million)
    - Reduction in tourism and transport administrative fees
    - Relief measures for airlines
  - Toronto, Canada ($12 million)
    - Administrative cost
    - Protection of jobs
    - SARS assistance office for the interests of employees

Lessons learned

What worked?

- Pandemic related stimulus programs helped rebuild the economy of affected countries. The recovery of consumer confidence in the economy as a result of these stimulus packages led to an uptick in business activity in the restaurant, travel, and retail sectors.[22]

What did not work?

- Although stimulus packages were effective, it can be said that these packages were costly, and the funds allocated for

CASE STUDY (cont'd)

them could have been used as investments in public services and for the general welfare of citizens.[22]

- Additionally, the increased deficit and inflation caused by the short-term actions may not have been worth the short-term benefits.[22]
- Policymakers were compelled to create and carry out economic policies in tandem with the SARS outbreak. This reactive strategy made it harder for nations to effectively reduce the economic effects of pandemics.

Recommendations

- In the event of a novel disease outbreak, there is an arsenal of both fiscal and monetary interventions that can be implemented.[28] To inform a response strategy for upcoming public health emergencies, an assessment of economic interventions by an international coordinating body—either a financial or public health body—is required. The findings of the evaluation will be critical in evaluating the possible short and long-term implications of possible economic interventions.
  - After the evaluation is completed, policymakers would benefit from the development of a comprehensive policy framework. This would enable decision-makers to respond swiftly to reports of a pandemic of a novel illness and focus their resources on implementation as opposed to reacting in parallel.
- Pandemic-related negative economic effects are never entirely preventable, and preventative and containment measures might not succeed. Thus, it is essential that public health experts and policymakers carry out an assessment of effective economic interventions that is utilized to develop a clear framework to respond on a public health emergency.

Vision for the future

A growing body of research is beginning to demonstrate the complex interrelationships of public health and economic crises.[29] Nowhere is the potential for a negative feedback loop with adverse health and economic impacts better encapsulated than in a global pandemic; where a health crisis creates economic strain, itself leading to adverse health outcomes impacting both pandemic recovery and further widening the health and economic disparities between populations, inter- and intranationally.[30] These impacts are exacerbated by the opportunities and challenges posed by an ever-globalized world. As financial markets continue to evolve a complex web of global interdependencies "contagion" is not limited simply to viruses but also to market disturbances.[31]

Preventing the next global pandemic before it could begin would of course prevents the negative economic implications of a public health crisis. Assuming that this goal is unrealizable in the near term, our vision for the future is the *mitigation* of the adverse economic implications of a pandemic, decreasing the severity of a negative feedback loop impacting public health. To achieve this goal, policymakers should consider the creation of a unified policy framework or template that can be applied to future pandemics.

Such a framework must be both granular enough that policymakers are not reinventing policies or procedures with broad applicability to pandemics de novo every public health crisis; any template must, however, be flexible enough to address the unique economic needs of a particular pandemic.[32,33] Policymakers should consider the applicability of a wide range of potential economic tools and consider the success of different international pandemic responses against the impact on individual nation GDP.[34] Policymakers must forecast the various phases of a pandemic to appropriately tailor their response to the unique supply and demand challenges experienced at these different phases. At-risk populations, industries, and macroeconomic forces must be recognized with the appropriate economic policy levers identified, such that the tools and framework are in place to ensure equitable access to the benefits.

Taking the example of the rescue and recovery plans or "living wills" required of many banks, international bodies, governments, and municipalities must consider not simply the content of a pandemic economic response framework but also its effective implementation.[35] Investment may be required in data collation, management, and economic modeling to achieve a better understanding of target populations in and out of stress scenarios. New budget authorities may need to be created to empower relevant supervisory bodies; lines of communication established and a unified approach to internal and external messaging developed.

A pandemic of sufficient magnitude and concomitant economic crisis will usually, or even optimally, require the deployment of central bank or legislative powers to stabilize the economy. Policymakers must identify where, how, and how much to intervene to mitigate the worst impacts of an economic crisis; the greatest challenge facing them is the *time* required to perform this assessment and implementation from scratch.[36] The work of economic policymakers must be informed by, and tailored to, the particular needs of the pandemic before them; any policy playbook must ensure effective communication channels between economic policymakers and public health officials. The creation of a unified economic policy approach with broad national and international applicability before the next pandemic will lead to greater efficiencies, allowing for a more tailored response, and save lives.

Actions

To achieve the vision of the future that we have set out, the creation of a unified policy framework that can be applied to future pandemics, several actions must be taken. These actions generally revolve around creating a strong evidence base to support such a framework. Most immediately, a database and well executed review of past policy interventions in response to public health emergencies should be created. Such a database and review would include both successful and unsuccessful responses, the granular impact of these responses, and contextual information (such as state and composition of the economy, type of public health emergency, and more). While it is not necessary that the co-ordinating body be either financial in nature or public health, it would need to be of sufficient scale and international reach to co-ordinate contributors across discipline and size of policy response, including, as an example, the now shuttered World Bank's Pandemic Emergency Financing Facility).[37]

It is important that the database and review which we propose be easily updatable and accessible, without biased information. This can best be done through the creation of a global collaborating center similar to those already used by the WHO to monitor emerging infectious diseases and other public health concerns. Such a global collaborating center also has the added advantage of ensuring that any cross-border economic effects are also taken into account—such as those stemming from trade restrictions and travel bans. However, it is equally important that all interventions and outcomes also take into account local contexts and not be so generalized that they are not useful at the country level, including the ability of individual countries to implement economic interventions (often dependent on the strength and power of their central bank); the reliance of the economy on travel; how the economy is set up; and more.

Any review and assessment as to the effectiveness of fiscal and monetary policies should also evaluate the implementation of these policies. A successful economic policy can easily become a failed one by virtue of ineffective execution. Any evidence database should also consider lessons learned on timing, public communication, intergovernmental or interagency co-ordination, and other operational considerations. Funding will be a key consideration, from creation, to maintenance, to implementation of a global set of standardized suite of policy tools and recommendations.

It is important to acknowledge however that lessons from the past will likely not be enough to combat the economic implications of future public health threats and emergencies. As such, thorough research is needed into the effects of hypothetical interventions. Of course, this is not possible via a natural experiment, so economic models and forecast techniques must be developed and improved upon to determine the efficacy of hypothetical interventions. In a similar vein, it is also important for countries to consider their current weaknesses. Using lessons learned from the past, the framework should include methods through which countries can examine their economic risk exposure in case of a public health emergency and offer methods for resolving such weaknesses. Alongside this, regular stress tests should be performed to ensure national and international levels of preparedness for public health emergencies, possibly coordinated by the aforementioned coordinating center.

Finally, it is also important that any framework on mitigating negative economic impacts of public health emergencies also consider the aftermath of public health emergencies. Many economic tools to combat the negative impacts of public health emergencies, such as quantitative easing and deficit spending, have long-term negative consequences including inflation and the need for austerity. This must be addressed and monitored, with long-term effects possibly considered and examined using forecasting methods.

References

1. Morrison JS, Reynolds C. Launching a new pandemic preparedness fund: a crack in the cycle of panic and neglect? *CSIS*; June 27, 2022. https://www.csis.org/analysis/launching-new-pandemic-preparedness-fund-crack-cycle-panic-and-neglect.
2. *Global Economic Prospects World Bank*. 2022.
3. Cuddington JT. Modeling the macroeconomic effects of AIDS, with an application to Tanzania. *World Bank Econ Rev*. May 1993;7(2):173–189.
4. *COVID Pandemic's "Acute Phase" Could End by Midyear*. WHO; 2022.
5. Gaskell J, Stoker G, Jennings W, Devine D. Covid-19 and the blunders of our governments: long-run system failings aggravated by political choices. *Polit Q*; August 11, 2020. https://www.ncbi.nlm.nih.gov/pmc/articles/PMC7436519/.
6. *How Influenze Affects Business the Arkansas Gazette*. Vol. 4. 1918.
7. Patterson KD, Pyle GF. The geography and mortality of the 1918 influenza pandemic. *Bull Hist Med Spring*. 1991;65(1):4–21.
8. Molinari NA, Ortega-Sanchez IR, Messonnier ML, et al. The annual impact of seasonal influenza in the US: measuring disease burden and costs. *Vaccine*. June 28, 2007;25(27):5086–5096.
9. Mauro Boianovsky GE. How economists ignored the Spanish flu pandemic in 1918–1920. *Erasmuc J Philos Econ*. 2021;14 (Special Issue on the Philosophy and Economics of Pandemics).
10. Karlsson M, Nilsson T, Pichler S. The impact of the 1918 Spanish flu epidemic on economic performance in Sweden: an investigation into the consequences of an extraordinary mortality shock. *J Health Econ*. July 2014;36:1–19.
11. David E, Bloom ASM. Does the AIDS epidemic threaten economic growth? *J Econom*. 1997;11(1):105–124.
12. Shelton NJ, McLean R, Theodore K, Henry RM, Bilali C. *Modelling the Macroeconomic Impact of HIV/AIDS in the English Speaking Caribbean: The Case of Trinidad and Tobago and Jamaica*. 1998.
13. John T, Cuddington JDH. Assessing the impact of AIDS on the growth path of the Malawian economy. *J Dev Econ*. 1994;43(2):363–368.
14. Cuesta J. How much of a threat to economic growth is a mature AIDS epidemic? *Appl Econ*. 2010;42(24):3077–3089.
15. Haskins C. *Rural Recovery after Foot-And-Mouth Disease*. 2001.
16. Thompson D, Muriel P, Russell D, et al. Economic costs of the foot and mouth disease outbreak in the United Kingdom in 2001. *Rev Sci Tech*. December 2002;21(3):675–687.
17. Blake A, Sinclair M, Sugiyarto G. Quantifying the impact of foot and mouth disease on tourism and the UK economy. *Tourism Econ*. 2003;9:449–465.
18. Frost M, Li R, Moolenaar R, Mao Q, Xie R. Progress in public health risk communication in China: lessons learned from SARS to H7N9. *BMC Publ Health*. 2019;19(3):475.
19. Abraham T. *Twenty-first Century Plague: The Story of SARS*. JHU Press; 2007.
20. *Summary of Probable SARS Cases with Onset of Illness from 1 November 2002 to 31 July 2003*. 2015.
21. Fan EX. *SARS:Economic Impacts and Implications*. Vol 15. Economic and Research Department Policy Brief; 2003.
22. *Asian SARS Outbreak Challenged International and National Responses: United States General Accounting Office*. 2004.
23. James K, Jackson MAW, Schwarzenberg AB, Nelson RM, Sutter KM, Sutherland MD. *Global Economic Effects of COVID-19: Congressional Research Services*. November 10, 2021.
24. Policy Response to COVID-19.
25. *About Severe Acute Respiratory Syndrome (SARS)*. 2013.
26. Qiu W, Chu C, Mao A, Wu J. The impacts on health, society, and economy of SARS and H7N9 outbreaks in China: a case comparison study. *J Environ Public Health*. 2018;2018, 2710185.
27. Gupta AG, Moyer CA, Stern DT. The economic impact of quarantine: SARS in Toronto as a case study. *J Infect*. June 2005;50(5):386–393.
28. Benmelech E, Tzur-Ilan N. *The Determinants of Fiscal and Monetary Policies during the COVID-19 Crisis*. National Bureau of Economic Research; 2020.
29. Gutman A. *The Great Recessions's Toll on Mind and Body Harvard Public Health Vol Failing Economy, Failing Health the Harvard T.H.* Chan School of Public Health; 2014.
30. *Tracking the COVID-19 Economy's Effects on Food, Housing, and Employment Hardships*. COVID Hardship Watch; 2022.

31. Larisa Yarovaya JWG, Brian L. *Financial Contagion during the COVID-19 Pandemic*. The FinReg Blog; 2020.
32. Bishop J. Economic effects of the Spanish flu. *Bulletin*; 2020. https://www.rba.gov.au/publications/bulletin/2020/jun/economic-effects-of-the-spanish-flu.html.
33. Filippo Annunziata CBM, Busch D, Clarke B, et al. *Pandemic Crisis and Financial Stability*. European Banking Institute; 2020.
34. Holsted E. *Negative Impacts on Gross Domestic Product Caused byPandemics*. Hubert Department of Global Health, Emory Rollins School of Public Health; 2021.
35. *Living Wills (Or Resolution Plans)*. 2021.
36. *Emergency Rental Assistance Program*. 2022.
37. *Pandemic Emergency Financing Facility*; 2020. https://www.worldbank.org/en/topic/pandemics/brief/pandemic-emergency-financing-facility.
38. Keogh-Brown MR, Smith RD. The economic impact of SARS: how does the reality match the predictions? *Health Pol*. October 2008;88(1):110–120.

CHAPTER

14

Health measures at points of entry as prevention tools

Nyri Safiya Wells[1], Charuttaporn Jitpeera[1], Mohamed Moussif[2], Peter S. Mabula[3] and Sopon Iamsirithaworn[4]

[1]Hubert Department of Global Health, Rollins School of Public Health, Emory University, Atlanta, GA, United States; [2]Casablanca International Airport, Ministry of Health, Casablanca, Morocco; [3]Emergency Medicine Department, Disaster Management, Safari City HealthCare Limited, Arusha, Tanzania; [4]Department of Disease Control, Ministry of Public Health, Bangkok, Thailand

Introduction

Points of entry (PoE) is defined as a passage for international entry or exit of travelers, baggage, cargo, containers, conveyances, goods, and postal parcels, as well as agencies and areas providing services to them on entry or exit.[1] Many of the practices conducted at PoE are guided by the International Health Regulations (IHR 2005).

The IHR was first adopted by the World Health Assembly in 1969 with the most recent revision in 2005. Historically, it has been the World Health Organization's (WHO) task to help its Member States (MS) limit and monitor the spread of disease.[2] These regulations exist as a legally binding instrument of international law that prioritizes international collaboration "to prevent, protect against, control, and provide a public health response to the international spread of disease."[3]

Historically, health measures at points of entry encompass a broad range of activities including health screening, vector control, and public health surveillance (PHS).[4] Health measures are meant to be used for prevention to avoid public health risks. Travelers and conveyors (e.g., shippers, truckers) are required to comply with government regulations at PoE. The IHR 2005 is the guiding source for MS to implement health measures.

Modernizing Global Health Security to Prevent, Detect, and Respond
https://doi.org/10.1016/B978-0-323-90945-7.00004-X

Background and significance

Points of entry and stipulations about conveyances constitute a major focus of the IHR (c.f., article IV IHR 2005 page 18 edition 3). Under the IHR regime, POE is to be formally designated by states' parties. Not all potential points where persons or goods enter a country are necessarily designated, given the porous nature of many land borders. At PoE, health measures are often put into place to handle routine activities and public health emergencies of international concern.[4] Management of public health events involves an intricate network of event identification, verification, risk assessment, and response.[5] Some authorities at PoE include agriculture departments, customs, immigration, airlines, emergency responders, law enforcement, rescue and fire departments, airport and port operators, and ground handling companies.

The IHR 2005 defines a PHEIC as "an extraordinary event that may constitute a public health risk to other States through the international spread of disease and may require an international response."[1] PHEICs are one of the most important cornerstones of the IHR because they initiate a sequence of events at PoE outside of routine health measures. Declaration of PHEICs also provides funding from WHO for countries that work to close certain capacity gaps inherent in global public health and initiate other supports.[5] Funding could also come from resources within the country itself including nonprofits and local government agencies.

Though the legal definition of PHEIC is clear, delays in the declaration have negative impacts on health measures at PoE. Delays can be attributed to miscommunications between authorities at PoE and late detection of the outbreak, itself. PHEIC declarations are a necessary part of the work done at PoE to control the occurrence of public health risks because they encourage timely evidence-based action, increase international funding for countries unable to provide it themselves, and limit the effects of emerging and re-emerging diseases.[5,6]

The standards and guidelines determined by authorities at PoE stipulate that health screening should be implemented for all travelers, specific travelers who have been to affected areas, or on particular travel routes.[2] Health screening can be implemented in the long term as part of a country's routine health measure or on a case-specific basis during public health emergencies. There are health screening measures at every stage of travel—before, during, and after boarding. Health screening at PoE typically occurs in two stages: primary and secondary. Primary health screening entails traveler observation and data collection on exposure history. Travelers with symptoms or exposure history will typically undergo secondary health screening measures. Secondary health screening concerns providing health assessments from healthcare or public health professionals. While health screening is a preventative measure, it is also a last resort effort for preventing disease importation or exportation. The likelihood of catching true cases is limited due to several factors explored below. However, health screening is important, and many countries cannot afford to remove these measures.[1,2,5]

Attitudes and behaviors of travelers

The rights of passengers were included within several articles and annexes of the IHR 2005.[1] However, few studies considered the impact that attitudes and behaviors could have on adherence to health measures at PoE. One study found that attitudes and behaviors of travelers are key to ensuring health measures have high uptake at PoE.[7] This study examined perceptions of health screening and the perceived severity of the influenza epidemic itself. Researchers found that demographic characteristics and perceived severity of illness are valuable to determine if travelers were willing to participate in protective behaviors. The results also found educational material and advice directed at travelers could be successfully tailored to subpopulations to increase participation.[7]

Designated points of entry

According to the IHR, States Parties must select airports and ports where the core capacities listed in the IHR Annex 1B are developed and categorize them as "designated." However, "designation" is not mandatory at PoE. State parties may justify designation after considering the volume and frequency of international traffic and the public health risks prevalent in countries of origin.[1]

Entry/exit screening practices at PoE

Health screening is one subcategory of health measures performed at PoE.[1] The goal of health screening is to avoid or delay the importation of infected cases or other public health threats.[1,2] Examples of existent screening practices include body temperature checks, physical exams, self-report questionnaires, visual checks, and passenger interviews conducted by officials.[1,2,4,5] Researchers found little evidence available about the effectiveness of implementing entry screening measures at PoE because it is difficult to ascertain incubating cases, including those who are asymptomatic, while in transit.[2] Existing evidence for the impact of health screening was deduced from the response to four officially declared PHEICs and other notifiable events (the 2003 SARS pandemic stimulated the IHR 2005 revision and occurred before the PHEIC concept came into being): (1) 2003 SARS, (2.) 2009 H1N1, (3) 2014 Ebola, (4) 2015–16 Zika virus epidemic, (5) 2018–20 Kivu Ebola epidemic, (6) 2020 SARS-COV-2, and (7) 2016–17 Yellow Fever.[8]

Severe Acute Respiratory Syndrome

The SARS 2003 outbreak began in China.[2] SARS was never officially declared a PHEIC but was the genesis of major revisions to the IHR 2005 and health measures at PoE. Several countries incorporated entry screening using body temperature thermometers and self-report symptom questionnaires with abysmal detection rates (Table 14.1).[9,10] Low detection rates during the SARS outbreak were likely due to vague case definitions and health screening

TABLE 14.1 Health screening measures across public health emergencies of international concerns.

PHEIC	Health screening measures used	Outcomes
2003 SARS	**1.** Thermal thermometer checks **2.** Self-reporting questionnaires **3.** Travel restrictions/No-fly lists **4.** Contact tracing	**1.** Unknown how many true cases were found **2.** Self-reporting questionnaires subject to under reporting by passengers 2,15
2009 H1N1	**1.** Thermal thermometer checks **2.** Self-reporting questionnaires **3.** Medical tents for symptomatic travelers	**1.** Medical tents helpful addressing symptomatic patients regardless of positive H1N1 test 2,7,12,15,20
2014 Ebola	**1.** Thermal thermometer checks **2.** Self-reporting questionnaires **3.** Visual symptom checks by PoE staff **4.** Interviews conducted by healthcare personnel	**1.** Individuals considered cases were provided with laboratory testing in the later stages of the outbreak12
2015–16 Zika	**1.** Thermal thermometer checks **2.** Self-reporting questionnaires **3.** On-site medical examinations	**1.** Medical examinations were only offered to symptomatic travelers which likely missed several infected passengers because the incubation period ranges from 1 − 12 days meaning symptoms likely occurred after exiting PoE
2018–20 Kivu Ebola	**1.** Thermal thermometer checks **2.** Self-reporting questionnaires **3.** Visual symptom checks by PoE staff **4.** Contact tracing	**1.** Individuals considered cases were provided with laboratory testing in the later stages of the outbreak12
2016 yellow fever	**1.** Thermal thermometer checks **2.** Visual symptom checks by PoE staff **3.** On-site medical examinations **4.** Vaccine passports **5.** Retroactive reporting from healthcare providers	**1.** Visual symptom checks proved helpful in identifying symptomatic cases because yellow fever frequently causes jaundice of the skin and eyes 21

TABLE 14.1 Health screening measures across public health emergencies of international concerns.—cont'd

PHEIC	Health screening measures used	Outcomes
2019–22 SARS-CoV-2	1. Travel restrictions/no-fly lists 2. Thermal thermometer checks 3. Contact tracing 4. Self-reporting questionnaires 5. Retroactive reporting from healthcare providers	1. Health measures could be successful at delaying but not preventing, disease transmission at PoE 2. Travel restrictions are valuable during the early stages of an outbreak because they prevent case exportation29 3. Several studies found that mild or asymptomatic passengers are not often detected by health screening measures 4. Passengers who take antipyretics (antifever medication) complicate final figures when measuring efficacy of screening procedures 5. Healthcare system (i.e., hospitals, clinics, testing centers) remains the most effective way to locate true cases.

measures that depended on self-reporting questionnaires.[2] It is still unknown how screening measures affected the spread of SARS during the outbreak.

Several countries chose to implement travel bans and no-fly lists to keep outbreaks contained. During the SARS outbreak in April 2003, Hong Kong special administrative region (SAR) utilized no-fly lists to avoid potential SARS cases traveling to other countries.[11] The Department of Health in Hong Kong created lists of SARS cases and contacts that were shared with immigration and PoE authorities throughout the outbreak.[11] Cases and contacts on the list were unable to leave the PoE.[11] It is unknown how many SARS cases and contacts were prohibited from leaving the country or the impact this measure had on slowing the spread of the virus. Throughout the COVID-19 pandemic, travel bans have been utilized with varying strictness and unknown efficacy.

Many of the countries that implemented travel bans were only able to do so after the contagion was present in the country. During the COVID-19 pandemic, the United States and South Africa both used travel bans as a health measure at PoE. There are limited studies exclusively analyzing the efficacy of travel bans as it pertains to the COVID-19 pandemic. One study used modeling to understand the impact of the travel ban in Wuhan on COVID cases in other countries.[12] They found that after authorities in Wuhan initiated a travel ban in February 2020, there was a 77% decrease in COVID cases imported from China to other nations.[12] The researchers also noted that risk is directly linked to travel flow in individual countries, thus understanding the impact travel bans can have must be considered on a case by case basis. More studies are needed to better ascertain the implications of incorporating travel bans as a public health response at PoE.[6,11–14]

2009 H1N1 pandemic

The H1N1 virus was an influenza type A virus originating in Mexico in 2009. The 2009 H1N1 epidemic was declared a PHEIC by the WHO in April 2009.[2,15] Health measures at

PoE were implemented immediately in various practices in over 170 Member States and territories. By 2010, the pandemic impacted over 200 MS and territories with ≥17,483 deaths.[14] After the pandemic, WHO was able to conduct a survey, which demonstrated that health screening accurately detected four confirmed cases per every 100,000 travelers in 10 countries (six in the Western Pacific, two in the Americas, one in South-East Asia, one in the European Region).[3] Though reports like this are illuminating, there are few others to draw upon. This highlights the gaps in understanding how health measures work as prevention tools.

2012-14 Ebola in Western Africa

Ebola outbreaks have been declared a PHEIC twice in the past decade (once in 2014 and 2016).[2] The Ebola virus disease (EVD) outbreak in Western Africa in 2014 was declared a PHEIC Aug 2014. With the precedents presented by the 2003 SARS and H1N1 epidemics, Member States followed many of the WHO's recommendations at all PoE. The health screening utilized at PoE consisted of body temperature measurement, symptoms and exposure questionnaires, visual reviews, and interviews conducted by healthcare personnel (Table 14.1).[5,16,17]

Several studies examined the screening measures during the outbreak. One published by the European Center for Disease Control (ECDC) looked at the Ebola outbreaks and the efficacy of screening measures.[5,16] Ebola was a particular issue at ground crossings, whereas the 2003 SARS and 2009 H1N1 outbreaks were major concerns at airports. Temperature screening was a common measure implemented using noncontact infrared thermometers (NCIT) or thermal scanner cameras (TSC). NCIT measures skin temperature but not basal body temperature. Several studies showed that NCIT can detect passengers with increased body temperature, although that does not always lead to case detection. Both NCIT and TSC are comparable in price but NCIT requires more personnel for training and operation. The ECDC report also showed that NCIT is slightly more accurate than TSC because it can be used at a closer distance. ECDC reported sensitivity for NCIT is at 80%–90%, which means that between 1% and 20% of febrile passengers will be missed (false negative).[5,16] The specificity for NCIT was 75%–99%, meaning between 1% and 25% of nonfebrile passengers will be misidentified as febrile. ECDC recommended that accuracy for NCIT is increased by taking the average of several readings.[5,16]

In contrast, the report stated that TSC cannot be used as a screening measure alone because their readings must be interpreted in conjunction with officially approved thermal screening instruments.[5,16] Their value is placed in the fact that they can be used to screen large groups of passengers at one time, which can be helpful in high-density areas. False negatives are a large concern with health screening that depends on the use of TSC. The report stated that this can be partially mitigated by increasing the temperature threshold that qualifies as febrile.[5,16]

There remain questions about the efficacy of health screening at PoE for identifying true cases. There are limited data that measure true cases detected using thermal temperature screening for Ebola.[5,16] This can be problematic because there can be discrepancies between skin and body temperatures, leading to underestimates and misrepresentations. Both skin and body temperature are also influenced by environmental conditions which can lead to inconsistencies in readings.[5,16] More research is needed to understand how these screening measures can be modified to be more useful in the future to avoid missing true cases in real time.[16]

ECDC also compiled case studies from different MS (i.e., United States, Canada, and United Kingdom) that implemented health screening for the 2014 EVD outbreak.[5] The United States utilized health screening procedures at five airports that received over 94% of travelers from EVD epicenters in West Africa.[2,13] All passengers from Guinea, Liberia, and Sierra Leone were screened closely upon arrival with the use of temperature screening using NCITs and symptom questionnaires.[5] Passengers who did not pass temperature checks were symptomatic or needed further screening were directed to U.S. CDC quarantine officers for verification and questioning. After verifying the temperature readings, symptomatic individuals were referred to proper public health authorities for treatment. Passengers who passed the temperature checks were reminded of symptoms to watch for upon leaving the PoE.[5]

In Canada, temperature screening was the main screening measure used in addition to self-reporting questionnaires where passengers could document exposure and travel history. Symptomatic individuals were referred to the Canadian Quarantine Officers for further treatment.[5]

The United Kingdom published statements from the Chief Medical Officer ahead of implementing screening measures to inform the public. This was a notable example of public health authorities and PoE officials working together toward better risk communication.[5] Health screening was initiated at two major airports (Gatwick and Heathrow) and Eurostar terminals. The screening measures of choice were travel history questionnaires and exposure history. Passengers were asked to inform officials of their travel history, travel arrangements upon exit from the PoE, and exposure history. Symptomatic passengers were given access to medical treatment, whereas asymptomatic individuals were notified of important symptoms.[12]

2015–16 Zika virus epidemic

The Zika virus disease (Zika) epidemic occurred from 2015–16 and impacted several tropical regions in Central America, Pacific Island Nations, and Africa. WHO declared Zika a PHEIC in February 2016 which also engages IHR Article 22.[2] One study conducted in Taiwan found that out of 21,083,404 screened passengers, only five were confirmed Zika cases.[18] The health screening measures utilized during the Zika epidemic included self-report questionnaires, visual observation, temperature screening, and on-site medical examinations. Both the self-report questionnaires and visual observations were instructed by the official list of symptoms provided by WHO. Temperature screening using NCITs in conjunction with infrared ear thermometers.[2,18] There are limited studies outlining the number of true cases detected through health screening at PoE.[2,18]

2016-17 yellow fever

The 2016–17 Yellow Fever outbreak occurred from August 2016 to May 2017 in Democratic Republic of Congo (DRC), Angola, and China. Between 2016–2017, there were 109,000 documented severe infections and 51,000 confirmed deaths occurring largely in Angola and DRC.[19]

The strong trade and travel relationships between China and Angola were a factor in disease importation and exportation at PoE. China's response to the importation of Yellow Fever cases at PoE involved temperature screening, visual symptom checks, and retroactive reporting from healthcare providers.[15] In addition, there were existent efforts to monitor the occurrence of *Aedes* mosquitoes which carry the yellow fever virus.[15] Authorities found 11 cases of Chinese nationals infected with Yellow Fever after returning from Angola. All of these cases were reported as living in Angola's capital, Luanda, at some point during the outbreak and 10 reported being bitten by mosquitoes.[19]

The IHR 2005 already has provisions for vaccination protocol and Yellow Fever.[15,19] The IHR 2005 stipulates that an International Certificate of Vaccination or Prophylaxis (ICVP) should be provided to travelers as proof of vaccination. There are various entry requirements stipulated by IHR 2005 that MS are expected to adhere to including no ICVP required, ICVP required if inbound from a high-risk country (high-risk status declared by WHO or the MS), ICVP required for travelers transiting a high-risk country where transit has exceeded 12 h (variables relating to transit exist) and ICVP mandatory (from any country despite risk status). Yellow fever symptoms can be broad and similar to other illnesses so case detection is reliant on patient travel history. Many countries require documentation of vaccination status upon entry.[15,19]

2018–20 Kivu Ebola epidemic

The 2018–20 Kivu Ebola epidemic occurred in the Democratic Republic of Congo (DRC).[2,16] There were over 3400 reported cases in the region. WHO declared the Kivu Ebola epidemic a PHEIC in July 2019. The health measures used during 2018–2020 were also used during the previous 2012–14 Ebola in Western Africa. Because this outbreak occurred on a smaller scale than the previous outbreak, health screening measures were not implemented outside of Africa.[16] There were limited studies looking closely at the impact of health screening on case detection for this outbreak. This was largely attributed to authorities' ability to contain the outbreak quickly.[16]

Despite occurring on a smaller scale, the EVD outbreaks had an impact on daily life because disease transmission was heavily influenced by low resource settings with porous land borders. Many researchers noted that screening measures were unsuccessful in identifying true cases through entry screening measures.[2,5,16] This has been attributed to the use of local shamans or other alternative medical practitioners that made it difficult to understand where disease origin began and the incubation period for Ebola being too long for screening measures to catch. The incubation period for EVD is anywhere from 2 – 21 days, though the average is measured as 8–10 days.[5] The incubation period introduces another complicated factor that health measures at PoE are unable to account for in real-time.[2,5,16]

MERS-CoV

The Middle East respiratory syndrome (MERS) outbreak occurred from 2012 – 13 in the Kingdom of Saudi Arabia (KSA). Though MERS was not officially declared a PHEIC, it quickly spread to other Middle Eastern countries and illuminated several facets of health screening at PoE. Case definitions were constantly changing at the start of the outbreak because MERS was a novel coronavirus. Thomas et al. conducted a cohort study looking at passengers who used self-report questionnaires, interviews, and travel advisories as health screening measures during the MERS outbreak.[17] They reported between Sep 24, 2012 and Oct 15, 2013, 77 travelers from Middle Eastern countries self-reported MERS-like symptoms. Of the 77 tested, two tested positive for MERS-CoV upon seeking medical care. They concluded that basing screening measures on self-reporting does not aid in detecting true patients.[17] Many patients that tested negative tested positive for other respiratory illnesses reinforcing the need for flexible case definitions and health screening methods upon arrival at PoE.[17] More studies are needed to understand the impact health screening at PoE had on disease prevention.[17]

2020 COVID-19 pandemic

For many countries, travel restrictions and advisories were the first health measures put into place. COVID-19 originated in Wuhan, China, and authorities quickly enacted a lockdown of the entire city, including PoE.[20–22] Despite this, several authors reported that most cases arrived at PoE before their symptomatic period which made it difficult for investigators to understand the scope. Several other cities in China followed Wuhan's example and began using lockdowns to decrease transmission.[20,21] For example, the Division of International Disease Control Ports at the Ministry of Public Health in Thailand used coordinated actions at PoE by involving several public health and travel sectors.[23] The Ministry of Public Health is the leading government sector that focuses on the activation of coordination and response mechanisms. The Communicable Disease Act and Emergency Decree is the legal frameworks in Thailand that allow different agencies to participate in decision-making and the empowerment of provincial Emerging Infectious Diseases committees.[23] Committees from each sector were initiated and facilitated multisectoral collaboration at the central and PoE levels. This included sharing information about cases and infectious disease dynamics through technology between sectors and levels to maintain the integrity of the data collected from various surveillance systems.[23]

Many studies found that international travel was monumental in the spread of COVID-19 during the early phases of the pandemic, and many screening measures were unsuccessful in preventing further transmission of SARS-CoV-2.[2,23,24]

One team examined the impact of international travel on the spread of COVID-19 in mainland China during the early phases of the pandemic.[20,21,25] They used COVID-19 incidence data and global airport network data to tabulate the importation and exportation rate of COVID-19 cases in mainland China. They found a significant correlation between airline travel and case exportation events. With Monte Carlo simulations, the investigators surmised that 64% of exported cases were in the presymptomatic period when they arrived. To determine the effectiveness of health measures at PoE, the researchers accounted for the variations in the incubation period for COVID-19. The incubation period was initially detailed as 1–5 days but was extended to 1–14 days.[20,26] The lockdowns conducted by Wuhan and Hubei decreased the disease exportation rate by 81%, and overall cases were prevented by 71%. Early detections in Wuhan made it possible to avoid transmission of the disease even, while passengers were in the asymptomatic period of infection.[27]

Following the outbreak in Wuhan, several countries began implementing health screening on small scales but eventually began using health screening techniques at all PoE. For example, the United States initially used health screening at only five airports: Atlanta (ATL), Los Angeles (LAX), San Francisco (SFO), New York City (JFK), and Chicago (ORD) during the early stages of the pandemic. This was updated to over 20 airports and shipping yards after the first wave of cases.[2,26,27]

A study by Dickens et al. examined strategies implemented at PoE in the face of the COVID-19 pandemic. Some included access testing and quarantine measures.[25] They found concerns about false negatives and polymerase chain reaction (PCR) testing kits. The researchers also suggested that quarantine measures should be enforced for passengers upon exit from PoE, and the efficacy of these quarantine measures depends on the length and location of quarantine.[25]

Mouchtouri et al. researched the health screening measures utilized during the COVID-19 pandemic. With case reports, news reports from the WHO and the European Center for Disease Prevention and Control (ECDC), and news articles, a statistical analysis was done.[20] They found that from Jan to Feb 2020, 26 countries reported 362 cases of COVID-19. Only 18 countries reported using health screening measures. Five countries did not use health screening during the study period: 1) Germany; 2) Finland; 3) Belgium; 4) Spain; and 5) Sweden. Fourteen (5.2%) of 271 imported cases were found through collective screening measures at other EU member states. An additional 11 were found through observational screening after arrival which increased the efficacy rate to 9.2%.[20]

They noted that contact tracing found another 15 secondary cases showing the effects of entry screening measures are enhanced by exit screening.[20] The rest of the cases (77.5%) were identified using the healthcare systems in each MS. The literature review also included an appraisal of entry screening practices, which demonstrated that health screening is resource-demanding with limited strengths.[20] One limitation of the study was that the researchers consulted mostly gray literature because complete figures were not released at the time of the study.[20]

Opportunities for improvement of health screening at PoE

There are concerns about the effectiveness of health screening during public health emergencies. Several articles examined the success of the past and current health screening measures at various PoE including international ports, ground crossings, and airports. Many questioned the limitations of health screening measures and considered if resources were better spent on other measures.

Singh et al. conducted a qualitative study examining health screening efficacy at ground crossings in Northern India.[28] They utilized the records from the WHO core capacity assessment tool and in-depth interviews with passengers as their data collection methods. The core capacity assessment tool is an excel spreadsheet document provided in the IHR 2005 meant to aid MS in assessing existing capacities and capacity gaps at airports, ports, and ground crossings.[1] The tool comes in multiple languages and measures the capacity to respond to PHEICs, quality of public health emergency contingency plans, vector control, and the ability to sanitize equipment and baggage.[1,28]

After using the tool, the researchers found that the implementation status was approximately 76%.[28] This value shows that there are gaps in implementation, and capacity is not being met.[28] Upon further investigation from both the tool and interviews, they deduce that the gaps are due to staff shortages, funding issues, and mishandling of chemical and nuclear waste. They concluded that gaps inherent at these ground crossings and lack of awareness and understanding from passengers increased the likelihood of disease transmission.[28]

Another study conducted in Tanzania, by Bakari et al. looked closely at barriers to implementing health screening measures under the IHR 2005.[29] They found several capacity gaps prohibited the proper implementation of health screening measures. First, there was limited precedent for Tanzanian PoE being designated as an IHR 2005 partner.[29] This is one of the core capacities highlighted by the IHR 2005. They also found that communication channels were clear at PoE, but many PoE do not have designated rooms for health screening measures to take place. Not only does this pose a health risk to passengers and authorities, but it makes it difficult to conduct public health screening. PHS is another mandatory aspect of the IHR 2005 that promotes better health outcomes. They proposed several solutions to these issues, which could be key in stopping the importation of cases if implemented. The researchers suggested more policy managers being stationed at PoE to help oversee the IHR 2005, public health policies, statutes, and guidelines are in practice.[29] These policy managers should also be cognizant of the delicate balance between health measures and travel and trade. Finally, they called on local policymakers to ensure the availability of resources for the execution of health screening measures at PoE.[29]

Another study examined the effectiveness of passenger screening for the SARS-CoV-2.[30] They found roughly 46% (95% CI = 36–58) of infected travelers were unlikely to be detected using current screening measures.[28] Researchers noted that efficacy is dependent on several factors including incubation period, the sensitivity of exit and entry screening, and the proportion of asymptomatic cases.[28] They concluded that airport screening is unlikely to detect a sufficient proportion of 2019-nCoV infected travelers to avoid entry of infected travelers.[28]

Gostic et al. used mathematical modeling to examine the effectiveness of health screening measures at PoE.[31] They found that the effectiveness of airport screening depends on factors like incubation period, available information on the pathogen, route of transmission, and the stage of the outbreak (i.e., beginning, peak, and end). The results detailed that for pathogens with short incubation periods, symptom screening was found to be more effective for avoiding disease importation.[31]

In contrast, diseases with long incubation periods (i.e., EVD, poliomyelitis, or HIV) were better mitigated with questionnaires preferably used during the initial stages of the epidemic. The most prevalent screening measure was temperature screening which they found to be effective in detecting febrile patients only 70% of the time.[31] They also found that PoE that rely on self-reporting questionnaires risk passengers not being truthful about their symptoms underreporting for other reasons.[31] This study occurred before the 2020 COVID-19 pandemic. They recommended that studies be performed to ascertain better measurement methods for the factors that influence case detection.[31]

Work by Chetty et al. included systematic reviews of the literature and found that thermal and body temperature screening was unsuccessful at stopping the spread of COVID-19.[27] They also consulted two separate modeling studies that assessed dermal temperature screening as a health screening measure and found that both individual and group dermal temperature screening would not aid in detecting enough true cases. This coincided with what several other studies have concluded about temperature screening procedures. Their results also stated that there was insufficient evidence that health screening measures are viable options for delaying community transmission and that international travel quarantine is a better option to significantly reduce case importations.[27] In countries where there are several comorbidities like tuberculosis (TB) and human immunodeficiency virus (HIV), shifting resources to health measures for those issues may be more conducive to promoting better health outcomes.[27]

Some in the public health community feel that health screening measures have the potential to divert meaningful resources away from other interventions. Bogoch et al. considered this point and made the argument that health screening measures at PoE in lower-income countries may not be a successful investment.[32] Efforts to stop case importations could be more effective in preventing large outbreaks.[32] They consulted incoming and outgoing flight data from Guinea, Liberia, and Sierra Leone, PHS data, and health screening measures practiced to evaluate the case-detection rates at PoE. They operated off the belief that each traveler had an equal likelihood of being infected and transmitting the virus. They found no true cases were discovered via screening measures. Instead, most cases were identified at local clinics or hospitals.[32] This could be attributed to strict travel restrictions that were put in place at the beginning of the outbreak in these countries where flights were cut by over 50% in each country. It is also possible that incoming passengers were in their asymptomatic stage; thus, temperature screening and self-report questionnaires were ineffective. They concluded that, during emergencies, low-income countries have decreased capacity that cannot sustain resource-intensive health screening measures. Declaration of a PHEIC does guarantee some funding, but this may not be enough for certain countries.[2,28,29]

The gaps identified in the previous section could be addressed by revisions to the IHR sections involving protocol for risk mitigation at PoE including crowd management and pretravel screening. Specifically addressing low case detection rates, troubles with data sharing, and communication between PoE authorities and travelers will help to close the gaps for health measures at PoE. Evidence suggests that health screening measures are only occasionally successful; however, governments continue to implement them to promote security and good will with the public.

CASE STUDY

The diamond princess cruise ship incident

Background:

- The Diamond Princess cruise ship is owned by the British-American company Carnival Corporation. In early 2020, the ship was quarantined off the coast of Japan due to an outbreak of SARS-COV-2 on board. This was one of the earliest outbreaks of SARS-COV-2 and WHO announced that more than half the known cases of COVID-19 initially came from the Diamond Princess.[33,42]

Observations:

- There were over 3700 passengers on board that were quarantined in their cabins after the outbreak was detected. A total of 691 confirmed cases were found onboard the vessel.[33,42]
- The captain of the ship was notified on January 25th, 2020 about a patient experiencing cough and fever by Hong Kong authorities. The captain waited 48 h before enacting any health measures or closing common areas.[33,42]

CASE STUDY (cont'd)

Key issues:

- Quarantine efforts were more successful for passengers who were confined to their rooms. The crew continued to congregate in common areas and seemed to work for the passengers but not for the crew who continued to work and have contact with passengers.[30,34]
- Unlike airplanes, ships are equipped with medical facilities onboard, and health insurance is available for passengers.[4]
- Epidemiological data were significantly delayed and hindered efforts to resolve the situation.[30,34]
- There were ineffective usage of PPE by staff in both contaminated and uncontaminated zones on the ship and lack of enforcement of health measures onboard the ship.[30,33,34,42]

Interventions:

- The ship docked in Tokyo Bay, and passengers were held on the ship for 27 days.
- February 3, 2020, the ship was split into red (contaminated) and green (uncontaminated) zones, and passengers were asked to stay in their respective cabins. However, these restrictions were not enforced strictly, and there was limited distinction between contaminated and uncontaminated zones in common areas.[30,34–36]

Lessons learned: (what worked, what did not work)

What worked?

- The diamond princess was critical in understanding the role true asymptomatic cases would play in the early stages of the pandemic because approximately half of the confirmed cases were found to be asymptomatic.[35]
- Studies conducted after the event found that quarantining passengers was successful in limiting transmission between living quarters but not totally successful stopping transmission in common areas.[35,36]

What did not work?

- Most of the transmission occurred in common areas likely due to overcrowding, continued gathering of the crew, poor ventilation, and limited sanitation capacity.[35,36]

Recommendations:

- Ship staff should be limited from entering multiple areas during an outbreak to stop cross-contamination and improve case investigation efforts.
- For safe and efficient health outcomes, standards and regulations provided by the IHR 2005 should be followed at every stage of a health crisis. Authorities should continue to develop management recommendations with global acceptance early on in epidemics. Accountability and efficiency at every level of public health and PoE outreach also are important to avoid crises like the Diamond Princess incident.

Vision for the future

The gaps identified in the previous section could be addressed by revisions to health screening, public health surveillance, and public health communication policies within the IHR to improve health measures at PoE. Communication from policymakers to frontline workers must be clear with an emphasis on collaborating to control the spread of disease. Policymakers should incorporate feedback from the field and improve screening measures through the practice. Using technical support on data collection would also be helpful to the screening performance.[37] For example, online health declarations decrease the risk of transmission, and the data links from thermal scanners to data collector channels insure symptomatic passengers are tracked throughout travel checkpoints.[20,32,33,37]

Increased risk perception and heightening travel advisories among passengers also promote adherence to public health guidelines. For example, the "Health Beware Card" was provided to inbound and outbound travelers in Thailand from risk areas, like Wuhan during the COVID-19 pandemic, to alert travelers and warn healthcare workers to initiate wearing full PPEs.[43] Data sharing between countries and airlines should be discussed and agreed upon together with the data authorities and security. All data should be transferred via official platforms to prevent losing information. Airline's should also send flight seat maps to aid retrospective case investigation and contact tracing of confirmed cases (Fig. 14.1).

Actions

- Travelers should be empowered to participate in health screening measures throughout their travels via communication on risk awareness through informative signage and text-based notification systems.
- Several researchers have found that most imported cases are found at clinics and hospitals upon arrival which makes the need for coordinated communication among PoE

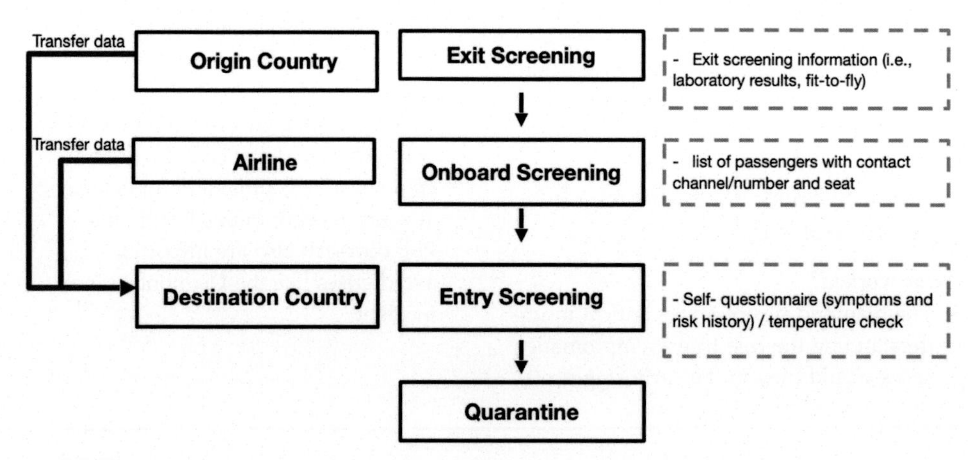

FIGURE 14.1 Idea framework of data transfer from the origin country to destination country.

authorities before, during, and after travel important for understanding the whole situation.[2,10]

- A syndicated approach where the primary authority for response to public health emergencies should be under the Ministry of Public Health with PoE authorities ready to prepare, supervise, and monitor actions.[33] The precedent for this type of syndicated approach has been set during several past public health emergencies including the COVID-19 pandemic.
- Public health authorities must use new technologies as they arise, like health and wellness apps and text message notification systems, to publish official information to support risk communication in the face of an emergency.[33] Helpful information should be updated in rapid response public health events.
- Public health screening measures are the backbone of current pandemic screening practice at PoE and should be synergized with other measures like contact tracing, epidemiological surveillance, quarantine measures, vaccination, and public health surveillance.

References

1. World Health Organization (WHO). *International Health Regulations.* 3rd ed. Geneva: The World Health Organization; 2005.
2. Mouchtouri VA, Christoforidou EP, An der Heiden M, et al. Exit and entry screening practices for infectious diseases among travelers at points of entry: looking for evidence on public health impact. *Int J Environ Res Publ Health.* 2019;16(23):4638.
3. World Health Organization (WHO). *Weekly Epidemiological Record.* 2010. Geneva.
4. World Health Organization (WHO). *Handbook for Management of Public Health Events on Board Ships.* 2016. Geneva.
5. European Centre for Disease Prevention and Control (ECDC). *Infection Prevention and Control Measures for Ebola Virus Disease: Entry and Exit Body Temperature Screening Measures.* 2014. Stockholm.
6. Wilder-Smith A, Paton NI, Goh KT. Experience of severe acute respiratory syndrome in Singapore: importation of cases, and defense strategies at the airport. *J Trav Med.* 2003;10(5):259–262.
7. Sharangpani R, Boulton KE, Wells E, Kim C. Attitudes and behaviors of international air travelers toward pandemic influenza. *J Trav Med.* 2011;18(3):203–208.
8. Durrheim DN, Gostin LO, Moodley K. When does a major outbreak become a public health emergency of international concern? *Lancet Infect Dis.* 2020:887–889. https://doi.org/10.1016/S1473-3099(20)30401-1.
9. Samaan G, Patel M, Spencer J, Roberts L. Border screening for SARS in Australia: what has been learnt? *Med J Aust.* 2004;180(5):220–223.
10. Chan E, Schloenhardt A. The 2003 SARS outbreak in Hong Kong: a review of legislative and border control measures. *Singapore J Leg Stud;* 2004:484–510. http://www.jstor.org/stable/24869491.
11. Chinazzi M, Davis JT, Ajelli M, et al. The effect of travel restrictions on the spread of the 2019 novel coronavirus (COVID-19) outbreak. *Science (New York, N.Y.).* 2020;368(6489):395–400. https://doi.org/10.1126/science.aba9757.
12. World Health Organization. *WHO Global Influenza Preparedness Plan: The Role of WHO and Recommendations for National Measures before and during Pandemics.* Geneva: WHO; 2005.
13. Zhang Y, Seale H, Yang P, et al. Factors associated with the transmission of pandemic (H1N1) 2009 among hospital healthcare workers in Beijing, China. *Influenza Other Respir Viruses.* 2013;7(3):466–471.
14. Schlaich C, Sevenich C, Gau B. [Public health measures at the airport of Hamburg during the early phase of pandemic influenza (H1N1) 2009]. *Gesundheitswesen.* 2012;74(3):145–153.
15. Li C, Li D, Smart SJ, et al. Evaluating the importation of yellow fever cases into China in 2016 and strategies used to prevent and control the spread of the disease. *Western Pac Surveill Response J.* 2020;11(2):5–10.

16. European Centre for Disease Prevention and Control. *Ebola Virus Disease Outbreak in North Kivu, Democratic Republic of Congo*. Stockholm: ECDC; February 22, 2021, 2021.
17. Thomas HL, Zhao H, Green HK, et al. Enhanced MERS coronavirus surveillance of travelers from the Middle East to England. *Emerg Infect Dis*. 2014;20(9):1562–1564.
18. Ho LL, Tsai YH, Lee WP, Liao ST, Wu LG, Wu YC. Taiwan's travel and border health measures in response to Zika. *Health Secur*. 2017;15(2):185–191.
19. Mouchtouri VA, Bogogiannidou Z, Dirksen-Fischer M, Tsiodras S, Hadjichristodoulou C. Detection of imported COVID-19 cases worldwide: early assessment of airport entry screening, 24 January until 17 February 2020. *Trop Med Health*. 2020;48(1):79.
20. Nhamo G, Dube K, Chikodzi D. Impact of COVID-19 on the global network of airports. *Counting Cost COVID-19 Global Tour Indust*. 2020:109–133.
21. Liu B, Sun Y, Dong Q, Zhang Z, Zhang L. Strengthening core public health capacity based on the implementation of the International Health Regulations (IHR) (2005): Chinese lessons. *Int J Health Pol Manag*. 2015;4(6):381–386.
22. Dickens BL, Koo JR, Lim JT, et al. Strategies at points of entry to reduce importation risk of COVID-19 cases and reopen travel. *J Trav Med*. 2020;27(8).
23. Wongsanuphat S, Jitpeera C, Konglapamnuay D, Nilphat C, Jantaramanee S, Suphanchaimat R. Contact tracing and awareness-raising measures for travelers arriving in Thailand from high risk areas of coronavirus disease (COVID-19): a cluster of imported COVID-19 cases from Italy, march 2020. Outbreak, surveillance, investigation and response. *(OSIR) J*. 2020;13(2):38–47.
24. Wells CR, Sah P, Moghadas SM, et al. Impact of international travel and border control measures on the global spread of the novel 2019 coronavirus outbreak. *Proc Natl Acad Sci U S A*. 2020;117(13):7504–7509.
25. The Centers for Disease Control. *Considerations for Health Screening for COVID-19 at Points of Entry*; 2021. https://www.cdc.gov/coronavirus/2019-ncov/global-covid-19/migration-border-health/considerations-border-health-screening.html. Accessed March 8, 2021.
26. Chetty T, Daniels BB, Ngandu NK, Goga A. A rapid review of the effectiveness of screening practices at airports, land borders and ports to reduce the transmission of respiratory infectious diseases such as COVID-19. *S Afr Med J*. 2020;110(11):1105–1109.
27. Gostic KM, Kucharski AJ, Lloyd-Smith JO. Effectiveness of traveler screening for emerging pathogens is shaped by epidemiology and natural history of infection. *Elife*. 2015;4.
28. Quilty BJ, Clifford S, Flasche S, Eggo RM. Effectiveness of airport screening at detecting travelers infected with novel coronavirus (2019-nCoV). *Euro Surveill*. 2020;25(5).
29. International Civil Aviation Organization. Annexes to the Convention on International Civil Aviation (ICAO). Geneva. Published 2020.
30. Xu P, Jia W, Qian H, et al. Lack of cross-transmission of SARS-CoV-2 between passenger's cabins on the *Diamond Princess* cruise ship. *Build Environ*. 2021;198, 107839. https://doi.org/10.1016/j.buildenv.2021.107839.
31. Singh R, Sumit K, Hossain SS. Core capacities for public health emergencies of international concern at ground crossings: a case study from North India. *Disaster Med Public Health Prep*. 2020;14(2):214–221.
32. Bakari E, Frumence G. Challenges to the implementation of international health regulations (2005) on preventing infectious diseases: experience from Julius Nyerere international airport, Tanzania. *Glob Health Action*. 2013;6, 20942.
33. Almuzaini Y, Abdulaziz M, Alhanouf A, et al. Risk communication effectiveness during COVID-19 pandemic among general population in Saudi Arabia. *Risk Manag Healthc Pol*. 2021;14:779–790. https://doi.org/10.2147/RMHP.S294885.
34. Yamagishi T, Kamiya H, Kakimoto K, Suzuki M, Wakita T. Descriptive study of COVID-19 outbreak among passengers and crew on Diamond Princess cruise ship, Yokohama Port, Japan, 20 January to 9 February 2020. published correction appears in Euro Surveill. 2020 Jun;25(24): *Euro Surveill*. 2020;25(23), 2000272. https://doi.org/10.2807/1560-7917.ES.2020.25.23.2000272.
35. Mizumoto K, Chowell G. Transmission potential of the novel coronavirus (COVID-19) onboard the diamond princess cruises ship, 2020. *Infect Dis Model*. 2020;5:264–270. https://doi.org/10.1016/j.idm.2020.02.003. pmid:32190785CrossRefPubMedGoogle Scholar.
36. Baraniuk C. What the Diamond Princess taught the world about covid-19. *BMJ*. 2020;369:m1632. https://doi.org/10.1136/bmj.m1632. Published 2020 Apr 27.

37. World Health Organization (WHO) Thailand. Joint Intra-action Review of the Public Health Response to COVID-19 in Thailand. Thailand. Published July 2020. p. 22-24.

Further reading

1. Wilder-Smith A, Osman S. Public health emergencies of international concern: a historic overview. *J Trav Med.* 2020;27(8).
2. Wickramage K. Airport entry and exit screening during the Ebola virus disease outbreak in Sierra Leone, 2014 to 2016. *BioMed Res Int.* 2019;2019, 3832790-3832790.
3. Parrish RG. Measuring population health outcomes. *CDC: Prev Chronic Dis.* 2010;7(4).
4. Bogoch II, Creatore MI, Cetron MS, et al. Assessment of the potential for international dissemination of Ebola virus via commercial air travel during the 2014 West African outbreak. *Lancet.* 2015;385(9962):29–35.

CHAPTER

15

Rights-based global health security through all-hazard risk management

Qudsia Huda[1], *Erin L. Downey*[2], *Ali Ardalan*[3], *Shuhei Nomura*[4] *and Ankur Rakesh*[1]

[1]**WHO Headquarters, World Health Organization, Geneva, Switzerland;** [2]**Harvard Humanitarian Initiative, Harvard T.H. Chan School of Public Health, Cambridge, MA, United States;** [3]**WHO Regional Office for the Eastern Mediterranean, World Health Organization, Cairo, Egypt;** [4]**School of Medicine, Keio University, Tokyo, Japan**

Introductions

Countries continue to face health security threats including those posed by infectious disease hazards, whether due to deliberate or intentional release or naturally occurring as exemplified by recent outbreaks of diseases with epidemic potential, such as Ebola, Lassa, Zika, MERS-CoV, plague, cholera, and influenza.[1] Pathogens know no borders and the COVID-19 pandemic demonstrated that our global health security is only as strong as its weakest link. Making the world safer for everyone also means scaling up action and innovations on many infectious diseases that kill 2.7 million people every year[2] such as HIV, TB, and malaria.

Critical to our understanding of global health security is the burden posed by multiple emergency events that occur or overlap simultaneously. Individual emergency events due to multiple hazards converge to create concurrent emergencies and have compounding impacts on the lives, livelihoods, and health of populations. Seasonal risks such as tropical cyclones (also known as hurricanes and typhoons) damage or destroy hospitals or force evacuations and leave many vulnerable without care, for example, three hospitals had to be evacuated due to Hurricane Ida[3] in the state of Louisiana at the height of the COVID-19 pandemic, while in the Philippines, crowded evacuation centers exacerbated infectious disease spread as a result of Typhoon Yolanda.[4] Seasonal risks continue to intensify, for instance wildfires that have devastated California, Australia, and Siberia with global predictions of extreme fires increasing

Modernizing Global Health Security to Prevent, Detect, and Respond
https://doi.org/10.1016/B978-0-323-90945-7.00007-5

to 50% by the end of the century.[5] The intensifying effect of wildfire's release of large quantities of carbon dioxide, carbon monoxide, and fine particulate matter into the atmosphere, all have devastating impacts on health, especially during a pandemic.[6] The compounding effects of multiple disasters and emergencies also have intensified the uncertainty of future challenges to global health security.

Concurrent emergencies are not just restricted to seasonality and pose a challenge to all-hazards emergency preparedness. In 2021, a 7.2 magnitude earthquake shook Haiti making the preexisting confluence of hunger, COVID-19, and security challenges more acute. In Myanmar, targeted violence against healthcare 1 year after Myanmar's military coup in 2021 reportedly harmed and intimidated healthcare workers, raided COVID-19 clinics, and blocked vital humanitarian aid, resulting in the decimation of the Myanmar health system amid a pandemic.[7]Nefarious forces have also exploited cyber security vulnerabilities at the intersection with healthcare during COVID-19, for example, in the Western Balkans four areas of health care experienced compounding threats: (1) governmental healthcare agencies; (2) hospitals and healthcare facilities; (3) research and development facilities; and (4) the use of new apps developed in the context of COVID-19.[8] Therefore, for robust health security, Governments need to work across prevention, preparedness, response, and recovery to deal with a confluence of risks and events that strain the ability of already burdened systems to protect their populations,[9] which was especially crucial during the COVID-19 pandemic[10]

Concurrent emergencies further compound the precarious situations faced by refugees, asylum seekers, and internally displaced people. As the fastest-growing populations in the world, they limited access to health services for all aspects of prevention and emergency services while being at risk for increased exposure to violence. Global, forcibly displaced population counts are estimated at 82.4 million.[11] Regional examples include migrating sub-Saharan populations that flee conflict, chronic violence, and other perilous situations into Northern Africa whose routes intersect with other migration routes, for example, the Western, Central, and Eastern Mediterranean routes. The closures of borders by countries during the COVID-19 pandemic suspended their ability to move and find means to work, availability of shelter, and access to health care.[12] The lack of access to the COVID-19 vaccine has further compounded their vulnerabilities. It is estimated that 100,000 million people have been pushed into poverty during COVID-19.[13–15] Together, concurrent emergencies have compounding effects on populations. These elements tend to intensify the staggering effects of ongoing emergencies and disasters on human, financial, and physical resources.

Global health security preparedness

While disease outbreaks and other acute risks to public health are often unpredictable and require a range of risk management measures, the basic principles of managing such health risks should be followed. Primary healthcare (PHC) access and delivery capabilities are a fundamental component of health systems and their resilience to resist, absorb, accommodate, adapt to, transform, and recover from the effects of hazards that could lead to a health emergency. The PHC services are the first level of contact for individuals, the family, and the community within the national health system. Investments in the PHC, as an investment in global health, are critical to keeping communities safe from infectious diseases including HIV,

TB, and malaria, and to help prevent, prepare for, and respond to new health security threats such as COVID-19. As a complementary component of PHC, essential public health functions (EPHFs) are usually seen as a list of minimum requirements for countries to ensure public health.[16] Together these core health systems components are essential during health emergencies. Health security preparedness, reflecting a broader overarching state for health systems, requires a common set of elements in all countries, for example, robust health surveillance systems including alert, monitoring, and reporting mechanisms that reflect robust strategies at every level within a country because pathogens exploit the weakest link of national preparedness.

The International Health Regulations (2005) (IHR) provide an overarching legal framework that defines all countries' rights and obligations in handling public health events and emergencies that have the potential to cross borders. The IHR (2005) is legally binding for 196 countries, including all 194 WHO Member States. It addresses global health security concerns by requiring that all countries have the ability to detect, assess, report, and respond to public health events.[17] Similarly, the Sendai Framework for Disaster Risk Reduction (DRR) 2015–30, which forms part of the 2030 agenda for sustainable development, reiterates the critical link between health and disasters and the responsibilities that all States have in all aspects of disaster management. The framework advocates for an equitable, all-hazards approach for comprehensive emergency and disaster risk management that integrates biological hazards such as epidemics and pandemics in disaster risk management policies, plans, and actions.[18] Together, the crucial concept of country resilience is reinforced by the need to reduce the risks of disaster for a comprehensive approach through prevention, preparedness, response, and recovery measures.[19,20]

Since the adoption of the IHR in 2005, a number of efforts have been launched to assess national preparedness capacity to detect, report, and respond to potential public health emergencies of international concern.[21] This includes the assessment of IHR capacities, through the State Party annual reporting (IHR-SPAR) process and voluntary external evaluation using the Joint External Evaluation (JEE) tool. In 2019, an international panel of experts developed the Global Health Security Index to provide an external assessment of similar state capacities.[22] These efforts are intended to assist Member States with guidance and frameworks to strengthen their health security preparedness at the national level while enabling a gradation of preparedness priorities for states that have different capabilities in place. However, more efforts are needed to address the risks of concurrent emergencies based on the identification of priority risks that the countries and communities are exposed to, using the whole-of-society approach.

All-hazard risk management approach for global health security

Galvanizing global health security requires a thorough understanding of the interconnected nature of risk from all-hazards. This is essential for better prevention, preparedness, readiness, response, and recovery, which facilitates health security at local, national, and global levels. Identification of risks at all levels must be based on evidence and then prioritized so that early and appropriate actions can be taken. Risk assessment is the process of determining risks to be prioritized for risk management, by the combination of risk identification, risk analysis, and evaluation of the level of risk against predetermined standards, targets, risks, or other criteria. Countries need to carry out strategic risk assessments through an inclusive approach in which a wide range of stakeholders participate in the identification of relevant hazards and the assessment of their level of risk within the country. The analyses of the level of exposure, vulnerability, and coping capacity to manage the identified risks before an event occurs will guide risk-informed programs that catalyze efforts to prevent, mitigate, detect early, prepare for, and to be operationally ready for a health emergency or disaster.[23] Strategic risk assessment followed by risk-informed and evidence-based actions are the key to creating enabling environments for health, ensuring that health and equity considerations are integrated throughout the processes of planning, and in investments and decision making at the global, national, and local levels. The Strategic Tool for Assessing Risks (STAR), a standardized methodology developed by WHO,[23] is one of the many approaches for conducting a strategic risk assessment. The assurance that health and equity are integrated throughout the processes of emergency and disaster risk planning requires that each individual or group of people is given the same considerations, resources, or opportunities for health as the whole of the population at risk. While equality means each individual or group of people is given the same resources or opportunities, equity recognizes that each person has different circumstances and allocates the exact resources and opportunities needed to reach an equal outcome.[24] Applied to emergency and disaster risk management, equity means that subpopulations of individuals or groups are not experiencing disproportionate risks or disadvantaged access to risk management measures from the greater population.

The COVID-19 pandemic highlighted that the existing levels of national pandemic preparedness and its multisectoral coordination have not been adequate.[25] Where such efforts existed, they have mostly been severely underfunded, and the costs associated with preparedness are a small fraction of the costs associated with response efforts and human losses incurred when an epidemic occurs.[26] Since the 2009 H1N1 influenza pandemic, various high-level panels and commissions have made specific recommendations to improve global pandemic preparedness, including improvements in the implementation of the IHR (2005). However, despite messages and general consensus that significant change is needed to ensure global protection against pandemic threats, very few of the recommendations have been implemented. Similarly, while the Sendai Framework for DRR includes pandemic risk in its purview, disaster risk reduction and capacity development have largely been separated from health-sector pandemic preparedness efforts in many countries.[27] COVID-19 provides an opportunity to address the gap in global governance, effective health systems, and sustainable financing for strengthening global health security through linkages to disaster risk reduction at all levels. However, this window of opportunity is shrinking rapidly.

As clearly demonstrated in the COVID-19 pandemic, emergencies and disasters including epidemics and pandemics have wide-ranging and often severe impacts across societies; they affect health and well-being, livelihoods, businesses and economies, and the continuity of essential services. Given the complex interdependencies between different sectors, reducing the risks and impacts of emergencies and disasters requires joined-up action across many sectors at all levels of society, under the leadership of government at the highest level of the national system, which is typically the Office of the Prime Minister, the Ministry of Interior or national disaster management authorities.[28,29] Therefore, global, regional, national, and local frameworks, plans, and guidance—including those developed by the World Health Organization (WHO) and other United Nations (UN) agencies—have emphasized whole-of-society actions by the health and other sectors to prevent, prepare for, respond to, and recover from emergencies and disasters.[18]

Cohesive efforts toward strengthening global health security require a methodology that is multisectorial and that leaves no one behind. WHO published the Health Emergency Disaster Risk Management (Health EDRM) framework guidance in 2019 which provides a common language and a comprehensive approach that can be adapted to the country and community context. When applied to health and other sectors, it harmonizes actions to reduce health risks and consequences of emergencies and disasters reflecting global policies and strategies like the IHR (2005) and the Sendai Framework for DRR. The Health EDRM Framework embodies a paradigm shift toward a risk-based, all-hazard, inclusive, and multisectoral approach. It is based on ethical principles that reiterate the value of implementing and sustaining the progress toward the Sustainable Development Goals (SDG) to effectively prepare for managing the risk of events like the COVID-19 pandemic and other concurrent hazards.[30] The framework highlights the need for making emergencies and disaster risk management "a shared responsibility" and "everyone's business" that builds upon the evidence and learning from good practices, including research and innovation.

Rights-based approach to risk management

Emergencies and disasters have a varied effect on individuals and communities. This could be attributed to the differential individual and group vulnerabilities that arise due to variable access to and management of resources that tend to result in vulnerabilities to all types of hazards.[31] Furthermore, emergencies and disasters may create a context in which human rights could be overlooked. However, it is pertinent that "the principles and values embodied in international and national ethics guidelines, as well as human rights instruments, must be upheld" as a core principle in disasters and emergencies (Fig. 15.1).[32]

Guiding frameworks such as the Sendai Framework for DRR (2015–30) and the IHR (2005) reiterate the importance of a rights-based approach to risk management for building stronger global health security. The Sendai Framework for DRR places people in the center and calls for the promotion and protection of all human rights and operates from the starting point that prevention and reducing disaster risk is also the means to protect and promote human rights. The guiding principle 19 (c) of the Sendai Framework for DRR explicitly states "19 (c). Engaging the risk of disasters is aimed at protecting persons and their property, health, livelihoods and productive assets, as well as cultural and environmental assets, while promoting and protecting all human rights, including the right to development." The Sendai Framework for DRR states that there should be a shift toward an "Inclusive disaster risk management," including the equality of rights, opportunities, and dignity of the individual while acknowledging the diversity that contributes to resilience of countries and communities. Similarly, the principles of the IHR (2005) are already structured to pay due diligence to human rights when public health laws authorize government authorities to take control of premises, facilities, and supplies, including health facilities and medical supplies, provided that reasonable

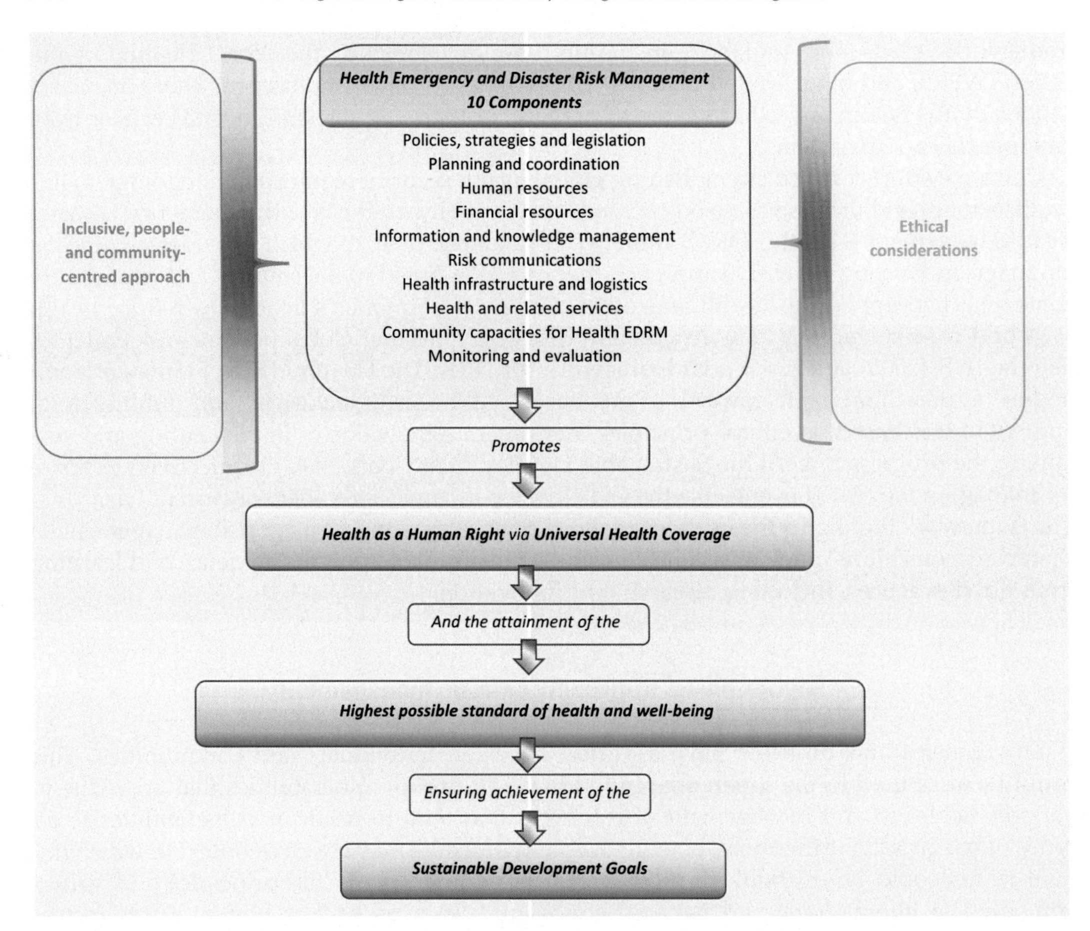

FIGURE 15.1 Health Emergency and Disaster Risk Management (Health EDRM) components upholding health as a human right.

compensation is paid. It is continuously stressed that countries shall treat all travelers with "respect for their dignity, human rights, and fundamental freedoms" taking into consideration the "gender, sociocultural, ethnic, or religious concerns" of travelers.[17]

Similarly, the Health EDRM Framework elucidates two principles that are most relevant to human rights namely, an inclusive, people- and community-centered approach and ethical considerations that include human rights.

Inclusive, people- and community-centered approach

Community members are central to effective Health EDRM, as it is their health, livelihoods, and assets that are at risk of any hazardous event including emergencies and disasters.[33] They are often well placed to manage their own risks through actions that provide protection to themselves, their families, and communities and are often the first responders to an emergency. Health EDRM employs an inclusive approach based on accessible and nondiscriminatory participation. It addresses the needs and capacities of people at greatest risk and disproportionately affected by emergencies and disasters, especially the poorest, as well as women, children, people with disabilities, older persons, migrants, refugees, and displaced persons, people with chronic diseases. Health EDRM framework emphasizes that policies and practices should integrate gender, age, disability, and cultural perspectives, in which the leadership of women, youth, and other at-risk groups should be promoted.

Ethical considerations

Decisions about priorities for reducing risks or responding to disasters include upholding health as a human right, as well as pragmatic, economic, political, and other considerations. Standards of ethics and international health law are central to the Health EDRM framework, driven by principles such as respect for persons, justice, solidarity, and cultural sensitivity. These principles enable ethical action with respect to Health EDRM policy, practice, communications, evaluation, and research and promote trust in interactions with affected communities.

Overall, the rights-based approach for risk management is aligned with the right to health in the WHO constitution: the "highest possible standard of health and well-being for all people who are at risk of emergencies, and stronger community and country resilience, health security, universal health coverage (UHC) and sustainable development."[34] Risk management approaches need to take this into account with an active involvement of the local government, civil societies, communities, and individuals. All risk management policies and practices should integrate gender, disability, age, and cultural perspectives in which the leadership of women, youth, and other at-risk groups should be promoted.[30]

Global health security and SDG in the COVID-19 era

Global health security and Health EDRM address core global health challenges because the world has realized that various hazards, including emergencies and disasters due to natural hazards, disease epidemics, refugee and migration, terrorism, and the effects of climate change, are threatening the health and well-being of individuals.[35,36] These events demonstrate there are fundamental weaknesses in critical global health functions that require collective global collaboration and solidarity, such as the management of cross-border externalities (e.g., pandemic preparedness, control of antimicrobial resistance), the provision of global public goods (e.g., vaccines, therapeutics, diagnostics), and effective leadership and stewardship of global systems (e.g., policies, guidance, standards, accountability, priority setting).[37] The SDG Report 2020, warned about the regressive impact of the COVID-19 pandemic on most SDG goals and the dramatic level of growing inequality.[38] The pandemic has not affected everyone uniformly, and its impact has fallen disproportionately on the most vulnerable along socioeconomic lines, threatening the SDGs' underlying ambition to leave no one behind.[39]

Existing inequalities exacerbated by COVID-19

Additionally, health is greatly influenced by social factors, that is, the social determinants of health. The groups, who already suffer from inequalities, such as the poor, women and children, ethnic minorities, migrants, refugees, the unsheltered, and those living in fragile states and conflict zones, are particularly affected by both the pandemic and its containment measures.[40,41] In fact, in the United Kingdom, during the first wave of the COVID-19 pandemic (January–September 2020), people from all ethnic minority groups had higher COVID-19 mortality than the white British group.[42] Most of the excess mortality risk for COVID-19 observed in some ethnic groups was explained by differences in residence, occupation, lifestyle, and certain preexisting conditions. COVID-19 risk of infection, serious illness, and death is fostered by inequality and reflects a complex overlap of social determinants of health.[43] Furthermore, the interplay of inequalities of these determinants had a complex role in impacting the mental health of populations during the COVID-19 pandemic. According to the latest meta-analysis, in 2020, the number of cases of major depressive disorder and anxiety disorders increased by 28% and 26%, respectively, compared to previous years.[44] There was a more significant increase in the prevalence of the disorders in women as compared to men. The gender difference in prevalence was even more significant between women and men than before the pandemic, providing supporting evidence that women are more vulnerable to the social and economic consequences of the pandemic. For example, UN Women has warned that women have suffered disproportionate job and income losses during the COVID-19 pandemic. According to International Labor Organization, globally, women's employment fell by 4.2% between 2019 and 2020, equivalent to 54 million jobs lost, while men's employment fell by 3.0%, resulting in 60 million jobs lost.[45] Moreover, past experiences with epidemics such as Ebola and Zika have suggested that the epidemics' social and economic impact on people's lives can change the nature and scale of violence against women.[46,47] When restrictions on movement and stay-at-home orders were issued to help limit the spread of COVID-19, there was a particular increase in the incidence of domestic violence among women.[48]

COVID-19 has further accelerated food insecurity in vulnerable national and regional food systems that are already stressed by increasing climate extremes and disasters.[49] The measures taken to contain the spread of the virus have disrupted supply chains with interruptions and delays in domestic and international logistics, and limited access to healthy food and nutrition due to the economic impact and the resulting high food prices.[50] In food and nutrition insecurity situations, the availability and price of food produced in a healthy and sustainable manner become even more challenging.

Inequality and geopolitical tensions inflamed by vaccine nationalism

One of the critical functions of the global health system is the provision of global public goods, which refer to assets that can be used by all people, including public infrastructure facilities, without anyone being left behind.[37,51] Therapeutic drugs and vaccines for infectious diseases are developed and manufactured by pharmaceutical companies, and each of them has intellectual property rights. However, when the spread of infection is related to the health security of people worldwide, they must have the nature of public goods and cannot be enjoyed only by rich countries and wealthy people.[52]

Developing a safe and effective COVID-19 vaccine has not been enough to end this pandemic. The vaccine must be available worldwide at a price that all countries can afford and distribute in a way that achieves equity.[53] Based on this recognition, the COVID-19 Vaccine Global Access (COVAX) Facility was established in April 2020 as an international framework for the fair and stable supply of COVID-19 vaccines.[54] The COVAX is led by Gavi, the Vaccine Alliance (Gavi), a public–private partnership that leads immunization efforts in low- and middle-income countries; the WHO; and the Coalition for Epidemic Preparedness Innovations (CEPI), which was formed after the Ebola epidemic in West Africa, along with UNICEF, as a key delivery partner. By the end of 2020, more than 180 countries and economies were participating in the COVAX. The COVAX scheme uses funds from donors and self-funding members to support the research, production, and distribution of the COVID-19 vaccine. Eligible low- and middle-income countries can receive the vaccine doses free of charge. All other donor countries will first receive doses in proportion to their population to ensure equity until 20% of the population has been vaccinated.

However, the global situation regarding access to COVID-19 vaccines has many problems and challenges.[55] Within months of COVAX being established, recruiting participants, and raising enough funds to prepurchase the vaccine doses, most of the initial global supply had already been purchased by rich countries.[56] They had increased their market power by entering into bilateral purchase agreements with COVID-19 vaccine manufacturers.[57] The experience of the 2009 H1N1 influenza pandemic, when rich countries bought up most of the global supply of the vaccines, is being repeated.[58] Such vaccine nationalism is a major threat to the achievement of global herd immunity and the restoration of social and economic activity.

Breaking the inequality and inequity

On September 23, 2021, world leaders attending the Global COVID-19 Summit hosted by the United States reiterated their commitment to ensuring equitable access to the COVID-19 vaccine for all countries through COVAX.[59] Before this summit, COVAX also called on donors and manufacturers to fulfill their commitments to COVAX, including lifting export restrictions and ensuring that manufacturers were transparent about their delivery schedules and queues. COVAX also encouraged donor countries that are already achieving high coverage rates to get ahead of the queue of vaccine manufacturers and give up their turn in line to COVAX-supported low- and middle-income countries.[56]

Today's world faces significant threats to health security, including terrorism, refugee and migration crisis, and the effects of climate change, and the COVID-19 pandemic further provides a stark lesson on the threats to health security that stem from inequality. In addition to the evident need for support to build public health capacity in areas suffering from long-term poverty and for vulnerable population groups, the realization of the SDGs' principle of leaving no one behind requires increased efforts to address the social determinants of poor health and well-being that generate inequalities. Strengthening global collaboration and solidarity in these areas of public health policy and practice, and moving the action forward, is the most critical health security challenge the world must now address.[43]

Global demand and supply on health

The Universal Declaration of Human Rights (adopted by the United Nations General Assembly in 1945)[59a] asserts that everyone has the right to a standard of living adequate for the health and well-being of himself and of his family (United Nations, b). The minimal corollary of such a right is the obligation on states, as signatories of the Declaration, to ensure the achievement of that right, and to satisfy treaty obligations under the 1966 International Covenant on Economic, Social and Cultural Rights (CESCR)[59b,c] which further specifies the right of everyone to the enjoyment of the highest attainable standards of physical and mental health (United Nations, a), where this is to be seen as a fundamental human right indispensable for the exercise of other human rights (UN Committee on Economic Social and Cultural Rights (August 11, 2000) General Comment No. 14: The Right to the Highest Attainable Standard of Health. Para 1.). Meeting such obligations is a chronic challenge for states, while being at the same time an essential element of global health security.

Although governments, typically through their Ministries of Health, are the lead for establishing and sustaining public health systems that prevent, detect, and respond to infectious disease threats in their country, the health workforce operationalizes this commitment. They follow the directives set forth from these obligations and transform them into practical reality through their profession. Therefore, the implementation of health security at all levels is heavily reliant upon the health workforce.

National policies for health emergency and disaster management aim to meet a country's universal obligations of access to health for its people even during crisis periods. The implementation of those policies, however, reflects geopolitical realities that create challenges in doing so. Examples of these challenges include the entrenched global inequities of access to health, porous borders, unprecedented counts of populations on the move, small island developing states affected by climate, and chronic situations of conflict and violence. Within these contexts, the demand for health services ranging from preventive care to emergency care and psychosocial support persists.

Disparities between demand and supply

The global demand on the healthcare workforce has steadily increased for decades.[60] In 2016, the World Bank projected that future health workforce demand for 165 countries will rise to 80 million by 2030 while the supply of health workers is expected to reach 65 million over the same period, resulting in a worldwide shortage of 15 million health workers.[60] The same report predicted that the lowest income countries are expected to experience the gravest access to health services, increases of comparative costs, and migration (exodus) of their health workforce.[60] Also in 2016, the WHO analyzed data from 130 countries and predicted a shortfall of the healthcare workforce of 18 million by 2030 including the same forecast for the low- and lower-middle-income countries being disproportionately affected.[61]

The International Labor Organization highlighted that the relentless demand upon the healthcare workforce is presenting multiple risks to their health, well-being.[62,63] The COVID-19 pandemic has stressed the health workforce in unprecedented ways, as evidenced by the escalation of deaths that in September 2021 were estimated to be 115,000 globally (range 80,000–180,000).[64] Equitable strategies for preventing deaths, preserving lives, and protecting the individuals that are performing frontline roles remain critical. Yet fundamental strategies such as providing sufficient PPE, unhampered access to vaccines, and addressing health workplace violence, have continued, if not been exacerbated, by the COVID-19 pandemic.[65,66]

The global supply of the healthcare workforce has important characteristics that require strategies to address this shortfall, for example, the World Nursing Report showed that from 191 countries, 90% of nurses are women.[67] The WHO also found that nurses account for 70% of the social and healthcare workforce globally.[68] The chronic problem of violence that occurs to health care workers has been increasingly documented and reported by governmental and nongovernmental organizations, including humanitarian organizations for the past 2 decades.[69–71] These trends exemplify why the protection of health care workers in general, and female health care workers specifically, is a crucial consideration when developing national and global strategies that aim to provide access to health for the whole of societies.

Implementing local-to-global networks at the national level

Local to global networks must maximize all their resources, especially given the increasing demands on health systems, which are simultaneously experiencing a reduction in the supply of healthcare workers. Emergency risk managers, including response agencies, have long known that system-wide elements of multisectoral coordination, collaboration, and communication are among the most critical strategies for immediate localized response and efficient use of resources. Successful implementation of these elements during emergencies and disasters heavily relies upon common goals, strong working relationships, and a willingness to share information, all well in advance of an emergency or disaster. Long-term investments in evidence-based strategies are necessary. All response sectors generally find many shortfalls in their response capabilities and must be willing to reassess, refocus, and reinvest in the strengthening of the health systems when a health crisis occurs.

CASE STUDY

Background

Public health experts have predicted a global pandemic for years.[72,73] The ongoing COVID-19 pandemic[74] reminded us that health systems are a critical part of community infrastructure to manage risks and serve urgent population needs. The health workforce is a central component of this crucial infrastructure and is essential for forming

CASE STUDY (*cont'd*)

and implementing equitable strategies that preserve human lives and livelihoods. The prepandemic prediction of the health workforce shortfall has left the world vulnerable to future pandemics.

Observations

- Noncommunicable Diseases (NCDs) kill 41 million people annually, equivalent to 71% of all deaths globally, but this number increases to 77% in low- and middle-income countries.[75]
- Among these NCDs, cardiovascular diseases account for most of the deaths annually (17.9 million), followed by cancers (9.3 million), respiratory diseases (4.1 million), and diabetes (1.5 million). Together, these four groups of diseases account for more than 80% of premature NCD deaths each year.[76]
- Chronic underfunding of NCDs during nonemergency exacerbates the inability to provide adequate NCD services during health emergencies.[75]
- Country capacities were profoundly ill-prepared to serve the NCD population's needs during the COVID-19 pandemic.[75]

Key issues

- Health service disruptions during COVID-19 in the WHO NCD assessment that reported of the 122 participating countries, 65% of disruptions were due to the cancellation of elective care, 46% were due to the closure of population-level screening programs, and 43% were due to the government or public transport restrictions hindering access to health facilities.[77]
- For healthcare worker-specific issues, 39% of disruptions were due to NCD-related clinical staff that deployed to provide COVID-19 relief, 33% were due to healthcare providers that were unable to provide services due to insufficient personal protective equipment (PPE), and 32% were due to countries that had insufficient staff to provide services.[77]
- For communities, fear and mistrust in seeking health care during COVID-19 were profound.[77]
- Differences among countries, no single risk assessment applies or can address all relevant gaps in a single health system; selection and use of risk assessments must be applied to a given country context.

What worked

- For the 122 countries within the WHO NCD report, the first and second mitigation strategies most often used to overcome disruptions (i.e., coping capacity) were "telemedicine deployment to replace in-person consults" and "triaging to identify priorities," respectively.[78]
- When all four country income levels were analyzed individually (i.e., low-income, lower-middle-income, upper-middle-income, and high-income),
    - "redirection of patients with NCDs to alternate health care facilities" was the second priority for low-income countries (and "telemedicine deployment ..." became an equal priority with "task shifting" as a third priority mitigation strategy to improve coping capacity).

CASE STUDY (cont'd)

- telemedicine was identified as a lower priority in low-income countries.
- "redirection of patients with NCD …" was the fifth priority mitigation strategy to overcome service disruptions for high-income countries.[78]

- Country capacity assessments, such as the WHO 2021 biannual NCD Country Capacity Assessment of service delivery during the COVID-19 pandemic, can inform national strategies for biological response to health emergencies.

What did not work

- Chronic underinvestment in the prevention, early diagnosis, screening, treatment, and rehabilitation of NCDs exacerbated the health service disruptions that occurred during the COVID-19 pandemic.[75]
- Lack of health services coping capacity for NCDs further stressed an already overburdened population health issue.[75]
- COVID-19 caused rapidly changing protocols and treatment policies for care-seekers and health service delivery, placing greater demand on the health workforce.[75]
- Planning shortfalls for NCD health services exacerbated the shortfalls of the emergency plans.[78]

Recommendations

- Risk assessment tools that identify and prioritize gaps in coping capacities for all-hazard risk management must be used as an essential component of accurately identifying country vulnerabilities. After risks are identified and prioritized, approaches for overcoming risks can be formulated.
- Systematic approaches to applying risk assessments to country emergency plans must be applied. Given the differences among countries, no single risk assessment applies or can address all relevant gaps in a single health system, and therefore the selection and use of risk assessments must be carefully tailored to a given context. For example, the Hospital Safety Index is a risk assessment used by health authorities and multidisciplinary partners to gauge the probability that a health facility will continue to be safe and operational in emergencies.[79] The Smart Hospital initiative builds on the Safe Hospital Initiative and focuses on improving hospitals' resilience, strengthening structural and operational aspects, and providing green technologies.[80]
- Coping capacity must be strengthened at the local level. All coping capacity strategies must incorporate rights-based approaches of access to healthcare and concepts of resilience building. In all cases, the health workforce is an operational resource that must be protected and supported as they implement their responsibilities.
- Public messages that anticipate information needs of people during emergencies and disasters are critical to managing the uncertainty that health emergencies create. If these are left unaddressed, they can lead to mistrust, fear, and anger and have cascading effects on populations that exacerbate the healthcare environment, for example, vaccination and PPE

CASE STUDY *(cont'd)*

reluctance, avoidance of the care environment, and violence. Misinformation, disinformation, and rapidly changing information fuel mistrust among the populations providing and receiving care and can exacerbate the mental health and psychosocial impacts for caregivers and care receivers, including their loved ones.

All countries have a profound responsibility to rededicate efforts to use and apply risk assessment tools that are particular to their country's context in the post-COVID-19 pandemic. The assessment of risks and coping capacities must include health services capabilities that may directly or indirectly be affected by health emergencies. As part of this coping capacity, the healthcare workforce capabilities must be included in the assessment. Increasing global health security, particularly during a biological event, can only occur when all countries and their respective subnational levels make a collective contribution to their universal shared responsibility of addressing their gaps in coping capacity for the security and health of their populations.

Vision for the future (including actions)

Resetting the health actions for a better-prepared world: Building a better future

Throughout history, humankind has been plagued by countless humanitarian emergencies resulting from all types of natural, biological, technological, and societal made hazards interacting with the exposure, vulnerabilities, and capacities of the affected communities. However, in recent years, the world has faced devastating impacts of emergencies, often concurrently due to conflicts, outbreaks, and disasters, in addition to the global COVID-19 pandemic.[81–83] These emergencies have affected population health both directly and indirectly through disruption of the health systems and other social services.

In the postacute phase or aftermath of emergencies in general and specifically pertaining to the COVID-19 pandemic, Ministries of Health and other health sector partners need to participate in multisectoral socioeconomic recovery planning and undertake in-depth health sector recovery planning using a comprehensive risk-informed and rights-based approach. This provides an opportunity to address weaknesses in health systems, "build back better"[84] and create systems that are more resilient and fit for purpose. Opportunities to address weaknesses in health systems can be lost, however, unless there is a well-planned and well-implemented strategy, an effective strategic risk and needs assessment, and advocacy to ensure that the health system and the health sector in general, and the strengthening of country and community capacities to manage health risks of all types of emergencies, are well-positioned both as a priority and as a condition for socioeconomic recovery. This may facilitate resource mobilization, increase investments, and budget allocations for the health sector. The COVID-19 pandemic has provided a global example of the consequences of not being preparedness and represents an opportunity to create a more resilient and fit-for-purpose health system that promotes and safeguards population health and global health security, advances progress toward UHC, and plays a central role in the building of resilience into communities.

The recovery phase provides an opportunity to rebuild the health system and other social services to ensure it is more resilient than the preemergency version. This is the core of the build back better principle, which is an approach to recovery that aims to reduce vulnerability to future disasters and build community resilience by addressing physical, social, environmental, climate, and economic vulnerabilities and shocks.[84,85] Postemergency recovery also provides an opportunity to better adapt models of care for future needs and to address distortions that existed in the system before the emergency, or which resulted from it.

Taking the building back better approach to postemergency recovery ensures that the rebuilt system is stronger, safer, smarter, and more resilient. This necessitates the identification and rectification of weaknesses inherent to the previous system. During the recovery process, health service provisions can be improved, for instance, by addressing previously neglected areas such as leadership, governance and coordination for all-hazards health emergency and disaster risk management, mental health and noncommunicable diseases, strengthening linkages between primary and secondary health services, reviewing the distribution of facilities against demographic changes, and constructing new facilities in areas with the greatest need.[86,87]

By taking a building back better approach, the resilience of the new health system can be improved by introducing disaster risk reduction measures in other sectors that also benefit public health such as building codes and land-use planning regulations. Damaged health facilities can be modernized or rightsized, and assets can be replaced with technologically up-to-date, environmentally sensitive, and climate-friendly alternatives (for example, by applying green hospital technologies for renewable energy, waste management, and water use reduction). Additionally, in postconflict settings, health can be used as an instrument for promoting and maintaining peace.[88] This includes conflict-sensitive programming, cooperation between health professionals across ethnic divides and the increase in state legitimacy that is generated by the provision of public goods, such as healthcare, that may contribute to the sustainability of peace, state-building, and conflict prevention efforts.[89] The desired result is stronger health systems that can manage risks while meeting the responsibilities for advancing UHC by providing quality, accessible, equitable, and affordable health services to the population.

To ensure that all partners work together and the opportunity to build back better is not missed, recovery should ideally begin at the planning stage with an analysis to identify opportunities, vulnerabilities, strengths and weaknesses in capacities, and key health system barriers, including bottlenecks to service delivery. The analysis should also include the different options and resources needed to address the identified issues.

The rights-based approach to risk management underscores the importance of Universal health coverage which implies that all people have access to the health services they need, when and where they need them, without financial hardship. However, COVID-19 has had a severe and significant impact on population health across the world, and revealed the lack of resilience of health systems globally, even in countries that have historically performed well on assessments of health security and UHC, that is, that all people have access to the health services they need, when and where they need them, without financial hardship.[90,91] The lack of individual country resilience has been exposed in many countries worldwide, extending from those with significant resources to those with few resources, including those of fragile, conflict, and violence-affected (FCV) contexts. As a result, the pandemic strained the global health systems that were sufficiently providing health services to their populations and further weakened the ones that were already frail before the pandemic.

Conclusion: A call to action

The lessons learned from the COVID-19 pandemic should guide the strengthening of global health security through a comprehensive risk-informed and rights-based approach. The goal of the recovery process therefore should be to build back better and transform health systems in such a way as to increase their resilience and equity. This will require concerted efforts at the national and global levels. First, there is a need for investments in health systems and efforts to strengthen them which are based on a PHC approach that is rooted in the principles of equity, social justice, and solidarity, and that ensures a broader role of people in health and well-being along the life course. Second, it will be vital to strengthen the EPHFs and have all-hazard emergency and disaster risk management plans and programs that emphasize capacity development for the implementation of the IHR (2005), the Sendai Framework, the Paris Agreement, and other relevant national, regional, and global frameworks and strategies, and a whole-of-society approach for better emergency preparedness based on country risk profiles. Third, facilitating and improving global information sharing and international and regional coordination and resource sharing for instance is necessary, for example, providing equitable access to PPE and vaccines. Finally, it is essential to learn from the current pandemic and the concurrent emergencies, and where necessary, to examine the adaptations that were made to protect health and maintain essential services safely before, during, and after emergencies, to determine which should become permanent features of the rebuilt and transformed health system.[92]

References

1. *Accelerating R&D Processes*; 2020. https://www.who.int/activities/accelerating-r-d-processes. Accessed November 1, 2021.
2. Global Health Security - The Global Fund to Fight AIDS, Tuberculosis and Malaria. Accessed November 1, 2021. https://www.theglobalfund.org/en/global-health-security/.
3. Achenbach J, Akilah J, Jacqueline D. Hurricane Ida forces three damaged hospitals to evacuate patients. *Washington Post*; 2021. https://www.washingtonpost.com/health/hurricane-ida-damages-louisiana-hospitals/2021/08/30/df90e5be-09c6-11ec-aea1-42a8138f132a_story.html on April 13, 2022. Accessed April 21, 2022.
4. Lacsamana B. Typhoon survivors left vulnerable to COVID-19. *Business World Online*; 2022. https://www.bworldonline.com/health/2022/01/05/421523/typhoon-survivors-left-vulnerable-to-covid-19/. Accessed April 21, 2022.
5. UNFCCC. Number of wildfires to rise by 50% by 2100 and governments are not prepared, experts Warn. *UNFCCC*; 2022. https://unfccc.int/news/number-of-wildfires-to-rise-by-50-by-2100-and-governments-are-not-prepared-experts-warn. Accessed April 21, 2022.

6. WHO. Wildfires. WHO. Accessed April 21, 2022. https://www.who.int/health-topics/wildfires#tab=tab_1.
7. ReliefWeb. "Our health workers are working in fear": targeted violence against health care one year after Myanmar's military coup, January 2022-Myanmar. *ReliefWeb*; 2022. https://reliefweb.int/report/myanmar/our-health-workers-are-working-fear-targeted-violence-against-health-care-one-year. Accessed April 21, 2022.
8. Achten N. Cyber threats during the COVID-19 outbreak and activities of national CERTs in the Western Balkans. *Geneva Centre for Security Sector Governance*; 2021. https://www.dcaf.ch/cyber-threats-during-covid-19-outbreak-and-activities-national-certs-western-balkans. Accessed April 21, 2022.
9. *Planning for Concurrent Emergencies—National Governors Association*; 2020. https://www.nga.org/center/publications/planning-for-concurrent-emergencies/. Accessed November 1, 2021.
10. WHO. *Preparedness for cyclones, tropical storms, tornadoes, floods and earthquakes during the COVID-19 pandemic. WHO*. 2020.
11. UNHCR. *Flagship Reports - Forced Displacement in 2020*. 2020.
12. Litzkow J. The impact of COVID-19 on refugees and migrants on the move in North and West Africa. *Mixed Migration Centre*. 2021:5.
13. Lippman B, Sutton R, Doby A, et al. *Covid-19: Understanding the Impact of the Pandemic on Forcibly Displaced Persons. Covid Collective Research for Policy and Practice*. 2022. https://doi.org/10.1017/S2045796020001079.
14. IOM-UNHCR-WHO. *UNHCR - Open Letter to G20 Heads of State and Government - UNHCR, IOM & WHO*; 2021. https://www.unhcr.org/news/press/2021/10/617bffc64/open-letter-g20-heads-state-government-unhcr-iom.html. Accessed November 1, 2021.
15. Reuters. *WHO calls for halting COVID-19 vaccine boosters in favor of unvaccinated. Reuters*; 2021. https://www.reuters.com/business/healthcare-pharmaceuticals/who-calls-moratorium-covid-19-vaccine-booster-doses-until-september-end-2021-08-04/. Accessed November 1, 2021.
16. WHO. Essential Public Health Functions. Accessed June 12, 2022. https://www.who.int/teams/integrated-health-services/health-service-resilience/essential-public-health-functions.
17. *World Health Organization. International Health Regulations (2005)*. 3rd ed. 2016.
18. WHO. *Everyone's Business: Whole-of-Society Action to Manage Health Risks and Reduce Socioeconomic Impacts of Emergencies and Disasters*. 2020.
19. Maini R, Clarke L, Blanchard K, Murray V. The Sendai framework for disaster risk reduction and its indicators—where does health fit in? *Int J Disaster Risk Sci*. 2017;8(2):150–155. https://doi.org/10.1007/S13753-017-0120-2.
20. UNDRR. *Sendai Framework for Disaster Risk Reduction 2015–2030*; 2015. https://www.undrr.org/publication/sendai-framework-disaster-risk-reduction-2015-2030. Accessed November 1, 2021.
21. Broberg MA. Critical appraisal of the world health organization's international health Regulations (2005) in times of pandemic: it is time for revision. *Eur J Risk Regul*. 2020;11(2):202–209. https://doi.org/10.1017/ERR.2020.26.
22. Global Health Security Index - The Nuclear Threat Initiative. Accessed December 25, 2021. https://www.nti.org/about/programs-projects/project/global-health-security-index/.
23. WHO. *Strategic Toolkit for Assessing Risks: A Comprehensive Toolkit for All-hazards Health Emergency Risk Assessment*. 2021.
24. Milken Institute School of Public Health. *Equity vs. Equality: what's the difference? George Washington. Online Public Health*; 2020. https://onlinepublichealth.gwu.edu/resources/equity-vs-equality/. Accessed April 21, 2022.
25. Haldane V, Jung AS, Neill R, et al. From response to transformation: how countries can strengthen national pandemic preparedness and response systems. *BMJ*. 2021;375. https://doi.org/10.1136/BMJ-2021-067507.
26. Future C on a GHRF for the, National Academy of Medicine Science. The case for investing in pandemic preparedness. *The Neglected Dimension of Global Security*. May 16, 2016. https://doi.org/10.17226/21891.
27. *Main Report & Accompanying Work - the Independent Panel for Pandemic Preparedness and Response*; 2021. https://theindependentpanel.org/mainreport/. Accessed November 1, 2021.
28. *GPMB Report*; 2021. https://www.gpmb.org/#tab=tab_1. Accessed November 1, 2021.
29. WHO. *WHO Guidance on Preparing for National Response to Health Emergencies and Disasters*; 2021. https://apps.who.int/iris/handle/10665/350838. Accessed December 23, 2021.
30. *Health Emergency and Disaster Risk Management Framework*. 2020.
31. Differential vulnerability - ESCWA. Accessed November 1, 2021. https://www.unescwa.org/sd-glossary/differential-vulnerability.
32. WHO. *Research Ethics in International Epidemic Response MEETING REPORT*. 2010:7.

33. WHO. *WHO COVID-19 Social Science in Outbreak Response. Community-Centred Approaches to Health Emergencies: Progress, Gaps and Research Priorities*. 2021.
34. WHO. *Constitution of the World Health Organization*; 1946. https://www.who.int/about/governance/constitution. Accessed December 25, 2021.
35. Moon S, Sridhar D, Pate MA, et al. Will Ebola change the game? Ten essential reforms before the next pandemic. The report of the Harvard-LSHTM independent panel on the global response to Ebola. *Lancet*. 2015;386(10009):2204–2221. https://doi.org/10.1016/S0140-6736(15)00946-0.
36. Shibuya K, Nomura S, Okayasu H, et al. Protecting human security: proposals for the G7 Ise-Shima Summit in Japan. *Lancet*. 2016;387(10033):2155–2162. https://doi.org/10.1016/S0140-6736(16)30177-5.
37. Schäferhoff M, Fewer S, Kraus J, et al. How much donor financing for health is channelled to global versus country-specific aid functions? *Lancet*. 2015;386(10011):2436–2441. https://doi.org/10.1016/S0140-6736(15)61161-8.
38. United Nations. *SDG Indicators*. 2021. https://doi.org/10.1007/978-3-319-95963-4_300147, 1095–1095.
39. The Lancet Public Health. Will the COVID-19 pandemic threaten the SDGs? *Lancet Public Health*. 2020;5(9):e460. https://doi.org/10.1016/S2468-2667(20)30189-4.
40. Paremoer L, Nandi S, Serag H, Baum F. Covid-19 pandemic and the social determinants of health. *BMJ*. 2021;372. https://doi.org/10.1136/BMJ.N129.
41. Campo-Arias A, Mendieta CT. Social determinants of mental health and the COVID-19 pandemic in low-income and middle-income countries. *Lancet Glob Health*. 2021;9(8):e1029–e1030. https://doi.org/10.1016/S2214-109X(21)00253-9.
42. *England - Office for National Statistics. Updating Ethnic Contrasts in Deaths Involving the Coronavirus (COVID-19), England - Office for National Statistics*; 2021. https://www.ons.gov.uk/peoplepopulationandcommunity/birthsdeathsandmarriages/deaths/articles/updatingethniccontrastsindeathsinvolvingthecoronaviruscovid19englandandwales/24january2020to31march2021. Accessed November 1, 2021.
43. The Lancet Public Health. COVID-19—break the cycle of inequality. *Lancet Glob Health*. 2021;6(2):e82. https://doi.org/10.1016/S2468-2667(21)00011-6.
44. Santomauro DF, Herrera AMM, Shadid J, et al. Global prevalence and burden of depressive and anxiety disorders in 204 countries and territories in 2020 due to the COVID-19 pandemic. *Lancet*. 2021;398:1700–1712. https://doi.org/10.1016/S0140-6736(21)02143-7.
45. Ginette Azcona AB. From insights to action: gender equality in the wake of COVID-19. *UN Women – Headquarters*; 2020. https://www.unwomen.org/en/digital-library/publications/2020/09/gender-equality-in-the-wake-of-covid-19. Accessed November 1, 2021.
46. Roesch E. Violence against women during Covid-19 pandemic restrictions OPEN ACCESS Protections for women and girls must be built into response plans. *BMJ*. 2020. https://doi.org/10.1136/bmj.m1712.
47. UN WOMEN. From insights to action: gender equality in the wake of COVID-19. *UN Women – Headquarters*; 2020. https://www.unwomen.org/en/digital-library/publications/2020/09/gender-equality-in-the-wake-of-covid-19. Accessed November 1, 2021.
48. Kourti A, Stavridou A, Panagouli E, et al. Domestic violence during the COVID-19 pandemic: a systematic review. *Trauma Violence Abuse*. 2021;24:719–745.
49. *2020 Global Nutrition Report - Global Nutrition Report*; 2020. https://globalnutritionreport.org/reports/2020-global-nutrition-report/. Accessed November 1, 2021.
50. The Impact of Disasters and Crises on Agriculture and Food Security: 2021. March 17, 2021.
51. Jamison DT, Summers LH, Alleyne G, et al. Global health 2035: a world converging within a generation. *Lancet*. 2013;382(9908):1898–1955. https://doi.org/10.1016/S0140-6736(13)62105-4.
52. Beyrer C, Allotey P, Amon JJ, et al. Human rights and fair access to COVID-19 vaccines: the international AIDS society–Lancet commission on health and human rights. *Lancet*. 2021;397(10284):1524–1527. https://doi.org/10.1016/S0140-6736(21)00708-X.
53. McAdams D, McDade KK, Ogbuoji O, Johnson M, Dixit S, Yamey G. Incentivising wealthy nations to participate in the COVID-19 Vaccine Global Access Facility (COVAX): a game theory perspective. *BMJ Glob Health*. 2020;5(11):e003627. https://doi.org/10.1136/BMJGH-2020-003627.
54. Gavi. *COVAX Facility Explainer Participation Arrangements for Self-financing Economies*. 2020.

55. Forman R, Shah S, Jeurissen P, Jit M, Mossialos E. COVID-19 vaccine challenges: what have we learned so far and what remains to be done? *Health Pol.* 2021;125(5):553–567. https://doi.org/10.1016/J.HEALTHPOL.2021.03.013.
56. WHO. *Joint COVAX Statement on Supply Forecast for 2021 and Early 2022*; 2021. https://www.who.int/news/item/08-09-2021-joint-covax-statement-on-supply-forecast-for-2021-and-early-2022. Accessed November 1, 2021.
57. Yamey G, Schäferhoff M, Hatchett R, Pate M, Zhao F, McDade KK. Ensuring global access to COVID-19 vaccines. *Lancet.* 2020;395(10234):1405–1406. https://doi.org/10.1016/S0140-6736(20)30763-7.
58. Wouters OJ, Shadlen KC, Salcher-Konrad M, et al. Challenges in ensuring global access to COVID-19 vaccines: production, affordability, allocation, and deployment. *Lancet.* 2021;397(10278):1023–1034. https://doi.org/10.1016/S0140-6736(21)00306-8.
59. UNICEF. Global leaders commit further support for global equitable access to COVID-19 vaccines and COVAX. *UNICEF;* 2020. https://www.unicef.org/press-releases/global-leaders-commit-further-support-global-equitable-access-covid-19-vaccines-0. Accessed November 1, 2021.
59a. United Nations. Universal Declaration of Human Rights (10 December 1948), Article 25. https://www.un.org/en/about-us/universal-declaration-of-human-rights.
59b. U. Nations. International Covenant on Economic, Social and Cultural Rights (16 December 1966), Article 12. https://www.refworld.org/docid/3ae6b36c0.html.
59c. UN Committee on Economic Social and Cultural Rights (CESCR). (August 11, 2000) General comment No. 14: the right to the highest attainable standard of health. Para 1.
60. Liua JX, Goryakin Y, Maeda A, Bruckner T, Scheffler R. *Global Health Workforce Labor Market Projections for 2030.* August 2016. https://doi.org/10.1596/1813-9450-7790.
61. WHO. *Global Strategy on Human Resources for Health: Workforce 2030.* 2016.
62. Forbes. Nearly 300,000 healthcare workers have been infected with covid-19 worldwide, threatening health systems. *Forbes;* 2020. https://www.forbes.com/sites/williamhaseltine/2020/11/17/the-infection-of-hundreds-of-thousands-of-healthcare-workers-worldwide-poses-a-threat-to-national-health-systems/?sh=1a54f4573499. Accessed November 1, 2021.
63. ILO. *Women Health Workers: Working Relentlessly in Hospitals and at Home;* 2021. http://www.ilo.org/global/about-the-ilo/newsroom/news/WCMS_741060/lang-en/index.htm. Accessed November 1, 2021.
64. World Health Organization. The impact of COVID-19 on health and care workers: a closer look at deaths. *WHO.* 2021.
65. WMA. WMA resolution on equitable global distribution of COVID-19 vaccine – WMA – the world medical association. *WMA;* 2020. https://www.wma.net/policies-post/wma-resolution-on-equitable-global-distribution-of-covid-19-vaccine/. Accessed February 18, 2022.
66. WMA. WMA resolution regarding the medical profession and COVID-19 – WMA – the world medical association. *WMA;* 2020. https://www.wma.net/policies-post/wma-resolution-regarding-the-medical-profession-and-covid-19/. Accessed February 18, 2022.
67. World Health Organization. *State of the World's Nursing Report-2020;* 2020:1–144. https://www.who.int/china/news/detail/07-04-2020-world-health-day-2020-year-of-the-nurse-and-midwife.
68. WHO. Delivered by women, led by men: a gender and equity analysis of the global health and social workforce—human Resources for Health Observer Series No. 24. *WHO.* 2019.
69. WHO. Stopping attacks on health care. *WHO;* 2021. https://www.who.int/activities/stopping-attacks-on-health-care. Accessed November 1, 2021.
70. MSF. Medical care under fire. *MSF;* 2013. https://www.msf.org/medical-care-under-fire. Accessed November 1, 2021.
71. ICRC. *HCiD - Home;* 2021. https://healthcareindanger.org/fr/. Accessed November 1, 2021.
72. Daszak P. Anatomy of a pandemic. *Lancet.* 2012;380(9857):1883–1884. https://doi.org/10.1016/S0140-6736(12)61887-X.
73. Morse SS, Mazet JAK, Woolhouse M, et al. Prediction and prevention of the next pandemic zoonosis. *Lancet.* 2012;380(9857):1956–1965. https://doi.org/10.1016/S0140-6736(12)61684-5.
74. World Health Organization. WHO Director-General's opening remarks at the media briefing on COVID-19-11 March 2020. *WHO;* 2020. https://www.who.int/director-general/speeches/detail/who-director-general-s-opening-remarks-at-the-media-briefing-on-covid-19-11-march-2020. Accessed January 19, 2022.

75. WHO. Rapid assessment of service delivery for NCDs during the COVID-19 pandemic. *WHO*; 2020. https://www.who.int/publications/m/item/rapid-assessment-of-service-delivery-for-ncds-during-the-covid-19-pandemic. Accessed April 21, 2022.
76. WHO. *Noncommunicable Diseases*. 2021.
77. WHO. *COVID-19 AND NCDs*. 2021.
78. WHO. *COVID-19 AND NCDs*. 2021.
79. WHO. Making Hospitals Safe in Emergencies. World Health Organization. Accessed April 21, 2022. https://www.who.int/activities/making-health-facilities-safe-in-emergencies-and-disasters.
80. Pan American Health Organization. Smart Hospitals. WHO/PAHO. Accessed April 21, 2022. https://www.paho.org/en/health-emergencies/smart-hospitals.
81. United Nations Office for the Coordination of Humanitarian Affairs (OCHA). *Global Humanitarian Overview 2019*. 2018.
82. World Health Organization (WHO). *Emergency Response Framework Second Edition*. World Health Organization; 2017.
83. *Union European, United Nations, World Bank. Joint Recovery and Peacebuilding Assessments (RPBAs): PreventionWeb*; 2017. https://www.preventionweb.net/publication/joint-recovery-and-peacebuilding-assessments-rpbas. Accessed November 30, 2021.
84. United Nations Office for Disaster Risk Reduction. *Build Back Better in Recovery, Rehabilitation and Reconstruction*. 2017.
85. World Health Organization (WHO). Building back better sustainable mental health care after emergencies. *WHO*. 2013.
86. Kamara S, Walder A, Duncan J, Kabbedijk A, Hughes P, Muana A. Mental health care during the Ebola virus disease outbreak in Sierra Leone. *Bull World Health Organ*. 2017;95(12):842–847. https://doi.org/10.2471/BLT.16.190470.
87. Epping-Jordan JAE, van Ommeren M, Ashour HN, et al. Beyond the crisis: building back better mental health care in 10 emergency-affected areas using a longer-term perspective. *Int J Ment Health Syst*. 2015;9(1):1–10. https://doi.org/10.1186/S13033-015-0007-9/TABLES/3.
88. World Health Organization. Global Health for Peace Initiative (GHPI). https://www.who.int/initiatives/who-health-and-peace-initiative.
89. Rubenstein LS, Senior JR, Program F. Post-conflict health reconstruction: new foundations for U.S. Policy. *US Institute of Peace*. 2009.
90. Crosby S, Dieleman JL, Kiernan S, Bollyky TJ. All bets are off for measuring pandemic preparedness. *Think Global Health*; 2020. https://www.thinkglobalhealth.org/article/all-bets-are-measuring-pandemic-preparedness. Accessed February 18, 2022.
91. WHO. *Universal Health Coverage*; 2015. https://www.who.int/health-topics/universal-health-coverage#tab=tab_1. Accessed February 18, 2022.
92. World Health Organization. Regional Office for the Eastern Mediterranean. *Implementation Guide for Health Systems Recovery in Emergencies Transforming Challenges into Opportunities*. 2020.

SECTION IV

Tools and techniques to modernize prevention, detection, and response to epidemics

Joseph N. Fair[1], Aamer Ikram[2]
Rana Jawad Asghar[3,4] and Marjorie P. Pollack[5]

[1]TriCorder Health, Tyler, TX, United States; [2]National Institutes of Health, Islamabad, Pakistan; [3]Global Health Strategists & Implementers, Islamabad, Pakistan; [4]University of Nebraska Medical Center, College of Public Health, Omaha, NE, United States; [5]PROMED, Brooklyn, NY, United States

In an era of rapid technological advancements and global connectivity, the tools and techniques available for preventing, detecting, and responding to epidemics have seen significant improvements. Advances in sequencing technologies, communication uplinks, online databases, and the active involvement of private technology companies and foundations have created the potential for laboratories worldwide to share high-fidelity sequence and epidemiological reports of communicable diseases in near real time. However, the COVID-19 pandemic exposed the limitations and challenges that persist despite these advancements especially in developing countries.

The COVID-19 pandemic revealed the existence of information silos, a lack of transparency due to national security or commercial concerns, and a lack of uniform command and operations centers in most political systems to effectively respond to pandemics. There were certain countries like Singapore that had a dedicated command center designed to respond to outbreaks at the pace they occur, making critical public health decisions promptly.

The pressing demand for enhanced biosafety and biosecurity measures has been underscored by certain noteworthy historical occurrences involving laboratory-acquired infections, as well as the persistent uncertainties surrounding the origins of COVID-19. These incidents have emphasized the critical importance of mitigating the risks associated with bioterrorism, dual use, or the unintended release of pathogens.

The development and widespread adoption of CRISPR methodologies have further added to the concerns surrounding intentional or accidental incidents, as genetically manipulated organisms pose a new and grave threat alongside naturally emerging infectious organisms. While functional sequencing centers have been established in many national laboratories, the upload of sequences to public databases for the development of diagnostics and therapeutics remains inconsistent.

A key obstacle to data sharing is the inequitable distribution of scientific recognition and funding. Historically, geopolitical factors and historical events have resulted in a concentration of scientists in wealthier, more developed nations, using data obtained from distant field sites to secure funding and advance their own careers and laboratories. This pattern has left frontline laboratories severely underfunded and disincentivized host countries from sharing their data without sufficient intellectual property protection.

To break this cycle and ensure the safety of all individuals, we must find ways to protect intellectual rights and incentivize data sharing. The upcoming chapters will address the need for laboratory enhancements, facilitated communications, and improved public health surveillance. The COVID-19 pandemic has brought to the forefront the importance of enhancing public health surveillance, as exemplified by the revision of the International Health Regulations (IHR) and the recognition of "Public Health Emergency of International Concern" (PHEIC). Additionally, innovative disease surveillance initiatives utilizing AI, web-crawling, social media, and participatory surveillance have emerged, leveraging nontraditional information sources.

This section will also emphasize the need to leverage "Big Data" effectively to drive population health impact and reduce inequalities. Advanced computing and analytics will be discussed as essential tools in addressing public health surveillance and outbreak response at a multiagency level. Furthermore, the section will address the phenomenon of infodemics, which arose alongside the COVID-19 pandemic. Infodemics refer to the unnecessary proliferation of information, including misinformation campaigns which may disrupt outbreak control efforts. Such campaigns have politicized nonpharmaceutical interventions and created distrust around vaccines, posing additional challenges to epidemic response.

By exploring the tools and techniques to modernize prevention, detection, and response to epidemics, this section aims to shed light on the gaps that need to be bridged and the innovative solutions that can empower global health systems to effectively combat future outbreaks.

Some of the approaches discussed in this section include early warning systems; rapid diagnostic tests; digital health technologies; big data analytics; genomic surveillance; vaccine development and deployment; One Health approach; public health communication; and international collaboration.

CHAPTER

16

Global laboratory systems

Lucy A. Perrone[1,2], *Francois-Xavier Babin*[3], *Sebastien Cognat*[4], *Juliane Gebelin*[3], *Emmanuelle Boussieres*[3], *Allegra Molkenthin*[5], *Barbara Jauregui*[5], *Koren Wolman-Tardy*[5], *Hanvit Oh*[6] *and Allison Watson*[6]

[1]University of British Columbia, Vancouver, BC, Canada; [2]University of Washington, Seattle, WA, United States; [3]Fondation Mérieux, Lyon, France; [4]World Health Organization, Lyon, France; [5]Mérieux Foundation USA, Washington, DC, United States; [6]Emory University, Atlanta, GA, United States

Introduction

Importance of diagnostics for clinical care and public health

Diagnostics and health laboratories are essential tools and healthcare services for effective patient care and public health. Every diagnostic test, whether performed in a doctor's office, pharmacy, or clinical or public laboratory, must be accurate and timely to be clinically useful for the care of patients and for public health. In high-income countries (HIC) such as the United States, it is estimated that 70% of clinical treatment decisions are based on diagnostic testing information.[1,2] Diagnostic services—whether "point-of-care" or laboratory-based—need to be reliable to provide the exact results for helping in correct diagnosis and in turn maintain provider and public trust. However, despite their importance, the recent 2021 Lancet Commission Report on Diagnostics revealed that just under half (47%) of the world's population have little to no access to diagnostics.[3]

Even in HICs, access to diagnostics is often geospatially and economically limited for rural, financially poor, and marginalized communities. The Commission's report estimates that reducing the diagnostic gap to only 10% for even just six major conditions would reduce the annual number of premature deaths in low-income and middle-income countries (LMICs) by 1.1 million (2.5% of total annual deaths in LMICs), and annual disability-adjusted life-year (DALY) losses by 38.5 million (1.8% of losses from all conditions). Clinical diagnostic tests should be available as an essential healthcare service, and importantly, prescribed and

Modernizing Global Health Security to Prevent, Detect, and Respond
https://doi.org/10.1016/B978-0-323-90945-7.00009-9

interpreted by physicians (e.g., early in the COVID-19 pandemic, serological tests were used too often and wrongly for diagnosis of acute infection).

Laboratory-based surveillance for pathogens is an essential public health service and compliments epidemiological and clinical data. Laboratory data are necessary for an effective public health response and inform health policy including alignment of necessary and appropriate health services with the needs of the population. International surveillance networks such as the Global Antimicrobial Resistance (AMR) and Use Surveillance System (GLASS) are critical tools for public health policy and action.[4] Outbreaks and public health threats such as those caused by Ebola, COVID-19, and AMR simply cannot be controlled without interconnected clinical and public health laboratory services. When they are functioning properly, supplied with the right tools, skilled workforce, and employing a quality management system enabling accurate, timely, and reliable results, health laboratories are an invaluable weapon in the public health response arsenal.

While the equal value of these two pillars of laboratory services is undeniable, consistent government investment in public health laboratories, which do not rely on patient payment for services or provider reimbursement, has historically taken a back seat to other health system priorities of governments (e.g., essential medicines), and this has had devastating consequences as illustrated by the recent COVID-19 pandemic. Following the international scare of the first SARS-CoV outbreak in 2003 and the revision of the International Health Regulations in 2005, steps must be taken now by nations and UN agencies to enable the development of effective laboratory systems and to enforce safety and quality standards of practice.

The decision taken to develop an enforceable instrument for pandemic preparedness and response by Member States at a special session of the World Health Assembly held in December 2021 marks a step forward in remedying gaps in national and international responses.[5,6] At this time of writing, it is unclear how this instrument may be used to increase and better evaluate laboratory readiness at the national and subnational levels, and the world awaits more details.

Essential functions of laboratory systems and networks

Predicting the pathogen and zoonotic source of the next infectious disease pandemic is difficult, if not completely impossible.[7] Only through capable diagnostics and effective laboratory systems can we detect new pathogens and respond to outbreaks. This impact of human, environmental, and veterinary health laboratories is most effective and synergistic when they are networked and working together as part of an integrated one-health-oriented laboratory system. This networking leverages individual facility strengths for greater health system capacity and creates efficiencies that, in turn, contribute to improved pathogen detection and public health response. This level of testing capability, coordination, and leadership is essential for an effective national and international response. The speed of global spread of the SARS-COV-2 virus that caused the COVID-19 pandemic has once again illustrated that building subnational, national, and global diagnostic readiness to detect pathogens that shorten the public health response time should be a paramount priority.

Despite the rapidity of the spread of the novel SARS-CoV-2 in late 2019, it is however helpful to reflect on the ways the world has gotten incrementally better at detecting and responding to novel pathogens in the last 25 years and reflect on the impact that certain strategic investments have had on shortening the public health response time. Significant bilateral and global financial investments in infrastructure, technology, and workforce training within priority health programs such as tuberculosis(TB), HIV, influenza, and malaria have helped strengthen the practice and quality of laboratory medicine and pathology services in LMICs.[8–10] Although limited in disease target scope, these investments in diagnostics and surveillance systems for these specific disease programs have in fact helped improve global detection and response capability, in particular because of investments in laboratory and epidemiology capacity. Investments in laboratory infrastructure, equipment, and workforce in many LMICs have had a measurable impact on the ability of those countries to provide better health services for their people and enable them to more fully participate in the global community's mission to detect and respond to pathogens of international concern in a more rapid and effective manner.

Recent outbreaks such as Ebola in West Africa and MERS in the Middle East have illustrated the value of these investments and which can and should be capitalized on. While the impact and value of this development are obvious, there remain serious challenges to expanding capabilities and sustaining progress because of financial sustainability, limited intersectoral coordination between human, animal, and environmental sectors, nonfunctional laboratory networking, gaps in operational overlap, continuing quality assurance, limited supply of skilled laboratory workforce, and lack of harmonization and collaboration between disease-specific programs—each of which greatly hampers effective system strengthening efforts.

It is notable that the SARS-CoV-2 virus was identified (23 vs. 86 days) and the sequence shared more rapidly (33 vs. 149 days) compared to the first SARS-CoV outbreak in 2002–03. The first PCR protocols were proposed only a few days after the sequence of the novel SARS-CoV-2 virus was shared in January 2020, and laboratory testing started less than 2 weeks after WHO was informed of the outbreak. This has been thankfully possible due to the lessons learned from previous outbreaks and, critically, the development of research and development systems and laboratory networks. Valuable public sequence data sharing platforms such as the Global Initiative on Sharing Avian Influenza Data (GISAID), which was launched in 2008 as a global science initiative and primary resource to provide open access to genomic data of pathogens, initially focused on influenza strains (due to urgent pandemic potential concerns of avian influenza) and have now pivoted to pathogenic coronaviruses such as SARS-CoV-2.[11] While this reflects an incremental success in the global public health response, most of the countries of the world are sorely underprepared to diagnose even common agents of disease, let alone novel or emerging pathogens. While HICs have enjoyed successful technological advancement in infectious disease detection and characterization, the highest risk pathogens of international consequence are most likely to emerge from zoonotic "hot spots" with high population density. This is where diagnostic capacity and system-strengthening efforts need to be strengthened the most.

Despite major investments by agencies and international aid programs such as the Global Fund and United States President's Emergency Plan for AIDS Relief (PEPFAR), many LMICs still have limited subnational laboratory and diagnostic services and can barely meet the needs of their own populations for routine health care. As detailed in the Lancet Commission Report on Diagnostics, nations need to develop and implement an Essential Diagnostics List, adapted from the WHO model list of essential in vitro diagnostics (EDL) according to their burden of disease, and integrated into their national health planning and financing.[12] To this day, well-outfitted HIV testing laboratories, clean and sparkling, sit next to dilapidated clinical laboratories that provide all other services including microbiology culture, hematology, and clinical chemistry. In the last 10 years, global value for and adoption of the Global Health Security Agenda (GHSA) and its nine core packages has spurred additional investments in public health laboratories beyond the large disease programs. However, the penetration of this support has largely stayed at the national level with investments made in centralized reference laboratories. A functioning laboratory system should exchange information, optimize laboratory services, and help control and prevent diseases and other health threats across the span of human, animal, and environmental health.

These systems only function well when sustainable investments are made at the country level in key areas: laws and regulations, infrastructure, workforce, information systems, quality management, and biosafety and biosecurity. Now the push for universal health care (UHC) is illuminating gaps in diagnostics and the need to remedy this to achieve UHC.[13] Laboratory system strengthening throughout the lower and interconnected tiers of national laboratory systems remains undersupported and must be addressed through significant, cross-cutting, system strengthening mechanisms and funding that is not tied to a singular issue or disease program. Laboratory services need to be recognized as a building block of strong health systems and extracted from the shadow of "medical products and technologies" where they currently reside in the current design of the WHO Building blocks (see Fig. 16.1).

The requirement by donors to measure the impact of their investments in a quantitative and discrete manner makes investments in system-wide strengthening measures (which can be challenging to measure quantitatively and require longer time frames to implement) unattractive and unjustifiable to those looking to make a legacy impact with their

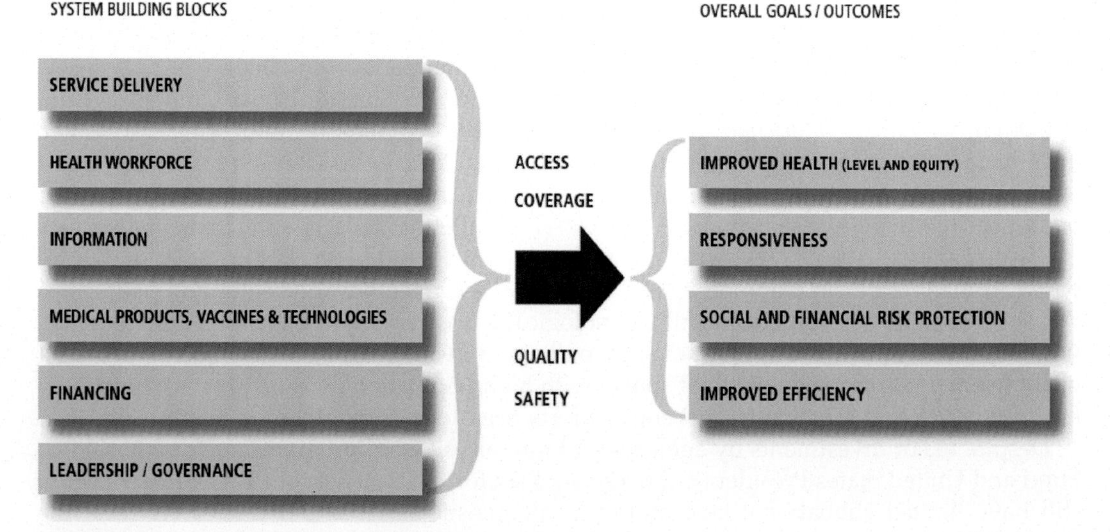

FIGURE 16.1 The WHO health system framework; the six building blocks of a health system: aims and desirable attributes. *Source: World Health Organization. Everybody's Business: Strengthening health systems to improve health outcomes—WHO's Framework for Action. Geneva: WHO, 2007.*

philanthropic money and justify spending to their constituents. The global commitment toward pandemic preparedness and response requires a long-range vision, faith, and leadership to invest in systems. Investments made today will take time to demonstrate impact.[14]

The time is now for sustained and system-wide investments in laboratory system strengthening

The IHR was revised in 2005 and entered into force in 2007 in response to the emergence or reemergence of pathogens, notably the emergence of the novel SARS-CoV virus in China in 2003. Increased threat awareness to the pandemic potential of these respiratory pathogens led to a scale-up of investments in national laboratory capacity development in all WHO regions, especially where ecological "hot zones" had been identified. Defining "laboratory preparedness" became paramount in the wake of avian flu and SARS; however, measuring national-level progress toward meeting the IHR occurred in various ways up until 2016 when the Joint External Evaluation (JEE) process and tool were widely adopted by member states.

A myriad of laboratory capacity evaluation tools have been utilized over the years and employed by siloed disease programs (e.g., Influenza, HIV, TB).[15] Laboratory assessments using these tools were often performed by a third party and conducted to assess the infrastructure, equipment, personnel, etc. needs to evaluate the capacity for diagnostic testing and surveillance, however information sharing about this capacity and harmonization of development investments across these programs was often lacking and uncoordinated. For example, in 2009, the WHO Laboratory Assessment Tool (LAT) was being used extensively to measure laboratory capacity largely concerned by the global influenza surveillance program. Then 2009 also saw the adaptation of the ISO 15189 standard and accreditation process into the new Stepwise Laboratory Quality Improvement Process Toward Accreditation (SLIPTA) checklist, driven largely by WHO AFRO and the global HIV program. Data from the WHO-LAT and SLIPTA checklists were utilized heavily to inform and justify technical and financial investments in HIV and influenza diagnostics and laboratories. Since 2008, the Global Laboratory Initiative (GLI) has provided a focus for TB within the framework of a multifaceted, integrated approach to laboratory capacity strengthening through a network of international partners.

It is valuable to appreciate that donors and nations have been making strategic investments and exhibiting good progress to improve laboratory quality and capacity using information collected through these tools and programs. However critical investments in systems has fallen behind. Member states' increasing value for ISO 15189 adoption for improved quality in clinical laboratory services and the rise of laboratory accreditation have contributed to stronger laboratories and systems over the last 20 years.

Launched in 2016, the JEE is a One Health-oriented, national-level evaluation framework and system assessment tool. Its implementation helped to objectify and systematize the process and types of information collected at the national level and inform investment decision making. There is a saying in public health that "what gets measured, gets done," and so it is critical that this information be utilized effectively. Singularly focused donor-driven initiatives and individual philanthropic pursuits seeking the eradication of a few select pathogens and diseases perpetuates a cycle of neglect for system-level investments that are sorely needed right now. The recent SARS-CoV-2 pandemic once again illustrates that the international community needs to be thinking and investing at the systems level.

Maintaining a pattern of siloed investments will continue to neglect detection and response *systems* which are ultimately responsible for an effective public health response; it is time to change the modus operandi.

CASE STUDY

RESAOLAB: Success factors for establishing and sustaining a regionally focused laboratory network

Background

The West African Network of Clinical Laboratories (RESAOLAB) is the first program established in the West African region aimed to support a coordinated approach to improving both national and regional laboratory capacity with the broader goal of strengthening health systems. RESOALAB was launched in 2009 as a regional collaboration between seven countries' ministries of health, and several West African-based healthcare stakeholders, including WAHO. The multinational network, which includes Benin, Burkina Faso, Guinea, Mali, Niger, Senegal, and Togo, is coordinated with support from the Mérieux Foundation and primarily funded by the French Development Agency (AFD). RESAOLAB addresses critical factors affecting the national laboratory system, from core national governance to operational performance at the facility level, taking a harmonized regional approach. RESAOLAB's activities are aligned with and integrated into the policies set by national, regional, and international health organizations.

Observations

Despite efforts to improve laboratory services in Africa, mainly focused on specific disease control programs, supporting and developing laboratory systems have historically been a low priority at the national level. Laboratory systems remain chronically under resourced, both in terms of personnel and budget, due to the lack of clear leadership for this sector. This leadership role must be assumed by the ministry of health. It should be granted regulatory authority through the designation of a specific and legally recognized department or unit within the ministry itself and locate at the national level to support and enable a functional, tiered laboratory system.[1] Furthermore, it is important to have a designated leader responsible and accountable for the dedicated department/directorate. One of the core functions of this entity is to develop and implement the National Laboratory Policy and National Strategic Plan, which establishes the priorities for strengthening the national laboratory system. Having clear leadership and a similar organization and management for laboratory systems within the region is a key factor for better collaboration and coordination, with positive impacts on patient care, infectious disease surveillance, and preparedness and response to epidemics. Since 2009, project RESOALAB's ambition has been to support ministries of health in establishing the leadership and organization needed to underpin an effective, tiered laboratory system.

Interventions

RESAOLAB interventions have been jointly defined and conducted with the ministries of health of member countries and

CASE STUDY *(cont'd)*

with the national laboratory departments/directorates to ensure they address national and regional priorities. As a strong lever of constructive collaboration, the network stakeholders meet twice every year at theme-based workshops to review and adjust planned activities. They share their experiences with laboratory governance (successes and difficulties) to inform and guide other members of the network. These meetings also provide an opportunity to address new scientific or regulatory developments. For example, the fourth edition of WHO's laboratory biosafety manual was featured during the 2022 RESAOLAB workshop in Togo.

The project focuses on two main areas.

1. Advocacy for the creation and recognition of the national health laboratory department/directorate
2. Financial and technical support for the operationalization of the national laboratory strategic plan

In each member country, RESAOLAB advocated for the creation of a dedicated and designated national laboratory department within the ministry of health. Once this department was created, the program focused on upgrading these departments to the level of a directorate to fully empower and give them the official and legal authority of a recognized decision-making entity, with adequate resources for effective governance. Step by step and over the years of the project, all members have established at least a national laboratory department and almost all have designated it as a directorate (5 of 7 countries). The last one was designated in 2021 in Benin. There is ongoing advocacy to provide these departments/directorates with adequate human and financial resources. With a governance structure in place, the RESAOLAB project went on to support member countries in establishing their national laboratory policy and its strategic plan as cornerstones for the improvement of their national laboratory systems.

The regional framework and collaboration among members have been key to strengthen advocacy in favor of laboratory directorates in member countries and harmonize the mandate and policies of these directorates/departments at the regional level, through experience sharing and emulation among countries. RESAOLAB activities support the implementation of the national strategic plans and target their priorities.

Main achievements include.

- Strengthened laboratory governance: Establishing a regulatory framework by developing and validating regulatory standards, manuals, and guidelines, such as national registration and licensing requirements of laboratories, a national laboratory biosafety manual, and manual for AMR surveillance. The regional scope has fostered a mutual assistance among countries, which share documents and work toward a regional harmonization of laboratory system governances.
- Improved quality of laboratory services following international standards: Over 200 laboratories benefit from external quality control to assess their performance and define action plans to address nonconformities.
- Enhanced initial training and continuing education: 14 training modules were developed through a collaboration involving member country experts Over 1000 laboratory technicians have been

(Continued)

CASE STUDY *(cont'd)*

trained. Scholarships have been funded for clinical lab-specialized training for doctors and pharmacists.

- A laboratory information system for infectious disease surveillance and patient care management: a free, open-source Laboratory Information System (LIS), LabBook, created by the Mérieux Foundation, has been installed in more than 150 labs across the network.

Lessons learned

What worked?

- RESAOLAB's main goal of establishing laboratory leadership was a success and the program helped to create a dedicated department for the governance of the national lab system in each country and a national laboratory directorate in most of them.

  The regional collaboration boosted advocacy efforts across borders and resulted in a coherent development of governance and coordination of laboratory systems at the regional level.
- The activities that are technically and financially supported by RESAOLAB are fully embedded within the national strategic plans and are a strong lever for their operationalization.
- The regional collaboration promotes scientific and technical exchanges and facilitates the implementation of some activities in less-resourced countries, through the sharing of experts, manuals, or guidelines for example.
- The national laboratory departments/directorates are in the process of being identified (and for some countries, this is already the case), by Funding and Technical Partners, like the Global Fund, as the key partners for questions and projects related to laboratory systems.
- During the COVID-19 pandemic, the national laboratory departments and directorates were the first line in developing COVID-19 laboratory response strategies because they were included in the national taskforce. They facilitated the mobilization of laboratory resources and expertise to back national responses and scaling up of testing capacity during the COVID-19 crisis.

  Their national anchorage facilitated the interactions with the other stakeholders of the ministry of health. They managed the coordination between the different international stakeholders to ensure the optimization of dispatch and use of reagents and supplies.
- RESAOLAB provided a specific online platform to exchange new scientific knowledge and data related to the pandemic between countries. Regular online meetings were organized to maintain regional contact despite the isolation requirements. They provided RESAOLAB members with the space to directly discuss and share national challenges and solutions during this period.

What did not work?

- The laboratory directorates and departments have not achieved the financial autonomy needed to fulfill their governance role and are still dependent on international funds.
- Despite advocacy efforts, all national laboratory departments were not upgraded to the level of the national directorate. Therefore, the development of

CASE STUDY *(cont'd)*

governance in all member countries is not at the same level. Laboratory departments were less included in the decision-making process compared to laboratory directorates, and they only had a consultative role. Furthermore, a more efficient decentralization of diagnostic capacities was possible in countries where a national laboratory directorate exists compared to a department, probably due to a stronger leadership over the national laboratory system. Preliminary results of the study on the network during COVID-19 identify the lack of a directorate as a limiting factor for effective response to the epidemic.

Recommendations

- The development of regional collaboration is a key factor for achieving the ambitious goal of organizing and managing national laboratory systems and tackling national priorities. This is only possible when there is clear leadership for the sector and an officially recognized body responsible for it in each country. The regional collaboration fosters emulation between countries, raises the overall level, and harmonizes the organization of laboratory systems across the region. Regional collaboration also promotes institutional, technical, and scientific exchanges, accelerating the development of national laboratory systems and alignment with the regional strategy.
- Dedicated financial and technical resources are required to make the strategic plan operational. Long-term and sustained central government financing for systems strengthening should be prioritized, with supplemental international donor funding sought for discrete health program work.
- Long-term programs (more than 10 years for RESAOLAB) embedded within the national strategic plans and contributing to their operationalization should be preferred for proper and sustainable actions and concrete impacts.
- Advocacy is still needed to designate a legally recognized, national laboratory department where one does not exist as a minimum at the ministry of health and to upgrade the current department into directorates. This must be completed by the advocacy for the allocation of adequate financial and human resources for both the functioning and operating budgets of the laboratory system.
- Adapting capacity-building program activities to national and regional policies and strategic plans is the best way to address national and regional priorities. This model, compared to more prescriptive projects where activities are defined outside of the national and regional policies, should be preferred as it promotes strong laboratory governance, local ownership, and a greater chance of long-term sustainability.

The structure of a regional network like RESAOLAB exemplifies concrete opportunities for strengthening laboratory systems through improved collaboration between countries, with international partners, and supports stronger coordination.

[1] World Health Organization. Regional Office for Africa. (2014). Guidance for Establishing a National Health Laboratory System. World Health Organization. Regional Office for Africa. https://apps.who.int/iris/handle/10665/148351

Actions

Our recommended actions detailed below address the major challenges that remain toward sustaining functional laboratory systems for effective disease detection and response. Specific recommendations are included herein.

Strengthen national governance such as dedicated national entities for laboratory services (e.g., national laboratory directorate)

Dedicated governance of laboratories with an organizational mandate and mission to regulate and support the laboratory system within national health agencies is the best way to ensure effective and appropriate laboratory services within a country. Laboratory and diagnostic services within countries are provided by different types of structures, both public and private, and are aligned with each tier of the health system. The typology can be broken down schematically as follows.

- Public laboratories connected to healthcare facilities of varying size, offering a range of lower to higher complexity of testing services;
- Laboratories with public health mandates for pathogen surveillance and characterization;
- Private laboratories offering a range of lower to higher complexity of testing services to paying clients;
- Research laboratories (commonly university-based) with or without clinical diagnostic activity.

Effective governance includes several core responsibilities that must be met with utmost competence and authority for enforcement.

- Regulatory: this entails establishing the regulatory framework for clinical diagnostics, that is, who can perform laboratory tests, under what conditions, with what legal liability in the event of an error in testing results. This also covers quality assurance for laboratory diagnostics, test performance, quality reagents and equipment, and the rules for their commercialization.
- Control: this involves the enforcement and monitoring of the regulatory framework through institutions that control and provide guidance, as well as impose sanctions in the event of noncompliance with the rules (for registration, quality, ethics, etc.).
- Policy: setting out a strategic vision for the laboratory sector's development and prioritizing actions to achieve this vision. The policy should be backed by strategic and operational plans.
- Coordination: coordinating the actions of the different entities and partners, whether national or international, public or private, to ensure consistency and complementarity within the sector. It is also a matter of ensuring effective coordination among human, animal, and environmental health laboratories as part of a One Health approach.
- Resource management: overseeing the use of public and/or private resources from technical and financial partners.

While this list is not exhaustive, it is intended to show the breadth of governance required and the importance of ensuring that all these aspects are well defined and supported by competent agencies in charge.

Like most health services, the way in which laboratory services for human health are organized within ministries of health varies according to the country and its history. In some countries, the laboratory is managed by the entities responsible for drugs and sometimes even more broadly responsible for all health products. The historical reason being that the forerunners of the laboratories were often the pharmacies where testing used to be performed.

In some other countries, the laboratory is part of the public health sector, under the responsibility of the epidemiology department. Sometimes this diversity is also apparent at the level of the scope represented. In some countries, a distinction is made between microbiology on the one hand and "pathology" on the other, which includes the analysis of noninfectious biological samples, whether solid or liquid or at the tissue (histology) or cellular (cytology) level.

For other countries yet, the distinction will be between solid and liquid samples, with anatomic pathology dealing with the analysis of solid samples and medical biology dealing with liquid samples on infectious and noninfectious components. *These differences in laboratory governance are vast, and each structure type has its advantages and disadvantages.*

Nevertheless, what is most important is to clarify these organizations and their respective responsibilities particularly in the context of international cooperation, to ensure that the terminology and the subjects of discussion are based on accepted standards and norms and that the specific organizational characteristics of the countries are considered.

Considering the importance of each of the components of governance, and the increasing complexity of the diagnostics landscape, it has become ever more crucial for ministries of health to establish well-defined mechanisms for the governance of laboratory services and to ensure that they have the necessary resources with dedicated teams and an operating budget. In addition to the urgent need to strengthen national-level governance and regulation, international oversight of biological research, and surveillance activities that pertain to pathogens of high consequence must be strengthened and be based on adopted international standards of biosafety and biosecurity.

Expand availability of workforce training and educational programs for laboratory professionals

Laboratory professionals need to have both technical and managerial competencies to deliver quality laboratory services. It is also crucial to be able to both explain the meaning of test results to clinicians and to ensure test reliability by maintaining complete control over the total testing process: spanning prescription, sampling, transport, analysis, reporting and archiving results, and quality processes—from biosafety to patient results traceability. All of this requires laboratory professionals to have the necessary competencies which encompass the multitude of testing aspects—from the biological fundamental knowledge, to technical aspects of testing, supervising others, quality assurance, and the medical implications to liaise with the clinicians. They must also ensure that the entire diagnostic workflow complies with quality requirements and that the results can be used in the course of care without endangering the patient. Needless to say, a diagnostic error can have disastrous consequences not only for the patient, but also for the community.

Personnel of this caliber are key to the successful utilization of laboratory services. There are currently efforts to standardize pathways, degree programs, and diplomas required to access the profession. However, it varies greatly from country to country. An attempt at standardization is currently underway in Europe with the development of a "laboratory medicine" specialty. Beyond standardization, it is extremely important to ensure that a critical mass of these specialists is trained to operate the laboratories and ensure the proper use of their services. Unfortunately, this type of foundational training does not exist in many countries, forcing those who wish to pursue this path to get trained abroad. An increasing number of laboratory technicians are being trained, but still not enough to meet the current demands, and they also need to be supervised by professionals who are able to liaise with clinicians and guide them in the use of laboratory results.

Furthermore, over and above technical training, it is very important that training in laboratory management and administration be provided. Ultimately, laboratories are people-driven organizations that require capable leadership to function properly. However, historically, competency-driven education and training in organizational leadership and management has been often overlooked for the more technically focused degree and training programs for medical biologists. It is for this reason that laboratory leadership and management development programs have been developed.[16–20] For example, the Global Laboratory Leadership Program (GLLP) was developed by WHO and other partners to improve laboratory coordination and management at the national and international levels across the One Health sectors of human, animal, and environmental health.[21] The best diagnostic tools, no matter how simple, are useless if they are not in the hands of competent personnel. Moreover, diagnostic testing is not a product, but a service relying on a sequence of actions, each of which must be perfectly executed and traced to ensure the quality of the process as a whole. It is therefore essential to ensure that laboratory professionals, medical biologists, "laboratory medicine" specialists, or the equivalent are trained in sufficient numbers and at a level that allows them to liaise with clinicians and public health professionals.

Require a systems approach to laboratory capacity building

Nearly every country in the world has adopted a tiered health care system whereby clinical and public health services are tailored to the national and subnational jurisdictions and available according to population needs at each level. Laboratory and diagnostic services follow accordingly with low-complexity tests available at the patient and community level, followed by more complex laboratory-based diagnostics available at higher levels.

Global investments in biomedical research and discovery have contributed significantly to advancements in public health and clinical diagnostics. The investments have helped drive technological advancements in health laboratories that have shifted the paradigms of clinical and public health and challenged many "gold standards" of surveillance and clinical diagnosis, which once held positions of unwavering respect. For example, the malaria blood smear, while practically useful in nearly all environment and community settings, is largely being replaced by easier-to-use, handheld lateral flow assays that can be done by a community health worker who has the ability to immediately prescribe antimalarial drugs and countermeasures such as bed nets. What used to take a day or two to diagnose and required families living far from the health clinic to sleep on makeshift beds set up next to a hospital, many miles from their homes, now can be done in their homes.

Powerful diagnostic tools and status-quo disruptions such as this example have evolved because the public and clinical health community and leaders challenged the current process and sought to do better for the benefit of patient care and public health. Advances in primary care such as for diabetes, maternal and child health, HIV, and TB are success stories because the health community pressed for change. However, similar advances have not occurred in the public health laboratory sector.

Diagnostic technologies have evolved tremendously over the last 3 decades with the development of molecular biology, flow cytometry, mass spectrometry, and the widespread use of computers for data processing. This has led to an extremely varied range of platforms and practices, from handheld rapid tests utilized at the point of care by community healthcare workers to fully automated, high-throughput platforms utilized in laboratories by highly trained laboratory professionals. The performance characteristics, reliability, and the utility of these diagnostics in various settings vary widely (appropriateness, turnaround time, accuracy, specificity, sensitivity, robustness, cost, throughput, etc.), and understanding them is essential to making an informed choice about test selection and utilization. In LMICs, these diagnostic choices are often driven by the technology availability and test cost, and importantly, often influenced by the international donor community supporting specific disease programs rather than a national level system strengthening approach deliberately accounting for national and local level burdens of disease and clinical and public health needs.

Establishing health sector development plans with prioritized objectives and activities to be implemented to achieve them is now a common practice in countries with limited resources to better coordinate, strengthen and optimize the sector. This translates in particular into setting a minimum package of activities to be implemented at all levels of the health pyramid, including laboratories. This means that a district hospital laboratory must be able to carry out all the tests specified in the minimum package of activities, and the same applies to the other levels of the pyramid. The development plans specific to the laboratory sector describe in greater detail the role and responsibilities of each level and how referral between levels works.

From the standpoint of technical and financial partners in health, this structuring helps to rationalize development aid, improve coordination of technical and financial partners around framework documents, and harmonize practices in the field to avoid disparities in the levels of facilities according to the support they may receive. Nevertheless, due to limited resources, especially in terms of expertise and laboratory staff, even the implementation of minimum packages of activities remains a challenge. Many facilities, although they have adequate equipment, are unable to implement certain activities due to a lack of human and/or financial resources. This is especially true when it comes to laboratory activities that require a high degree of expertise, such as molecular biology performed on open platforms, or bacteriology.

At the same time, in high-income countries, diagnostic resources are being concentrated around very large technology platforms and specimen transport networks. The paradigm is no longer that each healthcare facility or each laboratory must be able to perform a complete range of laboratory tests, but rather that the work is shared among laboratories, organized around facilities that handle sample collection and urgent tests, and other facilities with very large technical platforms to which samples are sent. This is true in both the public and private sectors and allows for significant economies of scale in human and material resources while providing quality services that are timely and reliable.

This calls for optimization of diagnostic services and to reexamine the organization of services, based on criteria such as the needs of the population, financial, technical, and managerial obstacles, and geographical accessibility. This entails addressing the laboratory system as a whole, that is, all the laboratories in a geographical area, and redefining the scope of their testing, based on needs and capabilities, rather than dogmatically adopting minimum activity packages for each level of the health pyramid. This also involves, in particular, working on the implementation and optimization of sample transport systems.

The COVID-19 pandemic and the ever-increasing demand for laboratory diagnostics, especially in the first phase of the pandemic for PCR testing, have greatly accelerated discussions and initiatives in this area. Furthermore, the importance of zoonoses, climate, and environmental issues shows that it is essential to consider the entire health ecosystem and include animal health and the environment. This also applies to the laboratory sector, where it is important to try to pool resources (harmonized activities including workforce training, surveillance systems, sample transport systems, etc.) to optimize the service provided and the sector's development. This systems-level approach has become essential.

Encourage transparency in communication and enable collaboration and information sharing within the international community

The SARS-CoV-2 virus sequence was shared with WHO and the international community through the Global Initiative on Sharing Avian Influenza Data (GISAID) a few days after the reporting of cases of viral pneumonia in Wuhan, People's Republic of China. Rapid sharing of the genomic sequence data allowed WHO and the scientific community to develop and share in-house testing protocols, and enabled the in vitro diagnostics industry to develop, manufacture, and distribute commercial assays in a rapid manner.[20] As a result, many countries acquired the capacity to detect this emerging virus at the early stages of the pandemic. The live virus was also rapidly available because of the widespread dissemination at the global level, minimizing the need for virus sharing for research and development purposes.

However, such rapid sharing of laboratory data and materials has not always been optimal. Information is often shared on a bilateral basis and ad hoc basis between countries or academic or research institutions, delaying proper risk assessment and development of medical countermeasures such as vaccines, diagnostics, and therapeutics. Delays in information sharing have many causes, from technical barriers to legal, financial, and political reasons.

Global mechanisms have been established to facilitate materials sharing in a timely and equitable manner, such as the Pandemic Influenza Preparedness (PIP) Framework, established in 2011 as a result of concerns identified during the avian H5N1 influenza outbreaks in the 2000s. Realizing the importance, WHO has also established a BioHub in 2021 to facilitate the voluntary sharing of pathogens with epidemic or pandemic potential in a safe and transparent manner.

Nevertheless, scientific and public health communities rely increasingly on digital information such as genetic sequence data. Such use triggers new legal and political challenges not properly addressed by existing global instruments, practices, and arrangements such as the Internal Health Regulations (2005), the PIP framework, or the 2010 Nagoya Protocol to the Convention on Biological Diversity. New and/or updated global mechanisms, such as a treaty on pandemic preparedness and response shall aim at better addressing access to such resources, in a fair and equitable manner through benefit-sharing measures.[22]

Build and utilize better tools to measure testing and response capacity at subnational levels

Following the revision of the International Health Regulations in 2005, a new monitoring and evaluation framework for the surveillance and response core capacities was established.[23,24] Laboratory capacity was identified as one of these core capacities to be measured through the various tools proposed. As part of this IHR M&E framework, the JEE tool includes a "National laboratory system" component that addresses laboratory testing for detection of priority diseases, specimen referral and transport system, effective modern point-of-care, and laboratory-based diagnostics, and laboratory quality system.[25] These indicators are measured on a scale from 1 (No capacity) to 5 (Sustainable capacity), and the tool includes contextual and technical questions as well as required documents to be collected.

Another tool, the WHO's LAT, emerged from internationally recognized standards and best practices and can be used to assess laboratories and national laboratory systems. It can be adapted to specific contexts and used in self-assessments or external assessments. However, these tools showed limitations to properly assess laboratory preparedness and readiness, notably for emerging diseases such as COVID-19.[26] The pandemic has shown the importance to decentralize testing capacity within the tiered levels of the health care system, within or outside laboratory facilities. Access to testing at the community level, fast regulatory approvals, procurement and delivery of new in vitro diagnostics (IVDs), as well as laboratory workforce retention and development should be among the elements to be better addressed in future tools. These tools shall also assess not only that structures and systems are in place, but also that they are operationally ready to respond to emerging threats. Readiness and response checklists have proven to be a useful approach to assess short-term needs when confronted with imminent threats.[27] Laboratory-focused intra or after-action reviews and targeted desktop or functional simulation exercises could help identify and address such operational weaknesses.[28]

Need for stronger international standards and enforcement of laboratory biosafety

Significant investments in laboratory biosafety improvement have been made in the last 25 years, notably in terms of infrastructure. Most of these investments have been based on the classification of pathogens in risk groups and associated biosafety levels. Although such an approach has proven to be effective in describing biosafety requirements and establishing regulations and oversight mechanisms, it has also shown some limitations, given the lack of consideration paid to the actual needs of the procedures and assays performed in the laboratories. This led to the adoption of hardly sustainable solutions in high-containment laboratories while neglecting basic occupational safety and health for laboratory workers and surrounding populations using low-containment facilities. The increased use of inactivated specimens and molecular technologies have, in many instances, challenged the classic and static association between a particular risk and a biosafety level, leading to a diversity of practices in the application of the biosafety level requirements from biosafety level 2 to the biosafety level 4.

Acknowledging the reduction in the use of propagative techniques for many infectious disease diagnostic activities, as well as the main causes of laboratory-acquired infections (i.e., human factors vs. engineering failures), WHO has published the fourth edition of the Laboratory Biosafety Manual, promoting a risk- and evidence-based approach for cost-effective, sustainable, and fit-for-purpose solutions.[29] Implementing such a new approach, moving away from the usual association of a particular risk group with a defined biosafety level, will require significant investments and updates of the national oversight and regulatory systems as well as an increase in the knowledge and know-how of the laboratory workforce on the implementation of a risk assessment framework.

At an international level, a large diversity of regulatory mechanisms can be observed, based on activities performed by the laboratories and the nature of the pathogens stored and handled. Regional and global initiatives are aiming at harmonizing and standardizing biosafety good practices and regulations, notably in the context of the implementation of the Biological and Toxin Weapons Convention (BWC) and related requirements under the United Nations Security Council Resolution 1540 (UNSCR 1540). On top of these international agreements, many manuals, codes, or instruments, such as the WHO laboratory biosafety manual or the OIE manuals and codes, as well as ISO standards 15190 and 35001, have inspired national biosafety legislations.[30]

However, adapting, implementing, and enforcing these global guidance documents through updated national legislation or soft laws such as guidelines remain challenging to fit at best the risks encountered given the resources available across the entire, multisectoral, laboratory systems.[31] This is an opportunity to advocate for a paradigm shift and implement an evidence-based risk management framework, keeping the laboratory workforce at the center of the establishment of a responsible life sciences culture.

Develop more resilient supply chains

The COVID-19 pandemic triggered unprecedented acute needs for laboratory and diagnostic supplies and reagents, as well as specimen collection kits. Global supply chains were stretched to their limits, resulting in competition between laboratories within and across countries to secure these limited resources. A diagnostics consortium was established by WHO and key partners such as the Global Fund and UNICEF to secure critical volumes for resource-limited countries that had limited access to the market. An allocation process was designed to ensure a fair and equitable distribution across the globe. Several bottlenecks limiting timely and equitable access to essential products have been identified, such as barriers to accessing biological materials needed to develop effective diagnostic tests or inefficient regulatory systems, exposing countries to low-quality devices.[32] Fast-track regulatory approvals developed by stringent regulatory bodies such as US Food and Drug Administration, the European Medicines Agency, as well as the Emergency Use Listing developed by WHO have certainly helped address some obstacles.

In the long-term, strengthened National Regulatory Authorities, improved management of donations, stronger forecasting, and supply planning, enhanced warehousing and distribution of diagnostics and increased use of open diagnostic platforms systems allowing multi-sourcing of reagents are among some of the key elements to be considered for better preparedness to the next pandemic.[33]

Summary of specific recommendations

1. Develop laboratory infrastructure, supply chain, and a workforce with improved governance, leadership, and financing mechanisms. Leverage and integrate better-existing laboratory and diagnostics resources using a One-Health approach.
2. Ensure representation of laboratory professionals in decision-making groups (e.g., laboratory directorate, hospitals (e.g., grand rounds), subnational working groups) and enable laboratory directorates to support all tiers of the system.
3. Support enabling regulatory environments through investing in governance-laboratory directorates to oversee the laboratory system. Empower and enable local leaders to act. Enable district-level responses through permissive policies.
4. Expand availability and access to laboratory workforce training and education programs.
5. Build communities of practice, promote, and strengthen professional societies with peers of shared goals, sharing of experience-peer networks
6. Promote transparency and timely information sharing. Enable information sharing by removing incentives to hide information for advantage
7. Ensure fair and equitable access to testing through UHC, which may create sustainable markets and stimulate investments. Invest at the community and district levels.
8. Invest in systems thinking and building. This includes functional specimen referral (including enabling financial reimbursement processes and addressing barriers). Develop supportive policies for collaboration between labs.
9. Enhance R&D and manufacturing capacities for affordable diagnostic technologies.

References

1. World bank country and lending groups. The World Bank. https://datahelpdesk.worldbank.org/knowledgebase/articles/906519-world-bank-country-and-lending-groups.
2. Strengthening Clinical Laboratories—Division of Laboratory Systems (DLS). *Centers for Disease Control and Prevention*; November 15, 2018. https://www.cdc.gov/csels/dls/strengthening-clinical-labs.html. Accessed January 18, 2022.
3. Fleming KA, Horton S, Wilson ML, et al. The Lancet Commission on diagnostics: transforming access to diagnostics. *The Lancet*. 2021;398:1997–2050.
4. Global antimicrobial resistance and use surveillance system (GLASS). World Health Organization. https://www.who.int/initiatives/glass.
5. Special Session of the World Health Assembly to Consider Developing a WHO Convention, Agreement or Other International Instrument on Pandemic Preparedness and Response. 2021. https://apps.who.int/gb/ebwha/pdf_files/WHA74/A74(16)-en.pdf.

6. WHO Director-General's closing remarks at the special session of the world health Assembly - 01 December 2021. *Who.int*; December 1, 2021. https://www.who.int/director-general/speeches/detail/who-director-general-s-opening-remarks-at-the-special-session-of-the-world-health-assembly-01-december-2021. Accessed January 18, 2022.
7. The Royal Institution. Are we ready for the next pandemic?—with Peter Piot. 2018. https://www.youtube.com/watch?v=en06PYwvpbI
8. Martin R, Barnhart S. Global laboratory systems development: needs and approaches. *Infect Dis Clin*. 2011;25(3). https://doi.org/10.1016/j.idc.2011.05.001.
9. Ridderhof JC, van Deun A, Kam KM, Narayanan PR, Aziz MA. Roles of laboratories and laboratory systems in effective tuberculosis programmes. *Bull World Health Organ*. 2007;85(5):354–359. https://doi.org/10.2471/blt.06.039081.
10. Nkengasong JN, Nsubuga P, Nwanyanwu O, et al. Laboratory systems and services are critical in global health: time to end the neglect?: time to end the neglect? *Am J Clin Pathol*. 2010;134(3):368–373. https://doi.org/10.1309/AJCPMPSINQ9BRMU6.
11. G20 Health Ministers recognize the importance of GISAID in regard to virus data sharing. GISAID. Accessed January 18, 2022. https://www.gisaid.org/.
12. Peruski AH, Birmingham M, Tantinimitkul C, et al. Strengthening public health laboratory capacity in Thailand for International Health Regulations (IHR) (2005). *WHO South East Asia J Public Health*. 2014;3(3–4):266–272. https://doi.org/10.4103/2224-3151.206749.
13. Decoster K. *UHC and Global Health Security: Two Sides of the Same Coin? Your Weekly Update on International Health Policies*; May 31, 2017. https://www.internationalhealthpolicies.org/blogs/uhc-and-global-health-security-two-sides-of-the-same-coin. Accessed January 21, 2022.
14. Craven M, Sabow A, Van der Veken L, Wilson M. Not the last pandemic: investing now to reimagine public-health systems. *Mckinsey.com.*; May 21, 2021. https://www.mckinsey.com/industries/public-and-social-sector/our-insights/not-the-last-pandemic-investing-now-to-reimagine-public-health-systems. Accessed January 13, 2022.
15. IHR Monitoring and Evaluation Framework. Who.int. Accessed February 16, 2022. https://extranet.who.int/sph/ihr-monitoring-evaluation.
16. Gopolang F, Zulu-Mwamba F, Nsama D, et al. Improving laboratory quality and capacity through leadership and management training: lessons from Zambia 2016–2018. *Afr J Lab Med*. April 30, 2021;10(1):1225.
17. Perrone LA, Confer D, Scott E, et al. Implementation of a mentored professional development programme in laboratory leadership and management in the Middle East and North Africa. *East Mediterr Health J*. February 1, 2017;22(11):832–839. https://doi.org/10.26719/2016.22.11.832.
18. Whoint. *Global Laboratory Leadership Programme (GLLP)*; July 23, 2021. https://www.who.int/initiatives/global-laboratory-leadership-programme. Accessed June 22, 2022.
19. Ascp.org. n.d. ASCP Certificate Programs. https://www.ascp.org/content/learning/certificate-programs#.
20. World Health Organization. *Laboratory Testing of 2019 Novel Coronavirus (2019-nCoV) in Suspected Human Cases: Interim Guidance*. World Health Organization; January 17, 2020. https://www.who.int/publications/i/item/laboratory-testing-of-2019-novel-coronavirus-(-2019-ncov)-in-suspected-human-cases-interim-guidance-17-January-2020. Accessed January 13, 2022.
21. Albetkova A, Isadore J, Ridderhof J, et al. Critical gaps in laboratory leadership to meet global health security goals. *Bull World Health Organ*. 2017;95(8).
22. Gian Luca Burci. International sharing of human pathogens to promote global health security - work in progress. *Asil.org*; July 20, 2021. https://www.asil.org/insights/volume/25/issue/13. Accessed January 20, 2022.
23. World Health Organization. *Laboratory Assessment Tool Annex 1: Laboratory Assessment Tool/System Questionnaire*; 2012. https://www.who.int/ihr/publications/Annex1_LAT.pdf?ua=1. Accessed February 9, 2022.
24. World Health Organization. *International Health Regulations (2005);: IHR Monitoring and Evaluation Framework*. World Health Organization; 2018. https://apps.who.int/iris/handle/10665/276651. Accessed February 16, 2022.
25. World Health Organization. *Joint External Evaluation Tool*; 2018. https://extranet.who.int/sph/sites/default/files/document-library/document/9789241550222-eng.pdf. Accessed December 15, 2021.
26. Stowell D, Garfield R. How can we strengthen the Joint external evaluation? *BMJ Glob Health*. 2021;6(5):e004545. https://doi.org/10.1136/bmjgh-2020-004545.

27. Tracie Eis B, Tipples G, Kuschak T, Gilmour M. Laboratory response checklist for infectious disease outbreaks-preparedness and response considerations for emerging threats. *Can Comm Dis Rep*. 2020;46(10):311–321. https://doi.org/10.14745/ccdr.v46i10a01.
28. World Health Organization. *Country Implementation Guidance: After Action Reviews and Simulation Exercises under the International Health Regulations 2005 Monitoring and Evaluation Framework (IHR MEF)*; 2018. https://apps.who.int/iris/bitstream/handle/10665/276175/WHO-WHE-CPI-2018.48-eng.pdf?sequence=1&isAllowed=y. Accessed December 2, 2021.
29. World Health Organization. *Laboratory Biosafety Manual*. 4th ed.; 2020. https://apps.who.int/iris/bitstream/handle/10665/337956/9789240011311-eng.pdf?sequence=1&isAllowed=y. Accessed January 13, 2021.
30. *Medical Laboratories – Requirements for Safety (ISO 15190)*. Geneva: International Organization for Standardization; 2020. https://www.iso.org/standard/38477.html. Accessed January 13, 2022.
31. World Health Organization. *WHO Guidance on Implementing Regulatory Requirements for Biosafety and Biosecurity in Biomedical Laboratories - a Stepwise Appraoch*; 2020. https://apps.who.int/iris/bitstream/handle/10665/332244/9789241516266-eng.pdf?sequence=1&isAllowed=y. Accessed January 13, 2022.
32. World Trade Organization. *Indicative List of Trade-Related Bottlenecks and Trade-Facilitating Measures on Critical Products to Combat Covid-19*; 2021. https://www.wto.org/english/tratop_e/covid19_e/bottlenecks_report_e.pdf. Accessed February 9, 2022.
33. The Rockefeller Foundation. Key Considerations and Strategies for Strengthening Covid-19 Diagnostics. https://www.rockefellerfoundation.org/wp-content/uploads/2020/12/LMIC-Covid-19-Guidance-Document_20-Nov-2020.pdf.

CHAPTER

17

Modernizing public health surveillance

Louise Gresham[1], *Wondimagegnehu Alemu*[2,3], *Nomita Divi*[4], *Noara Alhusseini*[5], *Oluwafunbi Awoniyi*[6], *Adnan Bashir*[7], *Affan T. Shaikh*[6,8] *and Scott J.N. McNabb*[6]

[1]PAX sapiens and School of Public Health, San Diego State University, Washington, DC, United States; [2]International Health Consultancy, LLC, Decatur, GA, United States; [3]Rollins School of Public Health, Emory University, Decatur, GA, United States; [4]Ending Pandemics, San Francisco, CA, United States; [5]Department of Biostatistics, Epidemiology, and Public Health, College of Medicine, Alfaisal University, Riyadh, Kingdom of Saudi Arabia; [6]Hubert Department of Global Health, Rollins School of Public Health, Emory University, Atlanta, GA, United States; [7]Health Information Systems Program, Islamabad, Pakistan; [8]Yale University, New Haven, CT, United States

Introduction

Worldwide, there are some 600 infectious disease outbreaks annually, of which 40–50 turn into country-level outbreaks or epidemics. Since 2007, six outbreaks have been declared a public health emergency of international concern (PHEIC) by the World Health Organization (WHO). Stopping a local outbreak before it becomes an epidemic and then a pandemic, necessitates rapid detection, reporting, and early response. Transforming local, national, and global surveillance to enable the rapid sharing of information between countries and with WHO can slow future epidemics.

Public health surveillance (PHS) is often referred to as the "foundation of public health practice, safeguarding communities through ongoing tracking of public health events threatening the health and well-being of populations."[1] PHS employs varied methods to collect data, analyze, interpret, generate, and disseminate actionable information.[2] Langmuir successfully applied the principles of surveillance to outbreaks preemptively, to define populations at risk for disease, determine interventions to counter infectious diseases, and monitor their impact.[3]

https://doi.org/10.1016/B978-0-323-90945-7.00002-6

Traditionally, disease occurrence is reported through a structured public health system with delays in time of disease onset to detection and time to report (from local to global). These measurable times become longer (or at least unpredictable) with fragile health systems and lack of transparency due to political pressures or fear of economic ramifications of reporting outbreaks. There is, however, a quantitative historical assessment of timeliness in infectious disease detection and public reporting of outbreaks that shows a positive trend toward shortened days to report. Chan et al. analyzed WHO Disease Outbreak Reports from 1996–2009 finding, overall, that the timeliness of outbreak detection improved by 7% each year.[4]

While the concept and principles of surveillance remain steadfast, the breadth of data sources and analytic tools, along with geospatial and mathematical modeling have tremendously advanced.[5] McNabb et al. outline new ideas on surveillance, capturing formal and unstructured data from diverse sources and interrelated systems, all the while setting realistic expectations of surveillance systems for daily use.[6] Multiple sources of information are necessary (and merited) as part of daily disease detection routines to find the "needle in the haystack." The rapid expansion in internet access is driving transparency in event-based surveillance by generating reports and alerts from massive amounts of unstructured information. The Program for Monitoring Emerging Diseases (ProMED) and Global Public Health Intelligence Network (GPHIN) are examples of producing timely threat alerts from unstructured data, curated by human experts, to complement traditional surveillance.[7]

COVID-19 prompted standardization of epidemiologic contact tracing, and genomic data resulting in higher quality, comparable, and representative data across select governments and health systems. Surveillance can be challenging in daily practice and, strikingly so, during health emergencies given the ethical dilemmas posed by privacy, equity, and common goods. One need only ponder that disease hotspots, many in low-income countries, are or will be the highest surveilled places in the world. Public health practitioners and policymakers need to anticipate ethical challenges as they utilize new technologies and prioritize resources to advance PHS goals.

Modern PHS takes advantage of novel data sources, scalable tools, and mathematical modeling and machine learning.[8] The explosion of data surrounding the COVID-19 pandemic highlights the need for human curation of high-volume data and diplomatic persuasion in data sharing. The aim is "real-time data" on public health events, achieved by employing multiple and diverse sources of data by the fastest means possible. Analysis of copious data collected from these sources is necessary to make sense of the information collected. Using artificial intelligence and data analytic methods may, for example, enhance scenario building and forecasting. Disease detection systems have increased incrementally in number and sophistication over time based on platform technologies, data analytics, laboratory diagnostics, and communication tools.

Literature review and gap analysis

A functional surveillance system is intended to be operational all the time, providing utility to counter daily threats whether they be foodborne, viral, antimicrobial resistance, or intentional threats. This kind of system allows communities (local to global) to pivot and respond when surveillance alarms are sounded.

WHO International Health Regulations (2005), global surveillance, local implementation

Modernizing PHS is one avenue to improve preparedness, prevention, and response, and led to revision of the 1969 International Health Regulations (IHR) to bolster the limited extant institutional and normative powers of WHO.[9] The current IHR (2005) is a global framework governing cross-border surveillance, thus establishing a surveillance system for PHEIC. However, implementation plays out at a country level. The IHR national focal points, organizations accessible to WHO and state parties, use a decision tree for formal notification to WHO of a PHEIC. The IHR (2005) spotlights several attributes of a surveillance system such as timeliness for reporting, timeliness of external notification of an event, and sensitivity of detection. Notably, it does not address representativeness, and the data quality relies on external validation and assessment.

Strengthening the capacity for surveillance depends upon its relevance to local populations and cultural context to counter endemic and epidemic-prone disease and the inequalities that exist broadly in health systems.[10] A well-established 2004 Center for Disease Control and Prevention (CDC) framework for Evaluating Public Health Surveillance Systems for Early Detection of Outbreaks, CDC outlined four elements of an effective surveillance system: purpose and objective, structure and processes, health events and gravity, and the resources required for high functionality.[11] Nonetheless, the framework does not include cross-sectoral and informal data sources, integration technologies. and laboratory technologies such as genomic sequencing, making the 2004 framework unable to convey the current urgency of a one health surveillance system.

To fill recognized voids, IHR (2005) supports new surveillance tools by widening the sources of human and animal information collected and analyzed to improve the sensitivity of the surveillance system (provisions imply 100% sensitivity as standard). Article 9.1 offers the ability to collect data from diverse sources, including the internet, nongovernmental organizations (NGOs), philanthropies, research organizations, and intergovernmental organizations.[12] Widening sources of data improve the core surveillance capacity of local and intermediate surveillance systems.

The importance of rumor surveillance is likely to increase as the international community considers a revised draft of the IHR 2005. Article 8 of the IHR (2005) Working Paper states, "WHO, in consultation with the health administration of the State concerned, shall verify rumors of public health risks which may involve or result in international spread of disease."[12] Expanding sources of data can circumvent a lack of state compliance with disease reporting requirements, as proven by the December 2019 ProMED and Global Public Health Intelligence Network (GPHIN) alerting of early pneumonia cases in Wuhan, China, before official reports were received by WHO. ProMED is a model of garnering deeper knowledge of conversational data by using expert moderators to review the many signals coming from nontraditional data sources before the distribution of epidemiologic data and critical situational awareness.

The strength of the IHR (2005) lies in its requirement for early detection based upon the appropriate capacity to identify and report outbreaks.[13] Burkle expressed that the intent of the IHR (2005) is "to improve the capacity of all countries to detect, assess, notify, and respond to public health threats has shamefully lapsed."[14] The economic disruption of an estimated $16 trillion from COVID-19 is a sobering business case to upgrade pandemic detection and response capabilities.[15] Understandably, there is pressure to assess the fit for purpose of the IHR (2005) for COVID-19 and future threats.

Implementation of the IHR (2005) depends on the strength of national core capacities, including public health surveillance.[16] A robust national surveillance system bolstered by functional subnational (provincial and district) systems that regularly report timely data is a prerequisite for successful prediction and early detection of public health events that inform action. Depending on the local context, and taking into consideration their global commitment, countries have adapted different models to build their national surveillance capacity. Case in point is in the WHO African region, where member states have resolved to use integrated disease surveillance strategy.[17]

An integrated framework for achieving the IHR

The building of modern One Health surveillance involves transparent, adaptive, and collaborative endeavors by human, animal, and environment stakeholders. The adoption of Integrated Disease Surveillance and Response (IDSR) framework "makes surveillance and laboratory data more useable, helping public health managers and decision-makers improve detection and response to the leading causes of illness, death, and disability in African countries."[18] The widespread use of IDSR confirms regional governance and national commitment to the integration necessary to broaden the scope of surveillance for emerging diseases and promote data sharing and communication.

The WHO African region adopted IDSR in 1998 and by 2017, 94% of African countries were implementing IDSR. One of the lessons learned during this long-time implementation of IDSR is that it should be aligned at all levels of the health information system.[14] This highly tested and proven PHS model proposes a more efficient and effective approach that will provide accurate, sensitive, specific, and timely PHS data, information, and messages where they are needed, when they are needed, and bypasses these limitations.[14,15]

Closing the gap with Global Health Security Agenda

Within days of the first confirmed Ebola case sparking a PHEIC in West Africa 2014, the Global Health Security Agenda (GHSA) was launched to accelerate implementation and compliance with the IHR (2005).[14] As demonstrated in the Ebola epidemic West Africa, there was a dire need to create a roadmap to build public health infrastructure and involve a whole-of-government and multi-sectoral effort to strengthen country-level capacity to counter health emergencies. This capacity to prepare for biological catastrophes, as measured by Joint External Evaluation (JEE) and GHSA2024, includes surveillance and accompanying security, finance, and development assistance.

WHO foresees a new international normative framework for pandemic preparedness and response (perhaps a pandemic treaty) and has agreed to create a negotiating body to explore the issues, instruments, and options available.[19] Alternatively, revisions of the IHR can be considered with the necessary clarifications to strengthen surveillance and data access.

A reliable and stable surveillance system demands State parties be dynamic in their approach to building capabilities, encouraging innovation, and updating public health laws. Surveillance, local to global, is at the core of IHR (2005) by providing direction for the four elements of a surveillance system, by defining a PHEIC, and by capturing a wide scope of communicable and noncommunicable disease events (natural, accidental, or intentional).[9] Preconditioned, the transformation of surveillance systems includes evaluation, for example, tracking outbreak timeliness metrics. Well-defined outbreak milestones (date of outbreak start, detection, report, lab confirmation, respond, and outbreak communication) capture key actions and progress across the timeline of an outbreak and allow for evaluation of the timeliness of a surveillance system.[20] Craven et al. of McKinsey & Company make a case for investing in preparedness and response and stated, "Adequate financing and the creation of a sustainable market will be needed for the establishment and continual maintenance of surveillance infrastructure. Countries should expect to spend about US$1–4 per capita annually on disease surveillance infrastructure and personnel."[15]

One Health surveillance

To help mainstream *one health* surveillance—thereby improving the ability to prevent, predict, detect, and respond to global health threats—the Food and Agriculture Organization of the United Nations (FAO), the World Organization for Animal Health (WOAH, founded as OIE), the United Nations Environment Program (UNEP) and the WHO came together to create an operational definition of One Health: "One Health is an integrated, unifying approach that aims to sustainably balance and optimize the health of people, animals, and ecosystems."[21] COVID-19 illustrated the importance of comprehensive surveillance that acknowledges the interdependency of the ecosystem of humans, animals, and plants, as well as reliance on quality laboratory systems and digital technologies.

Given this profound recognition of the interconnectedness between the health of animals, humans, and the environment, it is logical and timely to move toward the integration of disease surveillance systems. Current ministerial data systems are siloed by sector creating in many cases delays in information flow, redundancy of efforts, as well as missed opportunities for collaboration and joint outbreak investigations between sectors. Integration of multiple sector systems will require strengthened trusted relationships, major modification of workflows, and, realistically, diplomatic skills. Recognizing that this is not an easy adjustment, a step-by-step approach may be more acceptable. For example, sector surveillance leaders may be more willing to start sharing data if there was a predetermined list of diseases of common interest. Additional tools to help increase trust and transparency may include joint one health data dashboards comprised various sectors' data visualized on a single dashboard for simultaneous viewing. Accessible and shared information lends itself to improved transparency and accountability, as well as prompting warranted joint multisector investigations.

Enhanced surveillance to prevent, control, and respond to epidemics and pandemics

Accountable public health practice relies on data-driven decision-making, thereby guiding public health experts in the direction of modern, interoperable, and secure public health information systems. The mission of the newly formed WHO Berlin Epidemic and Pandemic Intelligence Hub launched in 2021, hosted by Germany and with other initiatives, is to provide the world better data and analytics to detect and respond to health emergencies. "Despite decades of investment, COVID-19 has revealed the great gaps that exist in the world's ability to forecast, detect, assess and respond to outbreaks that threaten people worldwide," said Dr. Michael Ryan, Executive Director of WHO's Health Emergency Programme.[22] This forward momentum, along with country capacity and buy-in, promotes equity in global decision-making, thereby leading to a more coordinated mechanism for one health pandemic preparedness and response. However, legal, governance, political, and workforce impediments remain and will obscure the ability to improve national and global surveillance effectiveness.

Digital public health tools

Digital connectivity has enabled the capture of data from the public for a variety of purposes including behavior predictions, monitoring disease trends, identifying risk factors, and early detection of public health events.

Electronic health records

Electronic medical record (EMR) systems, first introduced in 1992, had been anticipated to store sensitive patient data, but have not revamped healthcare intelligence and improved the quality of end-user experience as initially predicted.[23] Yet EMRs have evolved greatly through their implementation and still have the potency of saving trillions of dollars in a year, by employing state-of-the-art technologies like 5G, artificial intelligence, and blockchain.[24,25,26]

The National Notifiable Diseases Surveillance System (NNDSS) Base System (NBS) attempts to integrate information coming from EMRs to help local, state, and territorial public health departments manage reportable data and send notifiable data to CDC, but is used by only 26 health departments in 20 states including Washington, DC; CNMI; Guam; Puerto Rico; RMI; and US Virgin Islands. The WHO African region adopted IDSR in 1998 and by 2017, 94% of African countries were implementing IDSR. One of the lessons learned during this long-time implementation of IDSR is that it should be aligned at all levels of the health information system.[27,28]

An effective and efficient surveillance system is an essential building block for population health which improves health by capturing data and generating information that public health practitioners and stakeholders can use to improve the quality of their decisions and the effectiveness of their actions.[29]

Doctors around the world are still executing grassroots record-making via a pen, loading and storing the data offline in colossal stacks, henceforth, fundamental data are not readily accessible across specialties and not available when requisite.[21] Subsequently, there are missed correlations among the different diseases, which may have a short duration (acute), or may take years to cure (chronic).[21,30]

While some EMRs are in place and carry highly sensitive data, many proprietary EMRs operate in parallel and vary in design, coding language, and coding platforms. The United States has over 120 siloed, often antiquated, and fragmented public health surveillance (PHS) systems. The significant and persistent COVID-19 pandemic has revealed many liabilities and shortcomings in the current public health infrastructure. Also, has put public health professionals under extreme compression to cope with the understanding of the scope of diseases' distribution while maintaining the magnitude of public health surveillance for a multitude of preexisting maladies.[30]

The struggle to determine efficient and accurate data results in a lag of receiving data in a timely manner and places an extreme toll on the public health workforce at all levels. Even though various technologies are accessible to assist in the development of more effective public health surveillance systems, the constraint on existing public health surveillance system has hampered the ability to shift into the 21st century. The United States struggled to contain the virus and became the epicenter of the pandemic due to a lack of preparedness and insufficient containment efforts.[30]

The response to the COVID-19 pandemic in the United States has been inadequate due to a variety of factors, including limitations in the data fueled by inflammation surrounding the politicization of the spread of the virus. Delays in the data and accuracy limitations exacerbated hesitancy and lack of trust between public health professionals and the public.

Epidemic intelligence from open sources

Where the IHR (2005) architecture may not be fully functional, new tools can be added to enhance surveillance and interpretation of data. One illustration is Epidemic Intelligence from Open Sources (EIOS) using electronic public records, social media data, and data from smartphones and wearable devices. Led by WHO, EIOS aims to mitigate and, ideally, prevent public health emergencies by connecting experts around the world and providing them with the best possible solutions to detect, contextualize, analyze, assess, and share information for quick, evidence-based decision for public health action.

Discernibly, the EIOS initiative is created by and for experts across national and international organizations, networks, and government entities engaged in emergency preparedness and response. There is an urgency to retain the ability to generate alerts at the country level and bring initiatives down to the subnational level, not just build global capacity. Contrarily, it is possible to miss signals that can be grave in consequence.

EIOS builds on a growing global community of practice of multidisciplinary collaborators and an evolving fit-for-purpose system, "jointly working toward the EIOS vision of a world where health threats are identified and responded to so early and rapidly that they have zero impact on lives and livelihoods."[30] EIOS aims to create a strong public health information community supported by robust, harmonized, and standardized public health information systems and frameworks across organizations and jurisdictions. The EIOS system supports the IHR (2005) and refashions its scope with a WHO-led technology platform that uses, integrates, and visualizes multiple data sources across stakeholders and organizations. By analyzing volumes of data that provide deeper context to events that can disrupt communities and economies, we improve timely detection, verification, and communication so that appropriate action is taken to mitigate the threat. Risk analysis also highlights relevant risk factors for exposure and vulnerabilities. EIOS provides experts with "the information they need to facilitate an assessment of the hazard, the exposures and the context in which acute public health events do occur or could occur."[31]

The information explosion surrounding the COVID-19 pandemic clearly highlights the imperative of expert human data curation. The WHO emergency group noted that the human-curated ProMED alert triggered a review of pneumonia in EIOS "The EIOS system picked up the first article reporting on a cluster of pneumonia in Wuhan at 03:18 a.m. (UTC) on December 31, 2019 and has experienced an immense increase in data volume ever since. By the end of 2020, the EIOS system had tagged 26.3 million articles with the Coronavirus category, which amounts to about 85% of the 31 million articles that were imported by the system over the year. This is a sixfold increase from the five million articles that the system ingested in all of 2019—a clear testimony to the unparalleled flood of information accompanying the COVID-19 pandemic and an enormous challenge to professionals working in public health intelligence."[15]

In modernizing surveillance, we see EIOS as the current evolution of a long and trusted collaboration between the Global Health Security Initiative (GHSI), WHO, and the European Commission's Joint Research Centre (JRC). Beginning in 2008, GHSI acknowledged a broad and yet specific knowledge of the context of a public health event that ideally comes from different sectors or different data sources. Building on existing systems such as European Media Monitor (EMM) and Medical Information System (MEDISYS), GHSI and the European Commission funded and endorsed the Early Alerting and Reporting (EAR) platform in 2008 for monitoring and detecting all hazard events.[32] The complementarity of the EAR and HDRAS systems, and the commitment to join efforts and platforms across organizations were key drivers of the establishment of the EIOS Initiative in late 2017.[33]

For rapidly evolving and often unexpected situations, timely detection and assessment of health threats trigger a cascade of measures that may prevent or mitigate risk. EIOS focuses on the systematic collection and integrated analysis of publicly available sources of information—including media—that promote early detection of public health threats. The EIOS community of practice is supported by a maturing EIOS system that connects other systems and actors—including ProMED, HealthMap, and the GPHIN—and promotes and catalyzes new and innovative collaborative development.[30]

Go.Data

The Global Outbreak Alert and Response Network (GOARN) created and manages the outbreak investigation tool, Go.Data, to promote faster and more accurate outbreak investigations. Go.Data has multiple functionalities for case investigation, contact tracing, and visualization of chains of transmission. The software is available to ministries of health, academic institutions, and hospital responders for a wide range of outbreak scenarios. The software was tested with field data collection (also as a mobile application) when it was used in outbreaks of Ebola in Democratic Republic of Congo (DRC) with secure data exchange. It has been downloaded and implemented in over 35 countries to support the COVID-19 response. Go.Data is managed by the network of GOARN members and is coordinated by WHO.[34]

EpiCore

With the rise of digital disease detection across the globe, the need to validate and verify these early indications of a public health event increases. Innovations in disease surveillance can fill the gaps in traditional surveillance by gathering health information from informal sources, such as social media, word of mouth and local news, verifying that information, and then sharing it transparently. EpiCore is a network of human, animal, and environmental health experts who provide ground truths to validate rumors and informal data reports about disease outbreaks. EpiCore data also feeds into the EIOS system. This approach enables several successful aggregators in signaling alerts such as ProMED, HealthMap, and GPHIN to validate early signals of public health events.[35]

EU–ECDC epitweetr

Epitweetr uses a combination of signals and machine learning to build knowledge about infectious diseases and noncommunicable diseases. This R-based downloadable tool to detect public health threats early gives users a dashboard, alert page, geotagging, and a configuration page to look at underlying processes.[36] It is envisioned as an open-source all-hazards model to monitor trends by time pace and topic, developed for and by public health experts who can modify topics and keywords. An evaluation of epitweetr in November 2020 yielded 30% national-level accuracy of signal detection.[36] Epitweetr is part of the European Center for Disease Prevention and Control (ECDC) global threat detection applied to 70 unique topics, COVID-19-specific algorithms, and mass-gathering scenarios using data anonymization. As with EIOS, and participatory surveillance writ large, the innovative epitweetr leads to deeper questions on managing and accessing a wealth of data to create knowledge representations and reasonings to provide actionable intelligence on detected threats.

Participatory surveillance

Participatory disease surveillance refers to an active bidirectional process of receiving and transmitting data for action through direct engagement of the target population. Participatory surveillance, also referred to as participatory epidemiology in certain sectors, originated within the animal health community, and was used for public health surveillance for the first time in 2003 in the Netherlands.[37]

Participatory surveillance allows for timely two-way communication between health authorities and users of participatory disease surveillance systems, hence providing an important opportunity for health messaging and education. Participatory surveillance systems have been used to inform users about disease occurrence in their communities, to provide automatic messages, and to share appropriate and timely information on prevention and control measures. It also has been used to push out new resources to the community during an active outbreak. The system provides a valuable way for health authorities to message a large population of volunteer users, which may include hard-to-reach populations. Specific examples of useful information for the user include the location of vaccine distributors, availability of hospital beds for COVID-19, and mapping of disease activity close to the user.[38,39]

Participatory surveillance systems play a key role with COVID-19. The free national hotline (115 hotline) in Cambodia, was built in 2016 to grant direct reporting of public health events. This hotline was the principal means of providing the public with timely information about COVID-19. In fact, while on average the hotline received 600 calls per day with 20–30 requiring response from authorities, during COVID-19, this same hotline received 18,000 calls per day and is credited for identifying 90% of the early COVID-19 cases in the country.[40]

Participatory surveillance is also used to track events in animals. For example, in Thailand, community volunteers report suspected illnesses in domestic livestock through the submission of photos and simple data forms on mobile phones. Of note, in the first 6 months of implementation, more animal disease outbreaks were detected in two pilot districts of Chiang Mai province than were reported by all 25 of its districts over the entire previous year. Exemplifying its value, the system detected an initial case of foot and mouth disease in a cow that led to immediate wide-scale vaccination of all village cattle, preventing further spread. This rapid action is estimated to have saved the impacted village four million US dollars by preventing a ban on their milk.[41,42]

International mass gatherings are key use cases for disease surveillance and pandemic prevention. During the 2014 World Cup tournament in Brazil, participatory disease surveillance tools were tested for the context of such gatherings. The Brazilian Ministry of Health partnered with philanthropy and local tech partners to create and deploy "Saúde na Copa" (healthy cup), a smartphone app that allowed users to report health status or symptoms of illness daily throughout the games, a first attempt at using this approach in a mass-gathering setting. Encouraged by the success of this use case, the same partners created "Guardiões da Saúde" (guardians of health) for use during the 2016 Olympic and Paralympic Games in Rio de Janeiro.[43,44]

Vector-borne diseases have also emerged as a use case for participatory surveillance with the recent spread of chikungunya, dengue, and Zika viruses. Community-reporting applications such as "MoBuzz" and "Kidenga" have been deployed in Sri Lanka and across several states in the United States, respectively.[45,46]

Participatory disease surveillance provides flexible data systems that enable health authorities to adapt their system based on current data needs, enabling decision-making. Illustrative of the flexibility, the list of reported symptoms can be expanded if an emerging infectious disease is associated with additional symptoms not currently collected. Specifically, "Flu Near You," a system that tracks symptoms of influenza like illness across the United States began to collect new symptoms related to dengue and Zika and COVID-19 viruses as these diseases became more prevalent in the United States. Similarly, the 115 hotlines in Cambodia were adapted to enable contact tracing, a real need during COVID-19.[38,40]

Genomic sequencing

Genomic surveillance leverages next-gen sequencing and is viewed as necessary to be implemented worldwide to achieve effective mitigation and containment of COVID-19. It creates the availability of whole genome data and advances phylogenetic methods to detect variants that are phenotypically or antigenically different. We have seen intense collaborations between academic and public health authorities, such as with COVID-19 Genomics UK (COG-UK), providing breakthrough phylogenetic methods to produce reliable and actionable results to detect variants that are phenotypically or antigenically different.[47]

Having viral sequence data from animals and humans deepens the understanding of how an animal virus jumped species boundaries to infect humans. Genomic surveillance helps experts anticipate and plan mitigation and containment strategies for SARS-CoV-2 and other novel viruses. The detailed sequencing sheds light on the prevention of future zoonotic events by elucidating the infectiousness and transmissibility of SARS-CoV-2 in humans and animals.[48] The evidence also is suitable to paint a more definitive picture of theories of a pathogen's origin.

Successful genomic surveillance systems create an open space with publicly available data, immediately to inform real-time decision-making by public health officials and vaccine manufacturers. Genomic sequencing decodes a genome of a pathogen to better understand mutation and transmission from person to person. Vaccine development is premised upon the genetic detail available to test vaccine efficacy and variant changes, especially important with RNA viruses prone to replication errors than DNA viruses.[49]

Bioinformatic tools analyze sequence data and set up visualization linkages between pathogens and animals, humans, and places. Not to be forgotten, a modern surveillance workforce that can create and utilize bioinformatic tools needs to be in place and sustained to analyze extensive genomic sequences. Effective linkages between laboratories and public health institutes combine existing and new expertise at all levels—from global, to regional, to country, to local. As with many sources of surveillance data, analysis of genomic data can facilitate communication of uncertainties of an outbreak, in effect providing options to mitigate and control novel viruses. Increased international collaborative efforts may create the political will to make this highly sophisticated surveillance tool accessible to low- and middle-income countries.

Vaccine monitoring

Vaccine monitoring systems exist around the world at different levels of sophistication, intending to ensure vaccine safety. Vaccine safety and vaccine adverse effects, whether it is for measles or COVID-19, are monitored by government surveillance systems. In Europe, the European Medicines Agency (EMA) approves and monitors the safety of vaccines and pharmaceuticals for the European Union. The EMA uses a vaccine monitoring database called EudraVigilance, this system collects data on reported adverse effects from vaccines, researches these claims to determine if a link can be established between the vaccine and the adverse effect, and provides guidance to healthcare workers on vaccine administration.[50] In Saudi Arabia, vaccine safety is under the jurisdiction of the Saudi Food and Drug Authority (SFDA). In 2009, the SFDA established the National Pharmacovigilance Center (NPC) which handles the database where vaccine Adverse Effects (AEs) are reported by healthcare workers, pharmaceutical companies, and even vaccine beneficiaries.[51] In the United States, the CDC and the Food and Drug Administration (FDA) oversee vaccine safety and approval, respectively. Monitoring systems include the Vaccine Adverse Event Reporting System (VAERS) and Vaccine Safety Datalink (VSD). CDC uses VAERS data, considered passive surveillance, to analyze patterns and guide investigation on all adverse effects in the United States.[52] VSD, established in 1990, on the other hand, is the research arm of the CDC's vaccine safety monitoring system, and researches inquiries based on questions from healthcare workers.[53]

With the development of COVID-19 vaccines, the CDC decided to focus surveillance on vaccine breakthrough infections, referring to infections of fully vaccinated persons.[54] Most of the data collected are through passive surveillance systems that rely entirely on healthcare providers, public health officials, and communities to report. Adverse effects (AE) may not get reported because most people are uncertain about the causal link between a vaccine and the AE.[55]

Real-time vaccine surveillance systems can strengthen worldwide vaccination campaigns and resources, especially in low- and middle-income countries. Using this type of database can assist in tracking vaccination administrations, proper record keeping, and locating areas that have been missed in vaccination campaigns and assist with making sure that people in these areas are vaccinated.[56] Electronic immunization monitoring tools such as the National Immunization Management System (NIMS) or the Electronic Immunization Registries (EIR) were used in Pakistan to show that these types of real-time monitoring systems can be used to increase vaccination geographical coverage, especially in frequently missed areas in Pakistan, improving data-driven decision-making vaccination campaigns.[57]

At the start of the COVID-19 pandemic, the John Hopkins Coronavirus Resource Center (CRC) database tracked COVID-19 cases and deaths around the world in real time.[58] Previously, most governments only tracked their vaccine administration and shared their vaccination data with the WHO, hardly do we see this done in real-time. There are extant databases for tracking vaccination rates by country, and the vaccination rate per population of that country. Current COVID-19 vaccine administration rates can now be tracked in real time around the world. This is credited to the various databases such as "Our World in Data" that get vaccination data from governments and ministries of health from various countries and update daily a global data set on first and second doses of vaccination.[59] Databases such as these give not only the public but also health ministries and agencies around the world the ability to monitor vaccination rates on a global scale.

Contact tracing

Globally, many countries adopted contact tracing to interrupt chains of infection transmission of COVID-19.[60,61] According to the World Health Organizations guidelines, "at least 80% of new cases have their close contacts traced and in quarantine within 72 hours of case confirmation. At least 80% of contacts of new cases are monitored for 14 days."[62] The US and European CDCs recommended contact tracing as well, albeit with conflicting views. The US CDC stated, "When a jurisdiction does not have the capacity to investigate the majority of its new COVID-19 cases, case investigation and contact tracing may not be the most effective approach. At that point, jurisdictions should consider suspending or scaling down contact tracing activities and reimplementing strict mitigation measures (such as stay-at-home orders, business closures, and school closures) until transmission begins to decline."[63] The European CDC stated, "Contact tracing should still be considered in areas of more widespread transmission, wherever possible, and in conjunction with physical distancing measures. If resources are limited, high-risk exposure contacts of each case (close contacts) and contacts who are healthcare workers or work with vulnerable populations should be traced first, followed by as many low-risk exposure contacts as possible."[64] This contradiction in the guidelines can cause hesitancy among contact tracers. Contact tracing is optional in some countries such as the United Kingdom, Australia, Canada, Germany, and New Zealand, while others enforced it such as Russia, Singapore, South Korea, China, and Saudi Arabia (Table 17.1).[65]

CASE STUDY

Speed of detection and privacy

A systematic review of observational and modeling studies suggests that contact tracing is associated with better control of the COVID-19 pandemic.[66] Another systematic review included UK studies and concluded promising evidence of contact tracing and mass testing combined with social distancing and other preventative measures.[67] This case study reveals different contact tracing approaches in seven countries during COVID-19. It describes the population size, mode of technology, and policy enforcement measures of COVID-19 by country.

Background

- Contact tracing involves reporting individuals who contracted the virus or contacts of infected persons, which can be done manually where individuals enter their information or generate it automatically.[68]
- Some governments are restricting civil liberties of movement and enforcing the use of digital surveillance via several applications to contain the spread of the pandemic. Some countries have released apps with live maps of confirmed COVID-19 cases, location alerts, monitoring home isolation, government and self-reporting of signs and symptoms, and general health education.
- Some apps offer advanced services, including reporting physiological status, vital parameters, and virtual consultations. However, adoption rates varied across countries.[60,69,70]

Observations

TABLE 17.1 Examples of COVID-19 contact tracing initiated by country, 2020.

Country	Tool created in 2020	Technology	Mandated	Population in 2022	COVID-19 cases 12 Feb 2022[71]
China[72,73]	Alipay Health Code App	QR code	Yes, for all	1.4 billion[74]	106.863
Germany[73]	Corona Warn	Bluetooth technology	No	84.21 million[75]	12,126,991
Hong Kong[72]	StayHomeSafe	Wifi, Bluetooth and GPS—calibrated perimeter without tracking exact location	Yes, for 14 days for anyone arriving from overseas	7.59 million[76]	20,119
Japan[77]	COVID-19 Contact-Confirming Application (COCOA)	GPS-free Bluetooth enabled	No	125.85 million[78]	3,764,458
Saudi Arabia[79,80]	Tawakkalna app	QR code, GPS	Yes, for all	35.7 million[81]	724,525
Singapore[82]	Trace together app and tokens	Bluetooth. Geolocation not collected.	No, voluntary except for migrant workers	5.92 million[83]	449,570
United States[84]	Guidesafe (Alabama), AlohaSafe (Hawaii), COVID Notify (California), (Apple Google Bluetooth tool)	Most use Bluetooth	No	334.1 million[85]	79,228,628

(Continued)

CASE STUDY *(cont'd)*

Key issues

- Contact tracing has been facing criticism due to a lack of evidence of its effectiveness in controlling the spread of COVID-19. There are some limitations to contact tracing, such as incomplete or inaccurate recall of contact events and the reporting time, which can delay quarantine measures.[84]
- Privacy has been deemed a fair trade-off for a decrease in the potential effectiveness of contact tracing apps. This led to reluctance and hesitance to use such apps and lowered adoption rates.[60]
- In September 2020, a study in England revealed a code error in excel spreadsheet by health authorities, which caused a failure to trace 20% of cases in a timely manner. The random breakdown led to an increased number of cases and death.[86]

Interventions

- For decades, contact tracing has been the backbone of public health for controlling epidemics. The WHO recommends using contact tracing to break the chain of transmission through rapid identification of positive cases, followed by quarantine measures and required clinical attention.[87]
- According to a UK report, manual contact tracing would reduce the number of new infections by 5%–15%.[84]
- Countries developed contact tracing technologies while trying to ensure their citizens' privacy, but it remains a challenge to reach its full effectiveness. Some governments are enforcing using the apps, while others are keeping it optional.[60]

Lessons learned: What worked? What did not work?

- A balance between privacy and effectiveness is essential yet very challenging. Effectiveness requires certain measures such as enforcing quarantine, limiting activity, and notifying others about a positive case in the area, which can limit a person's privacy.[60]
- Most COVID-19 transmission occurred in the early stages before the onset of symptoms. The high transmissibility pattern before or near symptom onset with the short infectious period creates a need for more control strategies.[88]
- Although contact tracing is an effective public health tool for COVID-19 control, more evidence is crucial to understand how to optimize performance across different settings and contexts.[89]
- An agile contact tracing system and social distancing interventions can control COVID-19 transmission.[90]

Recommendations

- Take the epidemiologic profile and community needs into account before creating a contact tracing program. One size does not fit all.
- Encourage the use of contact tracing apps that protect the user's privacy.
- Modernize public health surveillance by using validated technologies.

Vision for the future

Surveillance must lead to immediate action and equitable sharing of the benefits of detecting outbreaks before they escalate. The world has seen a slow and inequitable response to the COVID-19 pandemic partly based upon the concentration of resources in a small number of countries. The authors envision creating capacities where there is a vast ability to generate and share data to warn the world about emerging threats. In this way, every country can collect, analyze, and use data over the continuum of disease prediction, detection, and response.

Expert high-level review panels on pandemic preparedness in 2021 coalesced around the need for a global surveillance and alert system. The Independent Panel for Pandemic Preparedness and Response (IPPPR) calls for a one health and shared data approach, leading to a global warning network with oversight by WHO. The G7 launched Global Pandemic Radar with a focused role on surveillance of pandemic prone disease. Further, the G20 and Pan European Commission both call for transforming surveillance and the powers of WHO: the Pan European review recommending an enhanced surveillance role of multilateral financial institutions.[91] Equitable collection and sharing of surveillance data creates distributed benefits and informs the use of governmental public health powers during a pandemic.[92] In the broad declaration of health emergencies to mitigate the spread of SARS-CoV-2, surveillance data were used by the governments for mandatory travel quarantines, bans on gathering, as well as business and school closures.[65] In like manner, equitable sharing of surveillance data helps reduce misinformation about any outbreak. And yet, while the world pauses to understand the commitment of the international community amid the shifting momentum of politicians—and considering IHR (2005) capacity has not yet been reached—action is now taking place to continue the modernization of public health surveillance.

The International Association of National Public Health Institutes (IANPHI), established in 2006, supports the IHR (2005) with increased technical capacity to detect and respond decisively to public health threats in low-resource countries.[84] Integration of one health data for analytics by the WHO Berlin Hub for Pandemic and Epidemic Intelligence, has the potential to generate valuable epidemic intelligence. Notably, the tripartite collaboration of FAO, OIE, and WHO has outlined a framework and considerations to establish a multidisciplinary and multisectoral (one health) surveillance system. The objective is a coordinated surveillance system for the timely detection of new and reemerging infectious disease events and the routine sharing of information across sectors.[93] Advanced analytics of one health data, as endeavored by the CDC and the WHO Berlin Hub, is drawing attention to the visualization of highly contextualized data and stimulation of political will for improved epidemic intelligence.[94]

Building national capacity

Surveillance cannot be left solely to global frameworks and structures. Country ownership of functional detection systems with capacities to verify alerts on their own and reduce the duration of a PHEIC is key to modernizing national disease surveillance systems.

Led by a senior health leader, the process of *owned* functionality may include agreeing on an established vision and innovation for surveillance, preparing an operational plan for change management, engaging relevant stakeholders to operationalize the plan, and communicating clearly on the process and implication of change to all levels. Implementation is dependent upon the availability of infrastructure and enabling system, to include a sufficiently trained and motivated workforce and sustainable financial support.

Official development assistance and philanthropy partners play critical roles in building, strengthening, and maintaining national surveillance systems. While policy-related decisions are made by governments, these partners (as stakeholders) provide technical support in assessing the situation, drawing a strategic and operational plan, advocating with senior government officials for increased funding, mobilizing additional resources to support implementation, as well as monitoring, evaluation, and learning. Development partners are expected to harmonize their plans and channel their support through implementing partners or work directly with government departments to increase a resource base. The support could include the provision of technical assistance and workforce development in surveillance and public health informatics. Partners could also use their influence to advocate for additional and sustainable financial resource mobilization from the government and other domestic entities. Partners should resist creating a parallel technical or funding structure to get the work done, as it undermines government efforts.

COVID-19 offers a perspective of comprehensive surveillance with the interdependency of humans, animals, and plants, as well as reliance on quality laboratory systems, genomic surveillance, and digital technologies. In analyzing the spread of SARS-COV-2 and how an outbreak became a pandemic, the chronology shows a surveillance system capable of predicting animal–human spillover and capable of detecting the spread and emergence of clinical cases of pneumonia of unknown etiology to generate global alerts.[9] The importance and increased acceptance of One Health strategies to detect pathogens with pandemic potential call for government investment to have systems and multisector workforces in place and utilized daily.

New generation of public health information systems

An efficient association of EMRs and interoperability among the currently present systems is required to achieve or maintain an effective public health surveillance system.[85,86] One such way is the full adoption and integration of IDSR in the United States, integrating all local, state, and national health departments while modernizing and updating public health surveillance.

Data should be entered once by human or veterinary healthcare providers or laboratory personnel into the secure web-based portal that will be developed in React Javascript. From here by utilizing deep learning-based artificial intelligence, automation processes will be developed. Through advanced data analytics techniques, data will be synchronously pushed in parallel to proprietary EMRs, laboratory information systems, and the One Health IDSR HIS.

All of these will operate in a parallel, concurrent cloud infrastructure with existing EMRs, veterinarians, and human and veterinary laboratory reporting systems. Instead of trying to harmonize and interoperate many isolated datasets having passed through various EMR systems, the One Health IDSR HIS architecture avoids all proprietary EMR impediments and provides a more effective, efficient, and secure mechanism of responding to outbreaks utilizing DHIS-2 and paves the way for implementation of IDSR.

A middleware platform will receive these data entered once from the secure web platform and provide concurrent, simultaneous reporting into the appropriate information systems (i.e., EMRs, LIMS, Vet Medical Record Systems and Laboratories, and the new One Health IDSR HIS).

The Middleware will then interact with robotic process automation (RPAs) developed individually for EMRs, LIMS, Vet Medical Record Systems and Laboratories. RPA will be utilized in workflow automation tools with a software developer who produces a list of actions to automate a task and interface to the back-end system using internal application programming interfaces (APIs) or dedicated scripting language.

There is an unrealized opportunity for improved data integration, sharing, and efficiency. Continuing without a One Health IDSR Health Information System (HIS) modernization plan and the pilot would exacerbate the ineffective and inefficient use of federal funding; local, state, and federal person-power; and modern information and communication technology (ICT) resources. The key to this framework is to create fast, responsive, and user-friendly applications.

The One Health IDSR HIS can be developed on the renowned DHIS-2 platform. Remembering that IDSR functions include *detect, confirm, analyze, and report*, the new HIS will perform these functions based on developed AI/ML. DHIS-2 is utilized in >70 countries around the world as their health information system.

Actions

In our vast interconnected space, a PHEIC has a consequential impact on the health of humans, animals, plants, and the environment across the world. In sum, the global community has a shared responsibility to avert this shared vulnerability.[95] According to IHR 2005, each sovereign country has an obligation to strengthen its national surveillance capacities to detect, verify, notify, and respond to any public health threats within the shortest possible time.[94]

Integrated surveillance systems with globally interconnected data systems for sharing information remain a futuristic goal. Nonetheless, it is a goal that government, academia, NGOs, and philanthropy should move swiftly toward, grounded in country priorities. Surveillance systems would ideally highlight five core principles (1) representative monitoring of human and animal populations, (2) appropriate laboratory confirmation capacity, (3) digitized data with unique health identifiers and privacy safeguards, (4) data transparency using standardized case definitions as well as informal data, and (5) adequate financing.[96] Morgan et al. suggest that National Public Health Institutes (NPHIs) could be mandated to "share information about transnational health threats with international bodies under the International Health Regulations (2005)" with full transparency and tracking of timeliness metrics, such as time to detect, report, investigate, and control disease outbreaks.[96] Experts and civil society organizations are being challenged to articulate and nurture new ideas to create a more coherent approach to integrated disease surveillance. There are multiple avenues to detect emergent patterns or unique disease occurrences, and all have value if there is an understanding of how data gets into a surveillance system, how it is validated, and how information is appropriately shared. Countries could build on current achievements using the integrated disease surveillance approach and sustain national surveillance capacities. National authorities could prioritize the following actions.

(1) Establish a one health integrated public health information surveillance and alert system with APIs connectivity and modern analytic capabilities
(2) Create collaborative commitments to data transparency and use of data sharing platforms
(3) Support the collaboration between extant government and academic epidemic intelligence hubs
(4) Fuller adoption of performance metrics, for example, outbreak timeliness metrics
(5) Expanded involvement of communities through well-managed, well-connected participatory/community surveillance systems to surveil public health threats daily

Surveillance initiatives must not rely solely on global governance, but rather be fully functional at the national and subnational interface to capture vital perspectives—boosting value regionally and globally. By strengthening local disease surveillance capacity and buy-in at a country level, the cycle of waiting for the official release of surveillance information from global structures can be broken with increased transparency and confidence based on each country's realities. In this way, we can be most successful at systematic reporting and aggregation of signaling alerts for the early identification of threats. Missing those signals can be catastrophic.

References

1. Pinner RW. Public health surveillance and information technology. *Emerg Infect Dis*. 1998;4(3):462–464.
2. Lee LM, Thacker SB. Public health surveillance and knowing about health in the context of growing sources of health data. *Am J Prev Med*. 2011;41(6):636–640.
3. Langmuir AD. The surveillance of communicable diseases of national importance. *N Engl J Med*. 1963;268:182–192.
4. Chan EH, Brewer TF, Madoff LC, et al. Global capacity for emerging infectious disease detection. *Proc Natl Acad Sci U S A*. 2010;107(50):21701–21706.
5. Hesse BW. Public health surveillance in the context of growing sources of health data: a commentary. *Am J Prev Med*. 2011;41(6):648–649.
6. McNabb S, Conde JM, Ferland L, et al. *Transforming Public Health Surveillance. Proactive Measures for Prevention, Detection, and Response*. 1st ed. Elsevier Inc; 2016:1–455.
7. Madoff LC. ProMED-mail: an early warning system for emerging diseases. *Clin Infect Dis*. 2004:227–232.
8. Kassler WJ. *Modernizing Public Health Surveillance*. Watson Health Perspectives; 2020.
9. Baker MG, Fidler DP. Global public health surveillance under new international health regulations. *Emerg Infect Dis*. 2006:1058–1065.
10. Halliday JEB, Hampson K, Hanley N, et al. Driving improvements in emerging disease surveillance through locally relevant capacity strengthening. *Science*. 2017:146–148.
11. Buehler JW, Hopkins RS, Overhage JM, Sosin DM, Tong V. *Framework for Evaluating Public Health Surveillance Systems for Early Detection of Outbreaks: Recommendations from CDC Working Group*. MMWR Recommendations and Report; 2004:1–11.
12. Samaan G, Patel M, Olowokure B, Roces MC, Oshitani H. Rumor surveillance and Avian influenza H5N1. *Emerg Infect Dis*. 2005:463–466.

13. Guerra J, Cognat S, Fuchs F. WHO Lyon Office: supporting countries in achieving the international health regulations (2005) core capacities for public health surveillance. *Rev Epidemiol Sante Publique*. 2018:S391.
14. Burkle FM. Global health security demands a strong international health regulations treaty and leadership from a highly resourced world health organization. *Disaster Med Public Health Prep*. 2015:568−580.
15. Craven M, Sabow A, Van der Veken L, Wilson M. *Not the Last Pandemic: Investing Now to Reimagine Public-Health Systems*. McKinsey & Company; 2021.
16. WHO. *Fifty-Eighth World Health Assembly Geneva: Revision of the International Health Regulations*. Geneva, Switzerland: World Health Organization; 2005.
17. Fall IISS, Rajatonirina S, Yahaya AA. Integrated Disease Surveillance and Response (IDSR) strategy: current status, challenges and perspectives for the future in Africa. *BMJ Glob Health*. 2019;4.
18. Integrated Disease Surveillance and Response (IDSR). Division of Global Health Protection. Global Health. CDC.
19. Burci GL, Moon S, Ricardo AC, Neumann C, Bezruki A. *Envisioning an International Normative Framework for Pandemic Preparedness and Response: Issues, Instruments and Options*. Geneva: Graduate Institute of International and Development Studies, Global Health Centre; 2021.
20. Crawley AW, Divi N, Smolinski MS. Using timeliness metrics to track progress and identify gaps in disease surveillance. *Health Secur*. 2021:309−317.
21. Food and Agriculture Organization of the United Nations. https://www.fao.org/one-health/en/.
22. *WHO, Germany Open Hub for Pandemic and Epidemic Intelligence in Berlin*. World Health Organization (WHO); 2021.
23. Evans RS. Electronic health records: then, now, and in the future. *Yearb Med Inform*. 2016;25(S 01):S48−S61.
24. Chamola V, Hassija V, Gupta S, Goyal A, Guizani M, Sikdar B. Disaster and pandemic management using machine learning: a survey. *IEEE Internet Things J*. 2020;8(21):16047−16071.
25. Shamsolmoali P, Kumar Jain D, Zareapoor M, Yang J, Afshar Alam M. High-dimensional multimedia classification using deep CNN and extended residual units. *Multimed Tool Appl*. 2019;78(17):23867−23882.
26. Yue X, Wang H, Jin D, Li M, Jiang W. Healthcare data gateways: found healthcare intelligence on blockchain with novel privacy risk control. *J Med Syst*. 2016;40(10):1−8.
27. McNabb SJ, Chungong S, Ryan M, et al. Conceptual framework of public health surveillance and action and its application in health sector reform. *BMC Publ Health*. 2002;2(1):1−9.
28. Fall IS, Rajatonirina S, Yahaya AA, et al. Integrated Disease Surveillance and Response (IDSR) strategy: current status, challenges and perspectives for the future in Africa. *BMJ Glob Health*. 2019;4(4):e001427.
29. Groseclose SL, Buckeridge DL. Public health surveillance systems: recent advances in their use and evaluation. *Annu Rev Publ Health*. 2017;38:57−79.
30. WHO. *Epidemic Intelligence Form Open Sources (EIOS): Zero Impact from Health Threats*. World Health Organization (WHO); 2021.
31. WHO. *Rapid Risk Assessment of Acute Public Health Events*. Geneva, Switzerland: World Health Organization (WHO); 2012:44, 2012.
32. Linge JP, Steinberger R, Weber TP, et al. Internet surveillance systems for early alerting of health threats. *Euro Surveill*. 2009;14.
33. Spagnolo L, AbdelMalik P, Doherty B, et al. *Integration of the Epidemic Intelligence from Open Sources (EIOS) System and the INFORM Suite: Enhancing Early Warning with Contextual Data for Informed Decision Making*. 2020:2020.
34. *GOARN. Go.Data*. World Health Organization (WHO); 2020.
35. Lorthe TS, Pollack MP, Lassmann B, et al. Evaluation of the Epicore outbreak verification system. *Bull World Health Organ*. 2018:327−334.
36. ECDC. *epitweetr Tool*. European Center for Disease Prevention and Control; 2020.
37. Marquet RL, Van Noort SP, Bartelds AI. Internet-based monitoring of influenza-like illness (ILI) in the general population of The Netherlands during the 2003−2004 influenza season. *BMC Publ Health*. 2006;14:19162.
38. Smolinski MS, Crawley AW, Baltrusaitis K. Flu near You: crowdsourced symptom reporting spanning 2 influenza seasons. *Am J Publ Health*. 2015:2124−2130.
39. Wójcik OP, Brownstein JS, Chunara R, Johansson MA. Public health for the people: participatory infectious disease surveillance in the digital age. *Emerg Themes Epidemiol*. 2014;11:1−7.
40. Development PfD. Case Study : Cambodia 's 115 Hotline Expanding Access to the Reporting and Detecting of Disease Outbreak.

41. Crawley A, Divi N. *Surveillance Technology in Thailand, Cambodia, and Tanzania: Case Study from Ending Pandemics*. Exemplars in Global Health; 2021.
42. Lazaro F. *Thailand, Tracking Animal Health to Prevent Outbreaks of Human Disease*. Public Broadcasting Service; 2019.
43. Leal-Neto OB, Dimech GS, Libel M. Saúde na Copa: the world's first application of participatory surveillance for a mass gathering at FIFA world cup 2014, Brazil. *JMIR Public Health Surveill*. 2017;3.
44. Leal-Neto OB, Dimech GS, Libel M, Oliveira W, Ferreira JP. Digital disease detection and participatory surveillance: overview and perspectives for Brazil. *Rev Saude Publica*. 2016:17.
45. Lwin MO, Jayasundar K, Sheldenkar A, et al. Lessons from the implementation of Mo-Buzz, a mobile pandemic surveillance system for dengue. *JMIR Public Health Surveill*. 2017;3:65.
46. Schmidt C, Phippard A, Olsen JM, et al. Kidenga: public engagement for detection and prevention of Aedes-borne viral diseases. *Online J Public Health Inform*. 2017;9.
47. Robishaw JD, Alter SM, Solano JJ, et al. Genomic surveillance to combat COVID-19: challenges and opportunities. *Lancet Microbe*. 2021:e481–e484.
48. Andersen KG, Rambaut A, Ian Lipkin W, Holmes EC, Garry RF. The proximal origin of SARS-CoV-2. *Nat Med*. 2020:450–452.
49. Krause PR, Fleming TR, Longini IM, Peto R, Briand S, Heymann DL, Beral V, Snape MD, Rees H, Ropero AM, Balicer RD, Cramer JP, Muñoz-Fontela C, Gruber M, Gaspar R, Singh JA, Subbarao K, Van Kerkhove MD, Swaminathan S, Ryan MJ, Henao-Restrepo AM. SARS-CoV-2 Variants and Vaccines. *N Engl J Med*. 2021 Jul 8;385(2):179–186. https://doi.org/10.1056/NEJMsr2105280. Epub 2021 Jun 23. PMID: 34161052; PMCID: PMC8262623.
50. European Medicines Agency Science Medicines Health. *Safety of COVID-19 Vaccines*. European Medicines Agency; 2021.
51. Albogami Y, Alkofide H, Alrwisan A. COVID-19 vaccine surveillance in Saudi Arabia: opportunities for real-time assessment. *Saudi Pharmaceut J*. 2021:914–916.
52. CDC. *Vaccine Adverse Event Reporting System (VAERS)*. Center for Disease Control and Prevention (CDC); 2021.
53. CDC. *Vaccine Safety Datalink (VSD)*. Center for Disease Control and Prevention (CDC); 2020.
54. Holtgrave DR, Vermund SH, Wen LS. Potential benefits of expanded COVID-19 surveillance in the US. *JAMA*. 2021:381–382.
55. Hazell L, Shakir SAW. Under-reporting of adverse drug reactions: a systematic review. *Drug Saf*. 2006:385–396.
56. Desmon S. *Improving Vaccine Campaigns through Real-Time Monitoring*. Johns Hopkins Center for Communication Programs; 2020.
57. Siddiqi DA, Abdullah S, Dharma VK, et al. Using a low-cost, real-time electronic immunization registry in Pakistan to demonstrate utility of data for immunization programs and evidence-based decision making to achieve SDG-3: insights from analysis of Big Data on vaccines. *Int J Med Inf*. 2021;149:104413.
58. JHU. *Johns Hopkins Coronavirus Resource Center Wins Research America Award*. John Hopkins University; 2021.
59. Mathieu E, Ritchie H, Ortiz-Ospina E, et al. A global database of COVID-19 vaccinations. *Nat Human Behav*. 2021:947–953.
60. Alanzi T. A review of mobile applications available in the app and Google play stores used during the COVID-19 outbreak. *J Multidiscip Healthc*. 2021:45–57.
61. Jnr BA. Use of telemedicine and virtual care for remote treatment in response to COVID-19 pandemic. *J Med Syst*. 2020;44:132.
62. WHO. *Public Health Criteria to Adjust Public Health and Social Measures in the Context of COVID-19*. World Health Organization (WHO); 2020:4, 2020.
63. CDC. *When to Initiate Case Investigation and Contact Tracing Activities*. Center for Disease Control and Prevention (CDC); 2020.
64. ECDC. *Contact Tracing: Public Health Management of Persons, Including Healthcare Workers, Having Had Contact with COVID-19 Cases in the European Union- Second Update*. Stockholm: European Centre for Disease Prevention and Control (ECDC); 2020, 2020.
65. Setó E, Challa P, Ware P. Adoption of COVID-19 contact tracing apps: a balance between privacy and effectiveness. *J Med Internet Res*. 2021;23.
66. Juneau C-E, Briand A-S, Pueyo T, Collazzo P, Potvin L. *Effective Contact Tracing for COVID-19: A Systematic Review. medRxiv*. Cold Spring Harbor Laboratory Press; 2020:2020.

67. Mbwogge M. Mass testing with contact tracing compared to test and trace for the effective suppression of COVID-19 in the United Kingdom: systematic review. *JMIRx Med*. 2021:e27254.
68. Braithwaite I, Callender T, Bullock M, Aldridge RW. Automated and partly automated contact tracing: a systematic review to inform the control of COVID-19. *Lancet Digit Health*. 2020:e607–e621.
69. Fulbright NR. *Contact Tracing Apps: A New World for Data Privacy*. Norton Rose Fulbright; 2021.
70. Sharma T, Bashir M. Use of apps in the COVID-19 response and the loss of privacy protection. *Nat Med: Nat Med*. 2020:1165–1167.
71. COVID Live – Coronavirus Statistics – Worldometer.
72. Utzerath J, Bird R, Cheng G, Ohara J. *Contact Tracing Apps in China, Hong Kong, Singapore, Japan, and South Korea*. Freshfields Bruckhaus Deringer; 2020.
73. Kostka G, Habich-Sobiegalla S. *Times of Crisis: Public Perceptions towards COVID-19 Contact Tracing Apps in China, Germany and the US. SSRN Electronic Journal*. Elsevier BV; 2020.
74. *China Population*. Worldometer; 2022.
75. *Germany Population*. Worldometer; 2022.
76. *Hong Kong Population*. Worldometer; 2022.
77. Nakamoto I, Jiang M, Zhang J, et al. Evaluation of the design and implementation of a peer-to-peer COVID-19 contact tracing mobile app (COCOA) in Japan. *JMIR mHealth uHealth*. 2020;8.
78. *Japan Population*. Worldometer; 2022.
79. Alassaf N, Bah S, Almulhim F, AlDossary N, Alqahtani M. Evaluation of official healthcare informatics applications in Saudi Arabia and their role in addressing COVID-19 pandemic. *Healthc Inform Res*. 2021:255–263.
80. Alsyouf A. *Mobile Health for COVID-19 Pandemic Surveillance in Developing Countries: The Case of Saudi Arabia*. Solid State Technology; 2020:2474–2485.
81. *Saudi Arabia Population*. Worldometer; 2022.
82. Cho H, Ippolito D, Yu YW. *Contact Tracing Mobile Apps for COVID-19: Privacy Considerations and Related Trade-Offs*. 2020.
83. *Singapore Population*. Worldometer; 2022.
84. IANPHI. Strengthening Public Health Institutes: Progress Report 2014-2015. International Association of National Public Health Institutes: International Association of National Public Health Institutes.
85. *United States Population*. Worldometer; 2022.
86. Fetzer T, Graeber T. Measuring the scientific effectiveness of contact tracing: evidence from a natural experiment. *Proc Natl Acad Sci USA*. 2021;118.
87. Coronavirus Disease (COVID-19): Contact Tracing.
88. Cheng HY, Jian SW, Liu DP, Ng TC, Huang WT, Lin HH. Contact tracing assessment of COVID-19 transmission dynamics in Taiwan and risk at different exposure periods before and after symptom onset. *JAMA Intern Med*. 2020:1156–1163.
89. Hossain AD, Jarolimova J, Elnaiem A, Huang CX, Richterman A, Ivers LC. Effectiveness of contact tracing in the control of infectious diseases: a systematic review. *Lancet Public Health*. 2022:e259–e273.
90. Yuan HY, Blakemore C. The impact of contact tracing and testing on controlling COVID-19 outbreak without lockdown in Hong Kong: an observational study. *Lancet Public Health*. 2022, 100374.
91. *Wellcome. Improving Global Pandemic Preparedness by 2025*. Wellcome; 2021:2021.
92. Edelstein M, Lee LM, Herten-Crabb A, Heymann DL, Harper DR. Strengthening global public health surveillance through data and benefit sharing. *Emerg Infect Dis*. 2018:1324–1330.
93. OIE WHO FAO. Taking a Multisectoral, One Health Approach: A Tripartite Guide to Addressing Zoonotic Diseases in Countries. In: *The Food and Agriculture Organization of the United Nations, the World Organisation for Animal Health*. The World Health Organization; 2019, 2019.
94. WHO. *WHO and Switzerland Launch Global BioHub for Pathogen Storage, Sharing and Analysis*. World Health Organization (WHO); 2021.
95. International Health Regulations. *Areas of Work for Implementation*. 2005, 2007.
96. Morgan J, Kennedy ED, Pesik N, et al. Building global health security capacity: the role for implementation science. *Health Secur*. 2018:S5–S7.

CHAPTER

18

Data for public health action: Creating informatics-savvy health organizations to support integrated disease surveillance and response*

Nancy Puttkammer[1], *Phiona Vumbugwa*[1], *Neranga Liyanaarachchige*[2], *Tadesse Wuhib*[2], *Dereje Habte*[3], *Eman Mukhtar Nasr Salih*[4], *Legesse Dibaba*[5], *Terence R. Zagar*[6] *and Bill Brand*[7]

[1]Department of Global Health, University of Washington, Seattle, WA, United States; [2]Division of Global HIV and TB, Global Health Center, United States Centers for Disease Control and Prevention, Atlanta, GA, United States; [3]US Centers for Disease Control and Prevention – Ethiopia, Addis Ababa, Ethiopia; [4]Hubert H. Humphrey Fellowship Program, Rollins School of Public Health, Emory University, Atlanta, GA, United States; [5]Ministry of Health, Addis Ababa, Ethiopia; [6]Independent Contractor seconded to the Division of Global HIV and TB, Global Health Center, United States Centers for Disease Control and Prevention, Atlanta, GA, United States; [7]The Task Force for Global Health/Public Health Informatics Institute, Decatur, GA, United States

Introduction

Public health is an information-dependent discipline. Effective detection and response to public health threats have always required quality information—data that are timely,

* Contribution of authors Nancy Puttkammer, Neranga Liyanaarachchige, Tadesse Wuhib, Dereje Habte, and Terence R. Zagar is subject to public domain.

Modernizing Global Health Security to Prevent, Detect, and Respond
https://doi.org/10.1016/B978-0-323-90945-7.00005-1

accurate, and complete. Failures to obtain, understand, and act on quality information are apparent in an information age when expectations for timely and abundant data are high. The variety, volume, and velocity of information required for emergency response place great demands on public health organizations to modernize data collection, analysis, visualization, and dissemination tools, and to provide sufficient staff with the right skills. The recent severe acute respiratory syndrome COVID-19 pandemic demonstrated that current public health systems lacked timely access to evolving information and epidemiology, as well as the tools to quickly share and exchange newly discovered data at all levels—local, regional, national, and global. Experience from the COVID-19 pandemic and other major outbreaks has highlighted the critical need for rapid and accurate collection, cleaning, analysis, and dissemination of health information. Achieving such speed and accuracy requires the digitization of a wide range of health information, including lab testing, treatment, vaccination status, vaccine safety surveillance, hospitalizations, and deaths, among other information. Achieving a transformation of health information from paper to digital forms requires more than well-designed and reliable information technologies, it also requires a clear vision of the desired outcomes, effective governance, and a skilled workforce.

Public health informatics is the scientific discipline that guides and supports improved capabilities in using health information and in designing effective health information systems. Informatics is about developing the capabilities to transform data into *actionable* information and knowledge for decision-making and problem-solving. Benefiting from informatics principles and methods requires a disciplined assessment of current capabilities and a clear articulation of the vision and desired goals for improved information systems, information processes, workforce capabilities, organizational outcomes, and, ultimately, health outcomes—ultimately leading to a robust public health system and healthy communities. This chapter discusses one framework—the informatics-savvy health organization (ISHO) framework—for achieving effective use of health information toward timely disease surveillance and response and improved care delivery and population health outcomes.

Literature review and gap analysis

Information needs for integrated disease surveillance and response

Data and information are at the center of mitigating threats to public health. The World Health Organization's (WHO) Integrated Disease Surveillance and Response (IDSR) strategy champions a single infrastructure for active and passive surveillance of health risks, involving multiple levels of the health system, rather than vertical, disease-specific surveillance initiatives and structures. The IDSR framework recognizes that improving data flow, quality, and use is critical to an effective response to both infectious and chronic disease threats.[1] The West African Ebola crisis of 2013–15 and the COVID-19 pandemic starting in 2019 each demonstrated challenges and opportunities in data and information management. Initially, the international response to the Ebola crisis was characterized by a lack of digitized data to track cases and their contacts by time and geography, availability of supplies, and treatment and community outreach activities. Gaps in digital connectivity and a framework for sharing data across sectors and across national boundaries hindered planning and response.[2] While the Ebola crisis spurred investment in data systems as part of the Global

Health Security Agenda,[3–5] COVID-19 and other recent emergent pandemics, including Zika and Mpox (formerly known as monkeypox), have highlighted the persistent gaps in data governance and data systems to enable effective detection and response. In the United States, challenges in data sharing across jurisdictions and the lack of standardized data disaggregated by risk groups made it difficult to identify disparities in COVID-19 infections, hospitalizations, and vaccination coverage, as they evolved over time.[6,7] Challenges with case reporting formats made it difficult to differentiate hospitalizations for conditions directly due to COVID-19 versus cases of hospitalized patients where COVID-19 was detected incidentally.[8] Globally, there remains likely underestimation of COVID-19 morbidity and mortality, as well as challenges in tracking emergence and progression of COVID-19 variants due to insufficient surveillance and reporting systems.[9–11]

Digital health investments in LMICs

Addressing such deficiencies in information management and use has led to several global policy initiatives aimed at increasing digital health capabilities. WHO defines digital health as *"The systematic application of information and communications technologies, computer science, and data to support informed decision-making by individuals, the health workforce, and health systems, to strengthen resilience to disease and improve health and wellness."*[12] Governments play a fundamental role in establishing an enabling environment for digital health.[12a,12b] A majority of low- to middle-income country (LMIC) governments have developed national digital health strategies,[12a,13,14] often with reference to the WHO-International Telecommunications Union (ITU) National eHealth Strategy Toolkit covering seven components: (1) leadership and governance; (2) strategy and investment; (3) services and applications; (4) standards and interoperability; (5) infrastructure; (6) legislation, policy, and compliance; and (7) workforce.[14a] In many cases, there is a recognition of the need for a national enterprise architecture and public policies covering data protection and confidentiality, unique person identification, and standards and security for interoperability.

In recent years, donors and funding mechanisms such as the United States President's Emergency Plan for AIDS Relief (PEPFAR), WHO, the United Nations Global Fund for AIDS, Tuberculosis, and Malaria (GFATM), the United Kingdom Department for International Development (DFID), the Bill and Melinda Gates Foundation (BMGF), and others have aligned investments around the Principles for Digital Development, including support of open-source digital "global goods."[15] These investments have covered interventions across WHO's digital health taxonomy.[15a] Digital health investments include large-scale deployments of electronic medical record systems (EHRs),[16–20] laboratory information systems (LIS),[21–23] logistics/supply chain management information systems (LMIS),[24–26] human resource information systems (HRIS),[27,28] community health information systems (CHIS),[29, 29a] health management information systems for aggregate health system data (HMIS),[30,31] and digital registries and surveillance systems.[3,32] Many digital health interventions scaled in LMICs now include mobile health capability and use data warehouses with dashboards and visualization to support data use for decision-making. However, the scale of adoption, maturity, sustainability, and integration of these tools is highly variable.[14,18]

In terms of digital registries and surveillance systems to support the IDSR approach, there has been promising advancement in several West African countries since the Ebola crisis. Starting in 2016 in Sierra Leone, the Ministry of Health and Sanitation (MoHS) collaborated with WHO and CDC to introduce a health facility-based electronic IDSR (eIDSR) solution using Android mobile devices, which was rolled out nationally at a relatively modest per-facility cost of $1021.[3–5] By 2019, 88% of health facility monthly surveillance reports were submitted by the mobile application.[3] An after-action review of a successfully contained Lassa Fever outbreak in Sierra Leone identified that the national eIDSR system played a key role in rapid reporting of the outbreak.[33] In Guinea, there was a similar national investment post-Ebola in digitizing routine surveillance data using District Health Information System Version 2 (DHIS2).[34] During 2018–20, completeness of weekly surveillance aggregate data reporting in Guinea's 38 health districts averaged 96.6% while timeliness averaged 71.3%; however, completion of individual-level case reporting via DHIS2 Tracker tended to be considerably lower (e.g., 48% for acute flaccid paralysis and 5% for acute diarrhea).[32] The presence of the DHIS2 system in Guinea at the onset of the COVID-19 pandemic enabled the country to rapidly adapt the system to collect COVID-19 related data, with minimal external technical assistance.[32]

Beyond West Africa, a review of digital tools potentially suitable for COVID-19 surveillance and response in 28 African countries identified 30 tools already in use as of early 2021.[35] While several were used for surveillance of a single disease, such as malaria or meningitis, a majority (17/30, 57%) supported integrated surveillance and included the capability for case-based surveillance including tracking of symptoms, hospitalization, exposures, events; geolocation and mapping; data synchronization with a central server using 3G, 4G, or Wifi; data aggregation; and dashboard capability. The DHIS2, Sormas, Go.Data and CommCare systems each supported a majority of critical functions, and have been most widely used in LMICs as part of the COVID-19 response. COVID-19 also spurred the development of digital vaccine certificates as digital innovation.[14]

Gaps in data systems for pandemic preparedness and response in LMICs

Despite the promising advancement in digital health ecosystems in LMICs, as described above, there remain important gaps in data governance, data systems, skilled workforce, and information quality and use, leading to persistent public health vulnerability.

- Governance structures for data systems remain weak.[2] While many countries have national digital health strategies, the quality and detail of these strategies are highly variable, and many plans are not costed.[14] Gaps in policies on data protection and data sharing lead to a lack of trust in security of data systems and inhibit potentially fruitful exchange of data between electronic health records used for clinical care and public health surveillance systems.[36]
- Many LMICs lack a coherent national enterprise architecture with shared data services. Enterprise architecture is defined as "a blueprint of business processes, data, systems, and technologies ... to support the workflow and roles of people in a large enterprise, such as a health system."[12] Having shared data services that use common security standards, communication protocols, and health terminologies can ensure that components can be leveraged rather than redeveloped for each health program need.[12b,37]
- Profound gaps in information technology infrastructure persist in many LMICs.[1] As of 2021, the International Telecommunications Union (ITU) estimated that 95% of the

world's population lived in areas covered by broadband, but 2.9 billion people remained offline based on lack of access to digital tools.[14] As a result, improving the digital connectivity infrastructure for every health worker and health facility is identified as the digital transformation priority requiring the greatest share of resources, approximately two-thirds to four-fifths of all investment required for digital transformation.[14]

- Information systems lack interoperability, or the ability to automatically exchange data. EHR, lab, and surveillance data systems are often unlinked, leading to delays in diagnostic data needed for appropriate clinical management and for public health surveillance. Absence of data standards, including for unique person identification, contributes to this gap.[12b] The OpenHIE framework for health information exchange[37a] and emergent work on machine-readable clinical guidelines based on common minimum datasets offer potential to advance interoperability,[38] but applications of these approaches at scale are limited.
- Many health information systems remain siloed, leading to both gaps in program overage and duplicate data capture that places great burden on frontline health workers.[1] Many EHR systems have been oriented to HIV epidemic control and lag in terms of reflecting the full array of features needed for primary care.[16]
- Limited integration of client-level data systems and aggregated data systems is typical, and contributes to inaccurate transcription and calculation errors, data lags, high burden of reporting.[39] Multiple reporting formats and manual processes are still too often required to summarize data for decision-making.[1]
- Local capacity to lead and manage data systems is often limited.[2,40] A promising training initiative in Kenya involved training of surveillance officers in 13 intervention counties with low performance on timeliness and completeness of routine surveillance using the DHIS2 system, finding that completeness and timeliness each improved by 14% over comparison counties following training.[41] However, sustainable models for local workforce development to support digital health interventions, including skill sets for software development, IT system administration, and frontline use of systems and data, are often lacking.[14]
- There is often underutilization of IDSR data, based on lack of skills and limited institutional culture of data use.[1,42] In Ethiopia, a study of data use from HMIS systems at the district level found low data use through the cycle of problem identification, problem prioritization, action planning, and implementation monitoring: while 52% of districts reviewed monthly performance, only 20% prepared action plans based on this review.[43] Although training can help to promote data use, it is often treated as a one-time strategy rather than institutionalized.[1]
- There remains limited integration of animal and human health data. While most surveillance efforts focus on humans, more than 60% of emerging diseases from 1940 to 2004 were caused by zoonotic pathogens.[1]
- There is poor inclusion of private health services in routine public surveillance and reporting systems.[1] Moreover, the private sector has low participation in providing market-based innovations for data collection, management, use, and visualization.[2] While open-source software communities offer promise for efficient learning about and

sharing of technical solutions, disease-specific categorical funding challenges digital health implementations from collaborating on software and technical solutions.

Vision for the future: Building informatics-savvy health organizations

Developing realistic digital health strategies and having the skills and methods to achieve them requires the effective application of informatics principles and methods, as well as strong, ongoing leadership support. There are no shortcuts to having the capabilities to transform data into *actionable* information and knowledge for decision-making and problem-solving. Implementation of digital health systems without an appropriate foundational framework can lead to duplication of investment and systems, the inability to exchange the data needed to meet both patient care and public health needs, and information technologies that do not serve the spectrum of public health needs, especially for rapid data collection and reporting of novel diseases. It is essential when optimizing digital health systems to first step back, and engage public health professionals at all levels to determine the data, workflows, and system functionality needed by those professionals as information users.[44,45] The COVID-19 pandemic has provided a critical opportunity for the reassessment of global and national digital health systems.

Role of informatics and data modernization in public health

The scientific discipline which supports the digital transformation in public health is public health informatics, commonly defined as "... the systematic application of information, computer science, and technology to public health practice, research, and learning".[46] A more operational definition is that public health informatics "... is the science of how to use data, information, and knowledge to improve human health and the delivery of health care services" (https://phii.org/how-we-do-it/defining-public-health-informatics/).

Informatics lies at the intersection of information, technologies, people, processes, and policies. All must work together for optimized organizational performance and results and to address information-related challenges such as those cited above.

A common informatics challenge is ensuring data quality; that is, ensuring that all data elements being captured, regardless of source, are timely (available when and where needed), accurate (faithfully representing the findings or measurements), and complete (no data elements missing from a record). A foundational strategy for achieving data quality is the adoption of health and other types of data standards—uniform ways for coding data elements so the data can be used by multiple information systems and for multiple purposes. Data standardization is thus critical to achieving another informatics challenge: system interoperability, or the ability to meaningfully exchange digital data between disparate information systems (systems can "talk to each other") while maintaining a consistent and accurate meaning.

Ensuring the confidentiality and security of the data is yet one more area in which informatics is concerned. Discussing each of these areas, and many more, is beyond the scope of this chapter but provides a basic foundation from which to discuss building informatics capabilities (the right skills and competencies in people and the right functionality in systems) and capacity (the right numbers of people and systems) in health organizations to be better prepared for future crises.

From understanding the data and technology forces impacting public health practice to simply knowing the right questions to ask to ensure appropriate allocations of resources, public health leaders, professionals, and organizations need to understand the role informatics can play in improving their organizational capacity to effectively manage, use, and safeguard information to respond effectively to emerging and present threats. The next section presents one approach for applying informatics principles and methods to rigorous assessment, planning, and implementation of digital health and data modernization initiatives.

Informatics-savvy health organization approach

Experience in both the United States and LMIC countries highlight three broad and interconnected areas for developing organizational robustness in public health informatics.

- An organization-wide vision, policy, and governance for how the organization uses information and information technologies as strategic assets.
- A skilled workforce; and
- Effectively designed and used information systems[47]

These mutually supportive capability areas are the three pillars of what is termed an *informatics-savvy health organization* (ISHO), "one that can obtain, effectively use, and securely exchange information to improve public health practice and patient and population health outcomes" (Fig. 18.1).[47]

An ISHO self-assessment tool was developed and tested by the Public Health Informatics Institute to enable health organizations to rigorously assess their current informatics capabilities (https://phii.org/module-4/self-assessment-tools/). The assessment uses a maturity framework, which can serve to benchmark and motivate evolution from lower to higher capability across technical, human, and organizational domains.[48–50] Specifically, the ISHO tool was adapted from the Capability Maturity Model developed initially by Carnegie-Mellon University Software Development Institute (https://resources.sei.cmu.edu/library/asset-view.cfm?assetid=6109). The model was selected because it was specifically developed for assessing capabilities related to software development processes, it had been subsequently tested across many process improvement applications, it provided a practical framework for developing a maturity questionnaire, and it provided for inter- and intraorganizational measurement consistency over time.

Each of the three pillars lists specific informatics core capabilities (see Appendix 1 and https://phii.org/module-4/self-assessment-tools/), with the idea that staff and stakeholders participating in the assessment would agree on the level which best represented their *lowest* capability level in each core area (since an organization's ability to act can be—and often is—limited by its weakest area). Ideally, ISHO assessments are government-led and include individuals from across the organization (since capabilities can differ widely among units within an organization), as well as stakeholders (since nongovernmental implementing

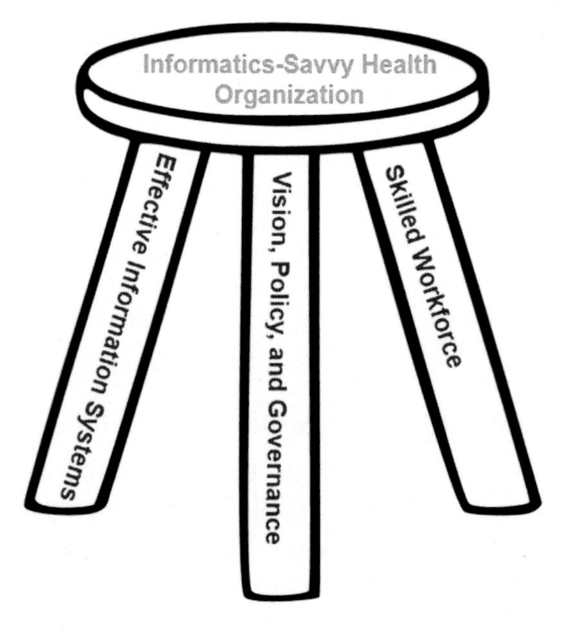

FIGURE 18.1 Three pillars of an informatics-savvy health organization (ISHO) framework. *Adapted by Dr. Herman Tolentino for the US CDC Public Health Informatics Fellowship Program.*

partners, funders, and multilateral development agencies each hold important perspectives in action planning and mobilization of resources toward shared goals).

The Public Health Informatics Institute tested and refined the overall model, the organizational self-assessment tool, and the process for conducting an organizational assessment in a variety of large and small, urban and rural, and state and local health departments across the United States. In 2018–22, the US CDC embraced and adapted the ISHO framework under its initiatives to strengthen national health information systems in LMICs as part of the President's Emergency Plan for AIDS Relief.[51] CDC adapted the framework for focus on HIV-related information systems, and for flexible use at multiple levels of the health system with different organizational types, including individual health facilities, district or regional health departments, and national Ministries of Health. The following case study from Ethiopia describes the ISHO assessment results for an LMIC.

ETHIOPIA CASE STUDY

Background

In 2018, 10 senior informatics leaders from Ethiopia conducted an ISHO assessment during the Intergovernmental Learning Exchange to Advance Data-driven Decision-making (I-LEAD), a training and capacity development workshop. Participants represented the Ministry of Health (MoH), Ethiopian Public Health Institute (EPHI), CDC Ethiopia, and ICAP (formerly the International Center for AIDS Care and Treatment Programs) at Columbia University. The CDC Health Informatics Team Lead coordinated the assessment. Steps in the assessment included:

ETHIOPIA CASE STUDY *(cont'd)*

- Virtual orientation of participants by CDC Headquarters Team
- Sharing of reference documents and assessment tools
- In-person review of above-site (or above-health facility) assessment tools
- Individual scoring by team members
- Compilation and averaging of the individual scores for each element
- Team meeting to review the deidentified average scores and reach on consensus of final scoring
- Prioritization of major gaps based on the ISHO scores
- Defining an implementation plan for the priorities

The team discussion and planning phase took 3 weeks, and the assessment process (individual and group work) took 2 weeks.

Observations

The final scoring decision for each element in the assessment relied on documented evidence. Prioritization considered the intensity of gaps on the ground, feasibility of available resources, technical capacity and timeline, and relevance for PEPFAR-supported HIS.

Priorities

Priorities for action that were incorporated in MoH, EPHI, CDC, and ICAP annual plans included.

- Governance: operationalize eHealth government structures, build effective partnerships, develop eHealth enterprise Architecture (EA) standards, develop data sharing agreements, achieve sustainable financing, and advance the security of personally identifiable information.
- Information systems: develop health information exchange (HIE), support standardized patient management systems, promote informatics practices, conduct an inventory of information systems, assess information system usability and effectiveness
- Health workforce: create eHealth/informatics workforce development strategy, create internship opportunities, incorporate informatics training in Field Epidemiology Training Program (FETP), develop job classifications, and adopt an e-learning portal.

Interventions and actions

Motivated by ISHO results from 2018, US CDC and the MoH in Ethiopia aligned priorities in the MoH national and PEPFAR annual plans and implemented the following interventions and activities:

Vision, policy, and governance

- Developed strong partnerships and collaboration between Ethiopia Ministry of Health, Addis Ababa City Administration Health Bureau (AACAHB), EPHI, University of Gondar, Jembi Health Systems, CDC, ICAP at Columbia University, and others to bring interventions to fruition.
- Developed the National Electronic Health Record (EHR) Standard.
- Adopted the ISHO assessment tool for nationwide above-site and site-level assessment.

(Continued)

ETHIOPIA CASE STUDY *(cont'd)*

- Adapted the Africa CDC Health Information Exchange (HIE) Policy and Standards Framework to the Ethiopia setting.

Effective information systems

- Advanced the capabilities of a national data warehouse for HIV case—surveillance.
- Enhanced the unique patient identification protocols for the deduplication of records within the data warehouse.
- Incorporated HIV recent infection testing into its routine surveillance systems and implemented a REDCap system to manage the data.

Skilled workforce

- Conducted Growing Expertise in E-Health Knowledge and Skills (GEEKS) training at the country level, with participation from public and partner organizations supporting eHealth.
- Conducted informatics training for health workers on IT project management and Informatics for Leaders (I4L) training for informatics and non-informatics professionals.

Lessons learned

ISHO assessment provided a stakeholder engagement platform for organizations involved in digital health and HIS. There have been increased efforts to collaborate and reduce duplication of efforts, hence pulling resources together in a more efficient and effective way. The collaboration opened an opportunity to discuss issues impacting HIS, specifically health workforce and governance gaps stalling systems strengthening. Ethiopia is planning a subnational ISHO assessment in 2023.

Virtual sharing of ISHO assessment tools helped the team to familiarize itself with the questions, however actual understanding of the questions differed among participants and this lack of clarity and standardization of data collection process prolonged the assessment process. Individual scoring was virtual, while consensus scoring was in-person which became a challenge; if done in-person, the process could take at most 4 h as opposed to weeks. The assessment would have been stronger and more generalizable with the involvement of a wider group of subject matter experts, although the resulting recommendations were still informative to the annual plans. The group did not identify documentation to justify all of the scores and noted that scoring was subjective for some questions.

Recommendations

Process: Assessment methods should be clearly defined to ensure accurate and reliable data collection, and in-person site visits are a preferred way of getting informationfrom end users when conducting site-level assessments. A more comprehensive analysis requires expanding ISHO assessment to national, subnational, and health facility levels, with stakeholder engagement key for selecting participating sites. Participation in the ISHO assessment should include subject matter experts at all levels of the organization to get generalizable results.

ISHO tools: By adopting the tools to a country-specific context, ISHO results can be used to measure progress of the eHealth national strategic plan through comparison of baseline and follow-up results.

Action

The ISHO framework connects assessment to action planning, as shown in Fig. 18.2.

Action Step 1: Gather the team. The involvement of a robust and representative group of stakeholders, including governmental leaders (Ministries of Health, Ministries of Telecommunications, and others), multilateral agencies like the WHO, bilateral funders, and implementing partners, brings multiple perspectives into each step. It is critical that the process of scoring, prioritization, and roadmap design all seek to develop consensus across multiple stakeholders. Using workshops or other convenings with expert facilitation to guide stakeholders toward consensus is helpful.

Action Step 2: Assess. ISHO assessment can be used to monitor the informatics capabilities in countries at above-site and/or site-levels. Above-site ISHO assessments focus on organizational units at district, regional, or national levels, which support planning and management of front-line health services delivery, but which do not directly deliver healthcare services. Site-level assessments focuse on capabilities at the facility level. Examples of facilities include clinics, hospitals, laboratories, and other "brick and mortar" structures where services are provided. Both levels provide an overarching snapshot of core informatics capabilities. ISHO assessments can be repeated over time to track evolution in maturity.

Action Step 3: Analyze and prioritize gaps. ISHO assessment uses a mixed method design where focus group discussions are used to review the justification for scoring (e.g., by reviewing written documents, plans, policies, job descriptions, or other evidence) and to develop consensus scores. Finally, a group process can be used to prioritize areas for action. Commonly prioritized gaps, evident in the Ethiopia case study as well as in other studies (see the Literature Review and Gaps section), are shown in Table 18.1.

Action Step 4: Building and maintaining the strategic informatics roadmap. The gaps and needs identified in the previous step, and actions to address the priority gaps, can then be documented in an informatics roadmap. The informatics roadmap is a strategic plan describing the goals, objectives, and tactics for advancing informatics capabilities and achieving the future vision. It is used to direct activities that foster an enterprise-wide strategic approach to addressing current deficiencies and aligning priorities. The roadmap also serves as a communication tool, a high-level document that helps articulate strategic thinking—the why—behind both the goal and the plan for getting there. It should guide strategic improvements and investments and must reflect an organization's unique informatics capacity, needs, and vision. A template roadmap outline showing sample sections with descriptions can be found at https://phii.org/module-5/post-assessment-resources/.

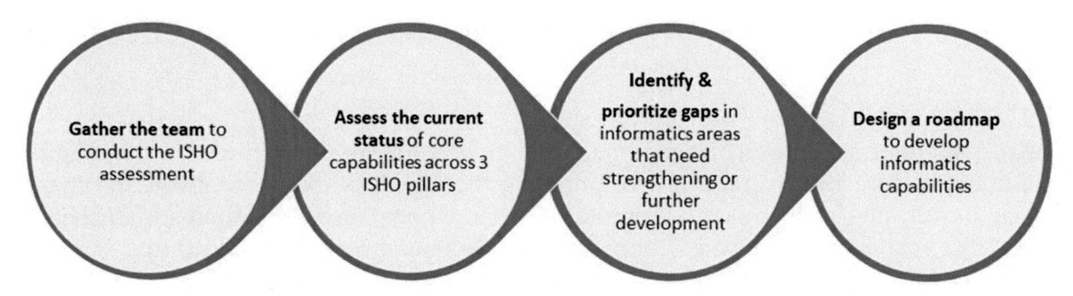

FIGURE 18.2 ISHO assessment leading to action planning.

TABLE 18.1 Typical gaps and priority action areas.

ISHO pillar	Priority areas	Explanation of needed investments or activities (some needs appear in more than one ISHO pillar)
Vision, policy, and governance	• Information and information system governance • National enterprise architecture • Patient privacy • Cybersecurity • Data sharing	• Need for a guiding national-level policy for management of patient data • Need to establish minimum cybersecurity requirements for workforce training and system implementation • Need to define standard data sharing conventions both within the country, but also outside the country (e.g., neighboring nations, WHO, CDC) • Need to determine when and how to use and manage cloud services as part of the digital health architecture, to improve data flow and data security • Need to identify and codify standards for health data terminologies, health information data exchange, digital messaging, data communications • Need for greater integration of human and animal health data
Workforce	• Leadership training in informatics • Project management (PM) tools and training • Informatics position descriptions	• Need to ensure awareness, motivation, knowledge, and skills in informatics planning and governance • Need to establish common project management practices and tools and provide training in best practices in informatics project management • Need to define specific informatics positions and associated position descriptions as a basis for career pathways • Need to increase local capacity to lead and manage data systems • Need to develop/provide training on effective use of data, reduce underutilization of data
Effective information systems	• Infrastructure • Cloud services • Health information exchange data and terminology standards	• Need to establish the requisite communications and power infrastructure to support digital health, through external partnerships • Need to determine when and how to use and manage cloud services as part of the digital health architecture, to improve data flow and data security • Need to identify and codify standards for health data terminologies, health information data exchange, digital messaging, data communications • Need for improved data sharing and integration between private health services and public surveillance systems

An ISHO-informed informatics roadmap may involve further detailed planning on digital health architecture and operations. Documenting the digital health architecture and maintaining it over time is a necessary underpinning to a national or organization-specific health informatics roadmap. The Open Group Architecture Framework (TOGAF) (https://www.opengroup.org/togaf) is one framework for developing and managing the enterprise architecture lifecycle. Enterprise architecture includes:

- An overview of the system
- A description of the system's goals and objectives
- A summary of known assumptions, constraints, risks, and issues
- Baseline business architecture
- Baseline application architecture
- Baseline data architecture
- Baseline technology architecture
- Baseline security architecture
- An overview of system operations and maintenance

Governance of digital health systems operations typically requires periodic evaluation of functional and nonfunctional capabilities as well as the implementation security of each key digital health tool or component.

Conclusion

Data and information are at the center of mitigating threats to public health. Developing realistic digital health strategies and having the skills and methods to achieve them requires the effective application of public health informatics principles and methods. No amount of advanced technology alone can effectively detect, prevent, or mitigate public health threats. That requires the most valuable resource of all: people. The human element includes both steadfast leadership to provide a vision and supporting governance, as well as a workforce willing to learn new ways of using and sharing information with others. The ISHO framework supports a rigorous process for identifying and prioritizing capabilities related to information as a strategic organizational asset, ensuring balance of effort and attention across governance, workforce, and technology pillars.

Supplemental Appendix 1: Capability maturity model response categories

Maturity level	General description
1. Absent	No capability is evident; "starting from scratch."
2. Initial	No organized, systematic efforts to build informatics capacity, only ad hoc efforts and isolated, individual heroics.
3. Managed	Some organized efforts begun or completed, but not systematically documented or institutionalized.
4. Defined	Systematic, ongoing efforts underway, but no overall method to measure progress or to ensure coordination.
5. Measured	Systematic, ongoing efforts underway to measure progress and ensure coordination.
6. Optimized	Systematic, ongoing efforts underway with quality improvement activities to align results with guiding vision, strategies, and performance metrics.

Supplemental Appendix 2: Core informatics-savvy health organization assessment areas

1. **Organization-wide informatics vision, policy, and governance** processes to include:
   - A well-articulated informatics visions, strategies, and metrics for how the agency obtains and uses information and Information Communication Technologies
   - An organization-wide (and potentially a nation-wide, continental-wide, or global-wide) approach to standards and interoperability
   - Policies to ensure privacy, security, and confidentiality
   - Defined governance and leadership structure for decision-making including the establishment of a focal unit and leader for informatics; and funding for positions and training
   - Effective collaborations with public and private partners, communities, and information technology departments
   - Sustainable practices responsive to rapid changes in the health environment
2. **Skilled workforce** to include:
   - Trained informaticians through competency-based academic programs or continuing education
   - Workforce strategies for continuously developing and improving informatics knowledge and skills (sustainability)
   - Staff within the organization who understand and apply informatics to their practice
   - Workforce recruitment and retention strategies (sustainability)
3. **Effective information systems** technology to include:
   - Effectively addressing the information needs, workflows, and practices of staff and programs
   - Development using sound project management principles and methods
   - Rigorously defined requirements, user-centered design, and sustainability design
   - Use of globally recognized vocabulary, messaging, and other standards for interoperability with internal and external systems
   - Effectively ensuring data integrity, and adhering to security protocols and procedures
   - Evaluation with end user and cost-effective outcome metrics
   - Sustainable implementation of HIS to include data, system, system interoperability, and easy system enhancement over time

Supplemental Appendix 3: Adaptation of ISHO framework for assessing national health information system maturity under the US President's Emergency Plan for AIDS Relief (PEPFAR)

Core essential element (CEE) and assessment question	Description
Pillar 1: Vision, policy, and governance	
CEE 1.1 eHealth/digital health strategy and costed plan *Question: Does the organization have a defined, documented, and adopted eHealth/digital health strategy and costed plan?*	WHO defines eHealth as " ... the cost-effective and secure use of information and communications (ICT) technologies in support of health and health-related fields, including healthcare services, health surveillance, health literature, and health education, knowledge, and research." Increasingly, we are using the term digital health, which is defined as "the use of digital, mobile, and wireless technologies to support the achievement of health objectives" and is inclusive of both mHealth and eHealth. Vision refers to a statement of what the organization seeks to achieve because of establishing a high level of information capability. The term strategy refers to a written "plan of action" for achieving specific goals or outcomes related to the organization's established information capability. Public health informatics is the science that underlies eHealth/digital health for public health and is defined as "the systematic application of information and computer science and technology to public health practice, research, and learning."
CEE 1.2 eHealth/digital health policies and legislation *Question: Do eHealth/digital health policies and legislation exist and are they enforced?*	Digital health policies and legislation refers to a written set of rules and regulations by a recognized authority in the country that oversees the implementation and use of digital health (e.g., Ministries of Health or Information & Communication Technology (ICT), Parliament). They may be established as part of the laws of a country through a legislative process or be drafted as national policies derived from existing legislation. Such policies and legislation govern the design, development, use, and disposal of digital technologies and often cover a broad range of areas including scalability, replicability, interoperability, security, and accessibility of digital health solutions. Such policies and legislation could derive from global institutions and be adapted for country-specific use, considering the local conditions, priorities, and interests, and are arrived at through a consultative process among key stakeholders.
CEE 1.3 eHealth/digital health governing body *Question: Does the organization have a governing body for eHealth/digital health planning and coordination that guides major decisions aligned with the strategic plan and following documented processes?*	Governance process refers to a formal process for decision-making. This may include a written plan that describes who participates in decision-making, a governance structure, such as committees or coalitions, and descriptions of how decisions are made.

(*Continued*)

—cont'd

Core essential element (CEE) and assessment question	Description
CEE 1.4 Financing and resource mobilization plans *Question: Does the organization have a financing plan that includes a resource mobilization plan to support staffing needs, infrastructure, and information systems?*	A sustained approach to funding may include activities undertaken to identify the potential sources of revenue (where will the money come from) and how the organization will seek the funds (Ministry of Health budget, donors, public–private partnerships) to support digital health and informatics activities. An approach may include the development of a funding plan that describes funding goals and includes measurable objectives or benchmarks, as well as action steps related to the funding strategy. It may also include an analysis of the financial, infrastructure, and human resources (both staff and volunteer) needs.
CEE 1.5 eHealth/digital health M&E plan *Question: Does the adopted eHealth/digital health strategy have a monitoring and evaluation Plan?*	eHealth/digital health monitoring and evaluation (M&E) plan is a document that measures progress made in a project and helps to track and assess the results of the interventions throughout its life. It includes the indicators, who is responsible for collecting them, what forms and tools will be used, and how the data will flow through the organization. An M&E plan for the digital health strategy should be developed at the same time as the strategy and should be used from the start of implementation of the targeted projects.
CEE 1.6: Data management, quality, and use policies and procedures *Question: Does the organization have policies and procedures in place for data management, quality, and use?*	Robust data management procedures include systematized plans and processes to collect, retain, protect, and enhance the value of data. Security and confidentiality protocols, data use agreements, and applicable statutes or rules may all inform or be included in data management procedures. Data quality assurance procedures include protocols to assess and ensure the accuracy, completeness, and timeliness of incoming and existing data.
CEE 1.7 Informatics focal point *Question: Does the organization have a focal point for informatics (e.g., an informatics unit, a chief informatics officer) with cross-organization responsibilities and authorities, including those related to the organization's information vision, strategies, and polices?*	Public health informatics is the science that underlies eHealth/digital health for public health and is defined as "the systematic application of information and computer science and technology to public health practice, research, and learning." It is one of the core sciences of public health. As an emerging discipline, it is often not well understood, and often misunderstood and confused with IT; informatics is not IT but IT plus. Informatics as a practice is increasingly seen as critical to the future capability of a public health organization that will enable it to handle its information challenges in the digital age and as such, it should be mainstreamed the same way as surveillance and laboratory sciences. For science-based organizations, establishing a focal area dedicated to informatics is one way in which organizations are working to address organizations' information needs.

—cont'd

Core essential element (CEE) and assessment question	Description
CEE 1.8 Personally identifiable information policies and procedures *Question: Has your organization established policies and procedures to ensure privacy, confidentiality, and security (storage, transmission, use) of personally identifiable information (PII)?*	Personally identifiable information, or PII, is any data that could potentially be used to identify a particular person such as a full name, date of birth, passport number, and email address or any other information that is linked or linkable to an individual such as medical, educational, or employment information. Patients expect that the PII they provide to a public health organization is held in confidence and thus should be protected by law. Mounting cybersecurity threats and the increased use of mobile devices, together with growing programmatic needs for patient-level data, call for increasing support to partners to mitigate these threats. Privacy refers to the right of an individual to keep his or her health information private and to control access to and use of one's information. Confidentiality is the protection of information against disclosure to unauthorized parties. Data integrity refers to the accuracy and consistency (validity) of data over its lifecycle indicated by an absence of any alteration in data between two updates of a data record. Data security is the practice of protecting data against unauthorized access and corruption to ensure privacy. Security rules require the implementation of three types of safeguards: (1) administrative, (2) physical, and (3) technical. The term procedures here is intended to cover the wide range of actions, defined and driven by written policy, and needed to ensure privacy, confidentiality, and security practices to achieve appropriate privacy protections. These mechanisms may include training, policies, procedures, and optimized technology attributes to protect data in electronic environments.
CEE 1.9 Information system standards *Question: Do the organization's information systems use nationally recognized terminology, exchange, and security standards?*	Terminology, exchange and security standards support data creation, sharing, and integration by ensuring a clear understanding of how data are represented. Standards help systems speak the same language. Adopting nationally recognized standards can decrease the time and resources needed for developing software, building interfaces, and supporting connectivity. Examples of internationally recognized standards for vocabulary include CVX, CPT, ICD, LOINC, and SNOMED. Messaging standards like HL7 and CDA often call for the use of specific vocabulary standards; for electronic laboratory messages, LOINC and SNOMED codes are routinely used for tests and results, while immunization messages use CVX and ICD codes for vaccines and administration methods. Data security standards, such as

(*Continued*)

—cont'd

Core essential element (CEE) and assessment question	Description
	ISO 27001 and NISP SP-800 series publications, are ways to ensure a standard set of security protections are available to an organization and systems that process, store, or access protected data. Some of the protections outlined in these standards are authentication, detection of cyberattacks, data encryption, patching, and network firewalls. Some of the most common cyber threats we face today are e-mail phishing; ransomware; loss or theft of equipment or data; insider, accidental, or intentional data loss; connected medical devices that may affect patient safety; and malicious system exploits by external government/state actors. These continue to evolve, and standards need to be continuously revised and updated accordingly.
CEE 1.10 Secure data exchange policies and protocols *Question: Has the organization adopted protocols, policies, frameworks, and procedures to support secure data exchange and sharing with internal and external partners?*	The concept here refers to compliance with the procedure, not compliance with the terms of the data-sharing agreements. The term procedure is intended to cover the wide range of actions needed to ensure compliance with data-sharing agreements. Data-sharing agreements are used to establish clear parameters for exchange between organizations or operational units within an organization. These are written agreements that may include: descriptions of allowable use of data, responsibilities of the parties to the agreement, the legal authority or business reason to share data, frequency of data exchange, provisions for reporting violations of agreements including breaches of privacy or security, privacy and security provisions, agreement on the purpose for the data exchange, and agreement on specific data elements to be exchanged.
CEE 1.11 Unique identifiers for PLWHA on ART *Question: Does your organization have a unique identifier to identify PLWHA on antiretroviral therapy (ART)?*	Unique identifier is a numeric or alphanumeric string that is associated with a single person living with HIV within a health register or a health records system. The unique ID is essential for the correct matching and linkage of patient records across systems. It is also necessary for the deduplication of patient records at the site and above the site level. A systematic generation of UID ensures that it remains consistent across every healthcare facility the patient visit and is ubiquitous.
CEE 1.12 GIS/geospatial capabilities *Question: Does your organization have GIS/geospatial capabilities?*	Geographic information system (GIS) is a framework for gathering, managing, and analyzing data. Rooted in the science of geography, GIS integrates many types of data. It analyzes spatial location and organizes layers of information into visualizations using maps and 3D scenes. With this unique capability, GIS reveals deeper insights into data, such as patterns, relationships, and situations—helping users make smarter decisions (ESRI).

—cont'd

Core essential element (CEE) and assessment question	Description
Pillar 2: Skilled workforce	
CEE 2.1 Informatics workforce strategy *Question: Does the organization have a workforce strategy or policy or guide that describes both the needed informatics capabilities and/or positions and the plans for recruiting, hiring, and/or developing existing staff to meet those needs?*	Strategies and action plans for human resources often include organization-wide efforts to meet organizational performance needs. Workforce planning strategy may include assessment, recruitment, training and development, retention, and succession planning. For informatics, the workforce strategy may include creating new positions or, because that is not always possible or desirable, training existing staff who have the interest and aptitude in informatics.
CEE 2.2 Informatics professional job classifications *Question: Does the organization support staff members across a broad range of job classifications to participate in informatics training?*	Human resources professionals must have job classification systems, position descriptions, and pay scales for all hiring situations. Because informatics is an emerging discipline, many agencies struggle to establish these positions within existing classifications. A major challenge is defining the informatics competencies, duties, and minimum requirements in ways that clearly distinguish them from IT classifications.
CEE 2.3 Staff informatics training support *Question: Does the organization support staff members across a broad range of job classifications to participate in informatics training?*	This question seeks to assess the availability of informatics training for individuals in a variety of job classifications. These positions can include those that support informatics capacity directly, as well as data analysts, epidemiologists, public health nurses, program managers, data quality specialists, and IT staff.
CEE 2.4 Academically prepared/highly experienced informaticians *Question: Does the organization have highly experienced or academically prepared informaticians in key roles at department and/or program levels, with backgrounds and training commensurate with their responsibilities?*	Highly experienced informaticians refer to those individuals that have the necessary combination of knowledge, demonstrated skills, and abilities to successfully contribute to effective informatics practice. While an academically prepared informatician may not be possible or feasible in many agencies, perhaps especially for local organizations, it is a good measure of informatics maturity within an organization. Key roles refer to the placement, availability, and access of individuals with informatics experience.
CEE 2.5 Information system capabilities among program-level staff *Question: Do staff members at the program level have the skills to effectively use information systems and tools, and knowledge of how to identify and document needed system improvements?*	Users of information systems need to know when those systems are not meeting their needs—requiring frustrating workarounds, inefficient workflows, or other problems—and be savvy enough to state or document their needs in sufficiently clear terms to serve as requirements for enhancements.

(*Continued*)

—cont'd

Core essential element (CEE) and assessment question	Description
CEE 2.6 Informatics capabilities among leadership/ management *Question: Do leaders/managers/supervisors in the organization have knowledge and skills of informatics principles, concepts, methods, and tools gained through education, training, or experience?*	Informatics principles, concepts, methods, and tools refer to the set of knowledge and skills necessary for leaders and managers to know how to lead the creation of an eHealth/ digital health/informatics vision, strategy, policies, governance, structures, and workforce for their organization; guide the development of standard-based and cost-effective information systems; and ensure investments are monitored and evaluated to optimize appropriate, efficient, and effective utilization of resources. For senior managers, it can mean providing governance/ oversight and clear purpose and direction for the project, understanding how requirements were gathered and vetted, whether end users were involved in the design, where the risks lie, and whether the system is delivering value. For those who manage the information system directly, it can include understanding the IT lifecycle, instituting sound requirements gathering and change control mechanisms, and being able to manage risks, problem solve, and ensure quality information is produced to support meeting program objectives.
Pillar 3: Effective information systems	
CEE 3.1 ICT hardware, connectivity, and infrastructure *Question: Does the organization have sufficient hardware, software, networking, connectivity, and ICT support infrastructure in place to ensure their operation and maintenance are in line with industry standards?*	Relevant hardware and infrastructure, including Internet connectivity and ICT support, are essential for the successful implementation of an HIS project. Well-organized agencies ensure there are adequate financial resources allocated to establish the relevant hardware and ICT infrastructure upon which the software will be implemented.
CEE 3.2 Software development life cycle *Question: Does the organization practice a standard software development life cycle (SDLC) process for requirements definition, system design, implementation, maintenance, and system disposition?*	A standard software development process may include some or all components to support the IT lifecycle, including initiation and concept, planning, requirements definition, design and development, testing, training and implementation, operations and maintenance, and disposition. A detailed requirements definition is particularly essential, as it includes understanding what the information system must do to support the program to meet its objectives. The output of requirements definition identifies, in very granular detail, the new product to be built or how an existing system is to be enhanced.
CEE 3.3 IT project management procedures *Question: Has the organization adopted and documented standard project management procedures for information technology projects?*	In this context, project management procedures refer to methods and strategies designed to oversee and guide the accomplishment of information system goals or projects. Typical project management components or steps coincide with a software development life cycle process which may include but are not limited to preparation, initiation, planning, execution, monitoring/controlling, and close-out. Project managers are responsible for coordinating and

—cont'd

Core essential element (CEE) and assessment question	Description
	conducting stakeholder communication and at times vendor contract management. Adhering to a standardized methodology of project management can help to mitigate risk, maintain timelines, manage realistic scope/project goals, and ensure success within a project.
CEE 3.4 Information system and services inventory *Question: Has the organization conducted an inventory of its information systems and the services/information they provide?*	In general, an information system is defined as any computerized database designed to collect, store, and process data for the purposes of delivering information and/or knowledge. In considering an inventory of information systems, it may be important for the organization to determine a uniform definition of information systems to be included and counted. An inventory of information systems and the services provided by existing systems can be a starting point toward a larger needs assessment to evaluate the degree to which information systems meet the needs of program staff and end users. Such assessments could include, identify, or enumerate the number and types of information systems that are in use, which standards are used, the current and possible future external and internal data exchange partners/users, technical capabilities, and resource needs. Such an activity may identify opportunities to reduce duplication or address multiple uncoordinated systems.
CEE 3.5 National-level aggregate data reporting system *Question: Does the organization have an aggregate data reporting system (ADRS) at the national level?*	Aggregate data reporting system (ADRS) at the national level: a tool for the collection, validation, analysis, and presentation of aggregate and patient-based statistical data, tailored (but not limited) to integrated health information management activities.
CEE 3.6 National electronic medical records (EMR) system is used to manage patient data for clinical care and clinical workflows and conforms to national standards at supported health facilities *Question: Does the organization have a national electronic medical records system (EMR)?*	National electronic medical records system (EMR) is used to manage patient data for clinical care and clinical workflows and conforms to national standards at supported health facilities.
CEE 3.7 National laboratory information system *Question: Does the organization have a national laboratory information system (LIS)?*	A laboratory information system (LIS) is used to manage data on lab orders, workflows, and reporting of results, and conforms to national standards at supported health facilities.
CEE 3.8 National data repository *Question: Does the organization have a national data repository (NDR)?*	National data repository (NDR) is used as a collection of deidentified, deduplicated patient-level data that is sourced and routinely updated from EMRs, LIS, and other patient-level systems that may be electronic or paper-based.

(Continued)

—cont'd

Core essential element (CEE) and assessment question	Description
CEE 3.9 National HIV case-based surveillance information management system *Question: Does the organization have a national HIV case-based surveillance (CBS) information management system (IS)?*	HIV case-based surveillance (CBS) information management systems capture, analyze, and present deidentified, individual patient-level data on key indicators that help monitor outcomes of HIV prevention and treatment programs. Individual patients are tracked from diagnosis until death.
CEE 3.10 Secure eHealth data exchange capabilities at the programmatic level *Question: Does the organization have capabilities to securely send, receive, and process electronic health data and/or messages between programmatic information systems?*	Electronic exchange of data or messages refers to the ability to send, receive, and process data that are electronically transferred from one information system to another. Generally, this does not include fax or email messages. Technical capabilities for electronic data exchange might include automated scripts for querying or extracting information from one system and securely transferring it to another. Internal data sharing might require enabling legislation or cross-program data use agreements. In some cases, it may be technologically possible to exchange data internally, but policy or programmatic hurdles may exist. Electronically processing information may refer to the ability to accurately match and merge records, reconcile differences, and automate deduplication processes.
CEE 3.11 Secure eHealth data exchange capabilities at health facility/clinical level *Question: Does the organization have the capability to securely send and receive electronic health data with health facilities/ clinical partners?*	Clinical partners may have a particular need to interact with public health data to meet reporting requirements and/or to leverage data for clinical decision support. For some areas of public health, clinical data exchange involves receiving health data through a unidirectional pathway, while others may have a need to securely exchange electronic health information bidirectionally. Unidirectional exchange is exemplified by electronic laboratory reporting where the external partner submits data to public health and may (or may not) receive an acknowledgment of that receipt. Alternatively, bidirectional exchange of data occurs when one system (e.g., a clinician's electronic health record system) submits data to or queries another (e.g., an immunization information system or registry), and that system returns a response that is incorporated by the requesting system.
CEE 3.12 Secure eHealth data exchange capabilities at external partner level *Question: Does the organization have the capability to receive and process electronic data and/or messages sent from external partners, including the private sector?*	Receiving a message means that the message reaches its intended target. Processing a message includes the ability to parse, store, and retrieve data. It also implies that the recipient can "read" or access the information contained in the message. Message processing capability also includes validating that the information contained in the message conveys an expected or appropriate value. Ideally, message receipt and processing would be automated, requiring minimal manual effort and human intervention.

—cont'd

Core essential element (CEE) and assessment question	Description
CEE 3.13 Organization-wide service sharing and integration *Question: Do programs share relevant services across the organization, such as an integrated provider registry, master facility list, master person index, terminology management, integration engine, and other applicable services?*	Shared or centralized services such as provider registries or master patient indexes that are leveraged across an organization can allow programs access to resources and tools, they would not otherwise be able to implement. Shared services can also facilitate a uniform and standards-based adoption of programmatic functions, while supporting common goals and processes.

For each of the specific CEEs, available assessment tools offer a range of closed-ended response options that are based on six levels of capability maturity. Only one response is selected for each question. It is not uncommon for organizations, even successful ones, to be at relatively low levels of this model. This reflects the organizational challenges that are inevitable with formally establishing new ways of working and then rigorously evaluating that work.

Supplemental Appendix 4: Ethiopia ISHO assessment results (2018)[a]

Capabilities: vision, policy, and governance	Ethiopia status
1.1. Vision and strategy: Does your Ministry of Health (MoH) have a documented informatics vision and strategy?	3
1.2. Information assets and needs: Has your MoH completed an assessment intended to describe its information assets and needs?	3
1.3 Governance process: Does your MoH have a governance process that guides the implementation of the informatics strategy?	3
1.4 Funding plan: Does your MoH have a systematic, sustained approach to funding informatics activities, including those to support staffing needs, physical facility, and information systems funding?	3
1.5 Stakeholder engagement (internal partners): Has the MoH completed an assessment to improve data exchange with internal stakeholders?	3
1.6 Stakeholder engagement (external partners): Has the MoH completed an assessment to improve data exchange with external stakeholders?	1

(*Continued*)

—cont'd

Capabilities: vision, policy, and governance	**Ethiopia status**
1.7 Data sharing agreement procedures: Has the MoH adopted procedures for establishing data-sharing agreements?	1
1.8 Privacy, confidentiality, and informed consent procedures: Has your MoH established policies and procedures to ensure privacy, confidentiality, and informed consent?	2
1.9 Informatics focal point: Does your MoH have an organizational focal point for informatics (e.g., an informatics unit, a Chief Informatics Officer) with cross-MoH responsibility and authorities, including those related to the MoH's information vision, strategies, and policies?	5
1.10 Effective relationship IT/informatics: Does your MoH have a strategy to support relationships with an information technology (IT) unit or services provider (internal or external) to support the achievement of informatics goals and objectives?	3
1.11 Collaboration with community partners to meet population health goals/objectives: Does your MoH effectively collaborate with community partners who have an interest/responsibility for population health assessment and/or management (NGOs, donors, etc.)?	3
Capabilities: Skilled workforce	
2.1. Workforce strategy: Does the MoH have a workforce strategy that describes its needed informatics capabilities and/or positions and have action plans for recruiting, hiring, and/or developing existing staff to meet those needs?	3
2.2. Job classifications for informatics professionals: Does the MoH have appropriate job classifications, including position descriptions and pay scales, for informatics professionals?	3
2.3 Training: Does the MoH support staff members across a broad range of job classifications to participate in informatics training?	2
2.4 Informatics professionals: Does the MoH have highly experienced or academically prepared informaticians in key roles at the department and/or program levels, with backgrounds and training commensurate to their responsibilities?	4
2.5 Informatics knowledge and skills (program level): Do staff members at the program level (e.g., epidemiologists, medical officers, public health nurses, data analysts) have the skills to effectively use information systems and tools, and the knowledge of how to identify and document needed system improvements?	2
2.6 Informatics knowledge and skills (program managers): Do managers/supervisors of large information system programs have knowledge and skills of informatics principles, concepts, methods, and tools gained through education, training or experience?	3
Capabilities: Information systems	
3.1. Software development process: Does the MoH practice a standard software development process for requirements definition, system design, implementation, and maintenance?	4
3.2. Project management: Has your MoH adopted and documented standard project management procedures for information technology projects?	3
3.3 Information systems inventory: Has the MoH conducted an inventory of its information systems and the services/information they provide?	4

—cont'd

Capabilities: vision, policy, and governance	Ethiopia status
3.4 Information systems usability: Has the MoH conducted an assessment of information system usability and effectiveness based on the needs of staff and programs?	1
3.5 Standards adoption and implementation: Does the MoH have information systems that use nationally recognized vocabulary, messaging, and transport standards?	1
3.6 Data exchange (internal): Does the MoH have the capability to electronically send, receive, and process data and/or messages internally between programmatic information systems?	2
3.7 Data exchange (external partners): Does the MoH have the capability to receive and process electronic data and/or messages sent from external partners?	0
3.8 Data exchange (clinical partners): Does the MoH have the capability to securely send and receive electronic health data with clinical partners?	0
3.9 Data management and quality assurance (internal): Has the MoH adopted procedures for data management and quality assurance for data housed in the Ministry of public Health's information systems?	4
3.10 Information technology systems plans and budgets: Does the MoH plan or budget for information technology systems maintenance?	4
3.11 Shared services: Do the MoH's programs share relevant services across the MoH, such as an integrated provider registry, master person index, integration engine, or other applicable services?	1

[a]The original PHII ISO assessment tool used a 0–5 scale, while the adaptation for PEPFAR programs in LMICS used a 1–6 scale..*

Acknowledgments

The chapter authors would like to acknowledge Debra Bara, MS, and Jim Jellison, MPH from The Task Force for Public Health/Public Health Informatics Institute (PHII) for foundational work in developing and testing the informatics-savvy health organization framework and toolkit.

References

1. Phalkey RK, Yamamoto S, Awate P, Marx M. Challenges with the implementation of an integrated disease surveillance and response (IDSR) system: systematic review of the lessons learned. *Health Pol Plann.* 2015;30(1):131–143. https://doi.org/10.1093/heapol/czt097.
2. Center for Strategic and International Studies (CSIS). *Can Digital Health Help Stop the Next Epidemic?* Washington, DC: Center for Strategic and International Studies (CSIS); 2019. https://www.csis.org/analysis/can-digital-health-help-stop-next-epidemic, 2019. Accessed June 29, 2023.
3. Martin DW, Sloan ML, Gleason BL, et al. Implementing nationwide facility-based electronic disease surveillance in Sierra Leone: lessons learned. *Health Secur.* 2020;18(S1):S-72–S-80. https://doi.org/10.1089/hs.2019.0081.
4. Njuguna C, Jambai A, Chimbaru A, et al. Revitalization of integrated disease surveillance and response in Sierra Leone post Ebola virus disease outbreak. *BMC Publ Health.* 2019;19(1):364. https://doi.org/10.1186/s12889-019-6636-1.
5. Sloan ML, Gleason BL, Squire JS, Koroma FF, Sogbeh SA, Park MJ. Cost analysis of health facility electronic integrated disease surveillance and response in one district in Sierra Leone. *Health Secur.* 2020;18(S1):S-64–S-71. https://doi.org/10.1089/hs.2019.0082.
6. Aliseda-Alonso A, Lis SB, Lee A, et al. The missing COVID-19 demographic data: a statewide analysis of COVID-19-related demographic data from local government sources and a comparison with federal public surveillance data. *Am J Publ Health.* 2022;112(8):1161–1169. https://doi.org/10.2105/AJPH.2022.306892.
7. Kauh TJ, Read JG, Scheitler AJ. The critical role of racial/ethnic data disaggregation for health equity. *Popul Res Pol Rev.* 2021;40(1):1–7. https://doi.org/10.1007/s11113-020-09631-6.

8. Klann JG, Strasser ZH, Hutch MR, et al. Distinguishing admissions specifically for COVID-19 from incidental SARS-CoV-2 admissions: national retrospective electronic health record study. *J Med Internet Res.* 2022;24(5):e37931. https://doi.org/10.2196/37931.
9. Adam D. COVID's true death toll: much higher than official records. *Nature.* 2022;603(7902):562–562. https://doi.org/10.1038/d41586-022-00708-0.
10. Allan M, Lièvre M, Laurenson-Schaefer H, et al. The world health organization COVID-19 surveillance database. *Int J Equity Health.* 2022;21(S3):167. https://doi.org/10.1186/s12939-022-01767-5.
11. World Health Organization (WHO). *Global Strategy on Digital Health 2020–2025.* Geneva, Switzerland: World Health Organization (WHO); 2021. https://www.who.int/docs/defaultsource/documents/gs4dhdaa2a9f352b0445bafbc79ca799dce4d.pdf.
12. World Health Organization (WHO). *Digital Implementation Investment Guide (DIIG): Integrating Digital Interventions into Health Programmes.* Geneva, Switzerland: World Health Organization; 2020. https://www.who.int/publications/i/item/9789240010567.
12a. Broadband Commission for Sustainable Development. Digital Health: A Call for Government Leadership and Cooperation between ICT and Health. https://broadbandcommission.org/wp-content/uploads/2021/09/WGHealth_Report2017-.pdf; 2017. Accessed July 2, 2023.
12b. Center for Disease Control (CDC). *Global Digital Health Strategy.* CDC; May 2022. https://vitalwave.com/wp-content/uploads/2022/09/GDHS-Full-Strategy-Final-20220524.pdf. Accessed June 29, 2023.
13. Silvestre E, Wood F. *Health Information Systems: Analysis of Country-Level Strategies, Indicators, and Resources;* 2020. https://lib.digitalsquare.io/handle/123456789/77348.
14. Transform Health. *Closing the Digital Divide: More and Better Funding for Digital Health Transformation of Health.* Basel: Transform Health; 2022. https://transformhealthcoalition.org/wp-content/uploads/2022/10/Closing-the-digital-divide-mainReport.pdf. Accessed June 29, 2023.
14a. International Telecommunications Union (ITU), World Health Organization (WHO). *National eHealth Strategy Toolkit.* Geneva, Switzerland: World Health Organization; 2012. https://www.itu.int/dms_pub/itu-d/opb/str/D-STR-E_HEALTH.05-2012-PDF-E.pdf. Accessed June 29, 2023.
15. Digital Impact Alliance (DIAL). *Principles for Digital Development.* Washington DC: Digital Impact Alliance (DIAL); 2019. https://digitalprinciples.org/principles/. Accessed June 29, 2023.
15a. World Health Organization (WHO). Classification of Digital Health Interventions v1.0. https://apps.who.int/iris/bitstream/handle/10665/260480/WHO-RHR-18.06-eng.pdf?sequence=1&isAllowed=y; 2018. Accessed June 29, 2023.
16. deRiel E, Puttkammer N, Hyppolite N, et al. Success factors for implementing and sustaining a mature electronic medical record in a low-resource setting: a case study of iSanté in Haiti. *Health Pol Plann.* 2018;33(2):237–246. https://doi.org/10.1093/heapol/czx171.
17. Kang'a SG, Muthee VM, Liku N, Too D, Puttkammer N. People, process and technology: strategies for assuring sustainable implementation of EMRs at public-sector health facilities in Kenya. *AMIA Annu Symp Proc.* 2017;2016:677–685.
18. Kumar M, Mostafa J. Electronic health records for better health in the lower- and middle-income countries: a landscape study. *Libr Hi Tech.* 2020;38(4):751–767. https://doi.org/10.1108/LHT-09-2019-0179.
19. Muinga N, Magare S, Monda J, et al. Implementing an open source electronic health record system in Kenyan health care facilities: case study. *JMIR Med Inform.* 2018;6(2):e22. https://doi.org/10.2196/medinform.8403.
20. Verma N, Mamlin B, Flowers J, Acharya S, Labrique A, Cullen T. OpenMRS as a global good: impact, opportunities, challenges, and lessons learned from fifteen years of implementation. *Int J Med Inf.* 2021;149:104405. https://doi.org/10.1016/j.ijmedinf.2021.104405.
21. Blaya JA, Shin S, Contreras C, et al. Full impact of laboratory information system requires direct use by clinical staff: cluster randomized controlled trial. *J Am Med Inf Assoc.* 2011;18(1):11–16. https://doi.org/10.1136/jamia.2010.005280.
22. He Y, Iiams-Hauser C, Henri Assoa P, et al. Development and national scale implementation of an open-source electronic laboratory information system (OpenELIS) in Côte d'Ivoire: sustainability lessons from the first 13 years. *Int J Med Inf.* 2023;170:104977. https://doi.org/10.1016/j.ijmedinf.2022.104977.
23. Njukeng PA, Njumkeng C, Ntongowa C, Abdulaziz M. Strengthening laboratory networks in the Central Africa region: a milestone for epidemic preparedness and response. *Afr J Lab Med.* 2022;11(1):1492. https://doi.org/10.4102/ajlm.v11i1.1492.

24. Fritz J, Herrick T, Gilbert SS. Estimation of health impact from digitalizing last-mile logistics management information systems (LMIS) in Ethiopia, Tanzania, and Mozambique: a lives saved tool (LiST) model analysis. *PLoS One*. 2021;16(10):e0258354. https://doi.org/10.1371/journal.pone.0258354.
25. Gilbert SS, Bulula N, Yohana E, et al. The impact of an integrated electronic immunization registry and logistics management information system (EIR-eLMIS) on vaccine availability in three regions in Tanzania: a pre-post and time-series analysis. *Vaccine*. 2020;38(3):562–569. https://doi.org/10.1016/j.vaccine.2019.10.059.
26. Vledder M, Friedman J, Sjöblom M, Brown T, Yadav P. Improving supply chain for essential drugs in low-income countries: results from a large scale randomized experiment in Zambia. *Health System Reform*. 2019;5(2):158–177. https://doi.org/10.1080/23288604.2019.1596050.
27. Dilu E, Gebreslassie M, Kebede M. Human Resource Information System implementation readiness in the Ethiopian health sector: a cross-sectional study. *Hum Resour Health*. 2017;15(1):85. https://doi.org/10.1186/s12960-017-0259-3.
28. Waters KP, Zuber A, Simbini T, Bangani Z, Krishnamurthy RS. Zimbabwe's Human Resources for health Information System (ZHRIS)-an assessment in the context of establishing a global standard. *Int J Med Inf*. 2017;100:121–128. https://doi.org/10.1016/j.ijmedinf.2017.01.011.
29. Agarwal S, Lasway C, L'Engle K, et al. Family planning counseling in your pocket: a mobile job aid for community health workers in Tanzania. *Glob Health Sci Pract*. 2016;4(2):300–310. https://doi.org/10.9745/GHSP-D-15-00393.

29a. Mechael P, Edelman K. *Global Digital Health Index: The State of Digital*. Global Development Incubator; 2019. https://saluddigital.com/wp-content/uploads/2019/06/Health-Enabled.-The-State-of-Digital-Health-2019.pdf. Accessed June 29, 2023.

30. Byrne E, Sæbø JI. Routine use of DHIS2 data: a scoping review. *BMC Health Serv Res*. 2022;22(1):1234. https://doi.org/10.1186/s12913-022-08598-8.
31. Etamesor S, Ottih C, Salihu IN, Okpani AI. Data for decision making: using a dashboard to strengthen routine immunisation in Nigeria. *BMJ Glob Health*. 2018;3(5): e000807. https://doi.org/10.1136/bmjgh-2018-000807.
32. Reynolds E, Martel LD, Bah MO, et al. Implementation of DHIS2 for disease surveillance in Guinea: 2015–2020. *Front Public Health*. 2022;9, 761196. https://doi.org/10.3389/fpubh.2021.761196.
33. Njuguna C, Vandi M, Liyosi E, et al. After action review of the response to an outbreak of Lassa fever in Sierra Leone, 2019: best practices and lessons learnt. *PLoS Neglected Trop Dis*. 2022;16(10):e0010755. https://doi.org/10.1371/journal.pntd.0010755.
34. Health Information Systems Programme (HISP) Centre at the University of Oslo. *District Health Information System (DHIS) Version (2.39)*. Oslo: University of Olso; 2022. https://dhis2.org/. Accessed June 29, 2023.
35. Silenou BC, Nyirenda JLZ, Zaghloul A, et al. Availability and suitability of digital health tools in Africa for pandemic control: scoping review and cluster analysis. *JMIR Public Health Surveill*. 2021;7(12):e30106. https://doi.org/10.2196/30106.
36. Dornan L, Pinyopornpanish K, Jiraporncharoen W, Hashmi A, Dejkriengkraikul N, Angkurawaranon C. Utilisation of electronic health records for public health in Asia: a review of success factors and potential challenges. *BioMed Res Int*. 2019;2019. https://doi.org/10.1155/2019/7341841.
37. Nadhamuni S, John O, Kulkarni M, et al. Driving digital transformation of comprehensive primary health services at scale in India: an enterprise architecture framework. *BMJ Glob Health*. 2021;6(suppl 5):e005242. https://doi.org/10.1136/bmjgh-2021-005242.

37a. Principles for Digital Development. *OpenHIE: communities building open standards for health information systems*. digitalprinciples.org; 2022. https://digitalprinciples.org/wp-content/uploads/PDD_CaseStudy-OpenHIE_v3-1.pdf. Accessed June 29, 2023.

38. Shivers J, Amlung J, Ratanaprayul N, Rhodes B, Biondich P. Enhancing narrative clinical guidance with computer-readable artifacts: authoring FHIR implementation guides based on WHO recommendations. *J Biomed Inf*. 2021;122:103891. https://doi.org/10.1016/j.jbi.2021.103891.
39. Kariuki JM, Manders EJ, Richards J, et al. Automating indicator data reporting from health facility EMR to a national aggregate data system in Kenya: an Interoperability field-test using OpenMRS and DHIS2. *Online J Public Health Inform*. 2016;8(2):e188. https://doi.org/10.5210/ojphi.v8i2.6722.
40. Khubone T, Tlou B, Mashamba-Thompson TP. Electronic health information systems to improve disease diagnosis and management at point-of-care in low and middle income countries: a narrative review. *Diagnostics*. 2020;10(5):327. https://doi.org/10.3390/diagnostics10050327.

41. Njeru I, Kareko D, Kisangau N, et al. Use of technology for public health surveillance reporting: opportunities, challenges and lessons learnt from Kenya. *BMC Publ Health*. 2020;20(1):1101. https://doi.org/10.1186/s12889-020-09222-2.
42. Nagbe T, Yealue K, Yeabah T, et al. Integrated disease surveillance and response implementation in Liberia, findings from a data quality audit, 2017. *Pan Afr Med J*. 2019;33. https://doi.org/10.11604/pamj.supp.2019.33.2.17608.
43. Endriyas M, Alano A, Mekonnen E, Kawza A, Lemango F. Decentralizing evidence-based decision-making in resource limited setting: a case of SNNP region, Ethiopia. *PLoS One*. 2020;15(7). https://doi.org/10.1371/journal.pone.0236637.
44. Gibson PJ, Shah GH, Streichert LC, Verchick L. Urgent challenges for local public health informatics. *J Publ Health Manag Pract*. 2016;22(Suppl 6):S6–S8. https://doi.org/10.1097/PHH.0000000000000479.
45. Lovelace K, Shah GH. Informatics as a strategic priority and collaborative processes to build a smarter, forward-looking health department. *J Publ Health Manag Pract*. 2016;22(Suppl 6):S83–S88. https://doi.org/10.1097/PHH.0000000000000452.
46. Yasnoff WA, O'Carroll PW, Koo D, Linkins RW, Kilbourne EM. Public health informatics: improving and transforming public health in the information age. *J Publ Health Manag Pract*. November 2000;6(6):67–75. https://doi.org/10.1097/00124784-200006060-00010. PMID: 18019962.
47. LaVenture M, Brand B, Ross DA, Baker EL. Building an informatics-savvy health department: Part I, vision and core strategies. *J Publ Health Manag Pract*. 2014;20(6):667–669. https://doi.org/10.1097/PHH.0000000000000149.
48. Carvalho JV, Rocha Á, Abreu A. Maturity models of healthcare information systems and technologies: a literature review. *J Med Syst*. 2016;40(6):131. https://doi.org/10.1007/s10916-016-0486-5.
49. Cresswell K, Sheikh A, Krasuska M, et al. Reconceptualising the digital maturity of health systems. *Lancet Digit Health*. 2019;1(5):e200–e201. https://doi.org/10.1016/S2589-7500(19)30083-4.
50. Gomes J, Romão M. Information system maturity models in healthcare. *J Med Syst*. 2018;42(12):235. https://doi.org/10.1007/s10916-018-1097-0.
51. US President's Emergency Fund for HIV (PEPFAR). *PEPFAR 2022 country and regional operational plan (COP/ROP) guidance for all PEPFAR-supported countries*. US Department of State; 2022. https://www.state.gov/wp-content/uploads/2022/02/COP22-Guidance-Final_508-Compliant-3.pdf. Accessed June 29, 2023.

CHAPTER

19

Analytics and intelligence for public health surveillance

Brian E. Dixon[1,2], *David Barros Sierra Cordera*[3], *Mauricio Hernández Ávila*[3], *Xiaochun Wang*[4], *Lanyue Zhang*[5], *Waldo Vieyra Romero*[3] and *Rodrigo Zepeda Tello*[3]

[1]Richard M. Fairbanks School of Public Health, Indiana University, Indianapolis, IN, United States; [2]Center for Biomedical Informatics, Regenstrief Institute, Indianapolis, IN, United States; [3]Mexican Institute for Social Security, Mexico City, Mexico; [4]Center for Global Public Health, Chinese Center for Disease Control and Prevention, Beijing, China; [5]Rollins School of Public Health, Emory University, Atlanta, GA, United States

Introduction

Following the COVID-19 pandemic, public health stands at a crossroads. One path is the familiar one, in which nations slowly resume usual operations including public health surveillance (PHS). We can return to "business as usual" that underinvests in prevention and public health systems. This is the pattern observed following previous significant public health threats such as SARS-CoV-1 and H1N1.[1] This is the path of least resistance as it does not require any change. Alternatively, we can take a different path that calls for an incremental change to invest in public health's human capital and infrastructure, including safe data sharing protocols, advanced computing and methods to transform data into knowledge. The alternate path requires new and innovative policies and social change, making it that much harder. It further requires a conscious effort to create and sustain scalable and secure intelligent systems at local and global levels that can improve health and well-being for populations.

Public health intelligence is defined as a specific public health discipline focused on the development and application of appropriate strategies, like early warning systems that can prevent the spread of new pathogens, and policies to improve the health of the population

Modernizing Global Health Security to Prevent, Detect, and Respond
https://doi.org/10.1016/B978-0-323-90945-7.00017-8

and reduce health inequalities underpinned by rigorous, robust, and timely evidence.[2] We view intelligence as an evolved form of PHS, aided by clear data sharing policies, advanced computing, informatics, and embedded institutional processes to ensure data are leveraged to drive impact and reduce inequality. Whereas current PHS processes (we refer to them as PHS 2.0) focus on producing reports that document disease incidence, prevalence as well as burden among populations, public health intelligence (PHS 3.0) emphasizes data-driven decision-making.

Existing PHS methods and processes might already be supported by information technology systems, including but not limited to online and offline data collection tools that feed into the District Health Information System (DHIS2), and other data aggregating tools, portals and dashboards that help centralize the data.[3,4] Although faxing and paper records remain the most common form of data collection in many nations, some data might be electronic as it is reported to the MoH or authority via mobile phones or computer systems in hospitals. This is wonderful, but PHS 2.0 is content to simply produce reports that document information which can be posted on websites or shared with the World Health Organization (WHO). A key aspect of public health intelligence is that epidemiologists, data scientists, data analysts, health system administrators, and ministers along with other decision-makers act on knowledge derived from the advantage of having real time data across the nation's health system. Instead of annual or monthly reports, intelligence focuses on identifying and answering questions about emerging phenomena and using these insights to advance health and well-being as events unfold. Intelligence also focuses on forecasting or predicting future variants of a novel pathogen, potential impacts of a public health intervention, and likely health impacts following a change in the environment.

The concept of public health intelligence is illustrated in Fig. 19.1; adapted from Choucair et al.[5] The figure illustrates that intelligence builds upon a strong foundation in public health informatics and information systems.[6] Existing efforts over the past 20 years to implement a variety of information systems and methods for strengthening the capture and management of digital health data are complemented by integrated approaches that leverage data and information to derive knowledge. That knowledge is then applied to public health action to advance health and well-being for populations within a community, state, nations, regionally, or globally.

Prior to the COVID-19 pandemic, several nations were already looking to move toward intelligent public health systems. In 2016, the U.S. Department of Health and Human Services, which includes the CDC, published a framework[7,8] known as "Public Health 3.0", and researchers in the United Kingdom[2] coined the term "Public Health Intelligence." Both groups describe what we perceive as the next evolution for PHS. The move toward intelligence is a natural evolution after a nation has established a basic infrastructure that captures and manages digital health data. Following the replacement of paper records with electronic versions, a country has a new challenge to manage large datasets and make sense out of information it previously could not derive from paper forms and basic disease counts. Before the pandemic, notions like virtual care, QR codes that open Web-based forms to capture medical data, and digital health passports were foreign concepts. Now, most countries require digital health declarations for travelers and queues for medical services are guided by information systems where patients virtually check-in like they do for airline flights. They might also leverage public-private partnerships to support their push toward Public Health

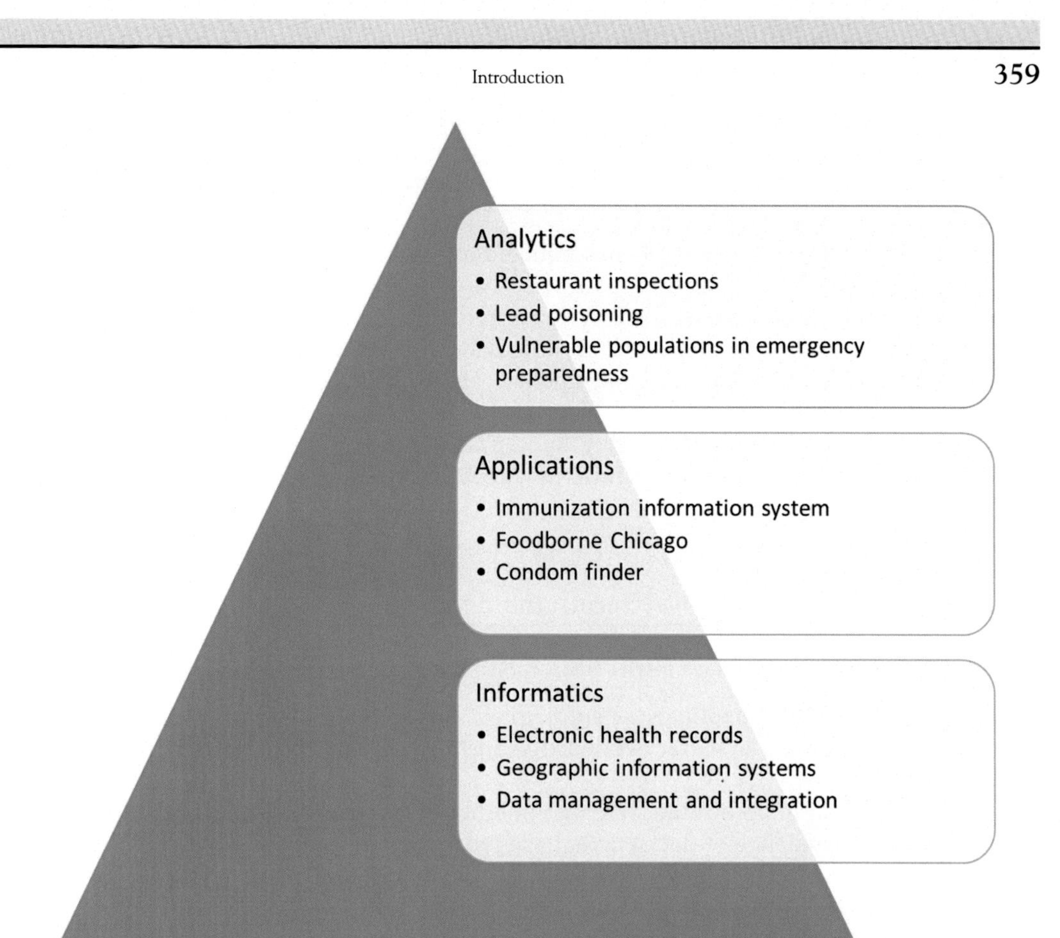

FIGURE 19.1 Intelligence evolves and advances PHS in an integrated way, building upon a strong foundation of robust informatics and information systems available for use by agencies to monitor and improve health and well-being for population.

Intelligence. However, it's not always clear what country authorities and other decision-makers *do* with this information, or if it drives impact at all. Most health systems globally are now primed for making the move toward public health intelligence.

Facilitating intelligence in public health requires, however, that ministries of health (MoHs) invest in human resources, health infrastructure, developing and maintaining robust information systems capable of managing the nation's ever-growing volume of complex digital health data, and implementing robust cybersecurity to protect health information. Although many nations currently have some digital health investments, including electronic medical records, digital health is not enough to create public health intelligence. Intelligence requires advanced computing infrastructure as well as health system (including MoHs) workers capable of transforming digital health data into knowledge that can guide surveillance efforts, policy decisions, and inform populations at large. Investing in a well-trained health informatics workforce, computing resources, data sharing protocols, and clear public communication strategies are critical, particularly at a time of overabundant information, much of which might be false or misleading (usually described as "Infodemic" explored in Chapter 20). These will be discussed at the end of the chapter in the policy recommendations section.

We note that in the U.S., the CDC has initiated efforts to upgrade infrastructure and data for PHS via its Data Modernization Initiative.[9] To date, the CDC has invested a total of $1 billion USD in infectious disease PHS systems and the PHS workforce through a series of grants to state, local, and territorial health jurisdictions.[10] However, as identified in a report from the Healthcare Information Management Systems Society, this amount is significantly less than the $36 billion required to achieve the vision of PHS 3.0.[11] All nations, including the U.S., will need to do more to increase the capacity of its technical and human infrastructure to integrate PHS activities across program areas and across infectious, chronic and injury divisions.

Key terms and definitions

This chapter uses several key terms from Kasthurirathne et al.[12] including.

- *Data science*: A multi-disciplinary scientific field that leverages various methods, processes, and algorithms to extract knowledge and insights from structured and unstructured data.
- *Big Data:* The field of computing research, methodology, and expertise on the extraction, analysis, and persistence of datasets that are too large and/or complex to be analyzed by traditional methods.
- *Analytics:* The discovery, interpretation, and communication of meaningful patterns found in data, as well as the application of data patterns for effective decision-making.
- *Artificial intelligence (AI):* A subdomain of computer science that focuses on the simulation of a human intelligence (or brain function) by a machine. AI is a wide domain that encompasses machine learning as well as other topics, such as logic, problem solving, and reasoning which are out of scope for this chapter.
- *Machine learning:* The ability of a computer system to learn from the external environment or a data source to improve its ability to perform a task. These approaches enable various algorithms to learn from data without need of any explicit programming. Machine learning is a subset of AI.
- *Deep learning:* An advanced form of machine learning that uses multiple layers within a neural network of algorithms to infer knowledge from data without explicit programming. Deep learning is also a subset of AI.
- *Cloud Computing:* On demand availability of data storage and computing power (e.g., processing) without direct, active management by the user. Cloud computing supports organizations with limited access to hardware and physical space expand their ability to manage Big Data and run analytics software.
- *Social Media:* Websites and computing applications that enable users to create and share content or to participate in social networking with each other. Most social media

platforms are open to the public, enabling anyone to become a user. Examples include Facebook, Instagram, Twitter, YouTube, and WeChat.

- *Blockchain:* A digitally distributed, decentralized, public ledger that exists across a network of computers. As a database, it stores information electronically in digital format. While best known for their role in cryptocurrency systems, such as Bitcoin, blockchains can maintain a secure and decentralized record of transactions, including health related events.
- *Internet of Things (IoT):* A system of physical objects like monitors, smartwatches, fitness trackers, and other medical devices that can receive and transmit data to the Internet. Data can be accessed by the individual who owns or wears the device, or data can be aggregated by healthcare providers as well as MOHs to monitor trends in patients and populations.

Literature review and gap analysis

Most nations now have some level of digital health infrastructure, such as electronic medical records, information networks, clinical decision support, and mobile health applications. Moreover, general trends in health care and information technology in high income countries are driving the development and use of precision medicine, which leverages Big Data and data science methods to analyze a large volume of data captured on cohorts of patients who seek treatment for specific diseases like cancer.[13] Analysis of Big Data includes leveraging genomic information integrated with clinical records and behavioral risk factors like diet, occupation, physical activity, etc. Increasingly, Big Data incorporates the growing volume of data available from consumers' wearable devices (often referred to as the Internet of Things or IoT) that track sleep, exercise, etc. Cloud computing resources in academic medical centers as well as ministries of health support the use of advanced computing techniques to analyze Big Data. However, in medicine, cohorts are often small and consist of a select set of patients who receive treatment at a specific medical center. Machine learning and other forms of AI help researchers create models to classify patients into distinct groups that can help predict outcomes based on the genetic, clinical, and behavioral risk factors.

Public health agencies in some countries are also leveraging Big Data and cloud computing to support PHS. Several of these examples are highlighted below in the case studies featured in this chapter. Numerous applications of AI and data science methods were used to analyze the spread of COVID-19 and predict transmission in multiple countries.[14,15] Several applications of machine learning, for example, were used to analyze social media data to identify or predict localized outbreaks of COVID-19 in cities or regions.[16,17] Others used advanced visualization tools to inform health system leaders of COVID-19 cases and hospitalizations,[18] including the example from the Regenstrief Institute[19] in Fig. 19.2. Previously, Big Data from social media platforms were leveraged to inform predictive models for seasonal influenza, Ebola, and other communicable diseases.[20,21] Data from environmental sensors have also been analyzed using Big Data and AI techniques to model risk of exposure for health outcomes.[22] Efforts to date provide hope that data science methods like AI can be

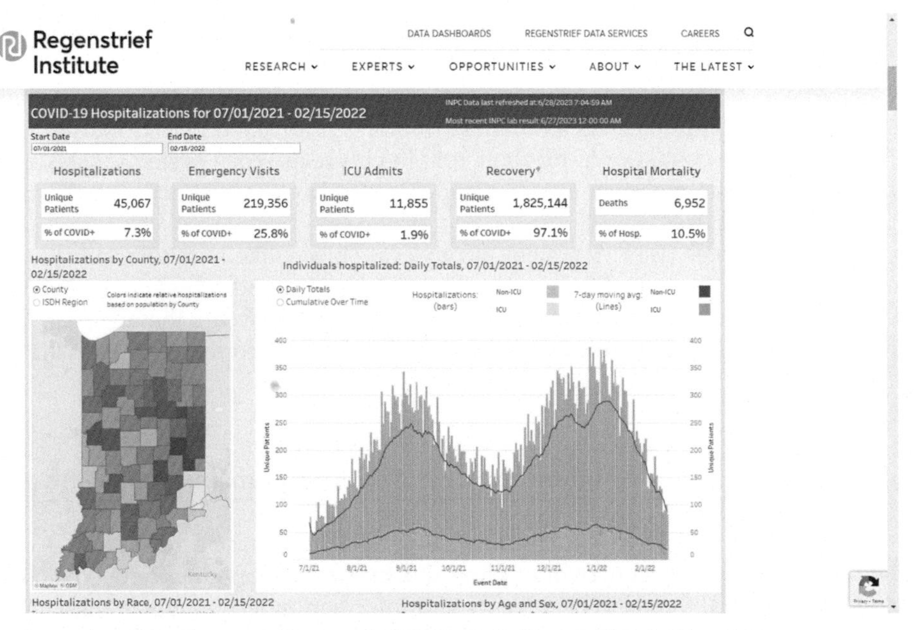

FIGURE 19.2 Screenshot of the Regenstrief Institute's COVID-19 dashboard. This dashboard leverages a statewide health information infrastructure that captures data on COVID-19 testing, cases, hospitalizations, and deaths in real-time using electronic medical records, public health case investigations, and mortality records.

operationalized in public health to analyze social media and other Big Data streams effectively and efficiently in order to augment traditional PHS methods.

Although there is significant optimism, real-world use of analytics and intelligent methods (e.g., AI) is currently limited in medicine as well as public health practice. Over the past decade, numerous peer-reviewed publications summarize various predictive models developed in medicine,[23,24] yet applications of AI and analytics are limited in medical practice. Many models cannot be generalized from one health system to the next one,[25] even though they perform reasonably well on local training datasets. This is a classic example of overfitting caused by noise (irrelevant or incorrect data elements included in the dataset) that does not generalize across other datasets. Moreover, the logic and details for most machine learning models, especially deep learning models, cannot be explained or interpreted by humans, so many clinicians may not be confident of the output and therefore do not use them in practice presently.[26,27]

The same is true for public health where AI and other intelligent methods are not widely deployed. One challenge is class imbalance, a situation where the disease or phenomena of interest is rare in the population limiting the number of positive cases on which to train the ML model. For example, the prevalence of HIV or Rabies may be significantly low across the general population. A model for detecting HIV cases may therefore not have enough 'signal' in the training data set to deliver adequate predictive performance.[12] Larger datasets as well as heterogeneous datasets from multiple regions or nations may be necessary to adequately train a generalizable predictive model. Moreover, there is concern about bias in

AI approaches, especially deep learning. Recent work highlights that when race and other demographic characteristics are co-variates in advanced algorithms, computers might exacerbate inequalities in population health rather than address them.[28,29] For example, researchers in the U.S. found significant racial bias in detecting influenza in social media data using three different machine learning methods.[30] Another challenge for social media platforms is representation. Although half of the world's population (approx. 3.6 billion people) are reported to use at least one social media platform, only 23% of the US adults use Twitter according to the Pew Research Center.[31] Twitter users tend to be young, affluent, and educated. Alternatively, in a study of COVID-19 symptoms and behaviors using Microsoft News in multiple countries,[32] researchers found users of that platform tend to be older than 50 years of age. Each platform therefore comes with its own set of potential biases in user populations. Public health must guard against bias in AI applications, and researchers need to work on developing equitable approaches that synthesize or integrate data from multiple platforms. Because public health seeks to serve all people, agencies should also monitor technology usage trends and adapt approaches to ensure appropriate representation across populations whose preferred channels change over time.

While innovations in analytics and surveillance accrue at an ever-increasing pace, they are seldom distributed homogeneously across the income spectrum both within and across country lines. Although programs like the Harnessing Data Science for Health Discovery and Innovation in Africa (DS-I Africa) by the US National Institutes of Health are currently developing infrastructure in low- and middle-income countries (LMICs), many advancements in analytics, particularly those enhanced by AI, machine learning or autonomous techniques and tools may remain only partially accessible to LMICs which often lack the necessary political, technological, human, or regulatory infrastructure to support such efforts (see **chapter on informatics enabled organizations**).

The underlying foundation for any form of analytics is good quality data. Therefore, while AI enhanced analytics may be able to circumvent or control for bad and unstructured data, full realization of PHS 3.0 will require deploying sustainable and long-term digital data collection solutions. This includes ensuring that as data are captured, they are codified and normalized using semantic standards, including but not limited to International Classification of Disease (ICD) codes, SNOMED CT concepts, and LOINC terms.[33] Furthermore, public health must look beyond traditional sources of data like disease counts and medical records. There now exist a variety of international, national, and local data sources that should be integrated into intelligence processes.[34] There are population-level data on social determinants as well as social media data on health behaviors and sentiment that can link to and augment existing data sources. Many data are readily available using application programming interfaces (APIs), including data on disease cases documented in DHIS2 and health care workers.[4,35] Additional data sources broaden the scope of intelligence activities, and they enable triangulation of persons, events, and time—the trifecta of surveillance.

Robust data management can only be achieved through good data governance and may require tripartite collaboration through country led public-private partnerships that engage with and adopt user-centric designs and solutions. Similarly, adopting flexible, interoperable technologies that allow online and offline data capture can be pivotal to ensuring visibility of health outcomes in particularly vulnerable populations like refugees and migrants. Technologies like blockchain might be used to verify delivery of health services, or delivery of essential goods in healthcare, with the added benefit of enhancing patient tractability and longitudinal analyses on populations.

Strong governance requires trust and deployment of robust security methods to protect the privacy and confidentiality of data.[36] National laws and regulations often define sensitive data and provide guidance on methods that health systems and public health agencies must use to ensure privacy and security of data, including Big Data.[37] Data sharing often requires data use agreements that permit sharing of sensitive data while ensuring privacy practices are enforced.[38] Public health should deploy robust data protections while balancing with the need to share data with partners in and out of the health system to maximize PHS intelligence.[37] It will not be possible to achieve PHS 3.0 goals without trusted partners with whom public health can collaboratively analyze Big Data and provide knowledge to decision-makers.

Case studies

In this section of the chapter, we summarize three case studies that illustrate various aspects of public health intelligence. Each case highlights the benefits of using intelligence-based approaches to address real-world PHS challenges in the digital age. Moreover, the case studies describe the infrastructure, including computing and human resources, necessary to make the shift to intelligence. Infrastructure can be developed over time, initially supporting a single use case then leveraged to support additional use cases as the nation evolves its health system.

Hospital-based Hepatitis C sentinel surveillance in China

Background

Global estimates in 2019 suggested 58 million people were living with chronic Hepatitis C Virus (HCV) infection with 15.2 individuals newly diagnosed. WHO further estimated nearly 300,000 deaths that year, most due to chronic liver disease (cirrhosis) and primary liver cancer (HCC). In 2016, the WHO developed a global viral hepatitis strategy and set goals to eliminate HCV, defined as a 90% reduction in new infection and 65% reduction in deaths, preventing more than seven million related deaths by 2030.[39] Implementation of the strategy and achievement of the goals face many challenges. For example, of those newly diagnosed in 2019, just 9.4 million (62%) were treated with direct antiviral agents (DAAs).

Hepatitis C is one of the serious public health problems in China. WHO (2022) estimates that there are about 7.6 million HCV infections in China, of which 4 to 5 million are chronic HCV patients. Despite national efforts to eliminate HCV under the guidance of the National Action Plan for Viral Hepatitis prevention and control, a certain percentage of chronic HCV infections are still unaware of their infection status. China started HCV case reporting surveillance as early as 1997, but the case reports only contained basic information. In order to guide the effective implementation of prevention and control measures for HCV, it is necessary to know more information about infections including the reasons and methods of HCV testing, epidemiological history, clinical and laboratory test results, etc. To this end, the China CDC conducted sentinel surveillance of hepatitis C cases based on hospitals through a study.

CASE STUDIES *(cont'd)*

Observations

- The "enhanced hepatitis C case reporting" was conducted in hospitals. In the hepatitis clinic of selected sentinel hospitals, doctors were required to complete the "Hepatitis C Case Report Form" for all confirmed hepatitis C patients, and input the content into a computerized Epidemiological Dynamic Data Collection (EDDC) platform.[40]
- The Report Form contents include demographic information, patient history, reasons for testing, liver enzyme levels (alanine aminotransferase ALT and aspartate aminotransferase AST), platelet (PLT) diagnostic tests (anti-HCV, HCV-RVA) and diagnosis (acute hepatitis C, chronic hepatitis C and hepatitis C cirrhosis).

Intervention

- Doctors, trained to use the system, feed the collected information on hepatitis C into the EDDC platform, which directly generates a database that analyzes each case's diagnostic classification, progression of liver fibrosis, and treatment received.
- Initially piloted in 43 hospitals across nine provinces beginning in 2017.
- Expanded to 61 hospitals across 12 provinces in 2018.
- Expanded to 83 hospitals across 15 provinces in 2019.
- Over a 3-year period, all confirmed cases of hepatitis C at these hospitals were analyzed using the EDDC platform.

The EDDC was established with the support of the China/WHO Health Technology Cooperation project "Development and Application of Online Customized Technology for Dynamic Collection of Surveillance data of Emerging (Re-emerging) Infectious Diseases." The platform has been designed on the advanced cloud computing service mode, and can realize the functions of questionnaire design, survey task release, survey form input, survey spatial location map as a service, survey data as a query and standardized data dictionary as a maintenance. Operators only need to use compatible browsers to access the platform and complete the entire information collection process of epidemiological thematic investigation tasks.

Lessons learned

Between 2017 and 2019, 16,241 cases of hepatitis C were reported in 83 hospitals in 15 provinces. The main findings[41] include: 59.77% of all hepatitis C cases were male and 75.96% of all cases were aged 45 years or older; 53.59% of HCV patients were tested through nucleic acid (HCV-RNA), while 46.41% of HCV patients only received antibody (anti-HCV) test; acute hepatitis C was diagnosed in 0.69%; 35.78% (3844/10742) of the cases had a risk of liver fibrosis and 19.51% (2095/10738) of cirrhosis.

The EDDC platform was applied at the hospital-based hepatitis C sentinel surveillance. The platform managed data collection and data agile diversity, timeliness, standardization and normalization of data, unified management and use of the characteristics of data. Since the release of the EDDC platform, the information center of China CDC had continuously improved the functions and performance of the platform. The platform had been successfully used in a number of large-scale national investigation projects.

CASE STUDIES (cont'd)

Recommendations

On the basis of case reporting of hepatitis C, it is necessary to further strengthen reporting of hepatitis C cases. The hospital based-hepatitis C sentinel surveillance has been entirely transformed by the EDDC platform. The study though helpful but is still in its early phase; the next need is to expand the surveillance area further, increase the sentinel hospitals with quality control.

Analytics to inform restaurant inspections in Chicago's Department of Public Health

Background

With a population of 2.7 million, Chicago is the third largest city in the United States. In 2011, Mayor Rahm Emanual launched *Healthy Chicago*, a comprehensive public health agenda with >200 strategies across 12 priority areas to improve the health of city residents. The plan sought to leverage neighborhood-level information and real-time digital health data to track, monitor, and protect health. The Chicago Department of Public Health (CDPH) is responsible for public health functions in the City of Chicago, including PHS. Specifically, CDPH has primary responsibility for foodborne illness surveillance as well as inspections of restaurants to ensure they follow all applicable city ordinances (e.g., laws) pertaining to food establishments. Like many public health departments, CDPH typically inspected restaurants once per year. Inspectors would physically visit each establishment to verify applicable ordinances were followed by owners, managers, and employees. For example, inspectors check temperatures of hot and cold food to ensure they are within the recommended ranges to keep food from spoiling.

Observations

- While CDPH was able to visit each restaurant once per year, the agency desired to visit establishments at risk of violations more than once per year. However, the agency could not visit all restaurants more than once per year.
- The *Healthy Chicago* plan aimed to develop data-driven processes to support the health and well-being of residents.
- CDPH desired to develop a data-driven method to predict which establishments might be at-risk of violating food ordinances so they could visit higher risk restaurants more frequently.

Intervention

- CDPH developed and implemented a predictive analytics solution that employed machine learning to assess foodborne outbreak risk among all restaurants.[5]
- History of prior violations along with a variety of other data, such as parking, vacant housing in the area, complaints of rats in nearby housing, and social media complaints by customers,[42] were fed into the predictive algorithm.
- The algorithm created a model to predict risk among all restaurants given a large number of co-variates (e.g., sanitation complaints, facility type, number of days since last inspection).
- Those establishments with predicted violations were prioritized for a secondary inspection.
- The predictive model is updated regularly allowing inspectors to visit establishments whose risk changes over time.

CASE STUDIES (cont'd)

Lessons learned

The project takes advantage of a robust infrastructure at the City of Chicago. Advanced computational resources and knowledgeable staff part of the Smart Data Project are available to help CDPH and other governmental city units implement predictive analytics solutions. Because of the partnership, CDPH staff received training on analytics and its use, building capacity for other public health intelligence projects. Based on their experience in this project, CDPH staff are working to leverage analytics elsewhere in the agency. The tool further complements and optimizes food protection processes so they can be allocated based on risk. The agency ensures that its use of intelligent methods is based on a shared understanding by the users (public health workers, customers), tasks, and environments. Ensuring solutions meet needs is a key lessons of informatics that is applicable across PH intelligence applications.

Recommendations

Analytics can and should be developed then scaled across PH agencies to support data-driven processes and decisions that will enable PH agencies to utilize scare resources (computing and human) efficiently.

A real-time early warning and alert system based on short-term disability claims registered at the Mexican Social Security Institute

Background

Timely detection of disease outbreaks is essential for public health action. During the COVID-19 pandemic, examples of successful interventions have been tied to a better understanding of the population at risk and health systems operations.[1–3] In Mexico, the health system is fragmented into several financers and providers. While beneficiaries frequently change between these systems, their medical data are rarely shared across providers. This results in belated national-level aggregated data and delayed decision making. The largest of these providers, the Mexican Social Security Institute (*Instituto Mexicano del Seguro Social-IMSS*), provides health care services in its own facilities and social security services to approximately 20 million workers in the formal economic sector and their families, which accounts for approximately half of the Mexican population. IMSS administers and tracks short-term disability claims (STDC). The allocation of this benefit is a medical-administrative action, requiring immediate notification and financial remediation by law. National level STDC data are accrued and aggregated centrally by IMSS, bypassing state or regional data processing. Given the requirements for the resolution of STDCs, IMSS provides an accurate and timely picture of the health conditions of workers within IMSS-affiliated businesses. In the context of the COVID-19 pandemic, this system was adapted and linked to other institutional medical-databases to provide real-time information of potential COVID-19 outbreaks on a more immediate and shorter timeline than health facility surveillance systems.

Observations

- IMSS has the mandate to provide medical, financial and social services to individuals and employers, and to support and guide them in the implementation of

CASE STUDIES (cont'd)

adequate risk reduction activities within work centers.

- IMSS has access to a nation-wide database of short-term disability claims with ICD-10 diagnoses encoded into most claims.
- Public Health Intelligence could be augmented by leveraging administrative, and/or economic data.
- >1 million businesses of all sizes are affiliated with the IMSS in Mexico

Key issues

- During the COVID-19 pandemic, COVID-19 health facility data reported by the federal government could be delayed by several days or even weeks in some locations, a significant delay inadequate for decision-making by businesses and economic sectors.
- After a rigorous lockdown implemented during the first COVID-19 wave, Mexico's economy and people needed a "reactivation strategy" that could ensure safe return to work for millions of people.
- As part of a "New Normality," IMSS sought to develop an automated, easy-to-scale early warning system for outbreak detection and communication strategy linked directly to business administrators.

Intervention

- By leveraging STDC records for individuals with COVID-19, as well as a range of other respiratory diseases like pneumonia, IMSS developed an algorithm that collects the total amount of RD-STDC of each business by epidemiological week.
- The algorithm is an adaptation of the Center for Disease Control's Early Aberration Reporting System C1 (EARS-C1)[6].
- If observed cases the following week exceed the a given level, an alarm is raised for the subsequent week.
- Once the alarm is raised, the system sends a warning signal to business administrators via an institutional platform, to which each affiliated business has access, and to IMSS Occupational Health experts who communicate directly to the business to begin field evaluations of work centers at risk of an outbreak.

Lessons learned

For an organization as large as the IMSS, efficient use of limited resources, like Occupation Health professionals during a pandemic, is paramount. The detection algorithm proved to be essential for IMSS to respond with a higher resolution and target specific businesses at risk of a potential outbreak.

While this algorithm, which could, has proven impactful, it will likely need revision as the pandemic unfolds. IMSS is already working on updating the algorithm in three ways: (a) establishing different detection methods for very low incidence diseases (fixed thresholds); (b) establishing an endemic channel analysis for the temporal detection method, wherein we can compare data from previous weeks to its equivalent epi week in previous years for infectious diseases of regional interest; and (c) establishing a spatial detection algorithm able to find clusters of diseases related to Occupational Health (cancer clusters for example), by analyzing incidence in discrete geographical areas and deploying a signal

CASE STUDIES *(cont'd)*

when that incidence is significantly higher than in other regions.

Finally, IMSS was able to leverage data that was otherwise unused for monitoring purposes with great effect (RD-STDC), and in the same line, help bridge known challenges within already existing PH data (health-facility data).

Recommendations

Early warning systems can be built upon varied sources of data, including those outside the care delivery system. Timely data is important, especially during a pandemic. Building an infrastructure that can promptly process incoming disease data and reduce lag in communication channels by informing decision makers directly can bring about high resolution action that much faster.

Vision for the future

In the future, we envision a world in which public health agencies employ robust digital health infrastructures that:

(1) enable collection of structured data through online and offline mobile devices;
(2) leverage appropriately designed instruments with standardized data ontologies;
(3) manage data securely and utilize robust governance processes that enable sharing of data while maintaining privacy and confidentiality; and
(4) feed data into cloud computing services that facilitate advanced analytics, including AI, to inform policy development, community-level interventions, and public communication strategies in real-time.

In the future, governments, and private institutions, alike will be able to take advantage of new technologies to facilitate efficient delivery and access to healthcare and other essential services in even the most complex settings. Concomitantly, public health institutions will be better able to understand and plan for the needs of the populations they serve as demands arise, while individuals will gain the benefits of digital identity, such as access to vital services, as well to participate socially and economically in an ever more digitized society. We envision that citizens will use digital technologies, including IoT devices, to manage their own health and share health data with public health authorities who can leverage these data to better manage the health and well-being of populations. We believe that each nation is capable of evolving PHS to Version 3.0 over time and that these efforts can be coordinated with the nation's digital health (e-health) strategy.

Actions

These efforts will be necessary to achieve the vision.

- Nations need to invest in data modernization, including investment in the nation's public health infrastructure, to enable PH agencies to leverage digital health data in their efforts to identify an unexpected or unusual public health event within its territory. Investments should be outlined in the nation's digital health (eHealth) strategy as recommended by WHO,[43] and nations should publicly report on the progress of achieving their digital health strategy goals.
  - Investments should include the adoption and use of data standards, including ICD (International Classification of Disease) and LOINC (Logical Observations, Identifiers, Names and Codes) to ensure that data can be harmonized across information systems deployed with the nation.
  - Investments should include appropriate digital collection and management of personal data regarding patients infected or affected by the public health event so that timely population health information can be provided to WHO and data can be linked across information systems in support of integrated disease surveillance and response (IDSR) approaches.
  - Nations should adopt advanced computing platforms that enable analytics for population health. These systems should be leveraged to identify potential disease outbreaks and trigger epidemiological investigation. This should be a core capacity for each member state, although LMICs may need to participate in regional computing hubs hosted by continental, multi-state collaboratives.
- Nations need to advance data sharing such that data captured by clinicians, social service agencies, and other entities within its territory can be integrated for population health at regional and/or national levels.
  - Improve existing systems, reduce manual data entry and unstructured data. A significant portion of health data is still collected manually, particularly in low- and middle-income countries. While the costs of manual data entry are well known, the burden of unstructured data is probably less visible, yet equally impactful. Multiple analysts suggest that 80%–90% of the world's data are unstructured,[44] making them either useless, or difficult to amend without robust analytical infrastructure. Existing data collection tools, standard instruments, and medical ontologies, as well as novel methods, need to be made available to health professionals to reduce paper-based data collection system, and promote digital solutions.
  - Data on individual patients should be linked across the health system so clinicians and ministries can identify early disease detection as well as treatment for disease for use during a public health emergency (e.g., IDSR).
  - Nations should execute data sharing agreements with regional entities (e.g., Africa CDC) and WHO so information on public health threats can be reported timely using digital health platforms established by regional authorities and WHO.
  - Information systems within nations should be interoperable, employing syntactic as well as semantic standards to enable data sharing within the territory. Nations

should similarly use standards to exchange digital health data with regional authorities and WHO.
  - With great data comes great responsibility. National-level Data Protection Regulatory frameworks must be established. While digitalization offers stronger data security than paper-based systems, it is by no means foolproof. In fact, from 2009 to 2020 (a period of significant digitalization), over 268 million healthcare records were either lost, exposed, stolen or were otherwise disclosed, with a clear increasing trend.[45] As nations evolve into PHS 3.0, and begin to leverage, analyze, and share more and more data, data privacy protocols need to keep pace to protect individuals and minority populations.
- Nations must determine how best to engage in and regulate social media. Many public health authorities have social media accounts, which they use to disseminate information on emerging health threats as well as important messages regarding health and well-being. Yet, social media platforms are also notorious for amplifying misinformation, a phenomenon observed numerous times during the COVID-19 pandemic.
  - Governments, including MOHs, must balance freedom of speech and private enterprise with the health and well-being of populations. Platforms that limit health literacy, encourage fearmongering, and blatantly spread misinformation may require greater regulation or oversight. Moreover, public health authorities need to collaborate with platforms who can help them disseminate evidence-based scientific information at appropriate literacy levels to at-risk populations. This is not easy but must be a component of digital health strategies.
- To support robust digital health infrastructures within a nation, MOHs should train public health informatics workers. Data and information science should be core competencies for public health informatics workers, and these workers should be deployed at local, regional, and national levels to support public health intelligence operations, including the implementation and maintenance of the nation's digital health infrastructure. This training should be a component of the nation's digital health strategy.
- The future of analytics and public health intelligence is here; it's just not evenly distributed. While digital health infrastructure is rapidly increasing, the rate of progress has been alluded to as the "new face of inequality,"[46] a symptom and effect of economic and social inequalities.
  - Although investment in modern infrastructure in LMICs will likely need donor support, the international community needs to guard against colonialism with respect to advanced computing and analytics for public health.
  - Algorithms can be biased, necessitating vigilance by MOHs and other health authorities to ensure predictive models help advance health for all, not just some, populations.
- As nations implement infrastructure and algorithms, appropriate technical safeguards and robust governance must be deployed, and scaled, to ensure privacy and confidentially of data are maintained in balance with use. Health data need to be available across the health system ecosystem, but identifiable data must remain private. When possible, data should be shared using the minimum necessary heuristic. Data should be used for PHS but only for appropriate purposes that are transparent with citizens and organizations involved in the capture, management, and analysis of health data.

- Big Data and AI may necessitate updated regulations as well as the use of updated data use agreements to ensure privacy and confidentiality while enabling use for PHS.
- Strong governance in PHS allows use of data and algorithms but maintains trust by patients, providers, and public health.
- In addition to transparency on how data are used, machine learning algorithms and AI approaches should be transparent. Results should be reproducible and interpretable. Solutions that use AI should further ensure that bias is minimized, especially when predicting or classifying characteristics of vulnerable subpopulations.

References

1. Dixon BE, Caine VA, Halverson PK. Deficient response to COVID-19 makes the case for evolving the public health system. *Am J Prev Med*. 2020;59(6):887–891.
2. Regmi K, Gee I. *Public Health Intelligence: Issues of Measure and Method*. 1 ed. Switzerland: Springer; 2016.
3. Kariuki JM, Manders EJ, Richards J, et al. Automating indicator data reporting from health facility EMR to a national aggregate data system in Kenya: an Interoperability field-test using OpenMRS and DHIS2. *Online J Public Health Inform*. 2016;8(2):e188.
4. Ahmed K, Bukhari MA, Mlanda T, et al. Novel approach to support rapid data collection, management, and visualization during the COVID-19 outbreak response in the world health organization African region: development of a data summarization and visualization tool. *JMIR Public Health Surveill*. 2020;6(4), e20355.
5. Choucair B, Bhatt J, Mansour R. A bright future: innovation transforming public health in Chicago. *J Publ Health Manag Pract*. 2015;21:S49–S55.
6. Magnuson JA, Dixon BE. Public health informatics: an introduction. In: Magnuson JA, Dixon BE, eds. *Public Health Informatics and Information Systems*. Cham: Springer International Publishing; 2020:3–16.
7. DeSalvo KB, Wang YC, Harris A, Auerbach J, Koo D, O'Carroll P. Public health 3.0: a call to action for public health to meet the challenges of the 21st century. *Prev Chronic Dis*. 2017;14:E78.
8. DeSalvo KB, O'Carroll PW, Koo D, Auerbach JM, Monroe JA. Public health 3.0: time for an upgrade. *Am J Publ Health*. 2016;106(4):621–622.
9. Centers for Disease Control and Prevention. *Data Modernization Initiative. Department of Health and Human Services, U.S. Public Health Surveillance and Data Web Site*; 2021. https://www.cdc.gov/surveillance/surveillance-data-strategies/data-IT-transformation.html. Accessed April 19, 2021.
10. Council of State and Territorial Epidemiologists. *CSTE Applauds Public Health Data Modernization Funding in President-Elect Biden's American Rescue Plan. PR Newswire*; 2021. https://www.prnewswire.com/news-releases/cste-applauds-public-health-data-modernization-funding-in-president-elect-bidens-american-rescue-plan-301209571.html. Accessed February 21, 2021.
11. *HIMSS. Public Health Information and Technology Infrastructure Modernization Funding Report*. HIMSS; 2022. https://www.himss.org/resources/public-health-information-and-technology-infrastructure-modernization-funding-report. Accessed August 22, 2022.
12. Kasthurirathne SN, Ho YA, Dixon BE. Public health analytics and Big data. In: Magnuson JA, Dixon BE, eds. *Public Health Informatics and Information Systems*. Cham: Springer International Publishing; 2020:203–219.
13. Velmovitsky PE, Bevilacqua T, Alencar P, Cowan D, Morita PP. Convergence of precision medicine and public health into precision public health: toward a Big data perspective. *Front Public Health*. 2021;9, 561873. https://doi.org/10.3389/fpubh.2021.561873.
14. Khanzada A, Hegde S, Sreeram S, et al. Challenges and opportunities in deploying COVID-19 cough AI systems. *J Voice*. 2021;35:811–812.
15. Hernandez-Avila M, Tamayo-Ortiz M, Vieyra-Romero W, et al. Use of private sector workforce respiratory disease short-term disability claims to assess SARS-CoV-2, Mexico, 2020. *Emerg Infect Dis*. 2022;28(1):214–218.

16. Shen C, Chen A, Luo C, Zhang J, Feng B, Liao W. Using reports of symptoms and diagnoses on social media to predict COVID-19 case counts in mainland China: observational infoveillance study. *J Med Internet Res.* 2020;22(5), e19421.
17. Mackey T, Purushothaman V, Li J, et al. Machine learning to detect self-reporting of symptoms, testing access, and recovery associated with COVID-19 on twitter: retrospective Big data infoveillance study. *JMIR Public Health Surveill.* 2020;6(2), e19509.
18. Dixon BE, Grannis SJ, McAndrews C, et al. Leveraging data visualization and a statewide health information exchange to support COVID-19 surveillance and response: application of public health informatics. *J Am Med Inf Assoc.* 2021;28(7):1363–1373.
19. Regenstrief Institute. *Regenstrief COVID-19 Dashboard.* Regenstrief Institute, Inc; 2020. https://www.regenstrief.org/covid-dashboard/. Accessed June 10, 2020.
20. Santillana M, Nguyen AT, Dredze M, Paul MJ, Nsoesie EO, Brownstein JS. Combining search, social media, and traditional data sources to improve influenza surveillance. *PLoS Comput Biol.* 2015;11(10), e1004513.
21. Jacobsen KH, Aguirre AA, Bailey CL, et al. Lessons from the Ebola outbreak: action Items for emerging infectious disease preparedness and response. *EcoHealth.* 2016;13(1):200–212.
22. Yang W, Park J, Cho M, Lee C, Lee J, Lee C. Environmental health surveillance system for a population using advanced exposure assessment. *Toxics.* 2020;8(3).
23. Ravizza S, Huschto T, Adamov A, et al. Predicting the early risk of chronic kidney disease in patients with diabetes using real-world data. *Nat Med.* 2019;25(1):57–59.
24. Klann JG, Anand V, Downs SM. Patient-tailored prioritization for a pediatric care decision support system through machine learning. *J Am Med Inf Assoc.* 2013;20(e2):e267–e274.
25. Dexter GP, Grannis SJ, Dixon BE, Kasthurirathne SN. Generalization of machine learning approaches to identify notifiable conditions from a statewide health information exchange. *AMIA Jt Summits Transl Sci.* 2020;2020:152–161.
26. Rasheed K, Qayyum A, Ghaly M, Al-Fuqaha A, Razi A, Qadir J. Explainable, trustworthy, and ethical machine learning for healthcare: a survey. *Comput Biol Med.* 2022;149, 106043.
27. Petch J, Di S, Nelson W. Opening the black box: the promise and limitations of explainable machine learning in cardiology. *Can J Cardiol.* 2022;38(2):204–213.
28. Kassem MA, Hosny KM, Damaševičius R, Eltoukhy MM. Machine learning and deep learning methods for skin lesion classification and diagnosis: a systematic review. *Diagnostics.* 2021;11(8).
29. Char DS, Shah NH, Magnus D. Implementing machine learning in health care—addressing ethical challenges. *N Engl J Med.* 2018;378(11):981–983.
30. Lwowski B, Rios A. The risk of racial bias while tracking influenza-related content on social media using machine learning. *J Am Med Inf Assoc.* 2021;28(4):839–849.
31. Pew Research Center. *Social Media Fact Sheet. Pew Research Center*; 2021. https://www.pewresearch.org/internet/fact-sheet/social-media/?menuItem=b14b718d-7ab6-46f4-b447-0abd510f4180. Accessed February 17, 2022.
32. Dixon BE, Mukherjee S, Wiensch A, Gray ML, Ferres JML, Grannis SJ. Capturing COVID-19-like symptoms at scale using banner ads on an online news platform: pilot survey study. *J Med Internet Res.* 2021;23(5), e24742.
33. Magnuson JA, Merrick R. Public health information standards. In: Magnuson JA, Dixon BE, eds. *Public Health Informatics and Information Systems.* Cham: Springer International Publishing; 2020:129–145.
34. Mensah E, Goderre JL. Data sources and data tools: preparing for the open data ecosystem. In: Magnuson JA, Dixon BE, eds. *Public Health Informatics and Information Systems.* Cham: Springer International Publishing; 2020:105–127.
35. Oluoch T, Muturi D, Kiriinya R, et al. Do interoperable national information systems enhance availability of data to assess the effect of scale-up of HIV services on health workforce deployment in resource-limited countries? *Stud Health Technol Inf.* 2015;216:677–681.
36. McKinley C. Data information and governance. In: Finnell JT, Dixon BE, eds. *Clinical Informatics Study Guide: Text and Review.* Cham: Springer International Publishing; 2022:221–226.
37. Hulkower R, Penn M, Schmit C. Privacy and confidentiality of public health information. In: Magnuson JA, Dixon BE, eds. *Public Health Informatics and Information Systems.* Cham: Springer International Publishing; 2020:147–166.

38. Fillmore AR, McKinley C, Tallman EF. Managing privacy, confidentiality, and risk: towards trust. In: Dixon BE, ed. *Health Information Exchange: Navigating and Managing a Network of Health Information Systems*. 2nd ed. Academic Press; 2023.
39. World Health Organization. *Combating Hepatitis B and C to Reach Elimination by 2030*. WHO; 2016. https://apps.who.int/iris/bitstream/handle/10665/206453/WHO_HIV_2016.04_eng.pdf. Accessed March 2, 2022.
40. Yujie M, Qi XP. The application of epidemiological dynamic data collection platform in public health investigation. *J Med Inform*. 2013;34(6).
41. Ding GW, Pang L, Wang XC, Ye SD, Hei FX. Analysis of baseline characteristics and treatment status of hepatitis C in sentinel hospitals from 2017 to 2019. *Zhonghua Gan Zang Bing Za Zhi*. 2020;28(10):844–849.
42. Sadilek A, Caty S, DiPrete L, et al. Machine-learned epidemiology: real-time detection of foodborne illness at scale. *NPJ Digit Med*. 2018;1:36.
43. World Health Organization. In: *National eHealth Strategy Toolkit*. Geneva, Switzerland: World Health Organization and International Telecommunication Union; 2012.
44. Harbert T. *Tapping the Power of Unstructured Data*. MIT Sloan School of Management; 2021. https://mitsloan.mit.edu/ideas-made-to-matter/tapping-power-unstructured-data. Accessed February 17, 2022.
45. *HIPAA Journal. Healthcare Data Breach Statistics*; 2021. https://www.hipaajournal.com/healthcare-data-breach-statistics/. Accessed February 17, 2022.
46. United Nations. *Don't Let the Digital Divide Become 'the New Face of Inequality': UN Deputy Chief*. United Nations; 2021. https://news.un.org/en/story/2021/04/1090712. Accessed March 2, 2022.

Further reading

1. Asadzadeh A, Pakkhoo S, Saeidabad MM, Khezri H, Ferdousi R. Information technology in emergency management of COVID-19 outbreak. *Inform Med Unlocked*. 2020;21, 100475.
2. Schomberg JP, Haimson OL, Hayes GR, Anton-Culver H. Supplementing public health inspection via social media. *PLoS One*. 2016;11(3), e0152117.
3. Yoon CH, Torrance R, Scheinerman N. Machine learning in medicine: should the pursuit of enhanced interpretability be abandoned? *J Med Ethics*. 2022;48(9):581–585.
4. Kasthurirathne SN, Grannis SJ. Analytics. In: Finnell JT, Dixon BE, eds. *Clinical Informatics Study Guide: Text and Review*. Cham: Springer International Publishing; 2022:227–239.
5. Kasthurirathne SN, Grannis S, Halverson PK, Morea J, Menachemi N, Vest JR. Precision health-enabled machine learning to identify need for wraparound social services using patient- and population-level data sets: algorithm development and validation. *JMIR Med. Inform.* 2020;8(7), e16129. https://doi.org/10.2196/16129.
6. Dixon BE, Holmes JH, Section Editors for the IMIA Yearbook Section on Managing Pandemics with Health Informatics. Managing Pandemics with Health Informatics. *Yearb. Med. Inform.* 2021;30(1):69–74. https://doi.org/10.1055/s-0041-1726504.

CHAPTER

20

Navigating a rapidly changing information and communication landscape amidst "infodemics"

Jacqueline Cuyvers[1], Aly Passanante[1], Ed Pertwee[1], Pauline Paterson[1,2,3], Leesa Lin[1,4,5] and Heidi J. Larson[1,6,7]

[1]Department of Infectious Disease Epidemiology, London School of Hygiene & Tropical Medicine, London, United Kingdom; [2]Health Protection Research Unit in Vaccines and Immunisation, London School of Hygiene & Tropical Medicine, London, United Kingdom; [3]Grantham Institute - Climate Change and the Environment, Imperial College London, London, United Kingdom; [4]Laboratory of Data Discovery for Health (D24H), Hong Kong Science Park, Hong Kong SAR, China; [5]WHO Collaborating Centre for Infectious Disease Epidemiology and Control, School of Public Health, LKS Faculty of Medicine, The University of Hong Kong, Hong Kong SAR, China; [6]Centre for the Evaluation of Vaccination, Vaccine & Infectious Disease Institute, University of Antwerp, Antwerp, Belgium; [7]Department of Health Metrics Sciences, University of Washington, Seattle, WA, United States

Introduction

Global health security is a persistent and pressing challenge in the 21st century, exacerbated by the rapid spread of infectious diseases and the parallel spread of misinformation in an interconnected world. As of March 2022, the COVID-19 pandemic had resulted in over 479 million cases and 6 million fatalities reported worldwide, with these figures continuing to rise on a daily basis.[1] Social media and online news sources offer unprecedented access to health information at a global scale. However, the speed of dissemination, uneven content quality, conflicting information, and lack of editorial control and

Modernizing Global Health Security to Prevent, Detect, and Respond
https://doi.org/10.1016/B978-0-323-90945-7.00023-3

regulatory oversight have fueled concerns regarding the impact that the contemporary information and communication environment may be having on a range of health outcomes.

These concerns came to the fore during the COVID-19 pandemic and the rollout of vaccination programs to address it. One 2021 study found that almost 8 in 10 people in the United States attached some credibility to at least one misinformation statement about COVID-19.[2] Belief in COVID-19 misinformation has likely had a negative impact on public health,[3] as it is associated with increased vaccine hesitancy,[4] decreased compliance with protective measures such as wearing masks,[5] and reduced trust in public health messaging[6] and authorities.[7] However, the problem extends well beyond the COVID-19 pandemic. During the 2014 Ebola outbreak in West Africa, misinformation about the disease and its transmission led to a significant delay in implementing public health interventions.[8] Similarly, the ongoing global measles outbreaks have been linked to vaccine misinformation and subsequent decreased vaccination rates.[9]

Concerns about the impact of the information and communication environment on the COVID-19 response led the WHO to declare in February 2020 that it was not just fighting a pandemic but also an infodemic. The term "infodemic", a portmanteau of "information" and "epidemic" was originally coined in the context of the 2002-04 SARS outbreak by David Rothkopf, who defined it as "a few facts, mixed with fear, speculation, and rumor, amplified and relayed swiftly worldwide by modern information technologies."[10] It has since entered common usage, reflected in its entry into the Oxford English Dictionary in 2020, where it is defined as "a proliferation of diverse, often unsubstantiated information relating to a crisis, controversy, or event, which disseminates rapidly and uncontrollably through news, online, and social media, and is regarded as intensifying public speculation or anxiety." In the context of the COVID-19 pandemic, the WHO narrowed the definition to "an infodemic is too much information, including false or misleading information in digital and physical environments during a disease outbreak."[11] This narrower definition has, however, been criticized as research suggests that audiences have developed sophisticated skills and strategies to navigate modern high-choice media environments and that consequently the problem of "information overload" may be overstated.

The "infodemic" concept has also been criticized for potentially conflating distinct problems under a single term. These include: the proliferation of user-generated content on social media; the parallel spread of misinformation and "fake news"; social media "bots" and other forms of "inauthentic" online behavior; state-sponsored disinformation; political polarization; populist distrust of experts; conspiracy theories; and the promotion of complementary and alternative medicines online, often for profit. While the term "infodemic" is probably unavoidable due to its popular currency, it bears emphasis that these problems do not stem from a single cause and are unlikely to have a single solution.

To address these challenges effectively, public health stakeholders must be proactive in identifying, assessing, and managing potential threats. This requires a comprehensive and coordinated response that embraces preventive measures, early detection and diagnosis, and effective treatments. Governments must also be prepared to respond resiliently to new outbreaks, anticipating changes in the epidemiological and technological landscapes, and addressing the social, political, and economic factors that affect health outcomes.

This chapter outlines the key challenges presented by the evolving information and communication landscape and offers key strategies on how to better understand and address health misinformation and "infodemics". Implementing these strategies could lead to more impactful engagement opportunities to improve global health communication, public health trust, and health outcomes.

Health misinformation and "infodemics"

The spread of health misinformation during infectious disease outbreaks is not a new phenomenon. During the Middle Ages, in the time of the bubonic plague pandemic in Europe, common belief held that the plague was transmitted through miasma, or poisoned air, could permeate one's skin through pores during a warm bath or through infected people's clothes, and was transmitted by cats who were then culled.[12] During the Influenza pandemic of 1918, misinformation about the origins of the outbreak pointed both to aspirin and biological warfare as causes.[13]

Given these historical precedents and the communicative opportunities afforded by contemporary digital technologies, it is not surprising that misinformation about the origins, prevention, and treatment of the COVID-19 pandemic was rife across blogs, forums, and social media.[14,15] Globally, social media usage increased by 20%–87% as the pandemic began to spread; so too did information seeking and unchecked information sharing.[3] Researchers found 1225 pieces of misinformation published online in English over a 4-month period, between January 1, 2020 and April 30, 2020, and of those, social media accounted for 50% while the other 50% came from a combination of sources such as news, tabloids, and individuals.[3]

The impact of misinformation during the COVID-19 pandemic was multifaceted. Initially, false or misleading claims about public health authority actions and policies created confusion and undermined effective communication. Additionally, the spread of misinformation targeting specific ethnic groups further strained public trust. In the early stages of the pandemic, researchers found that the most common form of misinformation was false or misleading claims about the actions and policies of public health authorities, followed by claims about the spread of the virus through communities, including content blaming particular ethnic groups for spreading the virus.[7] Subsequently, the proliferation of vaccine-related misinformation hindered vaccination efforts and contributed to increased vaccine hesitancy. Later, in the context of COVID-19 vaccination programs, misinformation around vaccines and vaccination became prolific and likely contributed to increased vaccine hesitancy and lower vaccine uptake than would otherwise have been the case. As COVID-19 vaccination programs were introduced, a surge in misinformation relating to vaccines and vaccination emerged. This proliferation of false information contributed to an increase in vaccine hesitancy and a lower uptake of vaccines compared to what would have been anticipated.[16] The dissemination of misleading claims regarding the safety and efficacy of vaccines, along with unfounded conspiracy theories, played a role in sowing doubt and uncertainty among the public. Consequently, the efforts to achieve widespread vaccine coverage and mitigate the impact of the pandemic were hindered by the influence of this misinformation. Recognizing and addressing these various forms of misinformation is essential to ensure the effective dissemination of accurate information and foster public trust in public health authorities.

Even prior to the COVID-19 pandemic, vaccine hesitancy had been identified as a key global health concern due to outbreaks of vaccine-preventable diseases like measles and whooping cough, which are responsible for the deaths of 2–3 million people annually. In 2019, the WHO identified vaccine hesitancy, defined as the "reluctance or refusal to vaccinate despite the availability of vaccines", as one of the top 10 threats to global health.[15] The COVID-19 pandemic has amplified the urgency of addressing vaccine hesitancy. While there have been over 13 billion doses of COVID-19 vaccines administered many eligible people chose not to be vaccinated, in some cases due to their belief in inaccurate, misleading, or false information about COVID-19 vaccines. A number of studies suggest a positive association between exposure to misinformation and vaccine hesitancy at a population as well as individual level.

An "infodemic" such as that witnessed during the COVID-19 pandemic creates a challenge to global health security in several interconnected ways. First, rumors and misinformation can lead to risk-taking behaviors that can harm individual or collective health. A Pew research poll in April 2020 observed that 50% of Americans found it difficult to determine what is true in the news about the COVID-19 pandemic; 64% reported seeing news that seemed fabricated; and 38% came across fabricated information that was later confirmed to be false.[17] Without clinical data to support their efficacy, some politicians promoted the use of the antimalarial drugs chloroquine and hydroxychloroquine as treatments/prophylaxis for COVID-19. This misinformation unfortunately led to mortalities.[18] More than any other unsubstantiated treatment, 44% of U.S. adults had heard of hydroxychloroquine as a possible treatment.[17]

Recent research suggests some reasons why misinformation can be so difficult to address. One reason is that misinformation has emotional and ideological determinants as well as rational ones.[19,20] Misinformation beliefs cannot be explained as a lack of "correct" scientific or medical information but rather as a result of complex belief systems shaped by internal and external factors.[21] An implication of this is that misinformation corrections can be ineffective if they are at odds with an individual's identity or worldview,[22] resulting in defensive responses[23] ranging from disbelief to denial to attempts at discrediting the sources of the correction.[21,24]

Moreover, when presented with scientific information, individuals process and remember both the verbatim, or rote facts, and the gist, or encapsulation of the meaning.[25] While it becomes harder to recall specific facts and figures over time, the gist of a message is often retained, especially if it resonates with an individual emotionally or fits with their existing worldview or beliefs.[26] Even when proven false and accepted as false, misinformation can still be damaging as it can impact an individual's reasoning in the future. This is known as the "continued influence effect".[21]

In addition to the problem of misinformation, conflicting messages during an outbreak can lead to mistrust in health authorities and undermine the public health response. Despite evidence, already available in March 2020,[27] that mask-wearing effectively reduced airborne transmission of COVID-19 risk, the US Centers for Disease Prevention and Control (CDC) continued to actively advise against mask-wearing until April 2020.[5] The US Surgeon General advised the public to stop buying masks as they were not effective in preventing infection.[28]

Some aspects of the academic research and publication process likely compounded these problems. There are strong institutional factors that incentivize quantity and rapidity of publication, as these are often taken to be markers of academic success. The growing practice of making studies publicly available through preprint services prior to acceptance by a journal, while it may increase the openness and transparency of scientific inquiry, also means that studies are often released to the public and journalists before they have been peer-reviewed.[29] Only a minority of articles published during the early stages of the COVID-19 pandemic were based on original research, while many preprints did not pass peer review or were later amended or retracted.[30] Nonetheless, many preprints were circulated and cited on social media.

For all these reasons, infodemics can intensify or lengthen outbreaks by increasing public uncertainty about the appropriate health behaviors to protect oneself and others.[31]

Health communication in a changing information landscape

Making best use of digital data

Despite the challenges that rumors and misinformation can bring, social media and other new communications technologies can also have a positive impact on improving public health outcomes. This is both because they are rich sources of data on public attitudes and behaviors and because they offer new opportunities for health communication.

Gunther Eysenbach and others have attempted to systematize the approach to collecting and analyzing digital data in a public health context. Eysenbach first coined the term "infodemiology" in 2002, which he defined as the "science of distribution and determinants of information in an electronic medium, specifically the internet, or in a population, with the ultimate aim to inform public health and public policy."[6,32] This definition has changed in scope and evolved over the last two decades. In its first iteration, the term focused on what was being published on the internet, or "supply-based infodemiology.[33] Infodemiology research at this stage focused on the quality and veracity of health content published online and its potential impact on public health. Shortly thereafter, the definition was expanded to include "demand-based" behavior, or what people were actively seeking online.[34]

Analysis of web content and search data has enabled researchers to identify information gaps and even predict emerging epidemics.[32] In 2009, researchers identified influenza outbreaks by monitoring an increase in relevant symptom search terms in specific geographic locations, which served as an early indicator of influenza, as this internet activity on average preceded the patient's visit to a doctor by a week.[35] Beyond outbreaks like SARS,[36] influenza,[37,38] and listeriosis,[39] numerous studies have investigated the relationship between search behaviors and the incidence of disease using online metrics and communication patterns as infodemiology indicators. This monitoring or surveillance behavior has come to be known as "infoveillance".[32] Sometimes, alternative terms such as "digital epidemiology" or "social listening" are used, although these are not strict synonyms.[40]

Whichever term is preferred, analysis of online conversations and posts from sources like blogs, social media, and forums is increasingly important in that this monitoring can provide researchers with early warning signs of public health misinformation about a topic so they can effectively enact countermeasures.[32] At the same time, monitoring search information seeking, such as Google search data or search behavior can provide an early warning of disease outbreaks or epidemics.[41] Compared to traditional primary research techniques such as interviews, focus groups, and surveys, this type of digital research can allow for near-real-time analysis of large-scale, naturally occurring data on health beliefs and behaviors.[32]

The collection and use of these data are not without challenges. Researchers should be aware of and work to mitigate potential issues with bias in the data, both in data collection and analysis. Firstly, attitudes expressed on social media data may not be representative of wider public attitudes,[42] as not all segments of the population are equally active online, and specific populations may be underrepresented.[40] Secondly, the choice of data sources and software tools can compound these underlying biases, especially where social media companies are selective about the data they make available to researchers. A good example is the arguable overuse of Twitter data in misinformation research, simply because Twitter has historically made its data more easily accessible to researchers than have other platforms. Thirdly, social media data are self-reported and subject to self-presentation biases, as users do not always posthonestly about their beliefs and behaviors. Finally, there is also the issue of selection bias in how social media data are collected and analyzed.[43] These challenges are not insurmountable, however, and we offer some specific recommendations on how to address issues of data bias and representativeness in the "actions" section below.

Leveraging social media platforms for health communication

Social media can also have a positive impact on improving public health outcomes through the communication opportunities that they afford.[44] Since their inception, social media platforms have served as tools to help people find health information, receive support from peers,[45] and complement or augment their offline health knowledge. This has helped patients reach a diagnosis earlier, improved self-management of chronic diseases, and led to improved mental health through increased psychosocial support.[44] Research has shown that social media activity and data volume increase in times of crisis, including epidemic outbreaks.[41]

During public health crises, social media channels can serve as powerful communication platforms for scientists and public health officials to quickly spread scientific information and public health communication. In the Czech Republic, early videos created by a data scientist about the protective benefits of mask-wearing raised public awareness of the scientific evidence. This YouTube video was viewed over 600,000 times and led a #masks4all movement, whose Facebook group had over 30,000 members handmaking masks to share with the public all in advance of the government's mask-wearing guidance of mandates.[46] In Vietnam, in February 2020, the National Institute of Occupational Safety and Health worked with a lyricist to write new handwashing lyrics over a popular pop song, "Ghen".[47] Choreographer Quang Dang then created a dance challenge to this song and shared it on TikTok under the hashtag #GhenCoVyChallenge. This video, which generated millions of views and thousands of responses is an example of a digital campaign that raised awareness of the importance of proper hygiene techniques to prevent the spread of COVID-19.[48]

However, leveraging social media for public health communication presupposes a public with access to computers, smartphones, and internet services, but also the ability to read, understand, and contextualize online content. Even in countries with high levels of education and internet penetration, like the United States, many people lack the levels of health literacy needed to make informed medical decisions.[49] As of a study in the United States, 53% of the population had "intermediate" levels of health literacy, with 12% "proficient", 22% "basic" and 14% "below basic", the group least likely to get their health information online or from written sources.[50] A recent systematic review of literature on the causes and impacts of the COVID-19 infodemic found that social media usage and low levels of health literacy were major causes (other factors identified included the speed with which COVID-19-related articles were published online, often in the form of preprint manuscripts that had not been peer-reviewed).[29] Improving health literacy could therefore help to mitigate the impacts of future infodemics, a point to which we return in the "Actions" section below.

CASE STUDY

Addressing public concerns on COVID-19 vaccines in East Asia

Topic and problem identified

Background

- Suboptimal uptake of COVID-19 vaccines across parts of East and Southeast Asia, such as Hong Kong, Thailand, and Singapore, has shown that vaccine hesitancy is a key impediment to the control of vaccine-preventable diseases (VPDs), despite advances in medical technologies. It highlights the challenges of overcoming vaccine hesitancy in the context of a world-devastating pandemic and resurgences fueled by COVID-19 variants.
- This case study focuses on the feasibility evaluation of COVID-19 vaccine chatbots.

Observations

- In response to this crisis, a novel solution was proposed to utilize conversational AI services (chatbots) to address common public questions, concerns, and misconceptions to combat misinformation (including fake news, rumors, and conspiracy theories) around COVID-19 vaccines and increase COVID-19 vaccine uptake with a focus on communities that have shown persistent hesitancy after vaccine rollout.
- Techniques discussed in this chapter, as well as social media listening and analysis of web search data, were used to identify the key questions, concerns, and misconceptions around COVID-19 vaccines in each of the study locations and target demographics.
- This case study focuses on a pilot study to evaluate the feasibility of COVID-19 vaccine chatbots.
- In addition to monitoring and assessing COVID-19 vaccine confidence and information needs in real-time, the intervention can support long-term mitigation of vaccine hesitancy by proactively immunizing the global community against vaccine misinformation—an innovative "prebunking" approach from inoculation theory.

CASE STUDY (cont'd)

Key issues

- Vaccine hesitancy can emerge, reemerge, and spread in any population for myriad complex reasons, including complacency against the threat of VPDs, a lack of confidence in vaccine safety and effectiveness, distrust in health authorities or political leadership, inconvenient access to vaccine information and vaccines, and exposure to vaccine misinformation.

Interventions

- The chatbots were modeled on preexisting chatbots and have two main features: Infodemic management, vaccine confidence booster, and booking services.
- We employed feasibility randomized control trials to assess the impact of the intervention on COVID-19 vaccine attitudes and acceptance, with a baseline survey performed before individuals had used the chatbots and a postintervention survey conducted 1 week after chatbot use.
- The novel solution was supported and facilitated with traditional research methods (e.g., social media analytics, formative consultations with existing COVID-19 chatbot developers) that helped derive strong evidence to better understand each person's reasons for vaccine hesitancy.
- The ultimate goal of developing vaccine chatbots was to test whether it was possible to generate a science-driven automated conversation that could promote vaccine confidence and uptake.

Lessons learned

What worked?

- The chatbots were able to disseminate accurate COVID-19 vaccine-related information to vaccine-hesitant communities in an easily accessible format and from a credible source. The chatbots enabled individual questions to be answered as soon as they came up, which can be more difficult with some traditional forms of health communication, such as conversations with a healthcare worker.
- Users were generally positive about their experiences with the chatbots. According to preliminary results, around four-fifths of participants in the pilot study would use the chatbot again.
- The chatbots will continue to be adapted based on user feedback and ongoing monitoring of COVID-19 vaccine-related social media conversations and search behaviors. This continuous monitoring and adaptation will allow the chatbots to remain timely and relevant.

What didn't work?

- The pilot study was conducted while the Omicron variant was spreading rapidly in the three study locations. In a context where the epidemiological situation was changing rapidly, it was challenging to ensure that the chatbot was programmed with the most recent and up-to-date information.
- The pilot study identified that chatbots can struggle with "off-topic" digressions. The chatbots were programmed to answer questions about COVID-19 vaccines specifically, whereas users were often additionally seeking information about related topics such as mask mandates or travel restrictions.

Recommendations

- Chatbots are potentially an important alternative information source that can

CASE STUDY *(cont'd)*

supplement other forms of health communication and can engage users who might feel judged when asking vaccine-related questions. More evidence is needed to assess if chatbots can effectively respond to concerns, rumors, and misinformation before they gain traction and/or can be utilized as a communication channel complementary to current patient engagement strategies.

- In order to be effective, chatbots need to be supported through ongoing analysis of social listening and web search data to ensure they are responding to current questions, concerns, and misconceptions in a timely and accurate manner.

Vision for the future

For all the reasons set out above, the current information and communication environment is an important contextual influence on a variety of health outcomes. It is not, however, the sole or even the most important influence in many cases. There are wider factors that influence public attitudes and behaviors, including: trust in political systems; personal experiences with health systems and providers; barriers to accessing healthcare; historical influences, such as experiences of racism and discrimination; cultural and religious beliefs; and the influence of community leaders or celebrities. It is beyond the scope of this chapter to address all of these issues, so here we focus specifically on how public health practitioners can help foster a more healthy information and communication environment in the future.

To mitigate the negative impacts of future infodemics, improve trust in public health messaging,[51] and improve health outcomes several steps need to be taken by governments and health systems. Public health officials need to engage in social listening real-time and have a team and plan in place to detect and address misinformation and rumors before they are widely shared. At the same time, health and media literacy must be improved at an individual and societal level, which can potentially be facilitated and expedited by new technologies.[52] To facilitate this, different frameworks have been proposed; however, the order of importance and implementation varies.[53,54] Interventions to combat misinformation must be carefully informed to understand the current and emerging topics and structured to overcome health literacy, and social and psychological barriers.

In the years to come, detecting, mitigating, and preventing misinformation must be streamlined and efficient, implemented by cross-functional teams at both micro and macro levels across cities, states, and nations. This vision is applicable both online and offline, using a collaborative approach to improve global public health education and outcomes.

Misinformation response units should work in collaboration with health departments, trusted community advocates, public health researchers, and communication experts to quickly understand misinformation topics and themes impacting specific communities and audience segments.

These teams will work rapidly to identify misinformation and challenges facing different audience segments and collaborate with community partners to tailor culturally appropriate, scientifically accurate messages to the different populations. They will use community-based response mechanisms that are both online and offline and deploy their messaging through ongoing community education and response that comes from trusted sources in that community.

The teams will use social listening for misinformation surveillance to closely track local, national, and global trends in misinformation on specific diseases. This research will track not only the local language of the community but also all languages spoken by or impacting members of the community, including foreign-language websites and social media. This social listening will use a methodology that aligns with the guidance provided by the Centers for Disease Control and Prevention's Social Listening and Monitoring Guide, which suggests monitoring not just what misinformation is spreading but also using search data to better understand what questions are being asked by the community and how attitudes and emotions are affecting health decisions.

The misinformation response teams will be coordinated with health departments to take a collective approach to rebuilding trust in public health messages and authorities and growing vaccine confidence. By partnering with community-based organizations and trusted community figures, they can rapidly disseminate culturally tailored information that is scientifically accurate and addresses the specific concerns of each segment with the help of leaders and representatives trusted by local communities. This information will be addressed by a two-pronged approach of using debunking and psychological inoculation to build resilience against the spread of misinformation. This approach relies on a coordinated approach underpinned by a comprehensive social listening and response system that works at speed to identify, respond, and communicate effectively, leveraging the trusted relationships already established within communities and demographic segments by their own members and leaders. This social listening will also work to monitor the efforts put forth by these teams, measure the impact, reach, and success of these campaigns within these audiences and segments, and quickly recalibrate where needed.

While this vision may be daunting, it has been successfully implemented to combat misinformation as a core function of public health at the micro and macro levels, achieving measurable results in communities such as New York City, U.S.A. The New York City Department of Health and Mental Hygiene formed the Misinformation Response Unit, which collaborated with more than 100 community partners to identify, create, and share culturally appropriate, scientifically accurate messages to different populations and segments within the community in partnership with trusted community leaders.

Actions to achieve the vision

Listening and learning from the community

Ongoing real-time social listening across online channels is a key activity needed to detect emerging misinformation narratives and allow them to be quickly countered with accurate health communication from trustworthy sources.[53] In order to do this effectively at scale, it is imperative to better understand the reach and influence of the misinformation. To facilitate this process, there is first a need to agree upon a codebook or shared public health taxonomy[55] and metrics upon which content and themes can be measured to detect a change in volume or sentiment.[56]

While social media data can provide valuable insights into public health issues and trends, it is also important to recognize that it may not necessarily be representative of the wider offline population. Therefore, it is important to consider other offline data sources in the context of online data, such as traditional survey data, clinical data, and administrative data, to supplement and validate findings from social media data.

Social listening tools can currently capture online content from public licensed sources such as blogs, forums, websites, and public-facing accounts on social media channels like Facebook and Twitter. However, as some of these channels have begun to shut down accounts and ban users for violations of their policies, including the spread of misinformation accounts, users and their content have moved to less restrictive social platforms such as MeWe and Parler,[57] which are not currently available for monitoring at scale. These, together with search data and emerging platforms and channels, would need to be factored into listening approaches.

While many infodemic studies rely on social listening as a technique, the variety of approaches, methods, and platforms[40,58] has further highlighted the need for increased accountability, coordination and guidance. The CDC has developed guidance on what tools are available for social listening, the channels and platforms they monitor, and how to conduct social listening, analysis, and reporting on an ongoing basis. Steps for conducting effective social listening include identifying existing monitoring tools—what the organization has versus what's needed. It sets out how to set up social and traditional media monitoring systems using Boolean queries to capture the datasets used for analysis in answering research questions before sharing guidance and templates on how to analyze and develop insights to provide regular reporting.[59] Further to this, the WHO has produced a landscape review and framework to enhance the effective use of digital social listening for immunization demand generation which provides methods and approaches in detail.[60]

Currently, global social media platforms like Twitter and Facebook do engage fact-checkers to combat misinformation and disinformation,[7] but these have little external review or oversight and are subject to interpretation. While Twitter employs internal fact-checkers and moderators, YouTube and Facebook are outsourced to third parties that fall under the International Fact-Checking Network, a nonpartisan fact-checking organization that is run by the Poynter Institute for Media Studies. The different moderation companies apply different moderation approaches and guidelines, and depending upon the sources they reference as guidance they are inconsistent in what is labeled misinformation or false; for example, on Facebook, mask-wearing as a preventative measure at the beginning of the pandemic was labeled false, then later factual.[61] Monitoring and platform-specific infoveillance could have greater oversight or consensus upon their monitoring and moderation.

Researchers also need to align on the theories that drive health decision-making as well as the viral spread of misinformation and vaccine hesitancy. Many of the studies to date have focused on the Health Belief Model and rumor theory, but consensus must be reached on how to measure the drivers and impact as well as plan for effective interventions. This can then be used to plan for public health policies more effectively through communication management and messaging that can more quickly and efficiently address misinformation that can lead to infodemics.

When conducting infodemiology social listening, researchers can take several measures to mitigate potential biases when analyzing social media data for public health insights. These include using multiple sources of data and analysis methods, using representative samples of social media users, cleaning and preprocessing the data, using data visualization, being transparent about the methods used, collaborating with stakeholders, and conducting the analysis ethically and responsibly. By adopting these best practices, researchers can reduce the impact of biases, increase the reliability and validity of their findings, provide more accurate insights into public health concerns, and identify misinformation quickly and accurately across platforms and user segments.

Accelerating health literacy and trusting the messenger

To increase health literacy more broadly it is critical to successfully measure, monitor, and engage online with timely interventions. Empirical studies have shown that empowering individuals with health literacy can contribute to reducing conspiracy beliefs and the spread of rumors.[62,63] While health literacy can be related to literacy and critical reasoning in general, there are ways to help expedite this learning. Improving health literacy alongside cultivating trust in the medical community and public health institutions could help enhance this process.[8] The nature of social media channels is interactive. Research has shown that these social connections and informal information seeking and sharing with online peers with similar health concerns positively impact individuals, even those with low health literacy, and improve health outcomes overall.[64]

Trust is an essential component for public health messaging and education to be effective. People need to trust the sources of the information they receive to believe and act upon it. The issue of trust has become even more crucial in the age of social media and the rapid spread of misinformation, leading to infodemics. The spread of misinformation and the public's lack of trust in authorities have significantly impacted public health outcomes, especially during the COVID-19 pandemic when people trusted the information about the safety, efficacy, administration, or even need for vaccination. Establishing trust through coordinated and culturally appropriate communication strategies is essential to leveraging the trusted relationships based within communities by their members and leaders. By identifying and collaborating with community partners, such as public health departments and trusted community advocates, misinformation response units can create tailored messaging for different populations and demographic segments. The messaging needs to be scientifically accurate, culturally appropriate, and delivered in a manner that resonates with the audience. Only through this comprehensive approach can we rebuild and sustain trust in public health messaging and authorities, ultimately improving public health outcomes.

Research has shown that engagement between patients and physicians through social media improves the patient's health literacy,[65] makes them more confident in their relationship with their physician, is better prepared for what to ask in an appointment, and increases willingness for seeking treatment and trusting healthcare providers.[66,67] Public health policies and communication strategies must be evidence-based, taking into account the latest research findings and acknowledging uncertainties. By relying on reliable data and continuously evaluating emerging evidence, public health authorities can enhance the credibility and effectiveness of their approaches, leading to improved public health outcomes. Looking into the online communities, social platforms could legitimize the accounts of healthcare professionals and organizations such as medical journals and promote their content ahead of the opinions of nonhealthcare professional, internet influencers, and celebrities contributing posts and content to healthcare discussions. By verifying the accounts of the healthcare experts, this would make it easier for the public to understand who to trust for health information. Further, while social channel advertising policies allow for governments and media to pay for the promotion of posts and contents, science organizations and publishers are prohibited from doing so,[53] leaving science and health information shared publicly in the hands of influencers rather than scientists or clinicians. Medical practitioners could also be encouraged to have a dialogue with patients to ask what they've heard or have questions about so they can fully answer questions and concerns or debunk misinformation before it becomes fully entrenched.

Empowering individuals to fact-check for themselves within the channel or platform they are already engaging in could overcome limitations inherent within these channels, which may be subject to algorithmic shaping of what content is shared and echo chambers, which could provide confirmation bias. Exposure to primary sources of information has been shown to be more impactful in informing health decision-making and increasing protective behaviors, acting as a moderator between knowledge and behavior, in particular for those with low health literacy.[68] However, this needs to be done within the network or platform, not relying on the user to go seek secondary sources for disconfirmation, as most users are not motivated to take the time to do so, and frequently credible sources are behind paywalls. By making additional resources available within the context of the user experience of a channel or platform through a chatbot or browser extension, confidence in the content and exposure to alternative viewpoints could help increase overall eHealth literacy and trust in the messaging.

Timely and relevant communication, measurement, and learnings

Timely and relevant communication tailored to the channel and issues needs to be part of strategies to address misinformation and rumors. Two communication strategies to combat vaccine misinformation are preemptive communication, or "prebunking" and reactive communication, or "debunking".[21] Prebunking involves providing users with the skills and knowledge to evaluate sources and identify misleading information when they subsequently encounter it, whereas debunking typically provides facts and counterarguments as a corrective to misinformation that is already in circulation.[69] Organizations need to develop risk communication strategies that are both informative and emotive, leveraging the communication theories that drive engagement across social media. Infoveillance, or social listening, is the first step in this process, to both develop the scales for risk assessment and intervention and also inform the topics and tone of content.[53,70] Messaging should rely on guidance from the community itself through social listening in order to be empathic in the communication style and address the unique communication preferences of the audience.[54] Further, health officials and organizations need to be honest about what they do not know, as the absence of information creates a content vacuum where fears and misinformation can thrive.[71] Effectively managing risk communication can lead to improved public trust in health organizations and institutions and better public health outcomes.[72,73]

In summary, the following action steps should be taken:

- Incorporate real-time social listening into routine public health activities.
- Implement best practices to reduce bias in data collection and analysis.
- Establish standardized approaches for information sharing and data comparison among agencies, countries, and over time.
- Adopt uniform methods for combating misinformation on social media platforms and within companies.
- Enhance the health and digital literacy.
- Empower individuals to independently evaluate and fact-check claims.
- Implement proactive strategies, like prebunking, to intervene earlier while maintaining preparedness to debunk when needed.
- Collaborate with community partners to develop targeted interventions that address the unique needs of specific communities, fostering trust.

Conclusion

This chapter has outlined some key challenges to health decision-making and public health communication as a result of the rapidly changing information and communications landscape. The current situation, where there is a large volume of online health information of decidedly mixed quality, especially when combined with low levels of health and media literacy, is a significant challenge to global public health. The large volume of online health content with misinformation and disinformation, alongside credible scientific content and public health messaging, has led to an infodemic, further destabilizing the public's trust in public health bodies globally and impacting health decision-making and compliance with public health measures. It is not an effective health communication strategy to just inform the public about disease awareness, prevention, or fill knowledge gaps. Researchers and public health officials need to listen, learn and communicate quickly and effectively through online and offline channels, in the language and medium of relevant communities, utilizing trusted community members and leaders to help improve the trust and adoption of communications.

Looking to the future, we envision a streamlined and efficient process for detecting, mediating, and preventing misinformation, implemented by cross-functional teams at both micro and macro levels across cities, states, and nations. These teams will use social listening for misinformation surveillance to closely track local, national, and global trends in misinformation on specific diseases, and they will work rapidly to identify misinformation and challenges facing different audience segments. By partnering with community-based organizations and trusted community figures, they can rapidly disseminate culturally tailored information that is scientifically accurate and addresses the specific concerns of each segment.

This approach relies on a coordinated approach underpinned by a comprehensive social listening and response system that works at speed to identify, respond, and communicate effectively, leveraging the trusted relationships already established within communities and demographic segments by their own members and leaders. By implementing prebunking, debunking, and inoculation theory, and improving eHealth literacy, we can overcome key infodemics and lead to greater vaccine confidence and improved public health outcomes. With the right tools, technology, and infrastructure in place, and through collaborative efforts between public health agencies, technology companies, and academic institutions, we can mitigate the negative impacts of future infodemics, improve and sustain trust in public health messaging, and ultimately improve health outcomes for all.

Acknowledgement

This work was supported by AIR@InnoHK administered by Innovation and Technology Commission.

References

1. WHO. WHO Coronavirus (Covid-19) Dashboard. Accessed March 25, 2022. https://covid19.who.int/.
2. Hamel LLL, Kirzinger A, Sparks G, Stokes M, Brodie M. KFF Covid-19 Vaccine Monitor: Media and Misinformation. Accessed February 13, 2023. https://www.kff.org/coronavirus-covid-19/poll-finding/kff-covid-19-vaccine-monitor-media-and-misinformation/view/footnotes/.
3. Naeem SB, Bhatti R, Khan A. An exploration of how fake news is taking over social media and putting public health at risk. *Health Inf Libr J*. 2021;38(2):143–149. https://doi.org/10.1111/hir.12320.
4. Fridman A, Gershon R, Gneezy A. COVID-19 and vaccine hesitancy: a longitudinal study. *PLoS One*. 2021;16(4), e0250123. https://doi.org/10.1371/journal.pone.0250123.
5. Hopfer S, Fields EJ, Lu Y, et al. The social amplification and attenuation of COVID-19 risk perception shaping mask wearing behavior: a longitudinal twitter analysis. *PLoS One*. 2021;16(9), e0257428. https://doi.org/10.1371/journal.pone.0257428.
6. Kricorian K, Civen R, Equils O. COVID-19 vaccine hesitancy: misinformation and perceptions of vaccine safety. *Hum Vaccines Immunother*. 2022;18(1):1–8. https://doi.org/10.1080/21645515.2021.1950504.
7. Breenen JS, Felix MS, Howard PN, Nielsen RK. *Types, Sources, and Claims of COVID-19 Misinformation*. RISJ Factsheet; 2020. https://reutersinstitute.politics.ox.ac.uk/sites/default/files/2020-04/Brennen%20-%20COVID%2019%20Misinformation%20FINAL%20%283%29.pdf.
8. Chou W-YS, Oh A, Klein WMP. Addressing health-related misinformation on social media. *JAMA*. 2018;320(23):2417. https://doi.org/10.1001/jama.2018.16865.
9. Benecke O, Deyoung SE. Anti-vaccine decision-making and measles resurgence in the United States. *Glob Pediatr Health*. 2019;6. https://doi.org/10.1177/2333794x19862949, 2333794X1986294.
10. Rothkopf DJ. When the buzz bites back. *Wash Post*; May 11, 2003. https://www.washingtonpost.com/archive/opinions/2003/05/11/when-the-buzz-bites-back/bc8cd84f-cab6-4648-bf58-0277261af6cd/. Accessed January 1, 2021.
11. Organization TWH. Infodemics. https://www.who.int/health-topics/infodemic#tab=tab_1.
12. Glatter KA, Finkelman P. History of the plague: an ancient pandemic for the age of COVID-19. *Am J Med*. 2021-02-01;134(2):176–181. https://doi.org/10.1016/j.amjmed.2020.08.019.
13. *The Flu Pandemic of 1918 and Early Conspiracy Theories*. September 29, 2020.

14. Zarocostas J. How to fight an infodemic. *Lancet*. 2020;395(10225):676. https://doi.org/10.1016/s0140-6736(20)30461-x.
15. Organization WH. *Ten Threats to Global Health in*; 2019. https://www.who.int/news-room/spotlight/ten-threats-to-global-health-in-2019.
16. Pierri F, Perry BL, Deverna MR, et al. Online misinformation is linked to early COVID-19 vaccination hesitancy and refusal. *Sci Rep*. 2022;12(1). https://doi.org/10.1038/s41598-022-10070-w.
17. Mitchell A, Oliphant B, Shearer E. *About Seven-in-Ten U.S. Adults Say They Need to Take Breaks from Covid-19 News*; 2020. https://www.pewresearch.org/journalism/2020/04/29/3-majority-of-americans-feel-they-need-breaks-from-covid-19-news-many-say-it-takes-an-emotional-toll/.
18. Niburski K, Niburski O. Impact of trump's promotion of unproven COVID-19 treatments on social media and subsequent internet trends: observational study. *J Med Internet Res*. 2020;22(11), e20044. https://doi.org/10.2196/20044.
19. Oliver JE, Wood TJ. Conspiracy theories and the paranoid style(s) of mass opinion. *Am J Polit Sci*. 2014;58(4):952–966. https://doi.org/10.1111/ajps.12084.
20. Rossen I, Hurlstone MJ, Dunlop P, Lawrence C. Accepters, fence sitters, or rejecters: moral profiles of vaccination attitudes. *Soc Sci Med*. 2019;224:23–27. https://doi.org/10.1016/j.socscimed.2019.01.038.
21. Ecker UKH, Lewandowsky S, Cook J, et al. The psychological drivers of misinformation belief and its resistance to correction. *Nat Rev Psychol*. 2022;1(1):13–29. https://doi.org/10.1038/s44159-021-00006-y.
22. Ecker UK, Ang LC. Political attitudes and the processing of misinformation corrections. *Polit Psychol*. 2019;40:241–260. https://doi.org/10.1111/pops.12494.
23. Hornsey MF, Fielding KS. Attitude roots and Jiu Jitsu persuasion: understanding and overcoming the motivated rejection of science. *Am Psychol*. 2017;72(5):459. https://doi.org/10.1037/a0040437.
24. Trevors G. The roles of identity conflict, emotion, and threat in learning from refutation texts on vaccination and immigration. *Discourse Process*. 2021. https://doi.org/10.1080/0163853X.2021.1917950.
25. Reyna VF. Risk perception and communication in vaccination decisions: a fuzzy-trace theory approach. *Vaccine*. 2012;30(25):3790–3797. https://doi.org/10.1016/j.vaccine.2011.11.070.
26. Reyna VF. A scientific theory of gist communication and misinformation resistance, with implications for health, education, and policy. *Proc Natl Acad Sci USA*. 2021;118(15), e1912441117. https://doi.org/10.1073/pnas.1912441117.
27. Leung NHL, Chu DKW, Shiu EYC, et al. Respiratory virus shedding in exhaled breath and efficacy of face masks. *Nat Med*. 2020;26(5):676–680. https://doi.org/10.1038/s41591-020-0843-2.
28. Netburn D. *A Timeline of the CDC's Advice on Face Masks*. Los Angeles Times; 2021. https://www.latimes.com/science/story/2021-07-27/timeline-cdc-mask-guidance-during-covid-19-pandemic.
29. Pian W, Chi J, Ma F. The causes, impacts and countermeasures of COVID-19 "Infodemic": a systematic review using narrative synthesis. *Inf Process Manag*. 2021;58(6), 102713. https://doi.org/10.1016/j.ipm.2021.102713.
30. Aea G. The "infodemic" of journal publication associated with the novel coronavirus disease. *J Bone Joint Surg*. 2020;102(13):e64. https://doi.org/10.2106/jbjs.20.00610.
31. Reporter S. Iran: Over 700 dead after drinking alcohol to cure coronavirus. Al Jazeera. https://www.aljazeera.com/news/2020/4/27/iran-over-700-dead-after-drinking-alcohol-to-cure-coronavirus.
32. Eysenbach G. Infodemiology and infoveillance: framework for an emerging set of public health informatics methods to analyze search, communication and publication behavior on the internet. *J Med Internet Res*. 2009;11(1):e11. https://doi.org/10.2196/jmir.1157.
33. Eysenbach G. Infodemiology: the epidemiology of (mis)information. *Am J Med*. 2002;113(9):763–765. https://doi.org/10.1016/s0002-9343(02)01473-0.
34. Eysenbach G. *Infodemiology- Tracking Flu-Related Searches on the Web for Syndromic Surveillance*. 2006:244–248.
35. Ginsberg J, Mohebbi MH, Patel RS, Brammer L, Smolinski MS, Brilliant L. Detecting influenza epidemics using search engine query data. *Nature*. 2009;457(7232):1012–1014. https://doi.org/10.1038/nature07634.
36. Eysenbach G. SARS and population health technology. *J Med Internet Res*. 2003;5(2):e14. https://doi.org/10.2196/jmir.5.2.e14.
37. Polgreen PM, Chen Y, Pennock MD, Nelson FD, Weinstein RA. Using internet searches for influenza surveillance. *Clin Infect Dis*. 2008;47(11):1443–1448. https://doi.org/10.1086/593098.
38. Hulth A, Rydevik G, Linde A. Web queries as a source for syndromic surveillance. *PLoS One*. 2009;4(2):e4378. https://doi.org/10.1371/journal.pone.0004378.

39. Wilson K, Brownstein JS. Early detection of disease outbreaks using the Internet. *Can Med Assoc J.* 2009;180(8):829–831. https://doi.org/10.1503/cmaj.1090215.
40. Barros JM, Duggan J, Rebholz-Schuhmann D. The application of internet-based sources for public health surveillance (infoveillance): systematic review. *J Med Internet Res.* 2020;22(3), e13680. https://doi.org/10.2196/13680.
41. Tang L, Bie B, Park SE, Zhi D. Social media and outbreaks of emerging infectious diseases: a systematic review of literature. *Am J Infect Control.* 2018;46(9):962–972. https://doi.org/10.1016/j.ajic.2018.02.010.
42. Blank G, Lutz C. Representativeness of social media in great britain: investigating facebook, linkedin, twitter, pinterest, google+, and instagram. *Am Behav Sci.* 2017;61(7):741–756. https://doi.org/10.1177/0002764217717559.
43. Zhao Y, Yin P, Li Y, He X, et al. *Data and Model Biases in Social Media Analyses: A Case Study of COVID-19 Tweets.* 2022:1264–1273.
44. Smailhodzic E, Hooijsma W, Boonstra A, Langley DJ. Social media use in healthcare: a systematic review of effects on patients and on their relationship with healthcare professionals. *BMC Health Serv Res.* 2016;16(1). https://doi.org/10.1186/s12913-016-1691-0.
45. Ho YX, O'Connor BH, Mulvaney SA. Features of online health communities for adolescents with type 1 diabetes. *West J Nurs Res.* 2014;36(9):1183–1198. https://doi.org/10.1177/0193945913520414.
46. Mclean S, Etzler T, Kottasova I. Masks made Czech Republic the envy of Europe. Now they've blown it. *CNN;* October 19, 2020. https://edition.cnn.com/2020/10/19/europe/czech-republic-coronavirus-intl/index.html.
47. Cost B. Coronavirus spawns viral TikTok dance about washing your hands. NY Post. https://nypost.com/2020/03/04/coronavirus-spawns-viral-tiktok-dance-about-washing-your-hands/.
48. *Illmer Andreas Coronavirus: Vietnam's handwashing song goes global.* BBC News; 2020. Accessed 1 January 2021.
49. Norman CD, Skinner HA. eHealth literacy: essential skills for consumer health in a networked world. *J Med Internet Res.* 2006;8(2):e9. https://doi.org/10.2196/jmir.8.2.e9.
50. National Center for Education Statistics USDoE. *The Health Literacy of America's Adults Results from the 2003 National Assessment of Adult Literacy;* 2006. https://files.eric.ed.gov/fulltext/ED493284.pdf.
51. Larson HJ, Clarke RM, Jarrett C, et al. Measuring trust in vaccination: a systematic review. *Hum Vaccines Immunother.* 2018;14(7):1599–1609. https://doi.org/10.1080/21645515.2018.1459252.
52. Risk Communication and Community Engagement (RCCE) Action Plan Guidance COVID-19 Preparedness and Response. 16 March 2020. https://www.who.int/publications/i/item/risk-communication-and-community-engagement-(rcce)-action-plan-guidance.
53. Eysenbach G. How to fight an infodemic: the four pillars of infodemic management. *J Med Internet Res.* 2020;22(6), e21820. https://doi.org/10.2196/21820.
54. Mheidly N, Fares J. Leveraging media and health communication strategies to overcome the COVID-19 infodemic. *J Publ Health Pol.* 2020;41(4):410–420. https://doi.org/10.1057/s41271-020-00247-w.
55. Tangcharoensathien V, Calleja N, Nguyen T, et al. Framework for managing the COVID-19 infodemic: methods and results of an online, crowdsourced WHO technical consultation. *J Med Internet Res.* 2020;22(6), e19659. https://doi.org/10.2196/19659.
56. Tran HTT, Lu S-H, Tran HTT, Nguyen BV. Social media insights during the COVID-19 pandemic: infodemiology study using big data. *JMIR Med Inform.* 2021;9(7), e27116. https://doi.org/10.2196/27116.
57. Dickson E. Inside MeWe, Where Anti-Vaxxers and Conspiracy Theorists Thrive. https://www.rollingstone.com/culture/culture-features/mewe-anti-vaxxers-conspiracy-theorists-822746/.
58. Karafillakis E, Martin S, Simas C, et al. Methods for social media monitoring related to vaccination: systematic scoping review. *JMIR Public Health Surveill.* 2021;7(2), e17149. https://doi.org/10.2196/17149.
59. Prevention USDoHaHSCfDCa. Social Listening and Monitoring Tools. https://www.cdc.gov/vaccines/covid-19/vaccinate-with-confidence/rca-guide/downloads/cdc_rca_guide_2021_tools_appendixe_sociallistening-monitoring-tools-508.pdf.
60. Organization WH. *Finding the Signal through the Noise;* 2021. https://www.gavi.org/sites/default/files/2021-06/Finding-the-Signal-Through-the-Noise.pdf.
61. Clarke L. Covid-19: who fact checks health and science on Facebook? *BMJ.* 2021:n1170. https://doi.org/10.1136/bmj.n1170.
62. Lee JJ, Kang KA, Wang MP, et al. Associations between COVID-19 misinformation exposure and belief with COVID-19 knowledge and preventive behaviors: cross-sectional online study. *J Med Internet Res.* 2020;22(11), e22205. https://doi.org/10.2196/22205.

63. Bin Naeem S, Kamel Boulos MN. COVID-19 misinformation online and health literacy: a brief overview. *Int J Environ Res Publ Health*. 2021;18(15):8091. https://doi.org/10.3390/ijerph18158091.
64. Hayat TZ, Brainin E, Neter E. With some help from my network: supplementing eHealth literacy with social ties. *J Med Internet Res*. 2017;19(3):e98. https://doi.org/10.2196/jmir.6472.
65. Wicks P, Massagli M, Frost J, et al. Sharing health data for better outcomes on PatientsLikeMe. *J Med Internet Res*. 2010;12(2):e19. https://doi.org/10.2196/jmir.1549.
66. Van Uden-Kraan CF, Drossaert CH, Taal E, Seydel ER, Van De Laar MA. Self-reported differences in empowerment between Lurkers and Posters in online patient support groups. *J Med Internet Res*. 2008;10(2):e18. https://doi.org/10.2196/jmir.992.
67. Bartlett YK, Coulson NS. An investigation into the empowerment effects of using online support groups and how this affects health professional/patient communication. *Patient Educ Counsel*. 2011;83(1):113–119. https://doi.org/10.1016/j.pec.2010.05.029.
68. Kim S, Capasso A, Cook SH, et al. Impact of COVID-19-related knowledge on protective behaviors: the moderating role of primary sources of information. *PLoS One*. 2021;16(11), e0260643. https://doi.org/10.1371/journal.pone.0260643.
69. Swire B, Ecke U, Lewandowsky S. The role of familiarity in correcting inaccurate information. *J Exp Psychol Learn Mem Cogn*. 2017;43(December 2017):1948–1961. https://doi.org/10.1037/xlm0000422.
70. Chatterjee R, Bajwa S, Dwivedi D, Kanji R, Ahammed M, Shaw R. COVID-19 Risk Assessment Tool: dual application of risk communication and risk governance. *Prog Disaster Sci. Oct*. 2020;7, 100109. https://doi.org/10.1016/j.pdisas.2020.100109.
71. Vaezi A, Javanmard SH. Infodemic and risk communication in the Era of CoV-19. *Adv Biomed Res*. 2020;9:10. https://doi.org/10.4103/abr.abr_47_20.
72. Vaughan E, Tinker T. Effective health risk communication about pandemic influenza for vulnerable populations. *Am J Publ Health*. October 2009;99(Suppl 2):S324–S332. https://doi.org/10.2105/AJPH.2009.162537.
73. Abdelwhab EM, Hafez HM. An overview of the epidemic of highly pathogenic H5N1 avian influenza virus in Egypt: epidemiology and control challenges. *Epidemiol Infect*. May 2011;139(5):647–657. https://doi.org/10.1017/S0950268810003122.

CHAPTER

21

Countering vaccine hesitancy in the context of global health

James O. Ayodele[1], Joann Kekeisen-Chen[2], Leesa Lin[3,4,5], Ahmed Haji Said[2], Heidi J. Larson[3,6,7] and Ferdinand Mukumbang[8,9]

[1]WHO Office for the Eastern Mediterranean Region, Cairo, Egypt; [2]Rollins School of Public Health, Emory University, Atlanta, GA, United States; [3]Department of Infectious Disease Epidemiology, London School of Hygiene & Tropical Medicine, London, United Kingdom; [4]Laboratory of Data Discovery for Health (D24H), Hong Kong Science Park, Hong Kong SAR, China; [5]WHO Collaborating Centre for Infectious Disease Epidemiology and Control, School of Public Health, LKS Faculty of Medicine, The University of Hong Kong, Hong Kong SAR, China; [6]Centre for the Evaluation of Vaccination, Vaccine & Infectious Disease Institute, University of Antwerp, Antwerp, Belgium; [7]Department of Health Metrics Sciences, University of Washington, Seattle, WA, United States; [8]Ingham Institute for Applied Medical Research, Liverpool, NSW, Australia; [9]Sydney Institute for Women, Children and Their Families, Sydney, NSW, Australia

Introduction

Vaccination is one of the most successful and cost-effective health interventions in history, with wider benefits throughout a person's lifetime. However, hesitancy continues to limit vaccine acceptance globally with direct impact on the success of vaccination programs and the degree of protection against vaccine-preventable diseases.[1,2] Vaccine hesitancy is complex and context-specific, varying with time, population, geographical location, and vaccine type. This chapter highlights the drivers of vaccine hesitancy and recommendations for engagement on vaccine and vaccination issues locally and globally.

https://doi.org/10.1016/B978-0-323-90945-7.00013-0

Literature rewiew and gap analysis

Regional and national disparities in vaccine hesitancy

Available evidence shows that vaccine hesitancy is a global phenomenon. It has been reported in Europe, the Americas, Asia, Australia, and Africa, with continental, regional, and national disparities.[3]

The World Health Organization (WHO) identified vaccine hesitancy as one of the top 10 global health threats in 2019,[4] and a Gavi report shows that hesitancy was a growing challenge in 15 of 40 Gavi-eligible economies, with significant impact on immunization services.[5] As the number of individuals and communities questioning the use of vaccines and delaying or refusing vaccination increases globally, so will be the risk of the spread of vaccine-preventable diseases and limitations on the potential to reach the required thresholds for herd immunity.[6–8]

Recent studies[3] have shown that people in high-income countries tend to report less confidence in vaccines and vaccinations than those in lower income countries. These studies attribute the low measles vaccination rates in Europe, the United States, and the United Kingdom; the low hepatitis vaccination in France and Denmark; and the large-scale measles outbreaks in Western and Eastern Europe and the United States in recent years, to the low confidence in vaccines.

Increasing vaccine hesitancy has been associated with declining childhood immunization rates in the United States and has been associated with outbreaks of invasive influenza type b, varicella, pneumococcal disease, measles, and pertussis diseases among one-quarter to one-third of parents.[8,9] The US Centers for Disease Control and Prevention reports show that more than 1% of children born in 2015 did not receive vaccinations, and in California, there were more than 9000 cases of whooping cough in 2010; higher than any yearly figure since the introduction of the whooping cough vaccine.[8,9]

A Wellcome Trust survey[10] that used the Vaccine Confidence Index questions with more than 140,000 people across 140 countries shows that only 72% of people in North America and 73% in Northern Europe agreed that vaccines are safe. Only 59% in Western Europe, 50% in Eastern Europe, and just one-third in France agreed that vaccines are safe. However, the figures were much higher in South Asia (95%) and Eastern Africa (92%).

On March 11, 2020, WHO declared COVID-19 a global pandemic.[11] The emergence of this infectious disease has led to the development, manufacturing, and distribution of safe and efficacious COVID-19 vaccines, 10 of which have so far been authorized for emergency use by WHO.[12] However, willingness to take the vaccine has not been commensurate with the speed and enthusiasm with which the vaccines were developed.

There have been significant variations in willingness by country. A survey of 13,426 people in 19 countries shows that on average, 71.5% of respondents were willing to take a COVID-19 vaccine; willingness was highest in China (88.6%) and lowest in Russia (54.9%).[12] A report published before COVID-19[2] shows that Eastern Europe had the lowest vaccine confidence scores, with 41% of respondents in France and 36% in Bosnia and Herzegovina reporting that vaccines are not safe. The report shows that in Poland, vaccine refusal increased from 4893 in 2007 to 23,147 in 2016, in Moldova hesitancy was 26% in 2020, and in Bulgaria only 23% of respondents were not hesitant to vaccines.

A Gallup survey[13] among adults aged 15 and older in 116 countries in 2020 shows that nearly 68% of adults worldwide were willing to take the COVID-19 vaccine. Willingness was highest in Myanmar (96%) and lowest in Kazakhstan (25%). Countries with the 10 highest acceptance rates were Myanmar 96%, Nepal 87%, Nicaragua 87%, Thailand 85%, Iceland 85%, Denmark 85%, Laos 84%, Ethiopia 84%, Cambodia 84%, Bangladesh 83%, and Egypt 83%. The 10 countries with the lowest acceptance rates were Montenegro 33%, Cameroon 33%, Bulgaria 33%, Kosovo 32%, Jamaica 32%, Jordan 31%, Bosnia and Herzegovina 30%, Hungary 30%, Gabon 29%, and Kazakhstan 25%. This survey[13] confirms the lowest acceptance rates already reported in Eastern European countries and the former Soviet Union States, including Russia. It further shows that before the COVID-19 pandemic, Russia had one of the highest rates of vaccine hesitancy in the world and that in Africa, two-thirds of respondents in Gabon and Cameroon, and majorities in Senegal, Togo, and Namibia, said they would not take vaccines if offered. In the United States, a little above half (53%), as of September and October 2020, agreed to be vaccinated if the vaccine was free; however, by March 2021, the proportion of those who agreed to be vaccinated if the vaccine was free had increased to 74%.

Although the proportion of adults who report being vaccinated for COVID-19 or intend to be vaccinated as soon as possible in the USA continues to rise, a May 2021 report by the Kaiser Family Foundation shows that the proportion of individuals who will "definitely not" take the vaccine has remained the same (14%) since December 2020.[14]

Despite successes in vaccination programs, a growing number of Africans had reportedly delayed or refused recommended vaccines for themselves or their children before the COVID-19 pandemic.[15] In sub-Saharan Africa, one in four children missed the three doses of diphtheria-tetanus-pertussis-containing vaccines partly due to hesitancy.[15]

Vaccine hesitancy in Africa is driven partly by controversies, rumors, and misinformation endorsed by some high-ranking public figures.[16] One of such is the rumor that the polio vaccine was an American conspiracy to spread HIV or cause infertility among Muslim girls in northern Nigeria in 2003/2004,[17] which led the Governor of Kano State to boycott polio vaccination for 11 months.[18]

In a survey in 15 African countries by the Africa Centres for Disease Control and Prevention (Africa CDC),[19,20] 21% of respondents, on average, said they would not take the COVID-19 vaccine even if it was safe and effective. Willingness to take the vaccine was highest in Ethiopia (94%), Niger (93%), and Tunisia (92%) and lowest in Gabon (67%), Senegal (65%), and Democratic Republic of Congo (59%).

A nationally representative survey in five West African countries—Republic of Benin, Liberia, Niger Republic, Senegal, and Togo—shows only four of every 10 respondents were willing to take the COVID-19 vaccine.[17] In South Africa, a May 2021 report shows that 71% were willing to take the COVID-19 vaccine, an improvement over a January 2021 report of 67% by the Human Sciences Research Council and a February 2021 report of 64% by the Ipsos-World Economic Forum.[20] The report suggests that vaccine acceptance in South Africa was higher than estimates from the United States and France but lower than in China, Brazil, and the United Kingdom.

A review by Sallam[21] shows low COVID-19 vaccine acceptance rates in the Middle East, Eastern Europe, and Russia and high acceptance rates in East and Southeast Asia. The highest rates were in Ecuador (97%), Malaysia (94%), Indonesia (93%), and China (91%), and the lowest rates were in Kuwait (24%), Jordan (28%), Italy (54%), Russia (55%), Poland (56%), the United States (57%), and France (59%). Acceptance was as low as 28% among healthcare workers in the Democratic Republic of Congo.

Eighty percent of 472,521 adults in the 20 Latin American countries said they intended to take the COVID-19 vaccine.[22] Mexico (88%), Puerto Rico US (85%), Costa Rica (84%), Brazil (83%), and Honduras (81%) had the highest proportions of those willing to take the vaccine, while Haiti (43%), Paraguay (65%), Dominican Republic (66%), Uruguay (66%), and Venezuela (69%) had the lowest proportions.

In India, before the COVID-19 pandemic,[23] 80% of parents were willing to vaccinate their daughters with HPV vaccine.

Drivers of vaccine hesitancy

Vaccine hesitancy is driven largely by the beliefs and perceptions of individuals and groups. This section highlights key drivers of vaccine hesitancy observed globally, including concerns about safety and efficacy of vaccines, mistrust in the health system and pharmaceutical companies, religious and cultural beliefs, misinformation and misconceptions, inadequate information about disease outbreaks or pandemics, fear of adverse reactions, limited knowledge of healthcare workers about the vaccines they administer, and negative attitudes of healthcare workers to vaccination.[19,20]

Lack of confidence in the safety and efficacy of vaccines

Confidence in the safety and efficacy of vaccines is a major marker for acceptance or hesitancy and a big threat to the effectiveness of vaccination programs.[8] Concerns about the safety and efficacy of vaccines are a major driver of vaccine hesitancy among people of different races, nationalities, and expertise globally, including healthcare workers, many of whom think vaccines are dangerous and harmful and can cause life-long morbidities and mortality.[24] In the United States, for example, movements against childhood and other vaccinations continue to grow due to negative perceptions that vaccines are not safe and can cause autism.[10] Some vaccine-hesitant individuals think that vaccination will worsen the already fragile condition of elderly people or increase their vulnerability to adverse health effects including death. A study in Africa reported distrust in vaccines because of the perception that vaccines are not safe; 18% of respondents said vaccines are generally not safe and 25% said a COVID-19 vaccine would be unsafe.[19,20]

The low vaccine acceptance rates before the COVID-19 pandemic already mentioned in Eastern European countries and the former Soviet Union States, including Russia, were driven largely by the belief that vaccines are generally not safe and not effective.[25] As of 2018, less than half of Russians who had heard about vaccines (45%) agreed that vaccines were safe, and 24% disagreed, and as of December 2020, only 37% of Russians said they would take a COVID-19 vaccine if offered.[25] In a survey of nationally representative sample of 5629 adults in South Africa, the three leading reasons given for hesitancy by 29% vaccine-hesitant respondents were worries about side-effects (31%) and efficacy (21%) and lack of trust in vaccines (18%).[26]

Globally, the novelty of the COVID-19 pandemic and the accelerated development of the COVID-19 vaccines may have further exacerbated public concerns and heightened risk perception regarding vaccine safety and efficacy. Uncertainty about the immediate and long-term effects of the vaccines has made some people reluctant even when they know that vaccines can protect them against the disease. There have been controversies and misinformation about the perception that mRNA vaccines alter the DNA of recipients and can cause deformities in the future. Similar concerns were reported with Ebola vaccine in 2019; confidence in the vaccine reduced because of the fear of contamination and uncertainty about the long-term effects on the body.[27]

Reports of blood clots in late 2020 and early 2021 due to the COVID-19 vaccine among some recipients triggered new waves of vaccine hesitancy across the globe. Suspension of the use of the AstraZeneca and Johnson and Johnson vaccines in the United States and Europe for some weeks and a total halt in Denmark due to concerns about blood clots further increased hesitancy against the vaccine.[28] The emergence of new strains of SARS-CoV-2 and uncertainty about the length of protection that the new vaccines could offer have also triggered concerns.

Distrust in public health authorities

Negative experiences with healthcare systems have been linked to widespread distrust in health authorities globally because some people think that data about vaccines and disease outbreaks are misleading and unreliable.[29–31] Even among healthcare workers, distrust in pharmaceutical companies and government has been reported. One physician twitted[32]: "As a hospital doctor of 48 years, I support everything you say my friend. Vaccines were not developed for COVID, COVID was developed for vaccines. The lethal injection is a double whammy to depopulate, to track and trace. It is a power grab."

A cross-sectional study shows that despite campaigns to combat COVID-19 in urban communities in Nigeria, many still deny the existence of the disease and see it as a political strategy to aid corruption.[33] Lack of confidence in government health programs in Africa and other parts of the world has been attributed to the lack of community involvement in planning and implementation of control measures and the limited effort by governments to debunk misinformation that Africans are "immune" to the pandemic because of their climatic condition.[33,34]

A five-country survey in Africa shows that only 31% of respondents, on average, said they trusted their governments enough to take the vaccine.[17] In Senegal and Liberia, 83% and 78%, respectively, said they did not trust their governments would ensure that any vaccine is safe before offering it to citizens, and because of this, they did not want to take the COVID-19 vaccine.

Lack of trust in health authorities and governments has been due largely to perceived corruption, lack of transparency in vaccine procurements, and mismanagement of funds, which in turn translate into distrust of vaccination programs.[2] Some parents who normally give routine vaccinations to their children refused to give them the polio vaccination because they did not trust the health system about the safety of the vaccine.[35]

In the COVID-19 era, vaccine politicization and vaccine nationalism are new concepts contributing to hesitancy among groups and individuals, in addition to public health errors attributed to governments.

Inadequate knowledge and information about vaccines and vaccination

Limited knowledge and information about vaccines and vaccinations among the general population is a main driver of vaccine hesitancy, and it has the potential to reduce risk perception, reduce the ability of individuals to make informed decision, and to increase the spread of misinformation.[2] Vaccine manufacturers and healthcare authorities have not done enough to explain the science of vaccines, including the adverse reactions to healthcare workers and the public. There is a general assumption that healthcare workers and rapid responders have enough information to educate their clients about vaccines and vaccinations, but this is often not the case. Limited knowledge has been reported even among healthcare workers, making them to reject vaccination for themselves and their children and to advise others against getting vaccinated.[36]

Studies in Eastern European countries have shown that healthcare workers relied on social and traditional media for information about vaccines and vaccination, which sometimes provide negative information that could discourage vaccination uptake.[2] Among Latin American immigrant mothers of adolescent children in Northern Virginia, the lack of, or mixed knowledge, about vaccine-related issues, including, vaccine-preventable diseases, has been reported as a major barrier to vaccine uptake.[36] These mothers agreed that childhood vaccination was necessary but questioned the necessity for vaccination for their adolescent children.

Myths, misinformation and rumors about vaccines and vaccination

Myths, misinformation, and rumors about several issues such as vaccine safety and efficacy, politics, science, etc. are major drivers of vaccine hesitancy; they instill fear in people and negatively influence their intention to vaccinate.[37] These have been largely responsible for vaccine hesitancy globally.

Misinformation through social media platforms has been identified as a key driver of vaccine hesitancy in South Korea and Malaysia.[38] For example, the ANAKI (raising children without medication) online community has been a leading voice against childhood immunization in South Korea. In Cameroon, in the 1990s, the purpose of childhood immunization was misconstrued because of the belief that the vaccines were meant to sterilize girls. In northern Nigeria, communities were told that the polio vaccine was meant to reduce Muslim populations.[39] Similar concerns have been observed with cholera, tetanus toxoid, measles, polio, and other immunizations in Mozambique, Zimbabwe, Nigeria, Democratic Republic of Congo, and other parts of Africa.[38] In China, negative information about vaccines was found to be associated with delay or refusal of childhood vaccination.[38]

Misinformation spreads partly because people have insufficient information, lack understanding, or are ignorant about vaccines, but also due to mischief. Some individuals spread myths, misinformation, and rumors to discourage people from accepting vaccinations, relying sometimes on statements by public figures, health officials, scientists, or researchers that are quoted out of context or misinterpreted and made to sound as authentic opinions of the original speaker or writer.[37] Others spread misinformation just because they disagree with scientific claims about vaccines.

False claims about deaths and severe adverse reactions due to vaccination have contributed to vaccine hesitancy; an example is a video alleging that vaccines increased death rates among individuals aged 80 years and above in the United Kingdom.[40]

Low risk perception

Controversies and claims that disease outbreaks are often exaggerated have been a reason for low risk perceptions that fuel vaccine hesitancy in some communities. Individuals who deny the existence of a disease outbreak are often among those who spread misinformation and would think that any measures to control the spread of such disease, including vaccination, are unnecessary. Such denials were observed during the Ebola outbreaks in West Africa and in the Democratic Republic of Congo.[41]

A perceived lack of severity of measles infection was reported as a reason for complacency toward measles-mumps-rubella vaccine uptake in Europe, with some parents seeking vaccination only when there is an outbreak.[2] A study in India has shown that parents who believed that their daughters were at a low risk of contracting cervical cancer were most likely not to allow them to get the HPV vaccine.[36]

The argument around COVID-19 vaccine has been entrenched in controversies about the pandemic itself. Some people still deny the existence of the pandemic, saying that the whole idea of COVID-19 is a hoax or a plot by authorities and big businesses to enrich themselves through fictitious and outrageous procurement and supply contracts. Others think COVID-19 is a tactic by scientists and pharmaceutical companies to make huge profits from the sales of medicines and vaccines.

Inequities in access to healthcare services

Inequities in access to healthcare services are among the drivers of vaccine hesitancy in some communities. These have been reported[42] in the United States and the United Kingdom, where vaccination coverage has been significantly lower among non-Hispanic blacks, Hispanics, and non-Hispanic Asians than among non-Hispanic whites in the United States. Only 48% of African Americans and 66% of Latinos said they would definitely, or probably, take the COVID-19 vaccine. Only 14% of African Americans and 34% of Latinos trusted that a vaccine would be safe, and only 18% of African Americans and 40% of Latinos trusted that the vaccine would be effective. In the United Kingdom, 72% of black respondents and 42% of Pakistani and Bangladeshi respondents said they were unlikely or very unlikely to accept the COVID-19 vaccine.

Vaccine hesitancy among health workers of black and Asian origin in the United Kingdom and among black and Latino groups in the United States is rooted in their negative historical experience with the health systems and deep-rooted historical trauma, caused mainly by marginalization, structural and systemic racism and discrimination, previous unethical healthcare research among people of color, and under-representation of minorities.[42–44]

Systemic racism, manifested in residential segregation, is rising in Europe, and it perpetuates and amplifies mistrust in government and health officials because of how it affects health and access to resources that enhance health.[45] In the United Kingdom, the Bangladeshi and Pakistani communities are reported to be most affected by segregation.[45] In December 2020, vaccine hesitancy was higher among black (odds ratio 12.96, 95% confidence interval 7.34–22.89), Bangladeshi, and Pakistani (both 2.31, 1.55–3.44) populations than among counterparts from a white ethnic background.[46] A January 2021 report shows COVID-19 vaccination rates were substantially lower among people over 80 years of age in ethnic minority populations in England (white 42.5%, black people 20.5%), and deprived communities (least deprived 44.7%, most deprived 37.9%).[47] Vaccination rates were also lower among ethnic minority healthcare workers (70.9% in white workers *vs* 58.5% in South Asian and 36.8% in black workers; $P < 0.001$ for both).[48]

The "big pharma" conspiracy

One of the reasons for refusing to support vaccines and vaccination is the deep distrust of the pharmaceutical companies. Antivaxxers think mass vaccination programs are a conspiracy by the "big pharma" companies and government officials to enrich themselves; they claim that vaccines have dangerous adverse reactions and that governments shield pharmaceutical companies from liability.[10] Antivaxxers allege that vaccines have caused deaths in the past and that pharmaceutical companies do not conduct enough clinical and social trials to prove that their vaccines are safe and efficacious. They believe that this is the reason pharmaceutical companies are unwilling to accept liability for any adverse reactions due to vaccination. Many antivaxxers have questioned the speed with which the COVID-19 vaccines were developed and say it is evidence of connivance between researchers and pharmaceutical companies to produce and sell vaccines, make huge profits, and control the global trade.

Misperception of vaccination as a depopulation agenda

There is a perception in some communities, particularly in Africa and Asia, that vaccines are bioweapons containing contraceptive hormones that can stop women from having children.[49] This perception is often driven by the belief that vaccination is a "Western plot" to sterilize, limit fertility, infect, or kill people to reduce the populations and influence of selected communities, in connivance with governments and funding agencies.[8,50]

Accusations of using humans as "guinea pigs"

Some individuals have alleged that the whole idea of vaccines is to enable scientists to test their products on humans, using them as "guinea pigs" to advance their research ambitions.[29,42] This perception has caused resistance to vaccine trials and vaccination in some parts of Africa, Asia, Latin America, and even in Europe and America. In Europe and America, there were oppositions to the COVID-19 vaccine on the grounds that the vaccines were developed too quickly, making people vulnerable to becoming "guinea pigs" for researchers and pharmaceutical companies.

Preference for alternative solutions

Proponents of traditional herbal medicines, mostly in Africa, Asia, and Latin America, have often expressed their preference for traditional medical solutions. They believe that traditional herbal preparations are natural and safer and more efficacious than artificially manufactured medicines. Therefore, they are less inclined to use vaccines as a preventive measure against any disease.

Traditional medication may not have major side effects if administered carefully, but they do not have the long-term protection that vaccines offer, and some of them require additional protection with synthetic medicines. For example, yellow fever, polio, measles, and meningitis vaccines have been proven to provide long-term protection, which traditional medicines do not provide.

Other perceived alternatives include vitamins and other mineral supplements, which can boost the body's immunity to disease but are much less effective than vaccines. Vaccines require very little time to activate the body's immune system, while supplements require a longer time and may not be able to provide the reinforced immunity needed to survive a sudden outbreak.

Preference for natural immunity

Some individuals believe that the body's immune system is sufficient and stronger than vaccines to protect them against diseases. Such people are of the opinion that the body can develop natural immunity against a disease if sufficiently exposed to an infection.[23] An example is the tweet[49] that said: "There is no need to #TakeTheVaccine when the vast majority of people can fight coronavirus with their own immune system, and the 'vaccine' is not actually a vaccine but an experimental mRNA injection whose manufacturers have no liability if you suffer side effects or death." A study in Romania shows that urban, well-educated mothers refused vaccination more than other groups often because of their preference for natural immunity.[2]

Religious beliefs

Religion plays a vital role in vaccine acceptance and health-seeking behaviors generally. Some religious groups say vaccination is against their beliefs. For example, the Zion and Apostolic sects in Malawi reject the HPV vaccine, saying that some medicines, including vaccines, promote immoral behavior and cause infertility.[50] Some religious groups in North America have labeled the COVID-19 vaccine as the "mark of the beast" and warned their members not to take it.[51] Some other religious groups believe that with prayer they would be healed of any disease, and because of this would not require a vaccine. A large majority of respondents in Niger (89%), Liberia (86%), and Senegal (71%) believed that prayer was more effective than a vaccine in preventing coronavirus infection.[17] These beliefs, coupled with misinformation about the COVID-19 vaccine, have been a major driver of hesitancy impacting the decision to be vaccinated by some religious groups.[29]

Disagreement over "vaccination passport" (or green pass)

Evidence of vaccination against diseases such as yellow fever, measles, and meningitis has been used for centuries as precondition for travel to some countries. However, with COVID-19, the concept has become more controversial and is making trust in governments and vaccination programs to decrease.[2] Some individuals say the concept of a "vaccine passport" is coercive, authoritarian, unjustified, and an infringement of their personal freedom of movement. Others think that "vaccination passport" is a government agenda to place citizens under illegal surveillance, and that everyone should be allowed to decide whether they want to be vaccinated or not without the fear of losing their jobs or their freedom. Opposers of the concept of "vaccination passport" or green pass have been using social media platforms to push for resistance to vaccination using such hashtags as #Freedom, #Liberty, #GreatRevolt, and #Coronapassport.

In the United Kingdom, 34% of survey respondents expressed concerns that "vaccination passport" would be a breach of human rights because it could discriminate against those waiting for the vaccine, those who cannot have it, or those who do not want it.[48] Others think that vaccination cannot offer freedom from lockdowns and restrictions as promised by scientists and health authorities.

Critics have argued that the European Union COVID-19 vaccine green pass discriminates against countries that do not have sufficient access to vaccines, particularly vaccines approved by the European Medicines Agency (EMA).[37] It raises the dilemma of whether individuals in such countries who wish to travel to Europe should choose to be vaccinated with a vaccine that is not recognized by the European Union or wait until a vaccine approved by EMA becomes available. Officials in countries like Philippines and Hungary had challenged the restriction of the European Union COVID-19 green pass to only EMA-approved vaccines because their health workers and senior citizens were vaccinated with available COVID-19 vaccines, mainly during the early phases of the vaccination drive, which were non-EMA approved.[37]

CASE STUDY

COVID-19 vaccine hesitancy

Background

By July 18, 2021, over 190 million COVID-19 cases and more than four million COVID-19-related deaths had been reported globally. Vaccination against the SARS-CoV-2 virus and its variants is considered a panacea for the pandemic, especially with increasing evidence that COVID-19 vaccination reduces the risks of being hospitalized.[52,53] However, despite the steady increase in availability of the COVID-19 vaccine, hesitancy continues to increase, thus posing a barrier to achieving the high vaccination targets needed to achieve herd immunity in many parts of the world.[54] Countries are increasingly focused on the protection of their citizens and are neglecting their obligations and commitments to protect asylum-seekers, refugees, and foreign-born migrants living within their borders. The vulnerabilities of asylum-seekers, refugees, and foreign-born migrants are particularly exacerbated during the COVID-19 pandemic. Consequently, they are more likely to suffer the physical and mental health and socio-economic consequences of COVID-19. Such a disproportionate impact warrants this demographic to be considered a most-at-risk population.

Observations

- There have been strong collaborations, partnerships, and dedication among scientists and health authorities to develop potent vaccines against the SARS-CoV-2 virus and its variants as quickly as possible.
- The alleged unethical and immoral vaccine hoarding and vaccine nationalism by high-income countries has been blamed for the poor access to vaccines in low- and middle-income countries.[55]
- Inconsistent messaging around vaccines and vaccination has fueled disinformation, misinformation, rumors, and conspiracy theories, which in turn are fueling vaccine hesitancy.

Key issues

- Limited capacity of low- and middle-income countries to develop and manufacture the vaccine doses needed to vaccinate their populations.
- Unwillingness of high-income countries that own the vaccine intellectual property rights to grant patent rights for manufacturing in low- and middle-income countries.

CASE STUDY *(cont'd)*

- Naming of SARS-CoV-2 variants based on their country or region of first identification and naming of vaccines based on their country of manufacture even when the active ingredient was developed by the same parent pharmaceutical company. These have contributed to stigma and discrimination because of the variation in the perception of efficacy, safety, and effectiveness of these vaccines.[56]
- Key populations such as immigrants, asylum-seekers, refugees, and internally displaced persons are not being adequately considered in the vaccination program of some countries or territories.[52]
- The COVID-19 pandemic has been shrouded in controversies regarding its origin, originality, and existence, and these controversies are also fueling hesitancy to the vaccines.
- Some opinion and political leaders have openly made negative statements about the efficacy and safety of the COVID-19, vaccines while others have outrightly denounced the vaccines.
- Reports of rare cases of severe side effects such as blood clots following vaccination, especially in the early days of COVID-19 vaccination, which contributed to hesitancy even among healthcare workers.

Interventions

- The establishment of the COVAX Facility, a platform to ensure equitable distribution of vaccines, was a good idea that has helped provide some quantities of COVID-19 vaccines to some low- and middle-income countries. The COVAX humanitarian buffer mechanism to reach high-risk and vulnerable populations in fragile and humanitarian settings has been effective.
- The subsequent establishment of the Pfizer/BioNTech vaccine manufacturing plant at Biovac Institute and a few other manufacturing arrangements in Africa are helping to improve availability and distribution of COVID-19 vaccines in Africa.
- The ability of China and India, two of the most populous countries in the world, to manufacture their own vaccines locally has helped cater for the vaccine needs of more than one-third of the world's population.
- The explicit inclusion of African populations in clinical trials in countries like South Africa has contributed to confidence in some Africans about the vaccines.
- Donor-based approaches to vaccine equity through the COVAX Facility and other national, regional, or continental entities and donations from high-come countries have facilitated availability of vaccines in low- and middle-income countries.
- More research is being conducted by pharmaceutical companies to fortify available vaccines and make them more effective against emerging SARS-CoV-2 variants.

Lessons learned

What worked?

- The inclusion of key populations such as migrants, asylum-seekers and refugees in countries such as South Africa, England, Portugal, Italy, and Germany have improved uptake of the COVID-19 vaccines and improved the possibility of reaching herd immunity.
- Pool procurement and distribution of vaccines through COVAX and other

(Continued)

CASE STUDY *(cont'd)*

regional and continental entities such as the African Vaccines Acquisition Taskforce (AVAT) have helped bridge supply chain problems to some extent.
- Countries like China, India, and Russia have manufactured the COVID-19 vaccines for their populations and provided these for free to some developing countries, thus contributing to increases in vaccination rates globally.

What did not work?
- The inability of COVAX to secure enough doses of the vaccines and in time as expected was a drawback to the Facility's vision of equal and timely distribution and access to the vaccines.
- While asylum seekers and other migrant populations were considered by some countries in their vaccination programs, the exclusion of undocumented migrants remains a "blind spot."
- Donor-based approaches to vaccine equity that are grounded in old and failed models to promote vaccine equity.
- Sourcing vaccine doses from high-income countries incapacitates low- and middle-income countries and increased the cost of procurement and supply chain.
- The nonavailability of clinical trial data has made vaccines manufactured by some countries not to be approved by the WHO for emergency use.

Recommendations
- Consistent, coordinated, prompt, and community-focused messaging from relevant authorities such as the WHO, regional or national disease control agencies, and governments could help counter conspiracy theories, rumors, misinformation, and disinformation.
- Accurate, evidence-based, and timely information about COVID-19 vaccines, including clinical trial data, communicated as essential facts to the general population and health authorities can improve vaccine acceptance and uptake.
- More funding is needed for equitable procurement and distribution of vaccines to countries that have limited supplies.
- Strengthen vaccine research, development, and manufacturing capacity of low- and middle-income countries to ensure sustainable and timely universal access to COVID-19 vaccines and other essential vaccines.
- Vaccine justice should be promoted, which entails moving beyond aid models of vaccine donation, to rapidly achieving global consensus for the intellectual property waiver, democratizing vaccine intellectual property rights and know-how, and supporting low- and middle-income countries to build vaccine manufacturing facilities.[55]

Vision and actions for the future

The success of antivaccine campaigns has been attributed partly to inaction or inadequate response to rumors and misinformation about vaccines and vaccinations by concerned authorities.[57] Just like any other community development activity, combatting vaccine hesitancy requires using multiple strategies to reach different population groups.[25] This section highlights some of these strategies.

Invest in global vaccine hesitancy research

Substantial investments in research are required globally to effectively address vaccine hesitancy; most of the available research studies on the subject are from high-income countries.[15,29,58] Investments should be targeted at developing simple and appropriate research tools that are applicable to high- and low-income settings and can be incorporated into routine immunization programs.[15] More research on vaccine hesitancy will help in understanding the concerns of individuals and groups as well as the underlying barriers and community behaviors toward vaccines and vaccinations.[29] Such understanding is essential for the development of evidence-based strategies to address vaccine hesitancy.

In addition to the development of tools, it is essential to develop the capacity to use the tools for socio-behavioral, community engagement, and demand generation research around vaccine hesitancy issues. Such capacity is needed for researchers, academics, clinicians, healthcare workers, communication experts, and pharmaceutical personnel involved in the vaccine value chain.

Strengthen communication and community engagement on vaccines and vaccinations

Communication and community engagement have been a substantial part of the successes recorded in vaccination for measles, smallpox, polio, HPV, and other vaccine-preventable diseases globally.[2] Consistent, coordinated, and targeted communication and community engagement is needed more than ever to counter misinformation and increase trust in vaccines and vaccinations. It is essential to increase demand by increasing understanding of the science, safety, and efficacy issues raised by antivaxxers and vaccine-hesitant individuals.

Engaging community members on their attitudes and beliefs regarding vaccines and vaccinations will help in identifying specific concerns requiring targeted interventions.[52] Continued engagement on issues of healthcare programming will assure community members that their interests are being considered in healthcare planning and will help build trust and confidence in health systems. When communities are involved in discussions and actions about the research, clinical trials, and distribution of vaccines, and allowed to provide relevant feedback, they are likely to become more receptive to vaccines and vaccinations.[44]

Collaborations between religious leaders and public health officials can enhance understanding of vaccine hesitancy and help identify sustainable solutions. Religious leaders are trusted members of faith communities and could play an integral role in positively sensitizing their communities about health seeking behaviors regarding vaccines and vaccinations. To be effective, engagement with religious and community leaders should, however, be based on scientific research and anthropologic evidence so that consultations can be relevant to the needs of the targeted communities.

The critical need for training in communication and community engagement for everyone involved in supporting vaccination programs, including the healthcare workers, cannot be overemphasized. Healthcare workers, frontline workers, community workers, and caregivers need in-depth understanding of the social and scientific aspects of vaccines and vaccinations and how to communicate these to clients at facilities and in the community.[2,53] They are often trusted by community members and are better placed to communicate acceptably about the safety, efficacy, adverse reactions, rumors, and controversies about vaccines.

Engage the services of community influencers

The use of trusted community influencers, including personal physicians, celebrities, community and religious leaders, and other community leaders, whose opinions are respected and accepted within the community, is essential in addressing vaccine hesitancy. Evidence has shown that parents and other community members trust their doctors and community healthcare workers, listen to their opinions, and are inclined to comply with their suggestions.[2] In Africa, Asia, and some other parts of the world, religious and community leaders command large followership and their opinions count in decision-making by their followers. Using these leaders as vaccination messengers can help increase vaccination uptake at the community level.

Strengthen social listening on vaccines and vaccinations

Misinformation, myths, and rumors about vaccines and vaccinations over social media platforms are on the increase, and they significantly affect the decision to accept or reject vaccination by individuals and communities.[52] Social listening through digital engagement and traditional community feedback systems is very important in understanding and addressing concerns about vaccines and vaccinations from different sectors of stakeholders including community members, community and healthcare workers, caregivers, political and opinion leaders, celebrities, and community gatekeepers. Systematic and consistent harvesting of social data using available digital software and traditional platforms has been effective in recent times and can provide evidence needed for planning interventions on vaccine hesitancy. Social listening requires substantial but essential investments that can turn the tide against vaccine hesitancy globally.

Conclusion

This chapter has highlighted the key drivers and regional disparities in vaccine hesitancy. Hesitancy is a growing trend that poses an increasing threat to public health globally. Universal vaccination is essential to achieve the desired breakthroughs in the control of infectious diseases, outbreaks, and pandemics of similar magnitude to COVID-19. Achieving breakthroughs will require targeted and prompt actions by the different actors in the public health space, as hesitancy continues, and advocates use different media to spread their antivaccine messages and campaigns. There is a need for carefully designed and targeted programs that will address the concerns of the different stakeholders, increase understanding of, and trust in, vaccines and vaccinations, and minimize hesitancy globally.

References

1. United Nations. *Misinformation and Growing Distrust on Vaccines, "Dangerous as a Disease" Says UNICEF Chief.* UN News; June 28, 2019. Available at https://news.un.org/en/story/2019/06/1041571.
2. Obregon R, Mosquera M, Tomsa S, Chitnis K. Vaccine hesitancy and demand for immunization in Eastern Europe and Central Asia: implications for the region and beyond. *J Health Commun.* 2020;25(10):808–815. https://doi.org/10.1080/10810730.2021.1879366.
3. Wilder-Smith AB, Qureshi K. Resurgence of measles in Europe: a systematic review on parental attitudes and beliefs of measles vaccine. *J Epidemiol Glob Health.* 2020;10(1):46–58. https://doi.org/10.2991/jegh.k.191117.001.
4. World Health Organization. *Ten Threats to Global Health in 2019.* Geneva: WHO; 2020, 2020 https://www.who.int/news-room/spotlight/ten-threats-to-global-health-in-2019.
5. Leah Ewald and Felicity Pocklington. *Understanding Vaccine Hesitancy: Cha's Story. Gavi.* The Vaccine Alliance; 2021, 29 April 2021 https://www.gavi.org/vaccineswork/understanding-vaccine-hesitancy-chas-story.
6. Stein RA. The golden age of anti-vaccine conspiracies. *Germs.* 2017;7(4):168–170. https://doi.org/10.18683/germs.2017.1122. Dec.
7. Di Pietro ML, Poscia A, Teleman AA, Maged D, Ricciardi W. Vaccine hesitancy: parental, professional and public responsibility. *Ann Ist Super Sanita.* 2017;53(2):157–162. https://doi.org/10.4415/ANN_17_02_13.
8. Public Health. *Vaccine Myths Debunked. Health Guide*; 2020. Available at https://www.publichealth.org/public-awareness/understanding-vaccines/vaccine-myths-debunked/.
9. Jacobson RM, Sauver JLS, Rutten LJF. Vaccine hesitancy. *Mayo Clin Proc.* 2015;90(11):1562–1568. https://doi.org/10.1016/j.mayocp.2015.09.006. November 2015.
10. Alex K. *Vaccine Mistrust on the Rise in Developed Nations, Survey Shows.* BioSpace; 2019. Jun 19, 2019. Available at https://www.biospace.com/article/vaccine-mistrust-grows-in-developed-nations-survey-shows/.
11. World Health Organization. *Director General's Opening Remarks at the Media Briefing on COVID-19.* 2020. Available at https://www.who.int/director-general/speeches/detail/who-director-general-s-opening-remarks-at-the-media-briefing-on-covid-19—-11-march-2020.
12. World Health Organization. *10 Vaccines Granted Emergency Use Listing (EUL) by WHO*; 2022. Updated 7 March 2022 https://covid19.trackvaccines.org/agency/who/.
13. Lazarus J, Ratzan S, Palayew A, et al. A global survey of potential acceptance of a COVID-19 vaccine. *Nat Med.* 2020;27(2):1–4. https://doi.org/10.1038/s41591-020-1124-9. Feb. 2021.
14. Kirzinger A, Sparks G, Hamel L, Lopes L, Stokes M, Brodie M. *KFF COVID-19 Vaccine Monitor: July 2021.* KFF; 2021, 4 August 2021 https://www.kff.org/coronavirus-covid-19/poll-finding/kff-covid-19-vaccine-monitor-july-2021/.
15. Cooper S, Betsch C, Sambala EZ, Mchiza N, Wiysonge CS. Vaccine hesitancy – a potential threat to the achievements of vaccination programmes in Africa. *Hum Vaccines Immunother.* 2018;14(10):2355–2357. https://doi.org/10.1080/21645515.2018.1460987. Published online 2018 May 22.

16. Makoni M. Tanzania refuses COVID-19 vaccines. *Lancet*. 2021;397(10274):566. https://doi.org/10.1016/S0140-6736(21)00362-7.
17. Seydou A. *Who Wants COVID-19 Vaccination? in 5 West African Countries, Hesitancy Is High, Trust Low*. Afrobarometer; 2021. Dispatch No. 432, 9 March 2021 https://afrobarometer.org/sites/default/files/publications/Dispatches/ad432-covid-19_vaccine_hesitancy_high_trust_low_in_west_africa-afrobarometer-8march21.pdf.
18. Ghinai I, Willott C, Dadari I, Larson HJ. Listening to the rumours: what the Northern Nigeria polio vaccine boycott can tell us ten years on. *Global Publ Health*. 2013;8(10):1138–1150. https://doi.org/10.1080/17441692.2013.859720.
19. Africa CDC. *COVID-19 Vaccine Perceptions: A 15-country Study*. Ethiopia: Addis Ababa; 2021:71. Available at https://africacdc.org/download/covid-19-vaccine-perceptions-a-15-country-study/.
20. Africa CDC. *Press Release: Majority of Africans Would Take a Safe and Effective COVID-19 Vaccine*; 2020. Available at https://africacdc.org/news-item/majority-of-africans-would-take-a-safe-and-effective-covid-19-vaccine/.
21. Sallam M. Rates. *Vaccines*. 2021;9(2):160. https://doi.org/10.3390/vaccines9020160. Published online 2021 Feb 16.
22. Urrunaga-Pastor D, Bendezu-Quispe G, Herrera-Anazco P, et al. Cross-sectional analysis of COVID-19 vaccine intention, perceptions and hesitancy across Latin America and the Caribbean. *Trav Med Infect Dis*. 2021;41, 102059. https://doi.org/10.1016/j.tmaid.2021.102059. May–June.
23. Toth-Manikowski SM, Swirsky ES, Gandhi R, Piscitello G. COVID-19 vaccination hesitancy among health care workers, communication, and policymaking. *Am J Infect Control*. 2022;50(1):20–25. https://doi.org/10.1016/j.ajic.2021.10.004.
24. Julie R. *Over 1 Billion Worldwide Unwilling to Take COVID-19 Vaccine*. World; 2021, 3 May 2021. Available at https://news.gallup.com/poll/348719/billion-unwilling-covid-vaccine.aspx.
25. Kahn T. *Study Finds 71% of SA Adults Willing to Get Vaccinated*. BusinessDay; 2021, 12 May 2021. Available at https://www.businesslive.co.za/bd/national/health/2021-05-12-study-finds-71-of-sa-adults-willing-to-get-vaccinated/.
26. Vinck P, Pham PN, Bindu KK, Bedford J, Nilles EJ. Institutional trust and misinformation in the response to the 2018–19 Ebola outbreak in North Kivu, DR Congo: a population-based survey. *Lancet Infect Dis*. 2019;19(5):529–536. https://doi.org/10.1016/s1473-3099(19)30063-5.
27. The Irish Times. *Denmark Stops Use of Johnson and Johnson Covid-19 Vaccine*. The Irish Times; 2021, 3 May 2021 https://www.irishtimes.com/news/world/europe/denmark-stops-use-of-johnson-johnson-covid-19-vaccine-1.4554449.
28. World Health Organization. *Factors that Contributed to Undetected Spread of the Ebola Virus and Impeded Rapid Containment*. WHO; 2015. January 2015. Available at https://www.who.int/news-room/spotlight/one-year-into-the-ebola-epidemic/factors-that-contributed-to-undetected-spread-of-the-ebola-virus-and-impeded-rapid-containment.
29. Degarege A, Krupp K, Fennie K, et al. Human papillomavirus vaccine acceptability among parents of adolescent girls in a rural area, Mysore, India. *J Pediatr Adolesc Gynecol*. 2018;31(December):583–591. https://doi.org/10.1016/j.jpag.2018.07.008.
30. Stecula DA, Kuru O, Jamieson KH. *How Trust in Experts and Media Use Affect Acceptance of Common Anti-vaccination Claims*. Harvard Kennedy School Misinformation Review; 2020. January 14, 2020. Available at https://misinforeview.hks.harvard.edu/article/users-of-social-media-more-likely-to-be-misinformed-about-vaccines/.
31. Kurten D. *No Need to #TakeTheVaccine*. Twitter; 2021, 19 February 2021 https://twitter.com/davidkurten/status/1362814635343294466.
32. College of Physicians of Philadelphia. *Cultural Perspectives on Vaccination*. The History of Vaccines; 2018, 10 January. Available at https://www.historyofvaccines.org/content/articles/cultural-perspectives-vaccination. Accessed October 24, 2020.
33. Letšosa R. What has the beast's mark to do with the COVID-19 vaccination, and what is the role of the church and answering to the Christians? *HTS Teol Stud/Theol Stud*. 2021;77(4). https://doi.org/10.4102/hts.v77i4.6480.
34. Dubé E, Gagnon D, Nickels E, Jeram S, Schuster M. Mapping vaccine hesitancy—country-specific characteristics of a global phenomenon. *Vaccine*. 2014;32(2014):6649–6654. https://doi.org/10.1016/j.vaccine.2014.09.039.
35. Housset B. Distrust of vaccination: why? *Rev Mal Respir*. 2019;36(8):955–961. https://doi.org/10.1016/j.rmr.2019.06.011.

36. Harvard TH. Chan School of Public Health. Dealing with Parents' Mistrust of Vaccines. n.d. Available at: https://www.hsph.harvard.edu/news/hsph-in-the-news/dealing-with-parents-mistrust-of-vaccines/.
37. Breeze D. *Vaccines Were Not Developed for COVID, COVID was Developed for Vaccines*. Twitter; 2021. March 19 https://twitter.com/dickbreeze/status/1372952507916431363.
38. Ilesanmi O, Afolabi A. Perception and practices during the COVID-19 pandemic in an urban community in Nigeria: a cross-sectional study. *PeerJ*. 2020;8:e10038. https://doi.org/10.7717/peerj.10038.
39. Vanguard. *Porous Borders, Cause of Rise in COVID-19 Cases-FG*. Vanguard; 2020, 3 April 2020 https://www.vanguardngr.com/2020/04/porous-borders-cause-of-rise-in-covid-19-cases-fg/.
40. Gesser-Edelsburg A, Shir-Raz Y, Green MS. Why do parents who usually vaccinate their children hesitate or refuse? General good vs. individual risk. *J Risk Res*. 2016;19(4):405–424. https://doi.org/10.1080/13669877.2014.983947.
41. Lu P, O'Halloran A, Williams WW, Lindley MC, Farrall S, Bridges CB. Racial and ethnic disparities in vaccination coverage among adult populations in the U.S. *Am J Prev Med*. 2015;49(6):S412–S425. https://doi.org/10.1016/j.amepre.2015.03.005.
42. Wan W. *Coronavirus Vaccines Face Trust Gap in Black and Latino Communities, Study Finds*. The Washington Post; 2021, 24 November 2021 https://www.washingtonpost.com/health/2020/11/23/covid-vaccine-hesitancy/.
43. Razai MS, Osama T, McKechnie DGJ, Majeed A. Covid-19 vaccine hesitancy among ethnic minority groups. *BMJ*. 2021;2021(372):n513. https://doi.org/10.1136/bmj.n513.
44. Razai MS, Kankam HK, Majeed A, Esmail A, Williams DR. Mitigating ethnic disparities in covid-19 and beyond. *BMJ*. 2021;(372):m4921. https://doi.org/10.1136/bmj.m4921, 2021.
45. Robertson E, Reeve KS, Niedzwiedz CL, et al. Predictors of COVID-19 vaccine hesitancy in the UK household longitudinal study. *Brain Behav Immun*. 2021;94:41–50. https://doi.org/10.1016/j.bbi.2021.03.008.
46. MacKenna B, Curtis HJ, Morton CE, et al. *Trends, Regional Variation, and Clinical Characteristics of COVID-19 Vaccine Recipients: A Retrospective Cohort Study in 23.4 Million Patients Using. Opensafely*. MedRxiv; 2021. https://doi.org/10.1101/2021.01.25.21250356, 9 April 2021.
47. Martin CA, Marshall C, Patel P, et al. *Association of Demographic and Occupational Factors with SARS-CoV-2 Vaccine Uptake in a Multi-Ethnic UK Healthcare Workforce: A Rapid Real-World Analysis*. MedRXiv; 2021. https://doi.org/10.1101/2021.02.11.21251548, 13 February 2021.
48. Brazal AM. Inoculation now or later? Lower efficacy and vaccine passport concerns. *J Publ Health*. 2021;43(3):e527–e528. https://doi.org/10.1093/pubmed/fdab179.
49. Wong LP, Wong P, AbuBakar S. Vaccine hesitancy and the resurgence of vaccine preventable diseases: the way forward for Malaysia, a Southeast Asian country. *Hum Vaccines Immunother*. 2020;16(1):1–10. https://doi.org/10.1080/21645515.2019.1706935.
50. Scott D. *Vital Whistleblower Testimony on #covid #vaccine Reactions and Effects*. Twitter; 2021, 1 February 2021 https://twitter.com/Albion_Rover/status/1356284751431294985.
51. Wiysonge CS. *Vaccine Hesitancy, an Escalating Danger in Africa*. Think Global Health; 2019, 17 December 2019 https://www.thinkglobalhealth.org/article/vaccine-hesitancy-escalating-danger-africa.
52. Steffens MS, Dunn AG, Wiley KE, et al. How organisations promoting vaccination respond to misinformation on social media: a qualitative investigation. *BMC Publ Health*. 2019;19(1348). https://doi.org/10.1186/s12889-019-7659-3.
53. Nachega JB, Sam-Agudu NA, Masekela R, et al. Addressing challenges to rolling out COVID-19 vaccines in African countries. *Lancet Global Health*. 2021;9(6):e746–e748. https://doi.org/10.1016/S2214-109X(21)00097-8.
54. Mukumbang FC. Are asylum seekers, refugees and foreign migrants considered in the COVID-19 vaccine discourse? *BMJ Glob Health*. 2020;5:4085.
55. Tande AJ, et al. Impact of the coronavirus disease 2019 (COVID-19) vaccine on asymptomatic infection among patients undergoing preprocedural COVID-19 molecular screening. *Clin Infect Dis*. 2021. https://doi.org/10.1093/cid/ciab229.
56. Wiysonge CS, et al. Vaccine hesitancy in the era of COVID-19: could lessons from the past help in divining the future? *Hum Vaccines Immunother*. 2021. https://doi.org/10.1080/21645515.2021.1893062.
57. Harman S, et al. Global vaccine equity demands reparative justice — not charity. *BMJ Glob Health*. 2021;6:e006504.
58. Lu L, Lycett S, Ashworth J, Mutapi F, Woolhouse M. What are SARS-CoV-2 genomes from the WHO Africa region member states telling us? *BMJ Glob Health*. 2021;6:e004408.

Further reading

1. UNHCR. *UNHCR Calls on States to Remove Barriers to Access to COVID-19 Vaccines for Refugees*. Press Release; June 24, 2021. https://www.unhcr.org/en-us/news/press/2021/6/60d45ebf4/unhcr-calls-states-remove-barriers-access-covid-19-vaccines-refugees.html.

SECTION V

Moving to the best-protected global community

Mohannad Al-Nsour[1], *Chima J. Ohuabunwo*[2,3,4], *Rana Jawad Asghar*[5,6] *and Randa K. Saad*[7]

[1]Eastern Mediterranean Public Health Network (EMPHNET), Amman, Jordan; [2]Department of Medicine and the Office of Global Health Equity, Morehouse School of Medicine, Atlanta, GA, United States; [3]Hubert Department of Global Health, Rollins School of Public Health, Emory University, Atlanta, GA, United States; [4]Department of Epidemiology, University of Port Harcourt School of Public Health, Rivers State, Nigeria; [5]Global Health Strategists & Implementers, Islamabad, Pakistan; [6]University of Nebraska Medical Center, College of Public Health, Omaha, NE, United States; [7]Center for Excellence for Applied Epidemiology, Eastern Mediterranean Public Health Network (EMPHNET), Amman, Jordan

More than 2 years after the outbreak, the world is still dealing with the impact of the COVID-19 pandemic. During these "pandemic years," the public health profession and disciplines have greatly evolved to better understand the complexities of pandemic preparedness and successful response. Similarly, integration of the traditional branches of public health with some nontraditional disciplines has gained critical importance. Indeed, we have further learnt that effective pandemic preparedness and response does not rely on the traditional core areas of public health only. In this section, authors discuss some traditional and nontraditional core capacities and essentials for enhancing the global health security.

Global public health emergencies have consistently demonstrated that science and politics are not divergent but intertwined. The COVID-19 pandemic further highlighted the importance of the interface between science and politics, and the dire need for public health professionals or institutions to learn, understand, and be equipped with skills to navigate this relationship.

Generally, public health has always been striving to get its fair share from already limited health budgets in most countries. Health finances are predominantly allocated to the clinical and administrative aspect of health and usually a limited amount is reserved for preventive programs, especially for pandemic preparedness. In the last 2 decades, the role of big philanthropy has been added to global health and most recently to pandemic preparedness and response. This gives private entities vast leverage to set health agendas especially in low resource countries. Understanding the current challenges of this arrangement and devising a future strategy that acknowledges the immense contribution from these entities can strike a balance in setting up a fair and effective roadmap for global health security.

Trust in public health agencies weakened in some countries during this pandemic. There were many factors beyond these agencies' control, but how with integrity, professionalism, collaboration, effective communication, community participation and transparency this lost trust could be regained is an important learning for all public health professionals. Similarly, the need for a skilled multidisciplinary workforce has been highlighted specially to secure global health security. Also important is ensuring that key personnel have been well trained, retrained, and resourced to prepare, identify, and respond to future outbreaks and pandemics effectively.

Global health security does not mean a standalone structure on top of existing health systems but rather working within existing systems. Health security could only be assured by strengthening and modifying basic health systems. These health systems need to have a multifaceted hazard approach with intersectoral collaboration and integration.

The global health community rightly anticipated future pandemics and had set up multiple preparedness evaluation frameworks with indicators for national and international levels. Unfortunately, these have not proven to be good predictors of severity and impact of pandemics in different countries. However, with additional pandemic experiences, work has already been initiated to strengthen and improve these indicators for a better global health security.

CHAPTER

22

Science and political leadership in global health security

Natalie Mayet[1], *Eliot England*[2], *Benjamin Djoudalbaye*[3], *Ebere Okereke*[3,4] *and Wondimagegnehu Alemu*[5,6]

[1]National Institute for Communicable Diseases, A Division of the National Health Laboratory Service, Johannesburg, South Africa; [2]Rollins School of Public Health, Emory University, Atlanta, GA, United States; [3]Africa Centres for Disease Control and Prevention (Africa CDC), Addis Ababa, Ethiopia; [4]Tony Blair Institute for Global Change, London, United Kingdom; [5]International Health Consultancy, LLC, Decatur, GA, United States; [6]Rollins School of Public Health, Emory University, Decatur, GA, United States

Introduction

"Science", "politics," and "leadership" are heterogeneous and intricate, each with its domain of expertise, perspective, and worldview. Additionally, there is a paucity of scholarship particularly on the interface and impact of collaboration of science and political leadership in the context of the multitude of crises that COVID-19 presented. On the other hand, Lipscy noting the fact that severe acute respiratory syndrome coronavirus disease-2 (SARS-CoV-2) spreads worldwide benefitting from the globally interconnected environment observes that the COVID-19 pandemic has given the opportunity to "refocus scholarly attention on the politics of crisis."[1] The scale, volatility, and varied responses at local, national, and global levels provide the opportunity to reflect on some lessons learned in science and political leadership that will undoubtedly be a topic of prolific future scholarship.

We recognize that multiple nuances and preexisting sociological, economic, and geopolitical factors affect different responses to the pandemic. This chapter aims to identify how the interface of science and political leadership discourse has influenced the unpredictable response to the COVID-19 pandemic.

Historically, "science and politics have been dependent on each other" in influencing national research agendas, public policy, and resource allocation.[2] Science provides evidence to inform public policy, as it has a complementary role to play for the benefit of society. Political

Modernizing Global Health Security to Prevent, Detect, and Respond
https://doi.org/10.1016/B978-0-323-90945-7.00016-6

leadership, recognizing the role of scientific evidence in guiding policy formulation, justifies the need to finance research topics as part of generation of scientific evidence.[2] The COVID-19 pandemic has boosted this interdependency to an even higher level (e.g., to find solutions to the SARS-COV-2 virus including the development of testing kits, vaccines, and therapeutics were supported and facilitated including allocation of adequate funding by political leadership). The lessons learned from severe acute respiratory syndrome coronavirus 1 (SARS-COV-1) and Middle Eastern respiratory syndrome coronavirus (MERS-COV) including the economic burden resulting from the rapid spread to Europe and North America could have contributed to better engagement of political leaders in implementing science-based public health interventions.[3] On the other hand, political leadership acknowledging scholarly independence is a foundation of modern research, and the consequences of retracting this principle can lead to mistrust, self-interest, loss of livelihoods, and loss of life. Howe suggests to "stick to the science when science gets political."[4] In contrast, given that governments pay for scientific research and development, they may decide to finance research topics prioritized for political gain claiming public interest. Thus, topics prioritized by the scientist requiring funding fail to attract funding as the topic did not make it to the priorities of the politicians. Abbasi gives examples of how science was suppressed during the UK pandemic response including engaging government advisors in the scientific advisory group for emergencies.[5]

Understanding science and politics

We use a simplistic definition of science in the broader context. Encyclopedia Britannica defines science as "system of knowledge that is concerned with the physical world and its phenomena and that entails unbiased observations and systematic experimentation. In general, science involves a pursuit of knowledge covering general truths or operations of fundamental laws" or by Science Council "identification, description, experimental investigation, and theoretical explanation of natural phenomena regardless of discipline of science."[6,7]

The definition of politics is a highly contested concept with many political theorists and practitioners providing many definitions that often could be contradictory to each other or overlapping. Recognizing the difficulty of having a single agreed definition of politics, for this chapter we opted to endorse the definition espoused by Yukl "the process of influencing others to understand and agree about what needs to be done and how to do it, and the process of facilitating individual and collective efforts to accomplish shared objectives."[8] This definition is similar to the Greek definition of politics "the set of activities that are associated with making decisions in groups, or other forms of power relations among individuals, such as the distribution of resources or status."[9]

When deliberating on the relationship of science and political leadership, various terms such as interface, interdependency, interconnectedness, and intersectorality are often used interchangeably. In this chapter, we chose to use "interface" between science and political leadership as an all-encompassing term. The term intersectorality is more comprehensively described in Integrating Science and Politics for Public Health.[10]

The interface between science and political leadership requires a diverse set of competencies where scientists understand diplomacy and apply the principles of translational science and where political leadership realize the benefit of evidence in directing policy. However, caution should be exercised not to politicize scientific concepts to the extent that politicization threatens the political party's agenda, though there is no controversy in the scientific community.[2] This requires competencies that need to be willfully determined and incorporated into future leadership capacity building initiatives to achieve credible scientific and politically achievable collaborative goals for global health security without falling into the partisan political trap.

To adapt, survive, and thrive in the volatility, uncertainty, complexity, and ambiguity (VUCA) of current and future pandemics, science and political leadership require a new approach for leadership architecture to address the complexities of health crisis using a new set of tools and capabilities.[11] The pandemic brings the opportunity to reinvent and cement "ordinary capabilities" of both scientists and politicians and move to "alliance formation" for combating current and future pandemics.[12] We review experiences and the impact of the interface of science and political leadership on pandemic preparedness and response against a framework described by Schoemaker et al. of the six characteristics required for a VUCA environment: anticipation; challenging; interpreting; making decisions; alignment with stakeholders, and continuous learning.[13]

We navigate this chapter through a myriad of nuances in that not all scientists are leaders and not all politicians are in policy-making positions, they may or may not be part of institutions of influence and often have relationships and links with other organizations or funders that have influence.[12] However, this does not mean that the collaboration of science and political leadership does not exist. Sometimes the political leadership fails to recognize the weight the science-based evidence must advance policy decision making such as making resources available for preparedness to respond to public health emergencies of international concerns including pandemics. The multifaceted nature of the political and science interface has an impact at all levels of governance and influence on multiple stakeholders on polygonal domains (Fig. 22.1).

Literature review and gap analysis

With the COVID-19 pandemic, scientists, public health officials, and political leaders found themselves at a pivotal moment in history with the need to acknowledge the significance of their cohesion, cooperation, coordination, and collaboration at the interface of public service. To produce an effective and mutually engaging relationship for effective collaboration among politicians and public health professionals, political leaders and citizens needed to engage intimately with the flexibility to clearly understand the science on which public health actions are based and vice versa.[14] Science provides evidence-based data to inform policy change to public health practice by bringing on board appropriate interventions and deimplement inappropriate and outdated ones. However, the policy changes may not be realized unless the political mechanism functions effectively or sometimes may be delayed due to conflicting political interest or communication failure.[15]

Zielonka questions "who should be in charge of pandemics, scientists or politicians?" He argues that the answer is not as simple as one would like but intriguing with other players like economists, manufacturers, entrepreneurs, social scientists, and civil society organizations all influencing political decision-making.[14] The world views and domains in which each operates are distinct yet blurred. He further questions the ability of the politicians

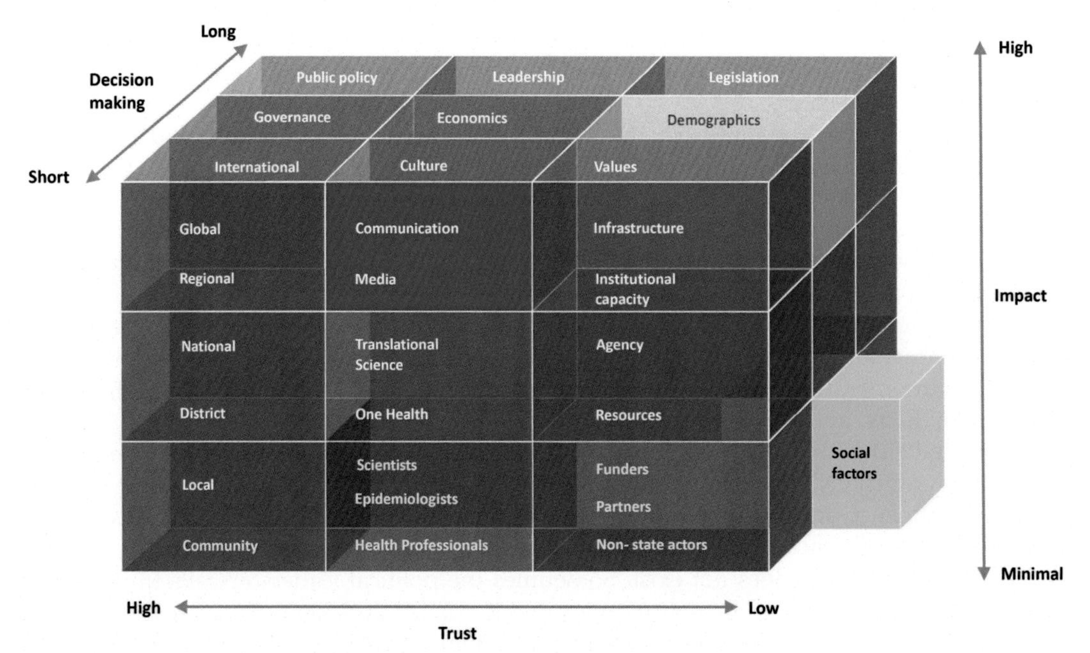

FIGURE 22.1 Science and politics interface cube.

and experts to come up with solutions that are "nonpartisan" and accepted by all. When exploring politics and science at an individual level, there are important similarities and differences in their purposes and decision-making processes.[15]

Politics is associated with power, power dynamics, and relationships within and between power(s); politicians navigate the realm of decision-making and often act to gain, transfer, or exert power to achieve specific interests.[13] Scientists, in contrast, are typically perceived as advocates of objectivity and empiricism. Power and relationships play a role in science, as in politics, but objectivity and empiricism are unique characteristics that manifest the separate identity for scientists and foster their independence from politicians.[13] There is a tendency for scientists to view themselves and their professional capacity as separate from mainstream society—at times they see their reasoning as immune from systemic problems and regard their analyses as apolitical.[13]

It is prudent to note that people and institutions in the scientific sphere come from and exist within a dynamic society, a society shaped by politics. Scientists are not neutral characters; nor are they rarified entities with only academics and geniuses. They are human beings with their own political beliefs, commitments, biases, and motives which they bring into their respective field, just like politicians and anyone else in the public.

At the system's level, the political mechanism is expected to be the vehicle through which public health science effectively initiates local, state, national, and international change.[14] However, scientific knowledge becomes politicized when it is used as a tool to garner support and authority.[2]

Science and public health play critical roles in governance as a nation's overall success is dependent on its health; but, many times politicians must concede and put greater value on elements other than on scientific deliberations.[15] The art of politics involves trade-offs and requires managing an array of economic, social, diplomatic, ideological, and personal aspects that have competing priorities and values.[15,16] Even if politicians want to follow science, many officials have limited science or public health training, and when presented with intricate, multileveled models, they do not always know how to deal with them appropriately. On the flip side, many scientists, including public health professionals, are not interested in joining politics to influence decisions from the inside. James Martin argues that scientists should aspire to political office and bring their evidence-based scientific insights and experiences to influence political governance. He writes that scientists should be more politically active to safeguard science and society.[17]

The interface of science and political leadership plays out on the global stage with the World Health Organization which was established by the United Nations as a specialized body "to act as the directing and coordinating authority on international health work" in 1948.[18] The organization is governed by the World Health Assembly, comprising delegates representing Member States. The International Sanitary Convention, adopted in 1892 following decades of negotiations and consultations, was first conceptualized in the mid-19th century when Europe was overwhelmed by the burden of cholera which interfered with trade and movement of people, intended as a tool to bring changes in global health governance. The State Parties of WHO adopted the International Sanitary Regulations in 1951, which were subsequently revised and renamed as International Health Regulations (IHR) in 1969. The major revision to IHR was initiated in 1995 to broaden the scope and make countries more accountable for its implementation. The IHR defines the rights and obligations of States Parties and the WHO Secretariat for handling public health emergencies with the potential to cross borders. It is a legally binding instrument that emboldens effective preparedness and response to public health emergencies by all member states.[19]

The IHR emphasized global solidarity over national sovereignty and situated the common global threat at the core of its framework; it broadened the WHO's influence in global health security and expanded its own scope of interest to include any events that would constitute a Public Health Emergency of International Concern (PHEIC).[19] The IHR transformed the relationship between international diplomacy and public health; its foundation was built upon rigorous scientific evidence and paved the way for a whole new outlook on global cooperation and collaboration. However, after the 2003 SARS outbreak, the reality that science and politics are inextricably connected became abundantly clear and the promise of the IHR (2005) revisions was met head-on with political interest, personal agendas, and disruption.[19] Infectious disease crises effectively expose what, when, and how leaders decide to allocate resources in support of science.[9] Among the major challenges was the lack of political commitment to invest in building National Core Capacity for IHR. The fact remains that none of the State Parties are in full compliance with the IHR obligations.[20]

The interference of politics has negatively impacted the functionality of IHR as an overarching convention to guide international response to PHEICs including the COVID-19 pandemic. This has resulted in WHO being criticized for inconsistent and unfounded methods of declaring PHEICs including for lack of transparency and acting on political agendas rather than global health well-being. In previous responses to the H1N1 pandemic,

Ebola Viral Disease outbreak in West Africa, and MERS outbreaks in the Middle East, WHO leaders were accused of providing little evidence and scientific reasoning for the decisions made.[21,22]

On the other hand, national and supranational public health agencies such as the US Center for Disease Control and Prevention (CDC), the European CDC, and the Africa CDC provided technical and scientific expertise as technical agencies of political bodies or as agencies funded by political bodies. COVID-19 became a "stress-test" for regional solidarity, more so in politics than science.[23] The coordination of responses varied with the EU experiencing "difficulties on many occasions to secure solidarity under its auspices in the first months, with the AU's response being better characterized as the swiftest amidst the pandemic."[24]

Scientists within the supranational institutions generally coordinate and lead evidence-based public health practice; public health surveillance; laboratory testing; applied research; epidemiological support; risk communication and social engagement with relevant stakeholders.[24] The interface among the scientists located in PHI largely informed the politicians and policymakers of appropriate interventions. This interface was largely based on established principles of trust in most instances and was highly valued and recognized on the African continent despite the limited establishment of public health institutes.[23,24]

At a local level, actions by member states demonstrated that at the heart of effective public health action and leadership was constituent receptivity. When those being led are not given the proper communication, in an appropriate manner, the myriad of health difficulties that arise—nonadherence and trivialization rule. If scientists in member states are part of global network and if a specific member state refuses to take certain approach, then scientists should be influential in changing the course of action by using their global and informed networks though the advocacy of open science community and collaboration. Abassi alleges that "science is being suppressed for political and financial gain" including by disproportionately overrepresenting "government advisors" in the "scientific Advisory Group for Emergency" instead of appointing scientists and independent public health experts.[5]

With the novel virus, the science was uncertain and the adage of "we are following the science" led to a plethora of scientific committees, ministerial advisory committees, scientific networks, and modeling consortiums often resulting in varied and divergent applications of "the science," sometimes used as a "shield and a general selling point to boost the legitimacy" of political actions.[25] The pandemic demanded immediate answers, not something that science is accustomed to with the scientific rigor of the peer review process. There was also a divergence of scientific opinion such as with the wearing of masks that was exacerbated by political stances as described earlier and also the use of the drug ivermectin among others.

Politicians changed stances supported by scientific evidence, and where policy makers had to "seek credibility for their approach by excessive deference to what they believe as science" they fell into what Muller termed "performative scientism." Interpretation is not linear and often relevant information is incomplete.[26] Acknowledging the limitations, we refer to the John Hopkins COVID-19 Behavioral dashboard that for the month of June 2022 the highest level of trust exists in counties like Sweden where there is concordance with trust in information from scientists at 71% and politicians at 66%. Similar but widening concordance is evident in the Philippines, 68% for scientists, and 52% for politicians. On the contrary, in South Africa, 46% of respondents trust scientists and a mere 7% trust politicians. The lowest level of trust at less than 5% exists in scientists and politicians in Ukraine, Poland, and the

Czech Republic. Interpreting this information in the geopolitical context of corruption and the war is relevant.[27] This demonstrates that misguided policymaking at the initial phase of the pandemic might have negative implications on the perception of the public in complying to effective public health implications.

Bollyky et al. following the analysis of SARS-CoV-2 infections and COVID-19 deaths for 177 countries and 181 subnational locations corroborates those higher levels of trust in "government and interpersonal trust has statistically significant associations with fewer infections." The authors attribute their views to higher levels of adherence to recommended behavioral interventions.[28]

A major issue during the COVID-19 pandemic that tested scientific evidence and political atmosphere in many countries involved the effectiveness of public health measures including Personal Protective Equipment (PPE) such as masks and vaccines against the SARS-CoV-2 virus. Despite strong scientific evidence that clearly showed masks are critical to minimizing the transmission of the virus between individuals, the issue has remained highly contentious and political. In the United States, there was a high correlation that members of the community leaning to the Republican party were against wearing masks and those leaning to the Democratic party and viewed as liberal were compliant to utilizing masks during the current pandemic. Green et al. in their paper "Elusive consensus: polarization in elite communication on the COVID-19 pandemic" showed that Democratic members of the US Congress were more likely to frame the pandemic as a public health threat than Republican members. In another study, Schoeni et al. reported that the adjusted percentage of adults with medical risk factors who wore a mask anywhere in the past week was lower for Republicans (87%) and Independents (91%) than for Democrats (97%).[29]

On the issue of vaccine acceptance, several studies across the world reported that politics and religion played significant roles in influencing public acceptance of vaccines against SARS-CoV-2. According to a Gallup survey reported in October 2021, 40% of Republicans in the United States did not plan to get vaccinated versus 3% of Democrats. The same report also highlighted the percentage of vaccinated people and their political affiliations. The report indicated that during the survey conducted between 13 and 22 September, 2021, 90% of Democrats had been vaccinated compared to 58% of Republicans. These reports underscore the high significance of politics in interpreting and utilizing scientific evidence toward the well-being of society.[30]

While the above is an example using the current COVID-19 pandemic, there are several public health challenges that have strong scientific evidence and yet the global society faces major challenges to get consistent political buy-in. One critical issue of planetary significance is climate change. Despite the mounting scientific evidence and its consequences in biodiversity loss, public health, community displacement, and other consequences, we continue to witness politicization of the issue and absence of robust policy changes that would have a lasting impact on society.

Many countries had pandemic preparedness plans on paper, these plans were drafted by scientists and signed off by politicians in compliance with IHR; Schoemaker et al. refer to this as "ordinary leadership."[13] When considering the characteristics of national responses to COVID-19, "high performing responses" are distinguished from "low performing responses."[31] Using lessons learned from managing previous outbreaks such as Ebola viral disease in West Africa, SARS in Singapore and MERS in Saudi Arabia and South Korea where partnerships between science and politicians were brokered, future threats are anticipated and scenarios enacted. This is accomplished either through real-life outbreak experience or from scanning the horizon to craft coordinated, collective, and collaborative evidence-based plans. Well-established multidisciplinary committees bring together leadership that anticipates using evidence of modeling teams, health economics, and social scientists.

CASE STUDY

Background

The Africa Centers for Disease Control and Prevention (Africa CDC), an autonomous body of the African Union (AU), has a mission to strengthen national public health institutions' capabilities to prepare for, detect, and respond to disease outbreaks and other health burdens. To achieve its mission, Africa CDC supports AU Member States to strengthen their national capacities and capabilities in key strategic priority areas in collaboration with relevant partners. Following the declaration COVID-19 outbreak as a Public Health Emergency of International Concern (PHEIC), the AU Commission Chairperson convened an Emergency meeting of Ministers of Health and established African Task Force on Coronavirus (AFTCOR) to serve as a Pan-African platform for preparation and response to COVID-19 pandemic on February 22, 2020, in Addis Ababa, Ethiopia. The continental task force endorsed a continental pandemic response strategy with the overall aim of Strengthening the capacity of member states to prevent, detect, and respond to the COVID-19 pandemic.[32]

The objectives of the strategy were to:

- Accelerate African involvement in the clinical development of a vaccine.
- Ensure African countries can access a sufficient share of the global vaccine supply.
- Remove barriers to wide delivery and uptake of effective vaccines across Africa.

Interventions

Accelerate African involvement in the clinical development of a vaccine

- The AU established Consortium for COVID-19 Clinical Vaccine Trials (CONCVACT) to serve as a coordinating body and enhance African research institutes' participation permitting sufficient data generation on safety and efficacy of the most promising candidate vaccines.[33]
- The consortium focused primarily on:
  - Research in vaccine development and promotion of translation of scientific findings to impact political and regulatory body engagement.
  - Facilitation of clinical trials for promising COVID-19 vaccine candidate.
  - Supporting the development of vaccine trial sites across Africa.
  - Accelerating posttrail regulatory approval.
  - Rolling out and uptake of vaccines and fostering Africa-based vaccine manufacturing capacity.[32]

Access to sufficient share of the global vaccine supply

- November 2020—African Vaccine Acquisition Task Team (AVATT) was established to ensure equitable access to COVID-19 vaccines for AU Member States and financing.
- In March 2021, AVATT has secured 400 million doses of the Johnson and Johnson single-shot COVID-19 vaccine with the support of the African Export-Import

CASE STUDY (cont'd)

Bank (Afreximbank). AU Member States are allocated vaccines, based on the size of their populations through a pooled procurement mechanism.
- April 2021, Partnership African Vaccine Manufacturing (PAVM) launched with the vision of manufacturing about 60% of its vaccines on the continent by 2040
- Senegal, Morocco, South Africa, Egypt, Algeria, and Rwanda have already signed agreements to develop vaccine manufacturing capacity.

Removing the barriers to delivery and uptake
- The African Union COVID-19 Response Fund has been established with the aim of raising resources for the continental COVID-19 response.
- The fund enabled Africa CDC to procure diagnostics and other medical commodities for distribution to Member States.
- The aim was to mitigate the socioeconomic and humanitarian impact of the pandemic on African populations.
- The Saving Lives and Livelihoods partnership between Africa CDC and Mastercard Foundation was launched in June 2021 to:
  - enable the purchase of COVID-19 vaccines for at least 65 million people and support the delivery to millions more across the continent.
  - lay the groundwork for vaccine manufacturing in Africa through a focus on human capital development and strengthen the Africa CDC.

Key observations

Leveraging the political, policy, advocacy, and convening power of AU, Africa CDC effectively coordinated the pandemic response. The high-level meeting convened by the AU commission engaged over 3000 political leaders and technical experts yielding the following key achievement.

- Access to COVID-19 Vaccines for Millions of Africans
- Strengthen Member States' vaccine rollout capacity
- Strengthen African institutions to drive a broader impact
- Ensure national, regional, and global health security

Lessons learned

What worked well?
- Leveraging the strong political leadership of the AU commission has helped to implement "the whole of Africa Approach" enabling the National Public Health Institutes (NPHIs) to play their critical role in the pandemic response.
- Development of continent-specific guidelines and strategies has enabled to attract more resources.
  - Africa Medical Supplies Platform (AMSP) offers direct access to vetted manufacturers of COVID-19 vaccines, drugs, medical equipment, and PPE
  - Over 34.5 thousand tons of medical supplies, equipment, and vaccine distributed to all 55 AU Member

CASE STUDY (cont'd)

States amounting to over 280 million USD
- AU COVID-19 Response Fund: Total pledged for COVID-19 response was $43,705,300 of which $37,233,883 ($33,292,410 in cash and $3,941,473 in kind) was received.

What worked less well?
- Insufficient well-trained and readied workforce to respond to PHEIC
- Insufficient capacity and capability of national public health institutions to prepare for and respond to public health emergency
- In adequate global collaboration to ensure equitable access to vaccines and other medical and nonmedical commodities for the pandemic response

Recommendations
- Urgent need for strengthening Africa's public health workforce
- Strengthen National Public Health Institutes
- Continuous advocacy to enhance global collaboration for equitable access to vaccines and other supplies

Vision for the future, challenges, and actions

The COVID-19 pandemic has not only impacted the lives of the global citizenry but has also given the opportunity for public health leaders, scientists, and politicians to reflect on how to work together collaboratively to respond to the pandemic using the "whole of Government and the whole of society approach."[34] It also provides a chance to examine the response in terms of what went well, what went not so well, and why to identify the strength and gaps, and prioritize actions to make the world safer.[5] How does one nurture trust among politicians, scientists, and communities in the context of multiple uncontrollable contextual factors? If we can leverage the pandemic to drive progressive change in strengthening the "three-legged stool" referenced by Turner as the "experts, the state and the public" it will require intentional deliberation for coherence to prevent the stool from wobbling.[35] Haldane et al. indicated that this will require transformative action propelled by adaptive leaders with new capabilities to architect an uncertain future. We briefly explore these leadership characteristics from lessons learned on the science and political interface in the pandemic.[31]

Challenges

As we navigate our way in and through the pandemic, science and political leadership will need to challenge the status quo of how health and social systems interact, they will need to review the need for more meaningful engagement and transparency and drive a research agenda that seeks to obtain an in-depth understanding of the outstanding challenges and some of the speculation of what worked well and not well.[28,29]

Protecting the independence of national and international organizations

Among the major challenges are the need to protect the independence of organizations that mainly advise the government on health emergency and security based on scientific merits. The case of the Centers for Disease Control and Public Health England could serve as a pointer to ensure that such institutions should maintain their ability to review and advise the public based on scientific merits available at the time.[5,36] The WHO constitution aims to achieve consensus and is alleged to value political correctness over scientific evidence to justify decisions, which resulted in WHO being accused of delaying PHEIC declaration during the COVID-19 pandemic and not announcing the need for travel restrictions timely with most countries making unilateral decisions by exceeding the WHO's travel guidance.[18]

Complying to International Health Regulations

The mechanism put in place to advance global health security is for each country to implement the IHR 2005, which requires building national core capacities to be able to prevent from, protect against, detect early, and respond rapidly to any PHEIC collaboratively through global efforts such as the Global Health Security Agenda framework.[37] The Joint External Evaluation result showed that the world was not prepared. The Independent Panel for Pandemic Preparedness and Response (IPPPR) report indicated that the political leadership failed to appreciate the scientific evidence and showed lack of commitment to prioritize and allocate funding for its implementation.[38]

As Campos and Reich write, managing the political dimension of policy implementation is critical, particularly in the area of preparedness for pandemics and health security.[39] Their six-dimensional framework to manage stakeholders which include interest group, beneficiaries, bureaucracies, financial decision, donors, and political leaders in the implementation of policy guidance and plans need to be managed skillfully nationally and globally if we have to avoid future surprises should a pandemic occurs.

The universal applicability of science will need to be contested, comparative political sociology between countries, within countries and governing the informal will require investigation.[14] While Offe asserts that female political leaders in New Zealand, Taiwan, Norway, Finland, and Germany have done "remarkably well"[40] this will have to be explored together with the assumption that both charismatic scientists and politicians had the greatest influence.[41–44] Strategic political and science leaders will need to become "savvy sense makers"[14] by synthesizing data from multiple sources, with multiple stakeholders in crafting options for testing multiple hypotheses.

Making decisions

In the context of public health crises, the similarities, and differences of the decision-making process between scientists and politicians should be appreciated. Political leadership in a pandemic will draw on a broad scope of decision-making tools that include economic, cultural, and ideological factors under the umbrella of "pandemic politics,"[9] while scientists will use objectivity and empiricism of science in search of truth.[8] The decision-making process is tested most in crises, decisions need to be made rapidly, on incomplete information, something that scientists are not accustomed to. Decisions were also influenced by political ideologies that existed before the universal crises with countries with neoliberal social democracies such as Brazil adopting the official position of "corona skepticism" and the United States and the United Kingdom with their "laissez-faire" response as compared to China who had a more "interventionist" approach.[8]

Alignment of actions with stakeholders

The whole-of-society and whole-of-government approach is essential and is demonstrated to be a key pillar of success in response (Victoria Haldane), but achieving universal consensus is rare.[31] It is increasingly being observed that the decentralization of the political and scientific decision-making process is enhanced by community participation and the understanding of the drivers behind divergent views, different perspectives, and the story behind the viewpoint. Social listening, adopted and implemented fairly late in the South Africa's COVID-19 response sought to provide the opportunity to listen, and direct initiatives should be a universal tool in soliciting input and seeking alignment, particularly for vulnerable societies.

Actions

The pandemic has presented a plethora of learning opportunities for scientists and politicians. We need to celebrate success and explore options for doing things differently in the new normal. If we are to engender the culture of learning we need to embrace feedback, overcome the existing vertical barriers of learning, and expand the skills of individual disciplinary approaches to a transdisciplinary mechanism necessary to effectively address complex public health challenges. This has been demonstrated very well during the pandemic as the fastest development of mRNA-based vaccine technology development, production, and distribution would not be possible without collaborative action by government, industry, academic, and regulatory partnership that transcends across many disciplines and administrative barriers at the global scale.

Rather than asking public health leaders to learn about politics on the job, the goal is to foster a new, more integrated account of public health science and policy interface that can be shared with students who will be the public health leaders of tomorrow.

The IHR (2005) decision algorithm is founded on rational action based on science, medical evidence, and global public health insight but does not address the need for skills and expertise required to navigate the increasingly political dimensions of outbreaks.[45] we posit that this is time to conduct a "capability audit"[10] of the skills and competencies required to navigate future pandemics.

There have been various attempts encouraging the nexus of competencies between science and politics such as the Madrid Declaration on Science that aims to promote the need to strengthen science diplomacy, making it part of foreign policy and international relationships.[46] Science diplomacy needs to be exploited to ensure that the leadership at supranational, national, and subnational levels align coordination of resources. The declaration advocates for capacity building through, interdisciplinary training that will enable diplomats, public officials, and scientists to cooperate. This declaration identified the need to establish new science diplomacy positions such as science advisors in key government and political institutions including foreign ministries and diplomatic missions.[36]

Conclusion

The COVID-19 pandemic has highlighted the growing awareness on codependencies of science, politics, and communities. This is a complex subject, and the authors acknowledge that this chapter could not do justice to the expanse of this topic. There is a lack of established measurement and evidence representing the concept of the political and science interface and this intersectoral relationship is nuanced by the varied intertwined consequences as a result of responses and actions determined by both of these groups.

There are, however, key lessons that emerge of the scientific and political interface in a public health emergency. These include the championing of trust relationships and partnership for transparency and mutual accountability in collectively anticipating threats, challenging our paradigm, interpreting the signals, and making evidence-based decisions that are aligned with jointly determined priorities in an environment where we can collectively grow together as humanity.

Pursuant to this addressing these lessons, we propose that the IHR (2005) be updated to address a comprehensive set of threats to humanity including the socioeconomic determinants of health and disease and that it should support and enable Member States to develop the skills and competencies required to handle and manage political environments, address public fears and opinions, and translate that into effective collective leadership.[10]

References

1. Lipscy P. COVID-19 and the politics of crisis. *Int Organ.* 2020;74(S1):E98–E127. https://doi.org/10.1017/s0020818320000375.
2. Modi N. Why Nature needs to cover politics now more than ever. *Nature.* 2020;586(Oct 6):169. https://doi.org/10.1038/d41586-020-02797-1.
3. Popescu SV. Historical perspective: lessons from SARS-COV-1. *Contagion.* 2022;7(5).
4. Howe (Producer). *Stick to the Science [Podcast].* 2020.
5. Abbasi K. Covid-19: politicisation, "corruption," and suppression of science. *BMJ.* 2020;371:m4425. https://doi.org/10.1136/bmj.m4425.
6. The Editors of Encyclopedia Britannica. "Science". *Encyclopedia Britannica;* 31 Dec. 2020. https://www.britannica.com/science/science. Accessed September 4, 2022.
7. Science Council. https://sciencecouncil.org/aboutscience/our-definition-of-science. Accessed September 4, 2022.
8. Yukl G. *Leadership in Organizations.* 9/e. Pearson Education India; 1981.
9. Wikipedia. Politics - Wikipedia.
10. Fafard P, Cassola A, de Leeuw E. *Integrating Science and Politics for Public Health;* 2022. https://link.springer.com/book/10.1007/978-3-030-98985-9.
11. Eden L, Hermann CF, Miller SR. Evidence-based policymaking in a VUCA world. *Transnatl Corp J.* 2021;28(3):159–182.
12. Whiteman WE. *Training and Educating Army Officers for the 21st Century: Implications for the United States Military Academy.* Fort Belvoir, VA: Defense Technical Information Center; 1998.
13. Schoemaker PJH, Heaton S, Teece D. Innovation, dynamic capabilities, and leadership. *Calif Manag Rev.* 2018;61(1):15–42. https://doi.org/10.1177/0008125618790246.
14. Zielonka J. Who should be in charge of pandemics? Scientists or politicians? *Pandemics Polit Soc.* 2021:59–74. De Gruyter.
15. van Wyk TD, Reddy V. Pandemic governance: developing a politics of informality. *South Afr J Sci.* 2022;118(5–6):1–6.
16. Martin J. Let science be a springboard for politics. *Nature.* 2017;546:577. https://doi.org/10.1038/546577a.

17. Hunter EL. Politics and public health-engaging the third rail. *J Publ Health Manag Pract*. 2016;22(5):436–441. https://doi.org/10.1097/PHH.0000000000000446.
18. World Health Organization. *Basic Documents Forty-Ninth Edition, Including Amendments Adopted up to May 31, 2019*. WHO | Basic documents; 2020.
19. Kickbusch I. *Governing the Global Health Security Domain* (Retrieved from Geneva, Switzerland). 2016.
20. WHO. *Global Health Observatory. Map Gallery Search Results (who.Int)*. 2021.
21. Gostin LO, Katz R. The International Health Regulations: the governing framework for global health security. *Milbank Q*. 2016;94(2):264–313.
22. Burkle Jr FM. Global health security demands a strong international health Regulations treaty and leadership from a highly resourced world health organization. *Disaster Med Public Health Prep*. 2015;9(5):568–580. https://doi.org/10.1017/dmp.2015.26.
23. de Melo DDN, Papageorgiou M. Regionalism on the run: ASEAN, EU, AU and MERCOSUR responses mid the covid-19 crisis. *Partecipazione E Conflitto*. 2021;14(1):57–78.
24. Fagbayibo B, Owie UN. Crisis as opportunity: exploring the African Union's response to COVID-19 and the implications for its aspirational supranational powers. *J Afr Law*. 2021;65(S2):181–208.
25. Stevens A. Governments cannot just "follow the science" on COVID-19. *Nat Human Behav*. 2020;4:560. https://doi.org/10.1038/s41562-020-0894-x.
26. Muller SM. The dangers of performative scientism as the alternative to anti-scientific policymaking: a critical, preliminary assessment of South Africa's Covid-19 response and its consequences. *World Dev*. 2021;140:105290. https://doi.org/10.1016/j.worlddev.2020.105290. PMID: 34580559; PMCID: PMC8457686.
27. Johns Hopkins Center for communication programs COVID Behaviors Dashboard. *COVID Behaviors Dashboard - Johns Hopkins Center for Communication Programs*; 2022. jhu.edu. Accessed September 22, 2022.
28. Bollyky TJ, Hulland EN, Barber RM, et al. Pandemic preparedness and COVID-19: an exploratory analysis of infection and fatality rates, and contextual factors associated with preparedness in 177 countries, from Jan 1, 2020, to Sept 30, 2021. *Lancet*. 2022;399(10334):1489–1512.
29. Schoeni RF, Wiemers EE, Seltzer JA, Langa KM. Political affiliation and risk taking behaviors among adults with elevated chance of severe complications from COVID-19. *Prev Med*. 2021;153:106726. https://doi.org/10.1016/j.ypmed.2021.106726. Epub 2021 Jul 16. PMID: 34280407; PMCID: PMC8284062.
30. Green J, Edgerton J, Naftel D, Shoub K, Cranmer SJ. Elusive consensus: polarization in elite communication on the COVID-19 pandemic. *Sci Adv*. 2020;6:eabc2717.
31. Haldane V, Jung AS, Neill R, et al. From response to transformation: how countries can strengthen national pandemic preparedness and response systems. *BMJ*. 2021;375:e067507. https://doi.org/10.1136/bmj-2021-0675.
32. Africa Union Africa CDC. *Africa CDC Consortium for COVID-19 Vaccine Clinical Trials (CONCVACT)*; August 2020. https://au.int/en/documents/20201005/africa-cdc-consortium-covid-19-vaccine-clinical-trials-concvact.
33. Africa Union Africa CDC. *COVID-19 Vaccine Development and Access Strategy*; August 2020. https://africacdc.org/download/covid-19-vaccine-development-and-access-strategy/.
34. Mosselmans M, Waldman R, Cisek C, Hankin E, Arciaga C. *Beyond Pandemics: A Whole-of-Society Approach to Disaster Preparedness*. 2011.
35. Turner S. The naked state: what the breakdown of normality reveals. *Pandemics Polit Soc*. 2021;43.
36. Rasmussen SA, Jamieson DJ. Public health decision making during Covid-19 — fulfilling the CDC pledge to the American people. *N Engl J Med*. 2020;383:901–903. https://doi.org/10.1056/NEJMp2026045.
37. World Health Organization Website. *Joint External Evaluation Report Average Score by Capacity*; 2020. JEE | Strategic Partnership for Health Security and Emergency Preparedness (SPH) Portal (who.int) https://extranet.who.int.sph/jee.
38. The Independent Panel for Pandemic Preparedness and Response. *COVID-19: Make it the Last Pandemic. (COVID-19: Make it the Last Pandemic*; 2021. theindependentpanel.org.
39. Campos PA, Reich MR. Political analysis for health policy implementation. *Health Systems & -Reform*. 2019;5(3):224–235. https://doi.org/10.1080/23288604.2019.1625251.
40. Offe C. Corona pandemic policy: exploratory notes on its "epistemic regime". *Pandemics Polit Soc*. 2021;25.
41. Aldrich AS, Lotito NJ. Pandemic performance: women leaders in the Covid-19 crisis. *Polit Gend*. 2020;16(4):960–967. https://doi.org/10.1017/S1743923X20000549.

42. Windsor LC, Reinhardt GY, Windsor AJ, Ostergard R, Allen S, Burns C, et al. Gender in the time of Covid-19: evaluating national leadership and Covid-19 fatalities. *PLoS One*. 2020;15(12):e0244531. https://doi.org/10.1371/journal.pone.0244531.
43. Lee E. Fact check: national beating back COVID-19 are female-led, but it's more complicated than just gender. *USA Today*; 2020. July, 28 beating-covid-19-female-led-coincidence/5451476002/.
44. Tyner K, Jalalzal F. Women prime ministers and COVID-19: within-case examinations of New Zealand and Iceland. *Polit Pol*. 2022;50:1078–1098. https://doi.org/10.1111/polp.12511.
45. World Health Organization website. *International Health Regulations*; 2022. https://www.who.int/health-topics/International Health Re International health regulations (who.int).
46. Using Science for/in Diplomacy for Addressing Global Challenges (S4D4C). (2019). The Madrid Declaration on Science Diplomacy. Madrid: S4D4C. The Madrid Declaration on Science Diplomacy – EU Science Diplomacy (s4d4c.Eu).

Further reading

1. Kwon D. *Science and Policy Collide during the Pandemic*. The Scientist; 2020. Website https://www.the-scientist.com/careers/science-and-policy-collide-during-the-pandemic-67882.
2. Mondal S, Van Belle S, Maioni A. Learning from intersectoral action beyond health: a meta-narrative review. *Health Pol Plann*. May 2021;36(4):552–571. https://doi.org/10.1093/heapol/czaa163.
3. Galston WA. For COVID-19 vaccinations, party affiliation matters more than race and ethnicity. *Brookings*; 2021. October 1 https://www.brookings.edu/blog/fixgov/2021/10/01/.
4. The Open University. What is politics? Open Learning. https://www.open.edu/-openlearn/society-politics-law/what-politics/content-section-2.4.
5. Navarro JA, Markel H. Politics, pushback, and pandemics: challenges to public health orders in the 1918 influenza pandemic. *Am J Publ Health*. March 2021;111(3):416–422. https://doi.org/10.2105/AJPH.2020.305958.
6. Reddy SP, Sewpaul R, Mabaso M, et al. South Africans' understanding of and response to the COVID-19 outbreak: an online survey. *S Afr Med J*. 2020;110(9):894–902. Retrieved from http://www.samj.org.za/index.php/samj/article/download/13049/9486.

CHAPTER

23

Influencing global health security through finance and philanthropy

Affan T. Shaikh[1,2], *Yashwant Chunduru*[1], *Sejal Waghray*[2] *and Julian Salim*[2]

[1]**Yale University, New Haven, CT, United States;** [2]**Hubert Department of Global Health, Rollins School of Public Health, Emory University, Atlanta, GA, United States**

Introduction

During the peak of the COVID-19 pandemic on July 6th′ 2020, the World Health Organization (WHO) lost the support of the United States, its largest benefactor, contributing US$893 million (20% of the total budget) in the previous 2-year budget cycle.[1] The second-largest donor behind the United States was not another member state (MS), but a nongovernmental organization, the Bill and Melinda Gates Foundation (BMGF), funded by some of the world's most powerful individuals. In that same budget cycle, the Gates Foundation contributed US$531 million (12% of the total budget). While, as a nonmember, the Gates Foundation could not set the WHO agenda, many questioned the influence and lack of transparency a nongovernmental agency could have.[2] Furthermore, the rise of NGO funding indicates a deeper issue of chronic underfunding of key institutions. Despite the WHO's key role in global health security (GHS), their annual budget is still less than that of a large teaching hospital in the US and just a fraction of that of the US Centers for Disease Control and Prevention.[3]

Modernizing Global Health Security to Prevent, Detect, and Respond
https://doi.org/10.1016/B978-0-323-90945-7.00028-2

FY22 operating budget (USD)	
US Centers for Disease Control and Prevention	$ 9,269,000,000.00
NewYork-presbyterian/Weill Cornell Medical Center	$ 7,240,896,211.00
Tisch Hospital	$ 6,444,113,234.00
Cleveland Clinic Main Campus	$ 6,240,384,003.00
World Health Organization	$ 6,121,700,000.00
Vanderbilt University Medical Center	$ 5,356,674,083.00

Definitive Healthcare, Hospitals with the Highest it Operating Budgets: Definitive Healthcare, 2022 WHO, Programme Budget 2022–2023. World Health Organization Congressional Research Service Sekar K. Centers for Disease Control and Prevention (CDC) Funding Overview. Congressinoal Research Service 2023

Global control of public health threats is dependent on both developing national capacity to build and maintain pandemic preparedness as well as ensuring sustainable financial and technical resources—simply stated governance and resourcing. The former gives the latter purpose and structure, but the latter sets agendas and drives programs. As we layer various stakeholders across local, national, regional, and global levels, the picture becomes increasingly complex, and the need to understand the influence of finance and philanthropy becomes more apparent.

Per the constitution of the WHO, "governments have a responsibility for the health of their peoples, which can be fulfilled only by the provision of adequate health and social measures."[7] Yet many MS struggle with their own national healthcare, let alone making contributions to the WHO. In Organization for Economic Cooperation and Development (OECD) countries, domestic spending for public health and preparedness is about 2.8% of GDP whereas a smaller proportion is distributed to disease detection and immunization.[8] On the other hand, the total government spending on health per capita in low- and middle-income countries (LMICs) is less than in OECD countries (US$35–85 per person/year).[8] The WHO maintains the Global Health Expenditure Database, which provides comparable data on health expenditures across 190 WHO MS.[9] The Pandemic saw the highest global spending on health, reaching US$9T, or 10.8% of global gross domestic product (GDP).[9] While health spending as a share of GDP (%) has been rising across the world since 2000, high-income countries will still account for 80% of health spending in 2020.[9]

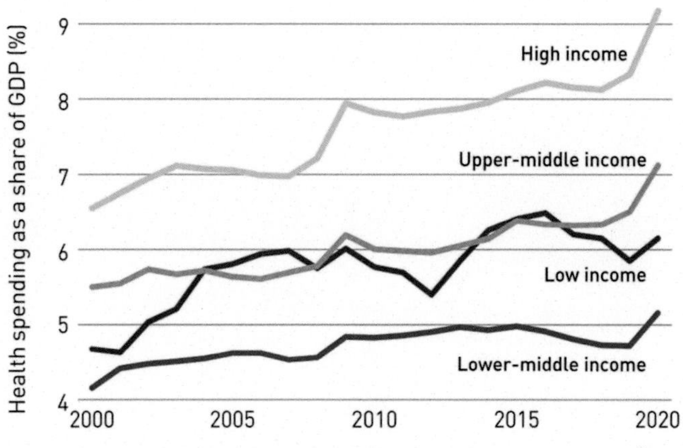

Data Source: WHO Global Health Expenditure Database, 2022[9]

It is important to allocate and sustain financial commitment to develop the core capacities of the International Health Regulations (IHR [2005]) from the central government, subnational levels of government, and the scientific community.[10] The math to fund GHS is clear. One estimate of the cost to build country-level capacity to address GHS gaps within the next 5 years could range from US$96 billion to US$204 billion.[11] Building public health core capacities offers 9–20x (multiple) returns on investment, especially in LMICs where benefits far exceed costs by factors. Sustained investments in prevention, detection, and response, underpinned by One Health, are estimated to save $15 billion annually from the prevention of outbreaks alone.[12] Yet, many countries have not yet established those resources, and a poor understanding of the associated costs has emerged as a barrier to effectively mobilizing assistance.[3]

"As the global community considers financing options, it must also examine where concessional financing will be needed to support global health preparedness investment".[8] In 2019, a total of US$374 million was disbursed for pandemic preparedness and response in LMICs, or less than 1% of all development assistance that went to other health-related measures (e.g., establishing national response plans and disaster management training).[13]

Notable increases in pandemic preparedness funding have been noticed following some severe epidemics.[13] After the 2009 H1N1 pandemic, the assistance for pandemic preparedness was about US$190 million. Further, in 2014 and 2015, after the Ebola outbreak in West Africa, about US$320 and US$370 million, respectively, were distributed for global pandemic preparedness.[13] Nonetheless, given the massive impacts of global health emergencies (e.g., the COVID-19 pandemic), that amount would not be sufficient to build resilience to prevent catastrophic consequences.

Financing and investing in public health systems strengthening, research and development (R&D), coordination, and contingency measures yield enormous advantages in protecting the world against the future pandemic. It is vital to comprehensively support the international community to upscale their financial capacity, allocation, and management during a worldwide public health crisis. In this chapter, we consider the influence of finance and philanthropy on GHS. Specifically, we start our conversation by considering what GHS finance is. We then dive deeper into the mechanisms used to fund GHS across public and private markets. Next, we explore some of the key issues around money, power, and politics when making GHS decisions. A case study exploring the eradication efforts around polio is presented that captures many of the key points described in the chapter. And finally, we present a vision for the future, rooted in the lessons of COVID-19, of how GHS can be secured through sustainable finance.

Literature review and gap analysis

The global health security dilemma

Global health security is an issue of paramount importance, and the question of who should fund it is one that has been debated for years. Some argue that it is a public good and that governments have a responsibility to fund the R&D of treatments and vaccines.

Others see the private sector playing a larger role, citing the obligation to safeguard economic security and the potential for profit. The debate is complicated by the fact that GHS is an issue that affects everyone. While some countries may have the resources to fund the necessary R&D, many do not, and if GHS is seen as a public good, it should be funded by everyone for the benefit of all, not just as a one-time investment but as an ongoing effort. Ultimately, country ownership of preparedness is critical to sustainable financing.[14]

Expenditures on global health security

Understanding current GHS expenditures can help us understand global priorities. The WHO's US$6.72B budget for 2022 and 2023 is broken into four segments: Base budget (including country, regional, and headquarter operating expenditures), special programs (i.e., UNICEF/UNDP/World Bank/WHO Special Program for Research and Training in Tropical Diseases; the UNDP/UNFPA/UNICEF/WHO/World Bank Special Program of Research, Development and Research Training in Human Reproduction; and Pandemic Influenza Preparedness Framework), the Global Polio Eradication Initiative (GPEI) (public-private partnership to eradicate polio), and emergency operations and appeals. The programmatic budget breakdown is presented below.

Program budget 2022–23	Allocations US$
Universal health coverage	1,930M
Health emergencies	1,250M
Healthier populations	455M
More effective and efficient WHO	1,333M
Polio eradication	558M
Special programmes	199M
Emergency operations and appeals	1,000M
Total	**6,725M**

2022–23 WHO Budget Programme Budget 2022–2023. World Health Organization.

U.S. Funding for Global Health Security, FY 2013 - FY 2023

(In Millions)

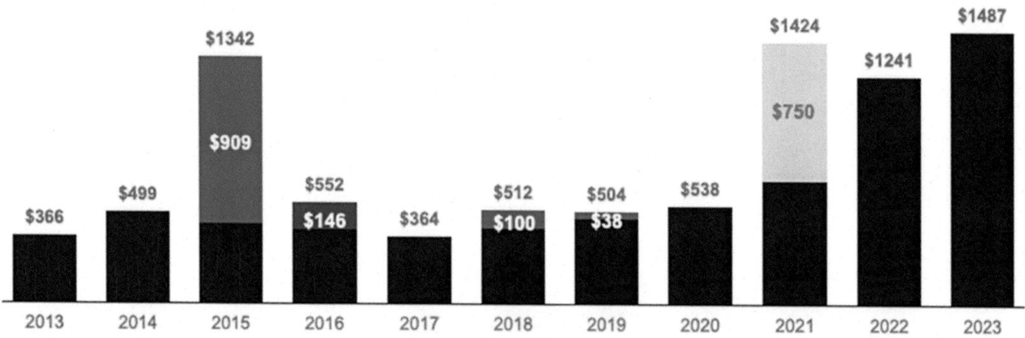

NOTE: Includes Global Health Security funding through USAID, CDC and DoD. Totals include base and supplemental funding. FY13 includes the effects of sequestration. FY23 is based on funding provided in the "Consolidated Appropriations Act, 2023" (P.L. 117-328) and is a preliminary estimate. FY21 and FY22 funding for GEIS at DoD includes $3.45 million and $8.438 million respectively in reprogrammed funding. FY23 funding for GEIS at DoD is not yet available; for comparison purposes, this amount is based on the prior year level. In FY15, Congress provided $5.4 billion in emergency funding to address the Ebola outbreak, of which $909.0 million was specifically designated for global health security. In FY16, Congress provided $1.1 billion in emergency funding to address the Zika outbreak, of which $145.5 million was specifically designated for global health security. In FY18, Congress provided $100 million in unspent Emergency Ebola funding for "programs to accelerate the capabilities of targeted countries to prevent, detect, and respond to infectious disease outbreaks." In FY19, Congress provided $38 million in unspent Emergency Ebola funding for "programs to accelerate the capacities of targeted countries to prevent, detect, and respond to infectious disease outbreaks." In FY21, Congress provided $9.4 billion in emergency supplemental global health funding to address the COVID-19 pandemic, of which $750 million provided through CDC was designated by CDC as global health security.

SOURCE: KFF analysis of data from the Office of Management and Budget, Agency Congressional Budget Justifications, Congressional Appropriations Bills, and U.S. Foreign Assistance Dashboard [website], available at: http://www.foreignassistance.gov. KFF personal communication from CDC, January 2023. • Get the data • PNG

KFF

Source: Kaiser Family Foundation[15]

At the country-level, financing GHS is dictated by several governance structures, including the IHR [2005], the Global Health Security Agenda (GHSA), and national needs around pandemic preparedness. Taking the US as an example, GHS funding has largely focused on addressing the emergence and spread of infectious disease threats, with budget fluctuations resulting from global public health emergencies such as Ebola, Zika, and COVID-19.[15]

Global health security represents only a small fraction (11%) of total US global health funding, with other key funding priorities including HIV, the Global Fund, maternal and child health, and malaria.[15]

U.S. Global Health Funding (in millions), By Sector, FY 2023

Total Funding = $13.0 billion

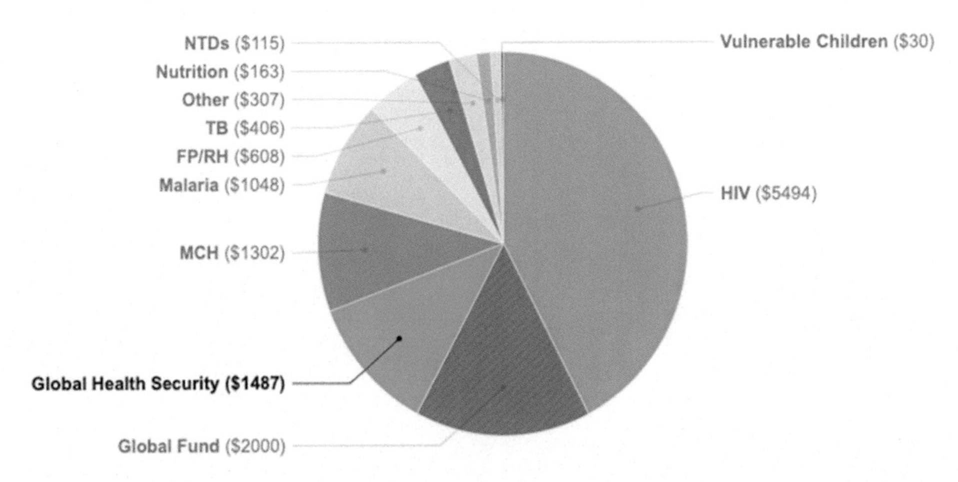

NOTE: Represents total known base funding provided through the State Department, USAID, CDC, NIH, and DoD. HIV includes funding through State/OGAC, USAID, CDC, NIH, and DoD. Global Fund includes funding through State. Malaria includes funding through USAID, CDC, NIH, and DoD. TB includes funding through USAID and CDC. Nutrition, NTDs, and Vulnerable Children include funding through USAID. MCH includes funding through USAID and CDC as well as contributions to UNICEF. FP/RH includes funding through USAID as well as contributions to UNFPA. Global Health Security includes funding through USAID, CDC, and DoD. "Other" includes funding through USAID, CDC, and NIH, as well as contributions to WHO and PAHO, and the Emergency Reserve Fund, which was created in the FY17 Omnibus bill to respond to contagious infectious disease outbreaks, and would be made available if there is an "emerging health threat that poses severe threats to human health." Some FY23 global health funding provided through CDC, NIH, DoD, and the Economic Support Fund (ESF) and Development Assistance (DA) accounts at USAID is not yet known; for comparison purposes, these amounts are estimated using prior year levels.
SOURCE: KFF analysis of data from the Office of Management and Budget, Agency Congressional Budget Justifications, Congressional Appropriations Bills, and U.S. Foreign Assistance Dashboard [website], available at: http://www.foreignassistance.gov. • Get the data • PNG

KFF

Source: Kaiser Family Foundation[15]

Let's further unpack GHS funding by looking at the IHR, GHSA, and pandemic preparedness.

International Health Regulations

IHR 2005 set out as a foundation for WHO MS to better collaborate and manage GHS concerns. IHR 2005 requires all countries to have the ability to detect, assess, report, and respond to public health threats.[1] IHR is a legally binding instrument to be administered by all WHO MS. The IHR (2005) intends to prevent, protect, and control the spread of disease globally and implement a public health response according to public health risks to avoid unnecessary barriers to international travel and trade.[16] Each MS is required to inform WHO about incidents that have the potential to be PHEICs. In implementing the IHR (2005), WHO is mandated to provide collaborative assistance to MS for the evaluation, assessment, and capacity building, sustainable technical and logistical assistance, and identifies the sources of funds needed to develop and maintain MS capacity.[17] To optimize these efforts, a large amount of financial support from various sectors is needed.

WHO, MS, and international development partners should urgently commit to provide financial support at the national, regional, and international levels for the successful implementation of the Global Strategic Plan.[18]

The strategic plan describes the approach of WHO to strengthening MS's ability to implement the main capacities required under the IHR (2005) to ensure national and global preparedness and response to public health events, including infectious disease emergencies.[19] It builds on and is aligned with existing international instruments (e.g., the Pandemic Influenza Preparedness Framework)[20] and regional approaches, networks, and mechanisms for health emergency preparedness and response. But in practice, the support from WHO is not optimal. The situations deemed to constitute a PHEIC often demonstrated the shortcomings of the instruments available. For instance, the IHR 2005 was insufficiently flexible in responding to the 2009 H1N1 and 2014 Ebola crises.[21] More than 11,000 deaths were directly attributed to Ebola, along with the broader economic, social, and health crises.[22] The weakness of infrastructure required for an effective response was exposed at all levels, driven by a lack of skilled human resources and prioritization and systemic underfunding.[18] Ebola showed that the inability to adapt the IHR's requirements in one country endangers the global ambition of being able to promptly and effectively respond to global public health emergencies.[18]

Given the many priorities in development and limited national budgets to achieve the goals of IHR (2005), external funding supports were needed to mobilize political commitment and federal action[23] to manifest more coordinated preparedness and response for preventing the problems caused by pandemics.

Philanthropies began to fill in the gap to help countries increase national capacities, but the majority of countries still failed to meet IHR capacities by the 2012 and then 2014 deadlines.[24] A new resourcing strategy was needed to bring countries to capacity.

Global Health Security Agenda

The year 2014 saw the largest emergence of Ebola in West Africa since the virus' discovery in 1976. Starting in Guinea, the virus quickly spread across porous borders into Sierra Leone and Liberia, taking a foothold in all three countries' capitals by July 2014.[25] A month later, with 932 deaths since the outbreak started, the WHO declared a PHEIC.[26] With this background, the US, WHO, OIE, FAO, and other partner nations launched the GHSA on February 13, 2014, with the intent to secure the world from infectious disease threats and unite nations to make new concrete commitments.[27] The GHSA is organized across 11 action packages that outline measurable steps required to meet IHR capacities. The US is committed to assist more than 70 countries over 5 years to achieve GHSA objectives specifically combating antibiotic-resistant bacteria, improving biosafety and biosecurity on a global basis, and preventing bioterrorism.[28]

Pandemic preparedness

The United States has traditionally taken a leading role in global health through high-profile tax-payer funded programs such as the US President's Emergency Plan for AIDS Relief, the President's Malaria Initiative, the Global Fund to Fight AIDS, Tuberculosis, and Malaria; Gavi, and the GHSA. However, these funds are not unlimited and are subject to political and economic changes. Yet this foreign assistance is a direct investment in global health.

From an economic standpoint, increasing the capacity of the pandemic response at the global, national, and local levels is interesting. The economic disruption caused by the COVID-19 pandemic could cost between US$9 and US$33 trillion—much lower from the cost projected to substantially prevent future pandemics,[29] which are estimated to cost between US$70 and US$120 billion for the next 2 years and US$20 billion to US$40 billion for each year after that.[30]

While global nations have a role in funding pandemic preparedness and health security, not every country can invest the same amount. Globally, 83% of government spending on health occurrs in high-income countries. At the same time, some LMICs, such as Somalia and the Democratic Republic of Congo, have total government spending of less than US$100 per person.[13] Expecting countries to contribute equally to these critical global health investments is not realistic. The amount of funding allocated to development assistance for health, which is the financial and nonfinancial contributions intended to improve or maintain health in LMICs should be significantly increased.

Following previous epidemics, significant increases in pandemic preparedness funding have been implemented. There was an increase in development assistance (US$90 million) for pandemic preparedness in 2010, after the 2009 H1N1 pandemic, and in 2014 and 2015 (US$100 million and US$120 million, respectively) after the Ebola outbreak in West Africa.[31] In general, the total development assistance for health allocated toward pandemic preparedness in 2019 was US$374 million, which is no more than 1% of all development assistance.[31] This condition shows that the global community has made efforts to overcome postpandemic health crises, illustrating that there was not serious attention to crisis preparedness.

Great Britain was predicted as the best-prepared country in the world to promptly respond and mitigate the pandemic.[32] However, when COVID-19 emerged in 2020, the UK had arguably one of the least effective responses among rich countries, where about 65,000 deaths were recorded until the end of 2020.[32] Even developed countries that are considered to have multibillion-dollar health funds and advanced pandemic plans still have gaps in certain aspects causing their investments to fail to produce a positive impact on their communities.

On the other hand, the experience of Taiwan can help picture potential paths forward to mitigate the spread of the pandemic. As of May 24, 2021, there were only 3755 active COVID-19 cases and a total of 29 deaths registered in Taiwan.[33] Taiwan's experience helped the government react promptly to the emerging crisis, classifying the unknown disease as "Severe Special Infectious Pneumonia" as early as January 15, 2019.[34] Further, after the SARS outbreak in 2003, the functions of the Center for Disease Control of the Taiwan Ministry of Health have significantly escalated by strengthening their efficiency, expanding the medical workforce for disease prevention, and elevating competence of information technology and skills of disease investigation and emergency operation.[35] Taiwan's success results from the massive investment in preparedness measures in governmental reorganization, the health insurance system, and public involvement. Taiwan is one of the fewcountries in the world that has not imposed lockdown on any cities; residents were allowed to perform daily economic activities as usual.[35]

The current COVID-19 pandemic highlights the urgent need to build a better-prepared global community by raising the urgency of pandemic preparedness on the global agenda.

Many countries around the world, including those considered to have better response capabilities, failed to implement early detection and respond to the SARS-CoV-2 pandemic. In addition, the death rate and economic damage became more massive because of the overlapping roles at various levels of government or between the public and private sectors.

Investments in pandemic preparedness are not currently a priority for global stakeholders; the investments that have been made in preventive measures have not seen success. World leaders need to determine more comprehensive strategies for pandemic preparedness investments to be able to accelerate response and strengthen the public health system. Infectious diseases will continue to emerge, and robust pandemic preparedness investment will prepare the world to respond better than we have so far to the COVID-19 pandemic.

Mechanisms of financing

Public funds

Calls for more potent international financing to curb new outbreaks before they become pandemics are growing louder following the slow-moving West Africa Ebola response in 2014. The WHO and World Bank Group took 2 months to approve and disburse funds during the Ebola outbreak. When funding arrived, the operational response was already disintegrated, and ad hoc.[36] The outbreak ultimately claimed more than 11,000 lives and led to US$53 billion in foregone economic output.[37] In an endeavor to help govern the immediate response of pandemic financing, The WHO's Contingency Fund for Emergencies (CFE) was established in 2015 to fund rapid reactions in the event of a new health emergency and expand responses when existing crises spike or during critical funding shortfalls.[38] The CFE can disburse up to US$500K within 24 h to finance immediate response activities and can scale as required after that.

In 2020, The World Bank Group allocated almost US$15 billion as a global COVID-19 support package to help LMICs strengthen health systems, pandemic interventions, and collaborative works with the private sector to minimize the economic consequences.[39] In Indonesia, the World Bank allocated US$300 million to the Government to support diversifying its financing sources to meet the unpreceded financial gap caused by the COVID-19 pandemic.[40] This funding resulted in the successful performance of the Indonesian Government to control the spread of COVID-19; there were 1.59 million total cases in 2021, or less than half of its prediction without financial supports.[41]

Diagnostic, treatment, vaccination, and improvement of the public health system are the primary funding areas for pandemic response. In 2020, diagnostics alone had received commitments of about US$250 million of the US$2 billion needed in LMICs. Nonetheless, the global initiative to deliver much-needed COVID-19 test kits to LMICs was still underfunded and faced challenges negotiating accessibility and affordability.[42] For instance, nearly 1 year after COVID-19 swept Indonesia, only 6.7 million people (approximately 2.5%) were tested in a population of nearly 270 million.[43]

Assuring global access to COVID-19 vaccines is vital to decrease death rates and increase population immunity. By November 15, 2020, a total of 7.48 billion doses of COVID-19 vaccine from 13 manufacturers were reserved—51% of them booked for high-income countries, which represented just 14% of the world's population,[44] compared to 33% acquired by LMICs accounting for 81% of the global population.[45] LMICs also deal with the most significant disparity between doses purchased and population (37% of the global population vs. 12% of purchased doses).[46] Leaving LMICs without adequate access to vaccines will cause massive economic hardship for the global community. Governments and manufacturers are required to provide much-needed assurances for equitable allocation of COVID-19 vaccines through accountability over the distribution arrangements.[44]

Funding for strengthening health systems has risen over time. However, it has narrowed as a percentage of total development assistance for health; it declined from 21% ($1.6 billion) in 1990 to 14% ($5.6 billion) in 2019.[46] The COVID-19 pandemic has driven a global focus on health systems and their capacity to handle multiple pressures. News headlines have been dominated by a shortage of personal protective equipment, inept testing supplies, hospital beds touching maximum capacity, and emergency construction structures to handle patient overcrowding.

The United States is considered as having the best health system and pandemic preparedness in the Global Health Security Index. However, it has, to date, reported the world's highest number of COVID-19 cases and deaths.[47,48] This anomaly reveals that a noble array of public and private laboratories, innovative pharmaceutical and technology companies, and a well-recognized national public health institute (NPHI) does not guarantee that quality implementation plans are in place. The United States ultimately relies on a fragmented, siloed healthcare system; each state funds and operates on its own plans for public health surveillance.[49]

There is an inadequacy of proper coordination and implementation of IHR 2005 core capacity that has so far hindered the country's readiness to precisely calculate the effect of COVID-19, resulting in a slowed response (e.g., testing and contact tracing).[49] The scarcity of centralized funding has also led to the inappropriate allocation of human and financial resources.[50] We need to strengthen in-country and far regions of the world health systems to reduce the risk of future pandemics threatening the global community and economy.

Private funds

Although funding from the public sector is generally larger, private sector funding of public healthcare has been growing, and, particularly with the COVID-19 pandemic, the private sector played a pivotal role in influencing, supporting, and funding public health. However, disparities exist as the bulk of private funding comes from developed nations for use in developed nations, while it is often LMICs that are disproportionately impacted by global public health events.

A Brookings 2017 report estimated that global private sector R&D spending on drugs, vaccines, and therapeutics totals about US$160 billion annually, primarily by pharmaceutical companies (US$156.7 billion) and venture capital firms (US$3.2 billion). However, only US$5.9 billion (3.7%) of the overall R&D spend globally is focused on developing nations, which contain over 70% of the world's population.[51] A number of mechanisms are used to access capital from private markets, including direct investments, advance purchase commitments, volume guarantees, and even crowdfunding.

Direct investments (equity/debt)

Direct investments can take the form of equity investments or debt financing for companies and organizations that are directly impacting public health or are public-health-adjacent. Direct investments still comprise a relatively small portion of public health financing, but they play an important role.

Advance purchase commitments

Advance purchase agreements, or advance market commitments (AMCs), represent another method for private sector actors to promote global health resilience. AMCs are a concept first presented by economist Michael Kremer, and they involve private donors or corporations making an upfront commitment to pay a large sum of money upon delivery of a certain viable drug or vaccine. The "buyer" commits money to purchasing vaccines at a high price per dose upon successful manufacture and delivery. Once the committed funds are exhausted, the drug manufacturer is obligated to supply the vaccine or medicine to LMICs at a low price. The buyer—a private donor corporation—essentially subsidizes the cost to LMICs and creates a supply of vaccines that would otherwise not be funded or economically viable, incentivizing the biopharma industry to "dust off abandoned, mediocre vaccine candidates.[52]

Gavi, the Vaccine Alliance, advocated for AMCs as a response to COVID-19 as early as May 2020 and launched COVAX, a $2 billion multinational funding mechanism, which included an AMC component as part of its efforts to ensure vaccine access to the most vulnerable LMICs.[53,54] Perhaps the most well-known AMC historically was a previous effort by Gavi with its pneumococcal AMC in 2009, which included a collective $1.5 billion of commitments from the BMGF, Canada, Italy, Norway, the Russian Federation, and the United Kingdom.[55] By agreeing upfront to buy large quantities of vaccines at established prices, the AMC mechanism provided pharmaceutical companies with an incentive to develop and produce suitable vaccines while also guaranteeing a sustainable price to provide coverage to the most vulnerable members of the population. According to Gavi, the pneumococcal AMC has been a "huge success over the last decade and has prevented 700,000 children's deaths in 60 developing countries."[56] According to a 2015 impact evaluation study conducted by Boston Consulting Group, the pneumococcal vaccine AMC facilitated by Gavi accelerated immunization coverage across 53 countries as of the date of the report and fully immunized 49 million children with three doses of PCV from 2009 to 2014. The study estimates that 230,000–290,000 deaths of children under 5 years have been averted through 2015 and that over 3 million deaths of children under 5 years will be averted by 2030.[55]

However, there can be "hold-up risk" with AMCs, particularly if proper markets are not created and there is unequal bargaining power among the parties. The hold-up problem is a market failure in economics where two parties can both benefit from a transaction, but it doesn't occur due to one party having outsized bargaining power. This exact issue was illustrated by the US Government's Project Bioshield Act, which was initiated in 2004. Under the original rules of the Act, government administrators had unilateral discretion on whether to purchase, how much to purchase, and at what prices, which left many companies that had invested millions in drug development without a customer.[52] As such, it's important to ensure that there are sufficient buyers. The economist Michael Kremer also advocated for multiple companies sharing in AMCs, lessening their individual burdens, and that LMICs are allowed to refuse the vaccine if circumstances change. There are also some concerns that AMCs might compete with public-private partnerships; however, AMCs only make payments once the vaccine is developed. They don't support the upfront costs of R&D, so there is still a need for public sector and grant-making. Coordination among industry, donors, and the public sector is critical for AMCs to work.

Volume guarantees and debt guarantees/other credit enhancements

Another mechanism used to fund worldwide impact initiatives, including GHS is volume guarantees, whereby donors or private actors will provide guarantees of quantity, allowing corporations to produce a greater supply of vaccines and medicines and sell them at a lower average price. Volume guarantees are similar to AMCs in that they provide third-party-funded economic incentives for industry actors to supply healthcare goods to LMICs; however, they drive impact via quantity first and not prices. An example of volume guarantees can be seen via the Gates Foundation's broad 2020 goal of providing contraceptives to 50% of the 225 million women that didn't have access to contraceptives but were seeking access to avoid pregnancy.[57] The Gates Foundation worked with donors from multiple countries and Merck and Bayer to produce roughly double the amount of contraceptive implants that the companies were originally planning to produce at about half the price. This essentially created a long-term fixed-price contract for the corporations, giving them visibility and incentive to produce at a higher production volume and lower price, given that any excess production units would be purchased by the donors or guarantors. According to a 2016 SSIR article, the price reductions "have already saved more than $240 million for global public health donors who procure products for the benefit of those most in need in developing countries," and total savings could top US$500 million by 2018.[57] Inclusive of the Merck and Bayer contraceptive guarantees, the Gates Foundation's Strategic Investment Fund lists a total of 16 volume guarantees in their portfolio to increase access to LMICs, five of which are listed as current. Other current and past volume guarantees from the Gates Foundation include rapid COVID tests with Abbott Laboratories and SD Biosensor made in 2020, single-pill HIV treatment pills with Aurobindo Pharma and Mylan Laboratories made in 2017, and polio vaccines with BioFarma in 2022 and Novartis in 2011, among others.[58] MedAccess is a UK-based volume guarantee organization that similarly seeks to promote positive health impacts in LMICs. They list nine current volume guarantee partnerships in their portfolio, including mosquito nets, access to malaria vaccines, HIV-syphilis dual testing, and access to COVID-19 supplies and vaccines, among others.[59]

Crowdfunding

Crowdfunding can also be a powerful tool to increase the private sector's and individuals' participation in global health funding. In the same vein as "venture philanthropy," certain concepts from the private market can be translated into the global health context, including crowdfunding. This idea of crowdfunding as a mechanism for financing philanthropy and global health initiatives was introduced by Özdemir, Faris, and Srivastava as a new approach for "philanthropists and citizens to cofund disruptive innovation in global health."[60] The authors argue that perhaps the existing model of biomedical R&D—research, proof of concept, and product development—does not work efficiently in health research, particularly in the developing world. Incremental advances and evidence of progress are often seen as the "safe" method to get funding, via traditional avenues like research grants or venture capital, which often require measurable metrics and milestones to be achieved. Over time, this may cull some of the "out-of-the-box" ideas and ground-breaking innovations that have the potential for substantial leaps in healthcare advancement. However, disruptive innovation, especially in healthcare, can depend on large creative leaps that often need to be made "without sufficient data and time to develop a fully-fledged proof of concept study."[60] The authors argue that this contributes to the issue that 85% of the $160 billion spent annually to fund biomedical research is deemed inefficient. The main reason, they contend, is "finding the right answers for the wrong questions": research findings that have little or no relevance for the communities who are meant to benefit.[60] Perhaps, then, crowdfunding philanthropic efforts and health initiatives can be a method to fund projects that the people want and need.

In the more traditional venture capital world, we are seeing VC opportunities becoming increasingly democratized, with individuals being able to write small checks into innovative start-ups they believe in through platforms like Republic or AngelList. However, individual citizens cannot effectively "cofund" health initiatives in their communities as they would be able to fund a new gadget or digital service on Kickstarter. "Crowdfunding 2.0" is a way to harness "economies of scale of massive populations" to address the right problems for their specific locales and reduce wasteful research. To emphasize the example used in the publication to illustrate this point, if every Indian citizen donated 1 rupee (or $0.012), over $17 million could be raised. To illustrate that same effect, but on a smaller scale, if 50% of a 100,000 population city contributed $5, about $250,000 could be raised. Crowdfunding presents an interesting opportunity for the public to support their priorities and bolster GHS.

Philanthropic contributions

In times of infectious disease emergencies, philanthropic resources can render an effective, efficient, and speedy means of funding new solutions.[61] As individuals, companies, and government agencies fight with COVID-19, philanthropy spent more than US$20 billion on COVID-19 in 2020 (private philanthropists gave more than $9.1 billion to response and relief efforts).[62] This spending included 177 funders, making 207 grants available for 42 recipient organizations globally.[62]

The Serum Institute of India (SII) has announced a total of US$150 million in partnership with Gavi, the Vaccine Alliance, and the Bill and Melinda Gates Foundation to stimulate the manufacture and delivery of up to 100 million doses of COVID-19 vaccines for India and other LMICs.[63] The partnership will fund at-risk manufacturing (which probably would not be funded by the government) of candidate vaccines from AstraZeneca and Novavax at a ceiling price of $3 per dose.[64] The funds will help SII increase its production capacity in preparation for the anticipated regulatory approval of vaccines.

Lately, philanthropy has been criticized from two conflicting directions: (i) Another entrepreneurial strategy to maximize company social value and leverage the future impact; and (ii) billionaire philanthropists are being "incriminated" in trying to solve problems that they have started.[61] Philanthropy is no longer considered a fair means of encouraging collaboration, advanced R&D, and intervention. Furthermore, philanthropic organizations are inhibited by a lack of accountability, transparency, conflicts of interest, and political regulation.[65]

Money, power, and politics of global health security

Influencing the global agenda

According to the Institute of Health Metrics at the University of Washington, nearly US$55 billion was spent on global health in 2020. From the below chart, it may be assumed that certain players have a larger influence than others based on the funds they contribute.[66]

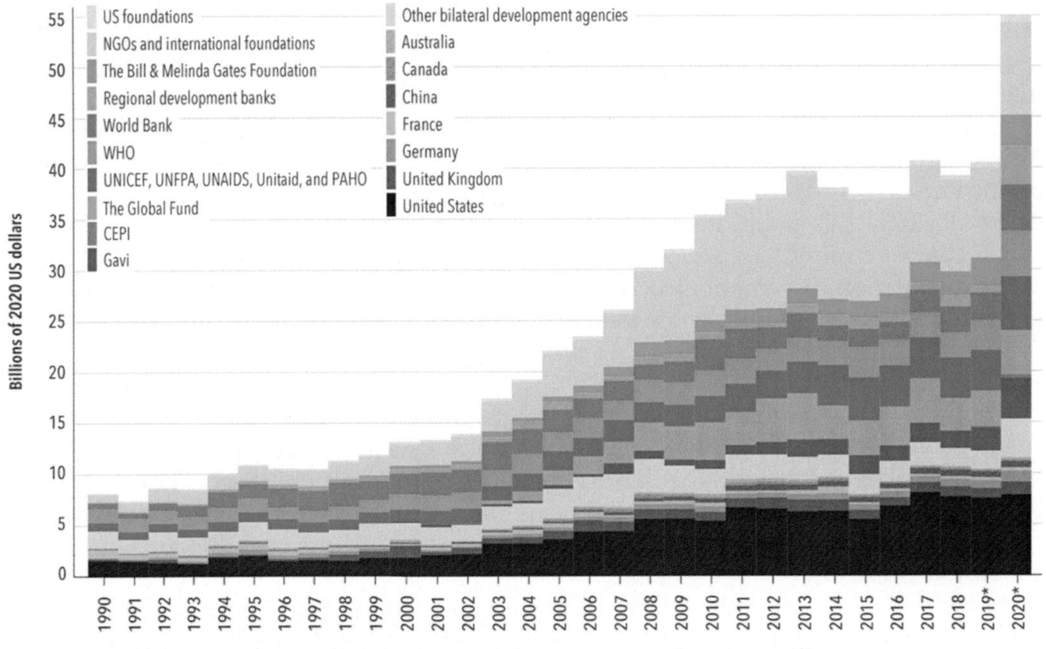

Source: Institute of Health Metrics, University of Washington[66]

WHO contributions and inclusivity

The WHO is funded through a number of different mechanisms. One such way is annual fees from member state known as "assessed contributions." This represents a percentage of a country's GDP that is agreed upon every 2 years at the World Health Assembly.[67] This percentage, unique to each country and also based on population size, has largely remained stable. The last increase in MS contributions of 3% was passed with much compromise in 2017.[68]

Funding from assessed contributions represents approximately 20% of the WHO's budget and is the most flexible funding the WHO has, allowing allocation across programs as needed. To ensure sustainable and consistent funding, there are calls to increase assessed contributions to represent up to 50% of the WHO budget,[69,70] worth up to US$4.4 billion.[71] This has been a key project for the current Director General, Tedros Adhanom Ghebreyesus, who in 2020 announced the WHO Foundation to help raise funds for the organization.[72] In 2021, the WHO Foundation mobilized approximately US$20 million for health programs (see more at https://who.foundation/).[73]

Aside from "assessed contributions," the WHO receives the remaining 80% of their budget from "voluntary contributions' from MS, NGOs, philanthropic organizations, and other private entities. These contributions, in particular, are typically earmarked for specific projects.[3] The result of this is that the WHO has very little discretionary control over how to allocate resources to the most pressing or emerging challenges or even address fundamental GHS concerns such as universal health coverage and pandemic preparedness.

Transparency and accountability

While tremendous good has come from philanthropic donations to WHO, the current predicament, where nongovernment agencies have taken a lead in financing critical agencies, leads to important governance questions around accountability, representativeness, and transparency. This is especially true when considering vulnerable populations, which are often left without a voice on the table. For their part, the Gates Foundation also recognizes the concern over being a leading funder of the WHO. The foundation's CEO, Mark Suzman, wrote in his annual newsletter, "It's not right for a private philanthropy to be one of the largest funders of multinational global health efforts [...] but make no mistake—where there's a solution that can improve livelihoods and save lives, we'll advocate persistently for it. We won't stop using our influence, along with our monetary commitments, to find solutions."[74] However, private funders can also change their mind about donations (i.e., the Trump administration withdrawing US$400 million),[75] or they could place conditions or demands on a public good. A private foundation's resources may also be tied to other capital markets. The Gates Foundation was once a large investor in fossil fuels, and the founder Bill Gates' personal fortune is tied to farmlands that sell produce to fast foods.[76,77]

The core of the WHO is driven by the central tenet that it is the governments' responsibility to provide their citizenry with healthcare locally as well as globally. The current financing structure delegates this responsibility to private entities that exist without the same transparency or accountability.

CASE STUDY

Background

The BMGF is structurally a merger between two organizations founded by the Gates family—the William H. Gates Foundation (for global health charitable giving) and the Gates Learning Foundation (formerly known as the Gates Library Foundation, which grew from providing internet access to low-income families in America to providing general education support). Collectively, both organizations rebranded as BMGF in

CASE STUDY *(cont'd)*

2000 with four main branches: the Global Development Program, the Global Health Program, the Global Policy and Advocacy Program, and the United States Program. Shortly after rebranding, a fifth branch for trusts/endowments was added to accommodate Warren Buffet's pledge of $31 million in Berkshire Hathaway stocks (Langley 2019). This trusts/endowments branch has since grown significantly, as "by mid-2015, 137 billionaires from 14 countries" were donating to BMGF's Giving Pledge (Martens and Seitz, 2015). With such powerful money from all over the world contributing to one of the most giving foundations globally, BMGF has a strong international presence. Today, BMGF's influence extends through its $50 Billion endowment as the "second-biggest donor to the WHO" and as the 12th biggest distributor of foreign aid if it were its own government.[78] The organization's restructuring has made it feasible for BMFG to employ a businesslike approach to philanthropy where return on investment is prioritized and financial return is critical to overall program retention.[79] In fact, "the BMGF is disproportionately supportive of vertical health programs, focusing on a range of individual diseases and a few syndromes rather than seeking to improve health systems."[80] One of the individual diseases that the foundation has chosen to prioritize is poliomyelitis, better known as Polio. More than just an individual disease they prioritize, polio was viewed by foundation cochair Bill Gates as a worthwhile investment into "innovations [that] could be applied to help us tackle other diseases in the future" and a change for "more excitement and more saved lives."[81] This business-model approach on investment impact and the foundation's scope for financial influence are highly correlated with the impact that BMFG has had on polio globally.

Polio "is a disabling and life-threatening disease caused by the [spread of the] poliovirus ... [infecting] a person's spinal cord, causing paralysis."[82] While polio was first recorded as early as 1580 BCE in ancient Egyptian art, it took until 1789 for polio to be formally studied and until the around the 1950s for vaccines to be introduced globally.[83] In 1988, polio "was present in more than 125 countries and paralyzed about 1000 children per day". Today, polio is only found in two countries: Afghanistan and Pakistan, with polio incidence decreased by 99% overall.[84] While growing immunization rates and ongoing efforts have helped the prevalence of polio rapidly decline, it is important to note the following key issues.

Observations

- Polio outbreaks in Nigeria, the Central African Republic, Guinea, Kenya, Cote d'Ivoire, Tajikistan, Congo, and more have been reported as recently as 2010. While these outbreaks have been controlled the risk of spread is ongoing.[84]
- BMFG notes that "if [they] fail to completely eradicate polio, within a decade we could witness a resurgence of 200,000 new cases annually."[84]
- The WHO published the polio eradication strategy for 2022–26. This global strategy discusses progress made by and current actionable items for the GPEI (a public-private partnership organization with BMFG and WHO—highly funded by BMFG—as two of the total six partners).[83]

CASE STUDY *(cont'd)*

Interventions

The BMGF is largely a financial supporter of polio efforts. One French political scientist claims that when individuals such as Bill Gates and Melinda French Gates have found a common cause "which they wish to promote in the world, they look for mutual assistance, and as soon as they have found one another, they combine."[85] BMGF found this "mutual assistance" through a variety of partnerships. The Center for Strategic and International Studies reports that BMFG drastically increased its support toward polio eradication efforts in 2007 with larger donations and "innovative financing mechanisms" to Rotary International, WHO, UNICEF, and the Islamic Development Bank,[81] and with GPEI when it became an official partner over 10 years ago.[86] In this partnership, BMGF contributes "technical and financial resources to accelerate [GPEI's] targeted vaccination campaigns, community mobilization, and routine immunizations". Moreover, at the 2022 World Health Summit, BMGF announced they will commit US$1.2 billion to "support efforts to end all forms of polio globally" and assist with the goals of the polio eradication strategy for 2022–26.[84] With the foundation dedicated to the eradication of polio, BMGF is earmarking its donations to go to specific partnering organizations and to only be used in specific ways.

With the WHO being a key stakeholder in polio eradication, it is imperative to note that the earmarking of funds is not limited to BMGF's partnership with GPEI but also extends to larger institutions such as the United Nations (UN). This makes BMGF's financial investment not only influential but also political, as "private funding runs the risk of turning UN agencies, funds, and programs into contractors for bilateral or public-private projects, eroding the multilateral character of the system and undermining democratic global governance."[87] The 990 forms for BMGF specifically outline the purpose for money being given to each organization, and each of these listed purposes aligns directly with BMGF's personal initiatives (such as that to eradicate polio). Given that vaccination uptake is critical for polio eradication, one of the most notable examples of the foundation's influence was in 2010. BMGF noted that vaccinations would be the most important issue in the upcoming decade and pledged US$10 billion specifically toward vaccination-based initiatives. Shortly after this moment, the WHO, World Bank, and several other smaller global health entities established "the decade of vaccinations".[88] Ultimately, BMGF's intervention on polio's eradication has been largely financial but the foundation's financial impact is influencing agendas across global health to align with their own.

Lessons learned

What worked?

Despite uncertainties behind the legitimacy of BMFG's influence, the significant reduction in polio cases has been a clear success. In fact, since GPEI was launched in 1988, "immunization efforts have reached nearly 3 billion children" and "the incidence of polio has decreased by 99%."[84] In 1988, 125 countries reported 350,000 cases of wild polio annually. In 2019, that number was reduced to 143 cases in only 2 countries (Afghanistan and Pakistan), with only 1 of 3 wild polio

CASE STUDY *(cont'd)*

strains left to be eradicated.[89] In order to see this success, strategies and areas of focus that were effective include, but are not limited to, vaccination campaigns with proper training for data collection and analysis, coordinated immunization systems as a foundation for all immunizations, strong surveillance and monitoring systems to effectively determine where to target polio campaigns, partnerships for accessible product development, containment policies to maintain reduced polio cases, and transition resource planning for sustainability among local communities.[84] It is important to note that BMFG listed their areas of focus across all parts of the holistic polio eradication program (from data to resource development to final transitioning). As a result, BMFG has maintained its value beyond just being a strategic partner or financial source but also as a crucial stakeholder for all stages of polio eradication efforts.

What didn't work?

While current efforts have been successful in reducing polio cases, it is imperative to note the foundational flaws in the approach. These foundational problems can be summarized through three categories: BMFG's philanthrocapitalism and venture-philanthropy-based approach to public health; BMFG's narrowed focus on innovation; and BMFG's difficulties adapting to its partnering organizations.

I. BMFG's philanthrocapitalism and venture philanthropy-based approach:

While a businesslike model of philanthropy is common, BMFG's extreme approach to this model has not been beneficial. Philanthrocapitalism and Venture Philanthropy are terms used to describe the approach to charitable giving from a corporate or for-profit perspective. "Philanthrocapitalist organizations like BMFG... are enthusiasts for data, measurement, and audit" and determine where to invest according to their profit or return rate.[90] Moreover, "BMFG's venture-philanthropy approach... is emblematic of the business models that now penetrate the global public health field... [and has] blurred boundaries between public and private spheres, representing a grave threat to democratic global health governance and scientific independence."[88]

II. BMFG's narrowed focus on innovation: BMFG's investment strategy has largely been to follow innovation. As mentioned earlier in the case of polio, Bill Gates narrowed his decision to invest in polio eradication programs because there was scope for applying similar innovative interventions to other diseases. However, Gates himself has acknowledged the shortcomings of this approach, indicating he "was pretty naïve about how long [the] process would take". Moreover, investing in innovation is time-consuming, and it is difficult to know which interventions would go on over time to save lives. This drive for innovative approaches has blinded BMFG to basic systemic problems that also need solutions. For example, in India, "the sheer number of [polio] doses that had to be distributed, twice a year, literally left no space in refrigerators for other vaccines against diseases such as measles." Moreover, polio funds being channeled into countries "led to local brain drains

CASE STUDY (cont'd)

into eradication and away from local and locally funded health priorities."[91] Ultimately, BMFG's narrowed focus on innovation inadvertently took away their ability to promote more time-conscious initiatives, did not support systems-based public health problems, and forced local health into following BMFG's agenda.

III.

BMFG's difficulties with adapting to its partnering organizations: BMFG relies on partnerships to effectively implement its programs. In the case of polio eradication programs, GPEI was one of their main partners among others. For polio eradication and other programs, the WHO has been a key partner. However, structural integration with the WHO has been a challenge as "many large donors have given priority to infectious diseases, to the extent that the WHO's polio program accounted for more than 20% of the program budget in 2016–17."[92] This is a significant problem as WHO's director-general, Dr. Tedros Adhanom Ghebreyesus, argued in favor of a budget that allows for much greater flexibility because "no organization can succeed when its budget and priorities are not aligned."[93] Outside of the WHO, success programs require "collaboration among many stakeholders across multiple socio-ecological levels ..., and governance and cooperation". Challenges such as politics, demand and supply, societal issues, poverty, mistrust, inequities, ongoing conflict, and more impact each of these stakeholders in a different way. In fact, many of these challenges are barriers in the goal to eradicate polio in Afghanistan and Pakistan.[89] While the most recent 2019–23 GPEI polio endgame strategy Plan now addresses potential strategies to address these different obstacles, the initiative to understand partnering organizations and contextual hardships should have been taken much sooner.

Recommendations

Overall, given the context of the polio eradication program, it is imperative that philanthropic organizations consider the following recommendations:

- Further diversifying the board of trustees to improve program accountability.

In 2022, BMFG made an unprecedented announcement to expand their trustees to include nonGates or Buffet family individuals. While this is a step in the right direction to increase accountability within the organization's decision-making, it is imperative to diversify the board beyond affluent, highly connected individuals so that programs are accountable not just to BMFG trustees, but to the individuals they are intended to benefit.

- Reprioritizing systems-based public health problems

BMFG's narrowed focus on innovation has prevented them from prioritizing sustainable solutions and short-term impact with a more systems-based agenda. By prioritizing healthcare infrastructure, resource capacity, local sustainability of programs, and surveillance initiatives, the foundation can diversify its investments instead of putting all its funds toward innovative and risky projects.

Vision for the future

In the future, we envision a world in which public health agencies are empowered with the resources to pursue GHS capacity building that is sustainable and adopts a long-term view, and where philanthropic agencies serve as partners, not agenda setters in meeting the public health needs of the world. Some national governments have developed detailed financial proposals to support pandemic preparedness strengthening, including the quality financing approach for R&D at the regional, national, and international levels. Countries have placed a domestic financing plan and resources to consolidate pandemic preparedness measures into their federal budgets.[43] For example, Liberia and Zambia have built NPHIs with a legal mandate to strengthen public health surveillance (PHS), infection prevention and control, laboratory capacity, public health capacity, the initial response to outbreaks, and the monitoring and evaluation of diseases with epidemic potential.[94,95]

Incentivizing countries to prioritize allocating funds to preparedness is a substantial way to convince the government to strengthen their public health systems.[96] IMF and the World Bank have regularly conducted country PHS processes to assess and identify economic and financial risks. These measures have successfully facilitated and increased the flow of incentives and other development assistance to better pandemic preparedness in regional, national, and international communities (Fig. 23.1).[39] It can be challenging to persuade stakeholders to prioritize a massive amount of their spending on preparedness and prevention measures because it is hard to claim credit for such an investment. However, previous efforts from such organizations have shown a commitment and initiative to move forward.

Sustainable financing and external partner commitment on mobilizing adequate resources to finance preparedness and increasing existing collective and bilateral commitments are essential concepts that have not received much coverage. Major international financing institutions (e.g., the World Bank) have not appropriately managed infectious disease risk assessments and responses under their legal framework.[39] There have been significant inclines in WHO Health Emergency Program (WHEP) budgets and expenditures in the last 5 years (about 75% are intended for the regional and country-level).[39] But the budgets and spending are not fully distributed at the beginning of 2 years, and actual expenditures at the ends of those periods used to be less than the approved budget.[39] Global collaborative supervision is needed to watch the implementation of WHEP and national health spending.

A significant reduction in the threat of pandemics can happen if countries have a big enough reason to invest in and strengthen their national preparedness systems.[96] Vietnam has shown the result of its investment by successfully implementing a comprehensive analysis of funding sources, agents, and users at the national and provincial levels. Vietnam integrated national health accounts with a systematic framework for mapping expenditures in health and the public expenditure tracking system, triangulating budget and financial records from various sources.[98] The government needs to be assured that it has the resources, capacity, and costs required to strengthen its national public health preparedness by adequate communication and financial support.

Preparedness helps manage the crucial phases of pandemics so that health systems and communities have a greater chance to recover and manage the impact of health crises. Investing in preparedness capabilities allows for lead time reductions of up to 67% (18 days)

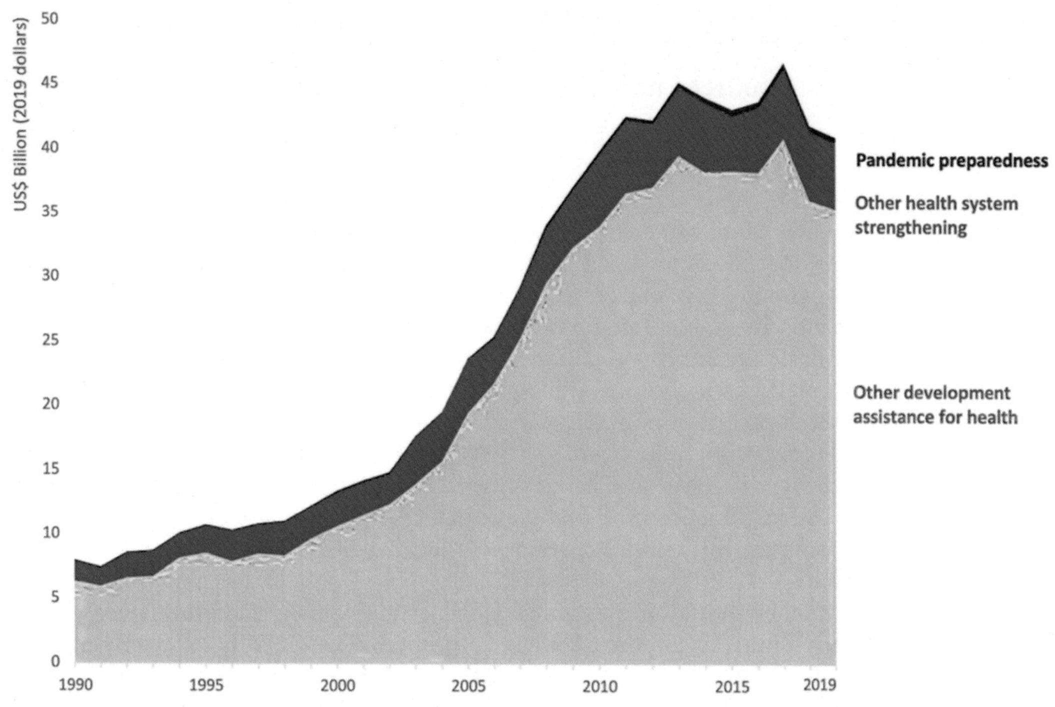

FIGURE 23.1 Pandemic preparedness fund of global health aid following some major outbreaks. *Council on Foreign Relations Bollyky TJ, Patrick SM. Improving Pandemic Preparedness: Lessons from COVID-19. 2020*

compared to a scenario without preparedness at significantly lower costs.[99] The global community is suggested to incorporate new scientific evidence, evolving epidemiology, and risk assessment into public health recommendations[100] to better mitigate the pandemic's impact and ensure humanitarian responses can address public health needs during a health crisis.[101]

Developing standardized monitoring and evaluation approaches for preparedness is less recommended during a global health crisis. Public health response should be focused on three priority areas: (i) Coordinated and consistent prevention orders across multiple jurisdictions; (ii) accelerated scale-up of testing; and (iii) improved healthcare capacity to respond.[102] In other words, when the world is experiencing strain on health care, economic, and social systems during a pandemic, priority assistance must be provided within the first hours following the disaster to increase the survival rate of the affected populations.

Despite improvements, significant gaps and difficulties exist in global pandemic preparedness measures. Some studies have assessed the systemic failures in the pandemic response, showing that only about one-fourth of countries in the world could sufficiently prevent, detect, and respond to public health emergencies.[103] Inadequate financing for pandemic preparedness, inflexible instruments for health crisis response, and slow but costly delivery of aid become other obstacles in a global pandemic situation. To support countries in better strengthening their capacity in pandemic preparedness and response, liable, transparent, sustainable, and honest cross-sectoral partnerships must be advanced.

Actions

Mobilizing domestic resources to build IHR core capacities

The government of New Zealand (NZ) is committed to enhancing resilience to its health crisis. Nine councils and organizations from Northland to Southland have been granted the 2017 Resilience Fund, an annual fund of $889,000 administered by the Ministry of Civil Defense and Emergency Management.[104] The grant is intended for national training programs, preparedness for schools and early childhood centers, improved waste management in emergencies, and planning for specific hazards.

The NZ governmentis also developing a national public alerting system, with criteria that placed a strong emphasis on improved collaboration and promoted consistent approaches. US$230,000 was allocated explicitly for the Integrated Training Framework, which provides a platform for the development of national standard training to enhance the competency of staff working in Emergency Operation Centers[104] These measures improve their understanding of hazards, boost preparedness, and build on world-leading skills and expertise. During the COVID-19 crisis, NZ experienced the first wave of COVID-19. However, after a period of elimination, NZ has subsequently only experienced a localized outbreak in the Auckland region in August 2020.[105]

A country's government should also establish a domestic financing plan that incorporates preparedness measures into its national budget allocation. The COVID-19 pandemic has increased the momentum in many countries to prepare and integrate preparedness programs. Those plans need to identify many competing interests and the priority investment areas for the whole public health systems (e.g., surveillance and economic capacity, well-equipped laboratories, and quality infection prevention strategies) to ensure meaningful progress and accountability for the public and private sectors as partners to help strengthen health security.

Mobilizing development assistance for preparedness

The spread of COVID-19 has proven that an infectious disease warning anywhere can be a threat everywhere. By investing in global health, the international community is doing more than combating COVID-19; they are fighting to ensure a future free of destructive pandemics. The United States, through USAID, has committed $2 billion to Gavi, the Vaccine Alliance, to support the procurement and delivery of COVID-19 vaccines to 92 LMICs and planned to disburse another $2 billion by 2022.[106] This grant is supporting fair access to COVID-19 vaccines for the world's most vulnerable and at-risk populations, including frontline health workers. A mutually beneficial approach involving governments and partners is demanded to allocate adequate resources to finance preparedness and have partners committed to existing joint and multilateral commitments. Collaborative work with the international community, including the WHO, governments, NGOs, and the private sector, to fight pandemics on the ground is essential to ensure countries have robust health systems and supplies and unify global response to the pandemic.

The international community needs to recognize preparedness capability as an indispensable part of the broader public health system strengthening and the UHC agenda. External funding and technical support must be leveraged for in-country capital investments for preparedness to minimize failure. The country's government should also strengthen its data, knowledge, and program accountability to increase the confidence of the international world to collaborate.

Incentivizing countries to prioritize fund allocations for preparedness

Investment in pandemic preparedness would no longer be solely the concern of the Health Minister from a macroeconomic criteria and analysis perspective.[96] A meaningful reduction in the threat of pandemics happens if countries choose to invest in and strengthen their integrated national preparedness systems.[96] However, to make such investments, governments need to be assured that the costs associated with that kind of investment are worth it and valuable for their future governance interests. There is a need to provide massive incentives for well-prepared and continuing pandemic preparedness investment and measures implemented by countries globally—not just by sending professionals for consultations and supervisions. This credit would increase the government's willingness to gain immense public attention, allocate more money for such a program, and identify potential weaknesses in underlying infrastructure and institutions that would benefit their future preparedness investment.

Mobilizing funding for research and development

The R&D around human resources must be a crucial part of heads of state and ministers of finance's decisions to increase domestic resource availability and facilitate future multisectoral participation. R&D evolution has been particularly evident during the COVID-19 pandemic, causing the accelerated development of vaccines and novel clinical trial designs.[107] National and international governments, donors, and funding agencies need to disburse more money for R&D in new product development and clinical research capacity as solutions for the emerging and reemerging global infectious disease outbreaks.

Health science today is also shifting rapidly and becoming more complicated, so no single researcher or single site can bring all the expertise to produce and validate health innovations or guarantee their safety.[108] Thus, efficient sharing of information between researchers, institutions, and countries has become even more critical than in previous eras. The development of international codes, federal legislation, and regulation of collaborative health R&D also needs to be strengthened in response to the potential health crisis in the future.

Increasing WHO assessed contributions

It is critical that we unshackle the WHO and allow it greater flexibility in budget allocations. The WHO is like a fire station that is being built during the fire, whose existing fire trucks are built for only one type of fire and not others. We've discussed a number of existing mechanisms to fund public goods. The exact approach will vary on national sentiment, but ultimately the funding of the WHO is an investment in a healthier, pandemic-free world. This can be accomplished in the short term by increasing "assessed contributions," ensuring that countries are providing the resources that the WHO needs to invest in GHS for the benefit of all.[109] The return on investment calculation demonstrates that both the public and private sectors stand to benefit in better prevention, control, and response capabilities. As such a pandemic preparedness tax, levied against corporations that benefit from the free flow of goods and services, can help provide a robust and consistent cash flow to the WHO.

References

1. *International Regulations (IHR)*; 2019. https://www.cdc.gov/globalhealth/healthprotection/ghs/ihr/index.html. Accessed March 12, 2022.
2. Cheney C. "Big concerns" over Gates foundation's potential to become largest WHO donor. *Devex*; 2020. https://www.devex.com/news/big-concerns-over-gates-foundation-s-potential-to-become-largest-who-donor-97377.
3. Carbonaro G. How is the World Health Organization Funded, and Why Does it Rely So Much on Bill Gates? 2023.
4. Hospitals with the Highest it Operating Budgets: Definitive Healthcare, 2022.
5. Programme Budget 2022–2023. World Health Organization.
6. Sekar K. *Centers for Disease Control and Prevention (CDC) Funding Overview*. Congressional Research Service; 2023.
7. *Constitution of the World Health Organization*. World Health Organization; 1946.
8. Glassman A, Smitham E. *Financing for Global Health Security and Pandemic Preparedness: Taking Stock and What's Next*. 2021.
9. *Global Spending on Health: Rising to the Pandemic's Challenges*. Geneva: World Health Organization; 2022.
10. International VH. Health regulations in the occupied Palestinian territory. *Lancet*. 2012;380(9851):1385–1386.
11. Eaneff S, Boyce MR, Graeden E, Lowrance D, Mackenzie M, Katz R. Financing global health security: estimating the costs of pandemic preparedness in Global Fund eligible countries. *BMJ Glob Health*. 2023;8.
12. National Academies of Sciences E, and Medicine. Global Health and the Future Role of the United States. The National Academies Press 2017.
13. Micah AE, Su Y, Backmeier S, et al. Health sector spending and spending on HIV/AIDS, tuberculosis, and malaria, and development assistance for health: progress towards Sustainable Development Goal 3. *Lancet*. 2020;396(10252):693–724.
14. *Delivering Global Health Security through Sustainable Financing*. World Health Organization; 2018.
15. Michaud J, Moss K, Kates J. *The U.S. Government and Global Health Security*. KFF; 2021.
16. WHO. Revision of the International Health Regulations, WHA58.3. 2005.
17. International Health Regulations. https://www.who.int/health-topics/international-health-regulations#tab=tab_1.
18. *Implementation of the International Health Regulations (2005): Report of the Review Committee on the Role of the International Health Regulations (2005) in the Ebola Outbreak and Response*. World Health Organization; 2016.
19. *Public Health Preparedness and Response: Implementation of the International Health Regulations (2005)*. World Health Organization; 2017.
20. *Pandemic Influenza Preparedness Framework for the Sharing of Influenza Viruses and Access to Vaccines and Other Benefits*; 2011. http://www.who.int/influenza/pip/en/.
21. Gostin LO, Katz R. The international health regulations: the governing framework for global health security. *Milbank Q*. 2016;94(2):264–313.
22. Broberg M. A critical appraisal of the world health organization's international health regulations (2005) in times of pandemic: it is time for revision. *Eur J Risk Regul*. 2020;0(0):1–8.
23. Suthar AB, Allen LG, Cifuentes S, Dye C, Nagata JM. Lessons learnt from implementation of the International Health Regulations: a systematic review. *Bull World Health Organ*. 2018;96(2):110.

24. Katz RDS. Revising the international health regulations: call for a 2017 review conference. *Lancet Global Health.* 2015;3:e352–e353.
25. Ebola Outbreak 2014-2016 - West Africa. World Health Organization.
26. *Statement on the 1st Meeting of the IHR Emergency Committee on the 2014 Ebola Outbreak in West Africa.* World Health Organization; 2014.
27. *FACT SHEET: Global Health Security Agenda: Getting Ahead of the Curve on Epidemic Threats*; 2014. https://obamawhitehouse.archives.gov/the-press-office/2014/09/26/fact-sheet-global-health-security-agenda-getting-ahead-curve-epidemic-th.
28. What is the Global Health Security Agenda? Centers for Disease Control and Prevention.
29. McKinsey. *Crushing Coronavirus Uncertainty: The Big "unlock" for Our Economies.* 2020:2020.
30. Craven M, Sabow A, Van der Veken L, Wilson M. *Not the Last Pandemic: Investing Now to Reimagine Public-Health Systems.* 2021:2021.
31. Financing Global Health. *Flows of Global Health Financing*; 2021. https://vizhub.healthdata.org/fgh/.
32. *Global Health Security Index: Building Collective Action and Accountability.* 2019.
33. *Number of Novel Coronavirus COVID-19 Cumulative Confirmed, Recovered and Death Cases in Taiwan from January 22, 2020 to May 30, 2021.* Statista; 2021.
34. Tu C-C. *Lessons from Taiwan's Experience with COVID-19.* Atlantic Council; 2021.
35. Lee W-C. Taiwan's experience in pandemic control: drawing the right lessons from SARS outbreak. *J Chin Med Assoc.* 2020;83(7):622–623.
36. Moon S, Sridhar D, Pate M, Jha A, Clinton C, Delaunay S. Will Ebola change the game? Ten essential reforms before the next pandemic. The report of the harvard-LSHTM independent panel on the global response to Ebola. *Lancet.* 2015;386(10009):2204–2221.
37. *Lessons Learned in Financing Rapid Response to Recent Epidemics in West and Central Africa: A Qualitative Study.* The World Bank; 2015.
38. Radin E, Eleftheriades C. *Financing Pandemic Preparedness and Response.* 2021.
39. *Pandemic Preparedness Financing Status Update.* The World Bank; 2019.
40. *Indonesia - First Financial Sector Reform Development Policy Financing: COVID-19 Supplemental Financing.* The World Bank; 2020.
41. *COVID-19 Dashboard by the Center for Systems Science and Engineering (CSSE)*; 2021. https://www.arcgis.com/apps/opsdashboard/index.html#/bda7594740fd40299423467b48e9ecf6.
42. Ravelo J. *Where is the Money for COVID-19 Diagnostics?.* 2020.
43. Sebaran P. *COVID-19.* 2021.
44. So AD, Woo J. Reserving coronavirus disease 2019 vaccines for global access: cross sectional analysis. *BMJ.* 2020;371(m4750).
45. Rouw A, Wexler A, Kates J, Michaud J. *Global COVID-19 Vaccine Access: A Snapshot of Inequality.* 2021.
46. Micah A, Dieleman J, O'rourke KF, et al. *Financing Global Health 2019: Tracking Health Spending in a Time of Crisis.* 2020.
47. Dalglish SL. COVID-19 gives the lie to global health expertise. *Lancet.* 2020;395(10231):1189.
48. *Joint External Evaluation of IHR Core Capacities of the United States of America: Mission Report June 2016.* World Health Organization; 2017.
49. Tromberg BJ, Schwetz TA, Perez-Stable EJ, et al. Rapid scaling up of covid-19 diagnostic testing in the United States - the NIH RADx initiative. *N Engl J Med.* 2020;383(11):1071–1077.
50. Marwaha J, Halamka J, Brat G. *Lifesaving Ventilators Will Sit Unused without a National Data-Sharing Effort.* 2020.
51. West D, Villasensor J, Schneider J. *Private Sector Investment in Global Health R&D: Spending Levels, Barriers, and Opportunities.* Brookings Institute; 2017.
52. Minkel J. Dangling a carrot for vaccines. *Sci Am*; 2006. https://www.scientificamerican.com/article/dangling-a-carrot-for-vac/.
53. What is the Gavi COVAX AMC? https://www.gavi.org/gavi-covax-amc.
54. Kremer M, Levin J, Snyder C. *Designing Advance Market Commitments for New Vaccines.* National Bureau of Economic Research; 2020.
55. Pneumococcal AMC Outcomes and Impact Evaluation. https://www.gavi.org/our-impact/evaluation-studies/pneumococcal-amc-outcomes-and-impact-evaluation.

56. What is an Advance Market Commitment and How Could it Help Beat COVID-19? 2020. https://www.gavi.org/vaccineswork/what-advance-market-commitment-and-how-could-it-help-beat-covid-19.
57. Bank D. *Guaranteed Impact*. Stanford Social Innovation Review; 2016.
58. Strategic Investment Fund Portfolio. https://sif.gatesfoundation.org/portfolio/?fwp_investment-type=volume-guarantee.
59. MedAccess Agreements. https://medaccess.org/our-agreements/agreements/.
60. Özdemir V, Faris J, Srivastava S. Crowdfunding 2.0: the next-generation philanthropy: a new approach for philanthropists and citizens to co-fund disruptive innovation in global health. *EMBO Rep*. 2015;16(3):267–271.
61. Woodcraft CAT, Munir K. *Pandemics and Philanthropy': An Opportunity in Crisis*. 2020.
62. *Philanthropic Response to Coronavirus (COVID-19)*; 2021. https://candid.org/explore-issues/coronavirus.
63. *Serum Institute of India. Product*; 2021. https://www.seruminstitute.com/products.php.
64. *Gates Foundation Invests $150 Million in COVID-19 Vaccine Production*. Philanthropy News Digest; August 10, 2020.
65. Dentico N. *SDG 3 – Philanthrocapitalism in Global Health and Nutrition: Analysis and Implications*. 2019.
66. *How to Boost Inclusive Investment Decision-Making in Global Health*. World Economic Forum; 2022.
67. How WHO is funded. https://www.who.int/about/funding.
68. Ravelo J. *8 Takeaways from the 70th World Health Assembly*. 2017.
69. Mazzucato M, Kickbusch I. *The WHO's Penny-Wise and Health-Foolish Members*. Project Syndicate; 2021.
70. Ravelo J. *Björn Kümmel: No Consensus on WHO Assessed Contribution Increase*. 2022.
71. Ravelo J. Countries Agreed to Sustainable Financing for WHO. What's Next? 2022.
72. *WHO Foundation Established to Support Critical Global Health Needs*. World Health Organization; 2020.
73. *WHO Foundation Annual Activity Report 2021*. WHO Foundation; 2021.
74. Suzman M. *Does Our Foundation Have too Much Influence? Here's How I See it*. Bill & Melinda Gates Foundation; 2023.
75. Ollstein A. *Trump Halts Funding to World Health Organization*. Politico; 2020.
76. Carrington D, Mathiesen K. *Revealed: Gates Foundation's $1.4bn in Fossil Fuel Investments*. The Guardian; 2015.
77. Glaser A. *McDonald's French Fries, Carrots, Onions: All of the Foods that Come from Bill Gates Farmland*. NBC News; 2021.
78. *The Gates Foundation's Approach Has Both Advantages and Limits*. 2021.
79. Brest P. Investing for impact with program-related investments. *Stanford Soc Innovat Rev*. 2016;14(3):A19–A27.
80. Butler C. Philanthrocapitalism: promoting global health but failing planetary health. *Challenges*. 2019;10(1).
81. Bristol N. *The U.S. Role in Global Polio Eradication*. 2012.
82. *What is Polio*; 2022. https://www.cdc.gov/polio/what-is-polio/index.htm.
83. *Polio Eradication Strategy 2022–2026: Delivering on a Promise*. World Health Organization; 2021.
84. Polio. https://www.gatesfoundation.org/our-work/programs/global-development/polio.
85. Mansfield HC, Winthrop D. In: *Alexis de Tocqueville, Democracy in America*. Press CUoC; 2000.
86. *GPEI Information*; 2022. https://www.cdc.gov/polio/gpei/index.htm.
87. Seitz K, Martens J. Philanthrolateralism: private funding and corporate influence in the United Nations. *Global Policy*. 2017;8(S5):46–50.
88. Birn AE, Richter J. *U.S. Philanthrocapitalism and the Global Health Agenda: The Rockefeller and Gates Foundations, Past and Present*. Global Policy Forum; 2017.
89. Alonge O. What can over 30 years of efforts to eradicate polio teach us about global health? *BMC Publ Health*. 2020;20(1177).
90. Reubi D. Epidemiological accountability: philanthropists, global health and the audit of saving lives. *Econ Soc*. 2018;47(1):83–110.
91. Fortner R. Has the billion dollar crusade to eradicate polio come to an end? *BMJ*. 1818;2021:374.
92. Reddy S, Mazhar S, Lencucha R. The financial sustainability of the World Health Organization and the political economy of global health governance: a review of funding proposals. *Glob Health*. 2018;14(119).
93. *Dialogue with the Director-General*. World Health Organization; 2018.
94. Rosenfeld EL, Binder S, Brush A, et al. National public health institute legal framework: a tool to build public health capacity. *Health Security*. 2019;18(S1).
95. *Who are the ZNPHI*; 2021. https://znphi.co.zm.

96. *From Panic and Neglect to Investing in Health Security: Financing Pandemic Preparedness at a National Level.* Washington, D.C.: World Bank Group; 2017.
97. Bollyky TJ, Patrick SM. *Improving Pandemic Preparedness: Lessons from COVID-19.* 2020.
98. *Money & Microbes: Strengthening Clinical Research Capacity to Pre- Vent Epidemics.* Washington D.C.: The World Bank; 2018.
99. Kunz N, Reiner G, Gold S. Investing in disaster management capabilities versus pre-positioning inventory: a new approach to disaster preparedness. *Int J Prod Econ.* 2013;00(00):1.
100. *Public Health Guidance for Community-Related Exposure.* Centers for Disease Control and Prevention; 2021.
101. *UNICEF/WFP Return on Investment for Emergency Preparedness Study.* Munich/Brussels: BCG; 2015.
102. Guest JL, Del Rio C, Sanchez T. The three steps needed to end the COVID-19 pandemic: bold public health leadership, rapid innovations, and courageous political will. *JMIR Public Health Surveill.* 2020;6(2), e19043.
103. Oppenheim B, Gallivan M, Madhav NK, et al. Assessing global preparedness for the next pandemic: development and application of an Epidemic Preparedness Index. *BMJ Glob Health.* 2019;4(1), e001157.
104. National Emergency Management Agency. *Funding to Improve Natural Hazard Resilience;* 2021. https://www.civildefence.govt.nz/resources/news-and-events/news/funding-to-improve-natural-hazard-resilience/.
105. Summers J, Cheng H-Y, Lin H-H, et al. *Potential Lessons from the Taiwan and New Zealand Health Responses to the COVID-19 Pandemic.* Vol. 4. Elsevier; 2020.
106. *COVAX Vaccine Roll-Out;* 2021. https://www.gavi.org/covax-vaccine-roll-out.
107. Lurie N, Keusch GT, Dzau VJ. Urgent lessons from COVID 19: why the world needs a standing, coordinated system and sustainable financing for global research and development. *Lancet.* 2021;397(10280):1229–1236.
108. Nass SJ, Levit LA, Gostin LO. *Beyond the HIPAA Privacy Rule: Enhancing Privacy, Improving Health through Research. The Value, Importance, and Oversight of Health Research.* Washington DC: National Academies Press (US); 2009.
109. Ravelo J. *Q&A: What Sustainable WHO Financing Means for Global Health Security.* 2020.

CHAPTER

24

Enhancing trust and transparency for public health programs

James O. Ayodele[1], Marika L. Kromberg Underwood[2], Duaa Al Ammari[3], Kara Goldstone[4] and Emmanuel Agogo[5]

[1]WHO Office for the Eastern Mediterranean Region, Cairo, Egypt; [2]Centre for Sustainable Healthcare Education, University of Oslo, Oslo, Norway; [3]College of Public Health and Health Informatics, King Saud Bin Abdulaziz University for Health Sciences, King Abdullah International Medical Research Center, Riyadh, Saudi Arabia; [4]Rollins School of Public Health, Emory University, Atlanta, GA, United States; [5]Resolve to Save Lives, New York, NY, United States

Introduction

Trust is a major pillar and a key contributor to the success of public health programs.[1] It makes policies and guidelines accepted by beneficiaries and communities and may contribute to program effectiveness and efficiency. Trust means gaining the confidence of stakeholders in the ability of health systems to deliver on promises and according to expectations. It is what makes patients believe in the integrity of healthcare services, including laboratory results, clinical diagnoses, pharmaceutical prescriptions, and the expected treatment outcomes. Without trust, there is no public health.

Experience has shown that trust is an ideal situation often missing in public health, and sometimes not considered in planning. Vinck et al.[2] have associated low trust in health institutions with the decreased likelihood of adopting preventive behaviors, as already observed with the acceptance of Ebola vaccines and seeking healthcare in public institutions in several countries.

Mistrust can occur even among healthcare workers. For example, health workers often perceive private institutions as profit-making entities and government establishments as facilities that are ill-equipped to deal with serious health conditions. Sometimes physicians would prefer to rely on personal skills and syndromic diagnosis instead of using laboratory results because they think laboratory data is not reliable.

Modernizing Global Health Security to Prevent, Detect, and Respond
https://doi.org/10.1016/B978-0-323-90945-7.00008-7

Public health is everybody's responsibility; therefore, every stakeholder must be involved in building trust in health systems.

Literature review and gap analysis

Determinants of trust in public health

People's attitudes to science are a product of several complex, interrelated, individual, and societal factors. Research has shown that trust is determined by how individuals or groups perceive competence, objectivity, fairness, consistency, sincerity, and faith,[3] and these perceptions often depend on their religious, cultural, educational, and social backgrounds and their beliefs. This section examines some of the determinants of trust in public health.

Capacity of healthcare systems and healthcare providers

Trust in health system has been positively correlated with patient satisfaction and found to strengthen compliance.[4] Trust is stronger when the health system is well capacitated with appropriate equipment and skilled providers. A health system that is adequately capacitated to provide comprehensive care at health facilities, at home and in the community, and to respond to emergencies, will positively influence decision-making on health-related behavior and gain the trust of the community.[4–6] Where the reverse is the case, there will be uncertainty and trust will be weak.

Community perceptions

The level of trust in the health system depends in part on how the community perceives the effectiveness of services in meeting their needs.[7,8] Such perceptions are often influenced by the quality of services and perceived sincerity or intentions of providers and health authorities. Effectiveness of and benefits from health services, ability to save lives, personal or group risk perception, and experience with or reputation of healthcare providers and the organizations they represent all influence community perception. For example, when a community has been exposed to the effects of counterfeit or adulterated medical products, they are more likely to distrust any subsequent offerings relating to that aspect of healthcare.

The educational level of community members, how much information they have about services, and where they get this information will influence their perception of, and trust in, public health. The perceived reliability of available information is often a function of the individual or groups providing it. If the community does not trust those individuals or groups, they will most likely reject whatever recommendations they make.

Politicization of healthcare programs to the disadvantage of the community will reduce their trust in authorities to provide effective solutions to their healthcare problems. Politicians sometimes play the give-and-take game with healthcare, tying healthcare projects to elections and their political agenda; communities that vote for or support them receive priority services while those that oppose them are neglected. And sometimes certain healthcare issues receive attention only when they affect top government officials or privileged members of the community.

Religious and cultural beliefs

Every society has its own beliefs, and these beliefs shape perceptions. For example, in Africa, Asia, and Latin America, there is a belief that "Western medicine" is out to suppress traditional indigenous health solutions, which have been in existence before the introduction of organized science.[9] And many think that the West is interested in selling their products and are not ready to invest enough in traditional medical research. Practitioners of traditional medicine, therefore, see pharmaceutical companies as competitors that are out to eliminate them from the local markets. This has been an object of controversy and a strong reason for mistrust in these settings.

Many preconceptions about public health are based on the cultural, religious, educational, and other socioeconomic backgrounds of individuals. For example, some people do not yet trust telemedicine technology to provide accurate consultation, diagnosis, and treatment, therefore, they still prefer to meet their physician or specialist physically. However, the COVID-19 situation has forced people to use virtual consultations because of the fear of getting infected at the healthcare facility. This has contributed positively and increased trust in virtual health services.[10]

Religion is a strong determinant of trust in public health and there is a clear contrast between religion and science.[11] Science believes in facts and evidence whereas religion is based on faith and followership. Some people believe that religion has some elements of science and should therefore accommodate science, whereas others think faith should override scientific evidence. These beliefs have great implications for trust in public health solutions. An example is the current controversy around vaccines in general including the COVID-19 vaccines. Some people think they don't need the vaccine because of their religious beliefs, others think their faith confers on them immunity against any disease and, therefore, see any attempt to mandate them to take a vaccine as fraudulent and an infringement on their personal rights.

Generational variations and messaging

Different age groups receive information from different sources with different types of messages. These differences in exposure and experience affect trust among age groups. For example, older people in Africa and Asia tend to trust traditional medical solutions more than the younger generation. They believe that they have lived healthier lives with traditional medicines than with modern medicine and therefore trust traditional medical practice more than modern medicine. However, the younger generation seems to be more trusting in modern medicine, perhaps because of the difficulty in explaining the science behind traditional medical practice and the poor documentation associated with it.

Evidence abounds of the inconsistent, incorrect, and contradictory messaging by governments during public health emergencies,[12] which have led to communication failures and mistrust. The inability of health authorities to adequately challenge misinformation spreading over social media about the COVID-19 pandemic, for example, has been a reason for the mistrust we currently witness in public health. Many government officials prefer to stick to "official channels," which often favor press releases, media interviews, and press briefings. Government officials, particularly the older generation, are yet to recognize the importance of social media in communication and for increasing trust in health programs.[13–16] They fail to realize that most young people get their information through social media and need to be targeted using the same media they seem to trust.

Gender roles and stereotypes

Gender stereotypes can affect trust in public health systems.[17] In societies where the patriarchal system dominates and women are not allowed to make decisions concerning their healthcare, women often have less trust in the healthcare system and may have very poor health-seeking practices. People who experience violations or discrimination while seeking healthcare, such as the transgender and gay communities, will be expected to trust less in the system.

Stereotypes also exist among health workers based on their training, the institutions they work for, or even their race or color. There is sometimes mistrust between service providers working for public and private institutions and between those working in rural and urban areas. Such stereotypes fuel mistrust in public health.

Reasons for mistrust

Failure of governments and health authorities

The government is expected to be the main custodian of healthcare provision, but this is not the case in many societies, especially in developing countries, hence there seems to be more trust in private than public healthcare service providers. For example, the Universal Health Coverage[18] was initiated to help advance access to quality healthcare for all, but service provision by government continues to deteriorate in many societies, making people to become personally responsible for their healthcare needs. This continues to deepen the lack of confidence and trust in government programs across the globe.[2]

In some societies, government investment in healthcare is perceived to be inadequate; medicines and other commodities are often inadequate and sometimes unavailable. Some communities witness the government destroying expired medicines, which are not made available to clients before their expiry dates, and this raises questions of why large quantities of medicines and consumables are allowed to expire while people need them.

Globally, the acute shortage of providers cripples healthcare services. In some communities, including some European countries, clients must book appointments many months ahead before they can access consultation by specialists in goverment facilities, sometimes for conditions that require urgent attention. Experience with government institutions is negative in some communities, making people perceive hospitals as places to go and die. Sometimes admission into the intensive care unit is seen as a death sentence, making some people prefer to die at a traditional care home than go to a government hospital. These inadequacies increase casualties and reduce trust in health systems.

Often there are accusations of misappropriation of funds that are allocated for healthcare provision by corrupt government officials, which raises a red flag on accountability by health authorities. When community members cannot explain the discrepancies between the living standards of senior government health officials and their official income, they question their wealth, believing that they are using their privileged positions to amass wealth at the expense of the community.

Distrust in governments and health authorities has been a major driver for rumors and misinformation about public health during the COVID-19 pandemic and for noncompliance to the prevention measures globally. Distrust has been reported to cause higher COVID-19 infection rates around the world.[19] Some individuals and groups think that COVID-19, the public health measures, and COVID-19 vaccines are all part of governments' deception. This has led to the creation of hashtags and slogans such as #COVID19Millionaires, #Arrest-COVID19Thieves, #SayNoToWitsVaccine, #KerajaanGagal (#FailedGovernment), #Kerajaan-Zalim (#CruelGovernment) and #KerajaanPembunuh (#MurdererGovernment), "political scam," "COVID-19 is a #Scamdemic," #Agenda21 (referring to conspiracy theories and totalitarian governments), "coronavirus is a lie," "conduit of global organized criminals who profit from the bloody pharma-trade," "we are not guinea pigs," "the infertility vaccine is masquerading as a cure for COVID-19," and "we do not trust anything with letter "C."[1,20,21] Although many of these hashtags and slogans have been deleted, they gained popularity on social media and acted as symbols for protests against the failure or inability of governments and health authorities to adequately address the health needs of their citizens.

Lack of transparency

Lack of transparency breaks trust in public health. For example, lack of transparency in budget spending by authorities or lack of transparency by pharmaceutical companies in the development of medicines, vaccines and supplies is causing distrust in health systems and institutions. Reports of public health scandals such as medical errors that are covered up, hidden agendas of pharmaceutical companies and industry partners, and corruption among public health managers and practitioners are very important reasons for distrust in public health.[3] Pharmaceutical and medical equipment companies are mostly privately owned and profit-driven. They have often been accused of not adhering to production and supply chain standards, supplying low-quality medicines and consumables, selling medicines to the public even when clinical trials show negative results, or repackaging and selling expired products to avoid losses.[22] These are very important reasons for mistrust.

Forgery and adulteration of pharmaceutical and medical products

The public health sector has been plagued with reports of forged, substandard, unregulated, and counterfeit products in recent years.[23] Sometimes regulatory certificates are forged to bypass government control or to avoid payment of taxes and duties. This has caused severe injuries, deaths, and other public health problems on many occasions, thus reducing community trust in the integrity of pharmaceutical and medical equipment manufacturers.

The disconnect between health authorities, health workers, and the community

Trust is often stronger at the microlevel, between clients and providers, and between healthcare facilities and the community where the facilities are located. However, as you advance to the top of the ladder, trust diminishes. This is because people tend to trust whatever is available and they can see or what originates from within their communities. Disconnection between policy and end-users sometimes results in the provision of low-quality or non-relevant services that do not meet the needs and expectations of the community.

In a survey of 961 adults aged 18 years and above in Beni and Butembo in North Kivu Province of the Democratic Republic of Congo, Vinck et al.[2] reported that communities trusted local authorities more than the provincial and national governments because they delivered services directly to community members. They trusted healthcare providers more frequently than the health authorities, highlighting the importance of a strong connection between health authorities, healthcare providers, and the communities to build sustainable trust.

When healthcare policies are developed in urban centers without the involvement of rural facilities there will be a disconnect, resulting in a lack of trust. Funding healthcare facilities in urban centers while neglecting rural centers will result in a shift in health-seeking behavior because of the lack of assurance of getting quality services at rural facilities.

Disconnection between health authorities and the community often gives room for rumors and misinformation, which provide a breeding ground for misunderstanding and mistrust. This has been observed with the Ebola outbreaks in Africa and the global COVID-19 pandemic.[24,25] Communication channels of governments and health authorities were not open enough for highly demanded interactive communication with the public.[2]

Unexplained public health occurrences or situations

There are many unknowns and many unanswered questions in public health, and these unknowns have been used as basis for the spread of misinformation, leading to mistrust in the public health system. Occurrences that are ignored or not adequately explained are a big risk to trust. For example, some medicines are labeled, "Not for sale in the United States or Canada" but are sold in Africa and Asia. And people ask why deaths due to HIV/AIDS are rare in the Western world. The perception is that the best quality medications are reserved for America and Europe while the lower quality ones are shipped to developing countries, sometimes as aid.

There have been questions of why adverse reactions to medicines are not being adequately addressed, for example, vaccine-derived polio deaths and failed measles vaccines in children; and why precautions and drug side effects are not being discussed openly.[26] There are unanswered questions about the duration of immunity that certain vaccines, including the COVID-19 vaccines, can provide. Many perceive these inadequacies as a lack of transparency on the part of health authorities and pharmaceutical companies.

At the beginning of the COVID-19 pandemic, there was no clear explanation about the disease and there were conflicting information and recommendations from different sources. Many recommendations about preventive measures were based on personal opinions, creating confusion and mistrust of health authorities and causing the pandemic to spread more widely.[27]

CASE STUDY

Responding to the 10th Ebola virus disease outbreak in DRC

Background

The 10th Ebola virus disease (EVD) outbreak was declared on August 1, 2018, in North Kivu Province of the Democratic Republic of Congo. In this region that had experienced over 25 years of war, crisis, and unrest, poverty was real and evident, healthcare services were very poor, or nonexistent in some communities, facilities were poorly resourced and some facilities had been looted or destroyed. Infrastructure had been badly destroyed with limited or no electricity supply for community members to connect to mass and electronic communication channels for essential information.

Observations

The sudden arrival of EVD rapid responders triggered suspicion among community members. Their mission was misconstrued and sparked mistrust in the communities. Community members felt they had been neglected for many years to suffer deprivation, destruction, and death despite the presence in the country of international organizations, including the United Nations. They felt that the government and international partners that could not prevent the killings were not capable of saving them from such a deadly disease. They, therefore, thought the responders were agents of government and international business owners who had been sent to exploit their mineral and material resources.

Key issues

- Community members did not believe there was any such thing as Ebola. They doubted the genuineness of the food distribution, treatment, and other services being provided as part of the response package. Many alleged that the food being distributed had been poisoned and that the blood transfusion, medicines, and vaccines offered at treatment centers were meant to infect, kill or make them infertile.
- Local politicians thought the whole idea of Ebola was a government tactic to delay elections. They, therefore, used the myths and misinformation around Ebola to create fear in community members, making them believe that only the politicians could protect them.
- Community members doubted the purpose of the medical equipment used for screening and treatment. For example, some of them thought the handheld thermal scanners had been programmed to alter their memory and thinking patterns so they could vote for certain politicians.
- These conspiracies and misperceptions led to resistance and affected contact tracing, monitoring, and treatment. Access was limited in some communities that had a heavy presence of soldiers and fighters. These in turn affected the effectiveness of response and prolonged the outbreak.

Interventions

- To gain trust, rapid responders embarked on aggressive engagement

CASE STUDY (cont'd)

with community members, using infographics, leaflets, short videos, animations, and direct messaging to provide information about the outbreak.

- The response team adopted different approaches, including one-to-one, house-to-house, and group meetings as the main channels of communication.
- To enhance the effectiveness of engagement, the response team hired and trained community members to serve as community relays who went into the communities to facilitate dialogue with community members.
- Consultations and advocacy dialogues were held with local politicians, religious leaders, influencers, and community leaders, some of whom became part of the task force of partners that provided training and information on Ebola and other public health issues. The association of media practitioners in the DRC was represented on the task force.
- A group of influencers, community chiefs, security personnel, healthcare workers, and other relevant stakeholders were brought in to help manage incidents at the community level and set up social listening mechanisms, including the use of the UNICEF U-report mobile app, which allowed them to interact with and get feedback from the community.
- The response team established partnerships and dialogue opportunities with local private business owners. They held education sessions and open-air outreaches in schools, hospitals, places of worship, markets, and other public places.
- Security personnel were deployed as part of the response team to facilitate engagement with community leaders and warring parties and secure access to communities that were inaccessible due to the presence of armed personnel.
- The team used personal testimonials of Ebola patients who had been treated and recovered to convince community members that they could recover if diagnosed and treated early.

Lessons learned

What worked?

- The use of one-to-one, house-to-house, and group meetings helped provide vital information to community members in the absence of electric power and limited access to mass electronic communication.
- The involvement of journalists in the task force helped equip journalists with the right information and enabled media alignment in messaging.
- Because of their involvement, the journalists facilitated daily talk shows on local radio stations, which provided opportunities for community members to ask questions about the disease and the outbreak.
- Setting up a group of influencers, community chiefs, security personnel, healthcare workers, and other relevant stakeholders that helped manage incidents at the community level and set up a social listening mechanism helped boost community engagement, and helped identify and counter misinformation, myths, and rumors.
- The use of evidence and testimonies from individuals who were treated and recovered, and protection for those who

CASE STUDY (cont'd)

were vaccinated helped in convincing community members that the response was genuine.

What did not work?

- In a conflict zone, the security of personnel depends largely on the good-will of the warring factions, which is not always guaranteed, community workers were attacked on some occasions, confirming the need for caution when working in such environments.
- Lack of trust in health authorities, the government, and the international development community made access difficult at the initial stage despite the genuineness of response efforts.

Recommendations

- In communities experiencing conflict, negotiating with local groups can be very helpful in gaining trust and enhancing access. Such negotiations should, however, be led by trained security personnel.
- Gain the trust of the community by identifying and engaging the real leaders to build strategic, sustainable partnerships before the commencement of any intervention. Once a dependable partnership is established, planning, and implementing with the community, rather than for the community, is very important.
- Build strong partnerships with the media so they can provide crucial support for the response. Training and educating them about public health issues will empower them with the knowledge and understanding they need to inform the public accurately and effectively.
- Demonstrate positive results through personal testimonials by beneficiaries to improve trust in public health response.
- Providing adequate, accurate, and timely information about an outbreak or a new disease is essential in securing community acceptance and support for any intervention.
- Involving local experts or community members who understand the local language and culture is essential in building trust when responding to a public health emergency.
- Continuous and substantial investment in quality and reliable healthcare services that meet the needs of community members can boost trust in public health because past positive experiences can motivate trust and acceptance of new interventions.

Vision for the future

It is possible to gain the trust of community members if careful attention and steps are taken to engage and involve them in designing interventions and making decisions on issues that affect their health. Openness, transparency, consistency, and clarity are crucial ingredients that foster trust in building a long-lasting, sustainable relationship with constituencies and stakeholders. Mistrust prevails when the concerns of community members are ignored or treated as secondary issues, or when there is a lack of clarity and openness. If the action items outlined in the following section are addressed, we will have a future of healthy people and communities who trust public health officials and programs. People and communities will be empowered to seek public health guidance, use their voices to engage in open dialogue, and take control of their health.

Actions: Trust-building in public health

Trust is an essential part of public health, and all parties must be involved in trust-building for the effectiveness of health programs. Actions should be consistent with words. This section examines the key ingredients for trust-building.

Consistent delivery of quality health services

Quality health service is a very strong booster of trust.[28] Clients will rate future expectations by the experience of the past. If services provided in the past fell short of their expectations, they will doubt the ability of service providers to meet their future needs. Health authorities must prove their credibility and competence by establishing healthcare systems that have the capacity to provide quality services. This is not an easy task, but it is essential to gain the trust of the public.[29]

To build trust in public health, communities need assurance that medicines, laboratory test results, diagnoses, and general patient care consistently meet quality standards and that services are inclusive and not discriminatory. This requires that authorities pay attention to strengthening regulatory and accreditation processes as well as to enforcing regulations to safeguard the integrity of health systems.[30]

Open conversations about public health issues

Trust-building begins with open conversations about health issues. Discussing public health issues openly may not appear profitable initially but it has potential for long-term benefits. Open conversations can bridge the gap between scientists, researchers, and the public and promote trust. It will provide opportunities to explain key terms and concepts to the public and to receive the feedback needed to improve programs. Every opinion should be considered seriously.

Open and transparent conversation has been applauded as one of the key strategies that helped the South Korean Government to have better control of the COVID-19 pandemic, as opposed to the situation with the Middle East Respiratory Syndrome (MERS).[22] During the MERS outbreak, the central government had initially refused to disclose necessary information, such as hospitals where patients were admitted, to the public thinking that it would create fear, panic, confusion, and a sense of insecurity. This nondisclosure and secrecy by the central government caused public outcry and tensions with local governments, which wanted to disclose essential information about the outbreak. For example, in opposition to the decision of the central government, the mayor of Seoul Metropolitan City held an emergency briefing to provide information to the public about the disease, the infection paths and hospitals where people were being treated.[22] This caused much friction between the central and local governments. However, the open conversation position of the metropolitan government was later accepted by the central government, and it helped individuals identify and assess their personal risks and the possibility of exposure to infection.

The South Korean Government[22] seemed to have applied lessons from the 2015 outbreak in managing COVID-19. The government provided essential information, and up-to-date statistics about the pandemic, including number of cases in the country, fatality rate, and other information that could help people avoid contact with infected persons. Using mobile apps such as the Corona Map helped citizens track the movement of infected individuals and to take appropriate caution. This open communication helped increase awareness about the disease, adherence to preventive measures, and provided information on how to get help if infected.

Openness can help create avenues for establishing trustworthy relationships between health officials and the public. It has been very useful for creating and sustaining dialogue with individuals and groups that had conflicting knowledge and information about the Ebola outbreak in West Africa and in the Democratic Republic of Congo, especially when the dialogue was relevant, socially contextualized, and adaptive to the continually changing nature of the outbreak.[2] Openness helped establish trusted relationships, which allowed individuals to willingly provide useful feedback from their communities. And such feedback helped in adjusting response strategies.

Another example of open communication has been observed in Singapore,[31] where government officials openly addressed the scientific uncertainties around COVID-19, such as the mode of transmission, the effectiveness of wearing masks, and the search for a vaccine. Even when the spread of the disease in the country seemed to be under control in early March 2020, the government warned that uncertainties around the virus, particularly the mode and extent of transmission globally and locally, meant that Singapore was also at risk and that number of cases could rise. Even if this created fear in the citizens it helped them understand the dynamics of the pandemic and to take caution appropriately.

Continuous engagement and involvement of the community in health planning and implementation

Trust between authorities and the community can only be built through open, consistent, and credible engagement on any issue affecting them, knowing that the community will not believe or accept a solution just because it is good or helpful but because they are convinced that it will benefit them.[19]

Healthcare design and decision-making should be community-focused and community-driven. To boost trust, health authorities should accommodate local community involvement in discussions because they are the end-users, and programs should address their needs. Strategies should consider the connections between the different sectors of the community and include plans for regular improvements through continuous monitoring and evaluation.

Involving the community in healthcare policymaking can bring healthcare closer to the community and increase understanding of public health programs and initiatives. Continuous engagement and involvement have the potential to empower community members as catalysts for the desired change.

Sensitization through one-to-one, house-to-house, and group interactions at different forums such as schools, places of worship, markets, community meetings, and through public health committees or community health groups made up of community members has enhanced the discussion of public health issues and uptake of services in different settings. Countries like South Africa, Uganda, and Zambia have used village health teams successfully, where members of health teams provide home care services and spend time to explain certain health conditions to community members.[32–34] This has helped reduce deaths due to diseases like HIV/AIDS, malaria, and even COVID-19.

For the effectiveness of engagement, it is important to use the community's own mechanisms, such as using recognized and accepted community and religious leaders, community influencers, local associations or groups, and gatekeepers to reach community members or groups. This is because communities trust their own systems and would listen to their own recognized leaders.

Another approach to involve the community is the use of community health workers and community relays who are appointed from among community members and trained to engage community members on public health issues.[35,36] Community health workers and community relays are a vital resource to trust-building in healthcare at the community level because of their closeness to the people and their mutual knowledge and understanding of health issues from the perspectives of the public health authorities and the community. Therefore, they can form an essential bridge between authorities and the community. For example, the use of community-based home care and counseling providers has helped increase trust in HIV/AIDS care and treatment in rural communities of Africa. Vinck et al.[2] have reported high-level adherence to some Ebola preventive behaviors, including avoiding direct, physical contact and compliance with hygiene measures because of direct engagement with local communities from the beginning of the outbreak in the Democratic Republic of Congo.

Continued engagement with the community can help avoid confusion, distrust, and misconceptions and help mobilize advocates for program support, particularly when there are oppositions.[37] It will help ensure that programs are tailored to fit specific community needs.

Clear, concise, and consistent messaging on public health issues

The effectiveness of public health interventions, including response to health emergencies, hinges on effective communication that is based on clear, concise, and consistent information from all sources.[19,29] Messages about existing and emerging public health issues should be provided to the public promptly, and such messages should be backed by scientific evidence and not the personal opinions of the messenger.

One-time communication is not enough, messages should be continuous, clear, concise, and consistent. Psychological research shows that the human mind is more trustful, more positive, and more receptive when information is repeated, clear, and simple in format and language.[37] Uncoordinated and inconsistent messaging by health authorities and healthcare workers will engender confusion and mistrust and force people to seek information from unreliable alternative sources.

Individuals obtain information about public health from different sources and the source can influence how they receive the information, which will, in turn, influence their perception and behavior. It is therefore important to use different platforms and channels to provide information about public health, based on the information-seeking characteristics of the target audience. The use of different channels has played a key role in the effectiveness of public health communication. Although resource-intensive, the use of diverse media outlets for communication proved effective in reaching populations during the West Nile Virus epidemic in New York City between 1999 and 2000.[29]

Social media has become an essential platform for the communication of political, social, economic, and health issues. Governments and health communicators should capitalize on the accessibility offered by social media platforms to reach different audience groups with public health messages. Other channels are equally important, such as the traditional media, village chiefs, community leaders, group leaders, faith leaders, opinion leaders, celebrities, and political leaders.[13–16]

Considering the social, economic, and political inequalities and cultural differences existing in communities, public health communication should be targeted and sensitive to the diversity of the target audience. It should address their specific information and communication needs per time, including language differences.[29,38]

The Government of Singapore effectively used different channels, including traditional and social media, for communicating about COVID-19.[31] They reached populations with updates and key information through weekly press conferences. They reached people in their own languages without using translators and interpreters and this was crucial in building trust and catalyzing effective beliefs about institutional behavior and competence in such a multicultural society. For example, when the Prime Minister addressed the nation on April 3, 2020, to announce the partial lockdown, he spoke in English, Bahasa, Melayu, and Mandarin and when infection clusters began to emerge in some of the foreign worker dormitories, the minister for communications and information conducted a dialogue in English and Tamil.

Very early in the pandemic, the Singapore Government recognized the role of WhatsApp in spreading information and quickly deployed this platform as a key medium for providing updates. As the first cases were reported in the country in late January 2020, the government established a meticulous contact tracing procedure and provided essential information about contact tracing records to the public through WhatsApp and other platforms. They provided 2–3 daily updates along with key prevention messages through government-designated platforms, including websites, WhatsApp, Twitter, Facebook, and Telegram. They countered misinformation through these same platforms. As the pandemic advanced, the government deployed other mobile apps that helped provide reliable information to the public to help them make appropriate decisions and behave in ways that would help reduce collective risk.

Public health messaging should be inclusive and sensitive to the needs of vulnerable groups, including the homeless, immigrant and displaced communities, communities in crisis, women and children, the poor and unemployed, racial minorities, indigenous groups, and people with disabilities.

It is good practice to provide information about anticipated shortcomings of any public health interventions to ensure that people understand what to expect and to prevent a backlash of resistance. When public health measures are likely to affect the community negatively, there must be clear communication about precautions and remedies as a way of saving lives and increasing trust in public health authorities and programs.[39]

Transparency

Honesty and transparency are vital tools in trust-building.[40] Any new therapeutic agent, procedure, or methods should be well explained to the public in a way that they can easily understand, including, but not limited to, the limitations of such products.[41] Therapeutic agents should not be imposed on populations. Marketing messages and advertisements should be well crafted to ensure that they do not impose undue pressure on people to use a particular product and do not cover up the negative aspects that the product may have. When there is a failure on the part of the authorities or pharmaceutical companies, or when they cannot keep their promises, they should be transparent enough to admit it and let the people know.

Health authorities, as carriers of public health messages, must show transparency and honesty in their dealings with the public and with each other for their messages to be believed and accepted. It is important NOT to present public health as the only solution, but to present every available option, including their pros and cons so that the public can exercise their right to make informed choices.

Investment in primary healthcare

Primary healthcare facilities are the closest to the community and should be the most resourced. However, in many developing countries they are the least resourced. When primary health care is inadequately resourced, individuals seek services at secondary and tertiary facilities, making those facilities become overwhelmed and inefficient because of the inability to meet the demand for services. This will reduce trust in the public health system. It is therefore important to invest adequately in primary health care if we must build trust in public health.

Paying attention to fair and equitable distribution of health resources will help increase trust in the government and health authorities. Communities that are underserved will be distrustful of public health institutions and will be tempted to seek solutions elsewhere for their health needs.[39] A survey among responders to the West Africa Ebola outbreak between May 2015 and June 2016 highlights the importance of consistent investment in community well-being and healthcare systems development as key to securing community trust in public health programs and healthcare providers.[42,43]

Investment in social listening

In today's digital landscape, it is vital to invest in real-time social listening systems that can collect and analyze data from various social media platforms, in multiple languages and across different geographical and political regions. Such analyses can help produce real-time maps that will enhance the ability of public health authorities and relevant stakeholders to understand the discussion and social dynamics around specific health issues and respond promptly and appropriately. Such real-time maps can provide vital information for developing communication and campaign strategies and toolkits to empower the public with accurate information for decision-making in line with recommended public health measures.

Social listening will allow authorities to access and filter inaccurate information and provide the public with the correct information based on scientific evidence, thus increasing trust in the government or relevant health authorities. It will allow authorities to learn about community-specific opinions, misconceptions, and fears before they become widespread and to develop targeted messages that will help them respond appropriately.[37]

References

1. DeBoer MJ. *Public Health, Public Trust, and Faith Communities*; 2021. Canopy Forum, 10 June 2021 https://canopyforum.org/2021/06/10/public-health-public-trust-and-faith-communities/.
2. Vinck P, Pham PN, Bindu KK, Bedford J, Nilles EJ. Institutional trust and misinformation in the response to the 2018–19 Ebola outbreak in North Kivu, DR Congo: a population-based survey. *Lancet Infect Dis.* 2019;19(5):529–536. https://doi.org/10.1016/S1473-3099(19)30063-5.
3. Wellcome. *Wellcome Global Monitor How Does the World Feel about Science and Health?* London: Wellcome; 2019. https://wellcome.org/sites/default/files/wellcome-global-monitor-2018.pdf.
4. Elisabetta L. Trust in health care and vaccine hesitancy. *Rivista di estetica.* 2018;(68). https://doi.org/10.4000/estetica.3553.
5. EU Expert Group on Health Systems Performance Assessment (HSPA). *Assessing the Resilience of Health Systems in Europe: An Overview of the Theory, Current Practice and Strategies for Improvement.* Luxembourg: EU; 2020. https://ec.europa.eu/health/sites/health/files/systems_performance_assessment/docs/2020_resilience_en.pdf.
6. World Health Organization. *Monitoring the Building Blocks of Health Systems: A Handbook of Indicators and Their Measurement Strategies.* Geneva: World Health Organization; 2010. https://apps.who.int/iris/bitstream/handle/10665/258734/9789241564052-eng.pdf.
7. Onyeneho NG, Amazigo UV, Njepuome NA, et al. Perception and utilization of public health services in Southeast Nigeria: implication for health care in communities with different degrees of urbanization. *Int J Equity Health.* 2016;15(12). https://doi.org/10.1186/s12939-016-0294-z.
8. Hu G, Han X, Zhou H, Liu Y. Public perception on health care services: evidence from social media platforms in China. *Int J Environ Res Publ Health.* 2019;16(7):1273. https://doi.org/10.3390/ijerph16071273.
9. Nchinda TC. Traditional and western medicine in Africa: collaboration or confrontation? *Trop Doct.* 1976;6(3):133–135. https://doi.org/10.1177/004947557600600315.
10. Thorburn S, Kue J, Keon KL, Lo P. Medical mistrust and discrimination in health care: a qualitative study of Hmong women and men. *J Community Health.* 2012;37(4):822–829. https://doi.org/10.1007/s10900-011-9516-x.
11. Benjamins MR. Religious influences on trust in physicians and the health care system. *Int J Psychiatr Med.* 2006;36(1):69–83. https://doi.org/10.2190/EKJ2-BCCT-8LT4-K01W.
12. Kim DK, Kreps GL. An analysis of government communication in the United States during the COVID-19 Pandemic: recommendations for effective government health risk communication. *World Med Health Pol.* 2020;12(4):398–412. https://doi.org/10.1002/wmh3.363.
13. Oak Ridge Associated Universities (ORAU). Is social media helping or hindering vaccination rates? Oak Ridge: ORAU, https://www.orau.org/impact/health-communication/is-social-media-helping-hindering-vaccine-rates.html.
14. Stein RA. The golden age of anti-vaccine conspiracies. *Germs.* 2017;7(4):168–170. https://doi.org/10.18683/germs.2017.1122, 2017 Dec.
15. Tate J. *Can Social Media Have a Role in Tackling Vaccine Hesitancy? the Health Policy Partnership*; 2019, 3 May 2019 https://www.healthpolicypartnership.com/can_social_media_have_role_vaccine_hesitancy/.

16. ScienceDaily. *Vaccine Misinformation and Social Media*. ScienceDaily; 2020, 17 February 2020 https://www.sciencedaily.com/releases/2020/02/200217163004.htm.
17. Rogers W, Ballantyne A. Gender and trust in medicine: vulnerabilities, abuses, and remedies. *Int J Fem Approaches Bioeth*. 2008;1(1):48–66. http://www.jstor.org/stable/40339212.
18. World Health Organization. Universal health coverage. Geneva: World Health Organization, https://www.who.int/health-topics/universal-health-coverage#tab=tab_1.
19. Institute for Health Metrics and Evaluation. *Lack of Trust Has Helped Fuel the COVID-19 Pandemic, New Study Shows*. Health Data; 2022, 1 February 2022 https://www.healthdata.org/news-release/lack-trust-has-helped-fuel-covid-19-pandemic-new-study-shows.
20. #SayNoToWitsVaccine. Twitter, https://twitter.com/hashtag/SayNoToWitsVaccine?src=hashtag_click.
21. Pauline P, Yin L, Amirul AR. *Hashtag Campaigns during the COVID-19 Pandemic in Malaysia Escalating from Online to Offline*. Singapore: ISEAS-Yusof Ishak Institute; 2021. December 2021 https://www.iseas.edu.sg/wp-content/uploads/2021/11/TRS21_21.pdf.
22. Moon MJ. Fighting COVID -19 with Agility, transparency, and participation: wicked policy problems and new governance challenges. *Publ Adm Rev*. 2020;80(4):651–656. https://doi.org/10.1111/puar.13214.
23. Buckley GJ, Gostin LO. *The Effects of Falsified and Substandard Drugs*. Washington DC: National Academies Press; 2013, 20 May https://www.ncbi.nlm.nih.gov/books/NBK202526/.
24. Bogart LM, Ojikutu BO, Tyagi K, et al. COVID-19 related medical mistrust, health impacts, and potential vaccine hesitancy among black Americans living with HIV. *J Acquir Immune Defic Syndr*. 2021;86(2):200–207. https://doi.org/10.1097/QAI.0000000000002570.
25. Cheung ATM, Parent B. Mistrust and inconsistency during COVID-19: considerations for resource allocation guidelines that prioritise healthcare workers. *J Medi Ethics*. 2021;47:73–77. https://doi.org/10.1136/medethics-2020-106801.
26. Holzmann H, Hengel H, Tenbusch M, et al. Eradication of measles: remaining challenges. *Med Microbiol Immunol*. 2016;205:201–208. https://doi.org/10.1007/s00430-016-0451-4.
27. Krause NM, Freiling I, Beets B, Brossard D. Fact-checking as risk communication: the multi-layered risk of misinformation in times of covid-19. *J Risk Res*. 2020;23(7–8):1052–1059. https://doi.org/10.1080/13669877.2020.1756385.
28. World Health Organization, OECD, World Bank. *Delivering Quality Health Services: A Global Imperative for Universal Health Coverage*. Geneva: World Health Organization, Organisation for Economic Co-operation and Development, and The World Bank; 2018, 2018 https://apps.who.int/iris/bitstream/handle/10665/272465/9789241513906-eng.pdf.
29. Ataguba OA, Ataguba JE. Social determinants of health: the role of effective communication in the COVID-19 pandemic in developing countries. *Glob Health Action*. 2020;13(1). https://doi.org/10.1080/16549716.2020.1788263.
30. Mate KS, Rooney AL, Supachutikul A, et al. Accreditation as a path to achieving universal quality health coverage. *Glob Health*. 2014;10(68). https://doi.org/10.1186/s12992-014-0068-6.
31. Wong CM, Jensen O. The paradox of trust: perceived risk and public compliance during the COVID-19 pandemic in Singapore. *J Risk Res*. 2020;23(7–8):1021–1030. https://doi.org/10.1080/13669877.2020.1756386.
32. Mulumba M, London L, Nantaba J, Ngwena C. Using health committees to promote community participation as a social determinant of the right to health: lessons from Uganda and South Africa. *Health Hum Rights*. 2018;20(2):11–17. https://www.ncbi.nlm.nih.gov/pmc/articles/PMC6293345/.
33. Karuga R, Kok M, Luitjens M, et al. Participation in primary health care through community-level health committees in Sub-Saharan Africa: a qualitative synthesis. *BMC Publ Health*. 2022;22:359. https://doi.org/10.1186/s12889-022-12730-y.
34. Gilmore B, McAuliffe E, Larkan F, et al. How do community health committees contribute to capacity building for maternal and child health? A realist evaluation protocol. *BMJ Open*. 2016;6, e011885. https://doi.org/10.1136/bmjopen-2016-011885.
35. Bhaumik S, Moola S, Tyagi J, Nambiar D, Kakoti M. Community health workers for pandemic response: a rapid evidence synthesis. *BMJ Glob Health*. 2020;5(6), e002769. https://doi.org/10.1136/bmjgh-2020-002769.
36. Gilmore B, McAuliffe E. Effectiveness of community health workers delivering preventive interventions for maternal and child health in low- and middle-income countries: a systematic review. *BMC Publ Health*. 2013;13:847. https://doi.org/10.1186/1471-2458-13-847.

37. WHO. *Vaccination and Trust: How Concerns Arise and the Role of Communication in Mitigating Crises*. Copenhagen: WHO Regional Office for Europe; 2017. https://www.euro.who.int/__data/assets/pdf_file/0004/329647/Vaccines-and-trust.PDF.
38. Adalja AA, Sell TK, Bouri N, Franco C. Lessons learned during dengue outbreaks in the United States, 2001-2011. *Emerg Infect Dis*. 2012;18(4):608–614. https://doi.org/10.3201/eid1804.110968.
39. Berger ZD, Evans NG, Phelan AL, Silverman RD. Covid-19: control measures must be equitable and inclusive. *BMJ*. 2020:m1141. https://doi.org/10.1136/bmj.m1141.
40. O'Malley P, Rainford J, Thompson A. Transparency during public health emergencies: from rhetoric to reality. *Bull World Health Organ*. 2009;87(8):614–618. https://doi.org/10.2471/blt.08.056689.
41. Fátima L, Adriana EC, Simon C, et al. Smart pharmaceutical manufacturing: ensuring end-to-end traceability and data integrity in medicine production. *Big Data Res*. 2021;24, 100172. https://doi.org/10.1016/j.bdr.2020.100172.
42. Depoux A, Martin S, Karafillakis E, Preet R, Wilder-Smith A, Larson H. The pandemic of social media panic travels faster than the COVID-19 outbreak. *J Trav Med*. 2020;27(3):taaa031. https://doi.org/10.1093/jtm/taaa031.
43. Ryan MJ, Giles-Vernick T, Graham JE. Technologies of trust in epidemic response: openness, reflexivity and accountability during the 2014–2016 Ebola outbreak in West Africa. *BMJ Glob Health*. 2019;4(1). https://doi.org/10.1136/bmjgh-2018-001272.

CHAPTER

25

Workforce development

Bernard Owusu Agyare[1], *Scott J.N. McNabb*[2], *Brittany L. Murray*[3], *Mabel K.M. Magowe*[4], *Peter S. Mabula*[5], *Chima J. Ohuabunwo*[2,6,7], *Affan T. Shaikh*[2,8] and *Laura C. Streichert*[9]

[1]Center for Global Health Science and Security, Georgetown University, Washington, DC, United States; [2]Hubert Department of Global Health, Rollins School of Public Health, Emory University, Atlanta, GA, United States; [3]Division of Pediatric Emergency Medicine, Global Health Office of Pediatrics at Emory, Emory University School of Medicine, Atlanta, GA, United States; [4]School of Nursing, University of Botswana, Gaborone, Botswana; [5]Emergency Medicine Department, Disaster Management, Safari City HealthCare Limited, Arusha, Tanzania; [6]Department of Medicine and the Office of Global Health Equity, Morehouse School of Medicine, Atlanta, GA, United States; [7]Department of Epidemiology, University of Port Harcourt School of Public Health, Rivers State, Nigeria; [8]Yale University, New Haven, CT, United States; [9]Public Health Consultant, San Diego, CA, United States

Introduction

Widespread disruptions caused by the rapid spread of the SARS-CoV-2 virus into a global pandemic affected nearly every aspect of life and activated the global health security (GHS) workforce on many fronts.[1] Clinical healthcare personnel, community health workers, epidemiologists, lab technicians, veterinarians, wildlife biologists, social scientists, and many other services have been at the forefront. Biomedical researchers developed breakthrough vaccines and treatments, while data scientists and IT specialists scrambled to develop and deliver real-time public health surveillance, dashboards, and predictive models. Experts in biosecurity, border control, and food safety were put on high alert to monitor supply chains and cross-border interactions. The need to respond at every level placed new pressures on already over-strapped budgets, with urgent decisions on funding reprioritization. The breadth of the GHS

Modernizing Global Health Security to Prevent, Detect, and Respond
https://doi.org/10.1016/B978-0-323-90945-7.00003-8

workforce is reflected in the transdisciplinary nature of those tasked to prevent, detect, and respond to emergencies.

Disasters can have significant impacts on the workforce supply because of displacement, disruption of workplaces, workers' ability to attend work, job losses, and shortages.[2] As the COVID-19 pandemic unfolded, it was met in many places with an insufficient, unprepared, and uncoordinated health security workforce that did not have sufficient capacity for the rapid surge response. Gaps in workforce preparedness as a result of staff shortages, lack of equipment, insufficient training, decentralized plans of action, unclear policies, and other issues led to the inability to mount a prompt and coordinated national and worldwide response.[3,4] Despite these "deep cracks" in global health systems, the GHS workforce, across disciplines and sectors, was galvanized to reach new levels of performance.

The World Health Organization (WHO) defines workforce development as "The enhancement of training, skills, and performance of health workers."[5] Workforce development for GHS goes further to include capacity building among professionals working across a broad spectrum of sectors and disciplines. The rationale for workforce development is woven throughout all major global health initiatives as central to emergency response and preparedness at every level.

Efforts to strengthen GHS and to build resilience in health systems involve preparing a workforce with competencies and skillsets identified as needed to perform their routine and special duties in the GHS architecture.[6] Breakdowns in health systems and services and the disruption of health intervention programs can compound the adverse effects of staff losses and or displacements.[4]

During COVID-19, the need for surge capacity required many professionals to be assigned or deployed to situations that required them to perform tasks outside of their training and normal job responsibilities. This put tremendous pressure on individuals and organizations alike and highlighted the need for broad-based competencies in systems thinking and collaboration skills and flexible organizational structures.[1]

Collecting and analyzing data using rigorous scientific methods and transparency and rapidly transforming it into information for broad consumption by diverse stakeholders is a powerful tool for building workforce capacity. The pandemic underscored the importance of working collaboratively across multiple sectors that span the natural, life, and social science arenas.[7]

The need for a workforce prepared to effectively prevent, detect, and respond to threats to GHS is a thread woven throughout this book. This chapter documents robust workforce development approaches for effective disaster and emergency response. We summarize the current state of thinking in a rapidly moving field and present a literature review and a gap analysis to shape a vision for creating a workforce of professionals with adequate knowledge, skills, and resources to perform their jobs effectively, safely, and with confidence, as well as a pipeline of new people entering careers in the expanding and diverse fields of GHS.

GHS workforce development efforts before COVID-19

The COVID-19 pandemic has led to a reexamination of how to align workforce development with health system requirements and lessons learned. Guidance is beginning to emerge in reports from WHO, published articles, and gray literature with the intention of implementing effective workforce development strategies to strengthen GHS.

Workforce development is a key pillar for a strong and viable global health system and an important component of capacity building for modernizing GHS to prevent, detect, and respond to public health threats. Previous epidemics have taught that, "There can be no health security without a skilled health workforce."[8] Drehobi et al., summarized four main factors that lead to a "lack of the right number of people with the right skills in the right place at the right time,"—the very opposite of what is needed. Although there are a number of variables, the four leaders fell under the categories of composition and number of workers, the competency of workers, contextual environment, and the work environment.[9] Effective workforce development must address all of these factors in a coordinated way, tailored to the specific need and context.[10,11]

Various intergovernmental frameworks and policies have addressed the inherent importance of the global health workforce and made special provisions for workforce development in their programs. The 2007 WHO Health System Strengthening (HSS) framework included a trained and motivated health workforce as a building block of a well-functioning health system.[12] Initially proposed to reach the targets of the United Nations Sustainable Development Goals (SDGs) in 2007, the HSS framework continues to be relevant.

Responding to a need for an overarching tool for measuring GHS after the 2003 severe acute respiratory syndrome (SARS) pandemic, the International Health Regulations (IHR-2005) were created and became legally binding on all WHO member states in 2007.[13]

The Global Health Security Agenda (GHSA) was adapted to implement the IHR and to develop capacities. Although the IHR-2005 urges member states to build, strengthen, and maintain required capacities and competencies and to mobilize the resources necessary, these directives have not been uniformly adhered to.[14] To address this, WHO provided strategic direction for scaling-up IHR learning activities and has made available the WHO-designed learning packages and activities based on needs identified locally, nationally, regionally, and globally. A prepared workforce is key to implementing the IHRs and disparities in IHR implementation are, in turn, reflected in workforce preparedness.

WHO's *Global strategy on human resources for health: workforce 2030* outlines measurable goals for workforce development. Their vision to "Accelerate progress toward universal health coverage and the UN Sustainable Development Goals by ensuring equitable access to health workers within strengthened health systems," reflects the long-term outcomes of workforce development on communities served and larger global health and security goals.[15]

GHS workforce development efforts after COVID-19

A lesson learned from COVID-19 is the need to embrace a coordinated multisector approach with all-hazard and hazard-specific measures to ensure preparedness for all types of emergencies at the community, national, and international levels.[7] For example, the expansion of the alliance between WHO, the Food and Agriculture Organization (FAO), and the World Organization for Animal Health (OIE) to a Quadripartite that includes the United Nations Environment Program (UNEP); the creation of a One Health High-Level Expert Panel

(OHHLEP); and the drafting of One Health Joint Plan of Action are concrete actions to promote cooperation across the animal, human, and environmental health sectors to accelerate integrated strategies to advance GHS.[16]

Considering COVID-19, WHO consolidated policy recommendations into an interim guidance note entitled, "*Health workforce policy and management in the context of the COVID-19 pandemic response*." This detailed report provides guidance for human resources for health (HRH) managers and policy-makers with validated interventions at the individual, management, organizational, and system levels.[1] Specifically, it targets system-wide change by addressing the requirements for mobilizing an intersectoral response, as detailed in Fig. 25.1.

The US Centers for Disease Control and Prevention (CDC), in partnership with collaborators worldwide, also play a role in developing training in GHS. Through the Field Epidemiology Training Program (FETP), currently spanning over 80 countries, the US-CDC aims to develop global epidemiology and public health surveillance workforce capacities.[17] Working with affiliates, such as Africa-CDC, and other organizations worldwide, the CDC's Division of Global Health Protection (DGHP) offers training activities that cover all the primary functions needed to prevent, detect, and respond to public health threats, regardless of the cause. The African Union, in partnership with Africa CDC and Emory University, has created an

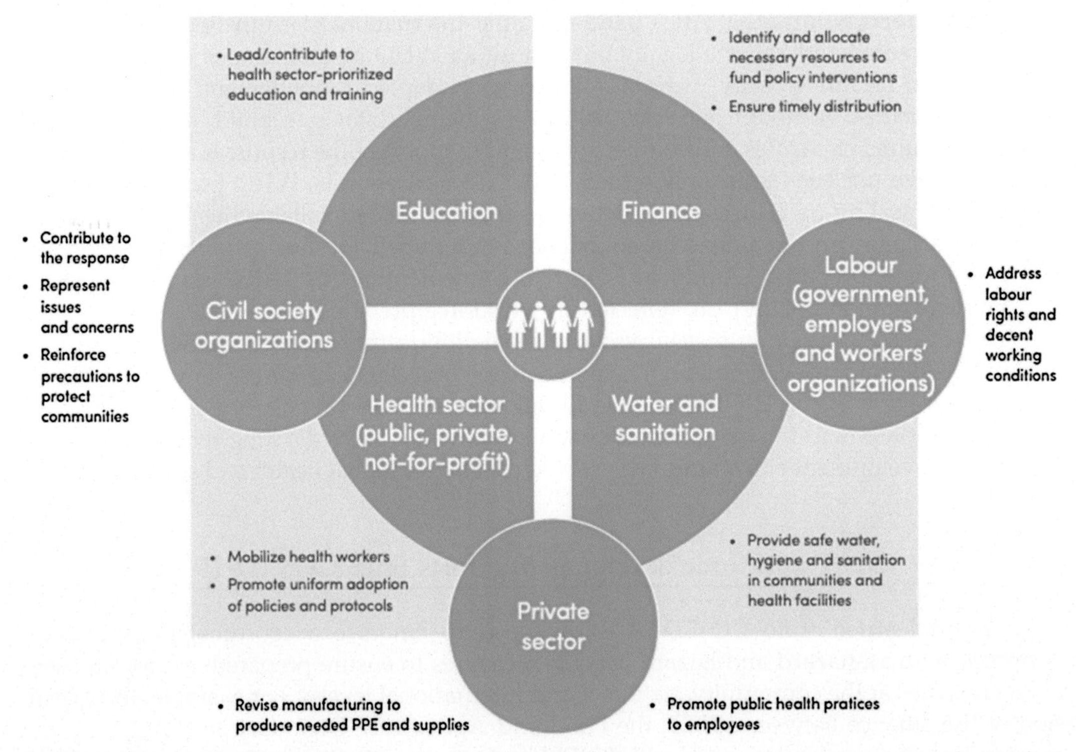

FIGURE 25.1 Mobilizing an intersectoral response to COVID-19 health workforce requirements. *World Health Organization. Health Workforce Policy and Management in the Context of the COVID-19 Pandemic Response, 2020. https://www.who.int/publications/i/item/WHO-2019-nCoV-health_workforce-2020.1*

Africa CDC Institute of Workforce Development (https://africacdc.org/institutes/africa-cdc-institute-for-workforce-development/). These are just examples of the many training initiatives with a focus on workforce development for public health.

During the COVID-19 pandemic, some healthcare systems responded by implementing protocols developed in response to previous health emergencies. An example is presented in the *Chapter Case Study*: A grassroots approach to engage local volunteers as part of Global Health Security workforce development in Uganda.

Community input and worker opinions on practices and priorities provide evidence to help target priority needs for information, training, and resources. For example, a survey of professionals working worldwide in One Health surveillance reported that "Obtaining access to data collected by other domains," was both the most frequent challenge and the most difficult barrier to improve. In that study, respondents from low-income or middle-income countries were more motivated to make improvements than stakeholders from high-income countries.[18]

Weaknesses in GHS workforce planning displayed during COVID-19 included barriers and facilitators to participation in response activities. A recent survey study provided a snapshot of the COVID-19 response activities and training needs of a cross-section of workers, including many self-affiliated with One Health. The results also indicated a role for One Health networks in training and worker activation, which can be used as a springboard for targeted training where primary users find it the most needed and effective.[19] More research is needed to fill the gap in evaluating the role of nonprofits, nongovernmental organizations, and others that support the operationalization of new collaborative initiatives.

Digital competencies are becoming increasingly important for the GHS workforce.[20] The COVID-19 pandemic catalyzed an explosion in telework and the use of online information and communication technologies.[21] Zoom, Microsoft Teams, webinars, video-conferencing, and other communication software could reach new audiences in synchronous and asynchronous formats at low expense. These formats also provide opportunities for evaluation and monitoring improvements in knowledge, skills, and performance for individuals and organizations alike. These platforms have also increased patronage of telemedicine, a phenomenon that has dramatically improved access to healthcare in certain geographical settings.

Online courses and videos have increasingly become a part of blended approaches to emergency preparedness training. For example, several self-learning packages are available at the WHO's Health Security Learning Platform and the OpenWHO platform provides several global health learning opportunities (openwho.org). Evaluation of a virtual training program to build capacity in COVID-19 infection prevention demonstrated course effectiveness on performance outcomes.[22] Nevertheless, the tremendous value and need for in-person meetings and training in the form of conferences, workshops, symposiums, and table-top exercises is still recognized and reflected in the return to live participation events.

Changes in how training and work are undertaken can have widespread intended and unanticipated effects that should be considered as training strategies move forward. The WHO and the International Labor Organization released a joint technical brief on the role of telework in the changing workplace and its potential impact on workers' health, safety, and well-being.[21] Among the benefits, virtual platforms offer broader engagement from global regions, reduce the time and cost of travel to individuals and organizations, and offer greater flexibility in programming. Some have noted this has led to an increase in international collaboration.[9] The WHO/ILO report acknowledged that telework settings may not meet typical occupational safety and health standards but offers suggestions for minimizing the physical and mental effects of working within a digital and socially isolated telework environment. Research is needed to determine the effectiveness of hybrid models in future strategies for recruiting, training, and retaining a strong GHS workforce.

To ensure the availability of a robust healthcare workforce, current healthcare workers must be supported, both professionally and personally, to excel in their roles at the moment.[23] This is key for workforce retention, as well as for the recruitment of new workers. During the pandemic, frontline healthcare workers including nurses, pharmacists, hospital staff, first responders, EMTs, and community health workers in both developed and developing countries shouldered the tremendous responsibility for providing care in an overtaxed healthcare system.[24]

The effect of the COVID-19 pandemic on nurses, specifically, illustrates how events can change the work environment for individuals and organizations. The demand for travel nurses, for example, is creating dramatic changes in nursing and reshaping the nursing profession. Many hospitals are losing nursing staff to resignations and retirements. Travel nurses demonstrate the value of an "all-hazard" approach and are equipped to operate appropriately in different situations.[25]

The COVID-19 pandemic showed the tremendous toll that such an urgent, intense, and ongoing emergency can take on the physical and mental health of workers.[24] For example, during epidemics and pandemics, healthcare workers are prone to experience a high workload with increased volumes of clinical duties, witness an increase in critically ill and dying patients, and have increased personal infection risks, all while experiencing overall situational anxiety. These workers are at high risk for mental health crises and burnout. Shortages or unavailability of PPEs and other logistics at health facilities posed a grave danger to health staff's safety and well-being. This was most visible among healthcare workers, especially doctors, nurses, and others on the frontlines who put their own safety at risk, and was even more prevalent in countries experiencing political or other forms of social instability.[26] The increasing number of violent incidents with public health and medical personnel has created unprecedented security concerns and new challenges for health systems. Reports are beginning to emerge of protective actions, such as increased penalties for committing assaults against health workers.[27] Supporting the physical and mental health of healthcare workers will help with workforce retention, as well as the recruitment needed to modernize GHS.[28,29] Curricula that incorporate the present experience with COVID-19 must be designed to train and prepare HCWs in advance for any future eventuality.

Gap analysis

A gap analysis of the current state of workforce development in the realm of GHS reveals what is working and where change is needed. We describe four topics where improvements would have an impact on modernizing workforce development considering lessons learned from COVID-19:

A central framework for the GHS workforce;

Robust data and public health surveillance systems;
Coordinated training using technical innovations; and
Building careers and capacity in GHS.

1. A central framework for the GHS workforce

   The lack of a clear understanding and definition for the GHS workforce hampers workforce development efforts. With the ever-increasing scope and complexity of existing and new health threats, the emergence of biosecurity/biosafety concerns, climate/global warming issues, and other factors, it is imperative to define the GHS workforce to codify the diversity and ensure that capacity-building programs are comprehensive and integrated. Traditionally, disease outbreak detection and response have been primarily within the ambit of public health staff and first responders. However, to better build capacity and prepare for future pandemics, the GHS workforce should be expansive to include diverse professionals important for GHS, like those from agriculture, food, animal health, defense, security, finance, environment, disaster management, border control, law enforcement, research, technology, communication, and education sectors.[30] In addition, the concepts and approaches of social sciences are increasingly being recognized for understanding the critical social context that drives human behavior.[31,32]

2. Robust data and public health surveillance systems

   Established systems for the dissemination of the data and information that workers need in real time can enhance worker capacity to prevent, detect, and respond to threats. To further health equity, health public health surveillance systems must be designed to include all stakeholders, including marginalized populations. Health system improvement must be guided by the concept that everyone, no matter where they are, has an equal right to health and to be counted in the data that inform decision-making. Strengthening the availability, quality, analysis, and use of health workforce data is important to direct political policy and to inform action at all levels.[33]

   Data are also needed to measure performance at the individual, organizational, and system levels and to reveal inequities across populations and between high- and low-resource nations. The development of standardized metrics for performance would enable longitudinal and comparative observations to measure the value of interventions to ensure a sufficient workforce with the necessary competencies. A rigorous approach to data systems and health public health surveillance is especially important in the global context of COVID-19 characterized by widespread public mistrust of science and health systems.[34]

3. Coordinated training that utilizes technical innovations

   No matter their work position, workers need to have the right competencies and skillsets to perform their routine and special duties in the GHS architecture. The top training needs for many public health workers have shifted from discipline-specific skills to more cross-cutting skills, such as systems thinking, communication, and coordination.[35] This requires confidence and positive attitudes toward flexibility and willingness to try something new. Fostering transdisciplinary and cross-sector exchange and the skills involved will build an understanding of different stakeholder issues and priorities. Despite recent guidance from the WHO and others involved in workforce development, there is still a need to address fragmented training approaches, review competencies, and develop comprehensive curricula. Ultimately, the creation of a Global Health competency model would facilitate improvements in education.[36] The integration of new technologies must also be considered in light of differences in digital access between low- and high-resource communities and nations.

4. Building careers and capacity in GHS.

   How and where people work in the broad range of disciplines and sectors related to GHS is changing at a rapid pace. Retention of skilled workers is essential in times of crisis. Recruitment of new professionals to careers in healthcare, public health, and other fields are needed to ensure a pipeline of people ready to confront present and future emergencies. In addition to anticipating the knowledge, skills, and support systems required of tomorrow's workforce, there need to be evidence-based strategies that include the system-level change in how professional development is created, implemented, evaluated, and supported.

Vision for the future

Workforce development and the optimization of the use of human resources are necessary to create GHS systems that provide equitable and high-quality care during normal times and emergencies. Based on the literature and the diverse experiences of the coauthors of this chapter, we describe a vision for overcoming the barriers identified in the gap analysis using lessons learned from COVID-19 and previous public health disasters.

Modernizing GHS begins with a recognition of the diversity of the frontline professionals involved in preventing, responding to, and recovering from the complex public health challenges of the 21st century. Creating a central framework for the GHS workforce would advance the coordination of efforts in research, practice, and policy for maximal benefit to both workers and the populations they serve. The recent adoption of a common definition for One Health by the newly formed Quadripartite alliance between the World Health Organization (WHO), the World Organization for Animal Health (WOAH, founded as OIE), the Food and Agriculture Organization of the United Nations (FAO), and the United Nations Environment Program (UNEP). The extent of multisectoral workforce response exemplifies a concrete step that can be taken to facilitate coordinated action.[16]

The future of workforce development will rely on training that is coordinated across multiple sectors that span the natural, life, and social science arenas. It will need to be driven by robust data, utilize technological innovations in communications, and be informed by research data and stakeholder input. The goal is a steady workforce of professionals armed with new competencies, and organizations with the overall capacity of human and other resources to respond rapidly and effectively to any incident. A key step in strengthening the capacity of the GHS is building flexibility into systems and eliminating silos in professional practice and training. In addition, to equitably serve all populations, there is a need to reduce disparities in capacities between high- and low-resource nations.

COVID-19 has taught us that strategies for scaling up need to be in place to address the increased workforce demand and surge capacity needs that arise in a global emergency.

One cannot establish a vision without considering the status of the public health workforce and a path to the future with clear steps for action.

Actions to achieve the vision

The COVID-19 pandemic clearly illustrated the need for a flexible healthcare workforce to respond to crisis situations. It unveiled weaknesses in healthcare workforce planning worldwide that limited the ability of systems to respond adequately to protect population health to the level needed. To improve health workforce development to ensure human resources are available for future emergencies, actions based on what has been learned through the COVID-19 pandemic and prior healthcare surges are necessary to move forward. Table 25.1 presents some examples of actions that will advance the vision for workforce development shaped by identified gaps.

A central hub for accessing health data and information, as well as training opportunities, would facilitate informed action and evidence-based decisions. Data-sharing protocols need to be in place to ensure transparency and inclusion of all relevant data in shared datasets. In addition, platforms for knowledge, lessons learned, and information and opportunities for interaction are needed for workers to find information outside of their typical domain. To ramp up facilitation for capacity building through workforce development, the WHO has released joint statements and reports that present solid guidance based on expert opinion and research. With so much information already available, it is inefficient for individuals and organizations to "reinvent the wheel." The WHO, through the mechanism of the IHR, could formulate actionable and measurable workforce development checklists for member

TABLE 25.1 Summary of identified needs for GHS workforce development with a vision for success and practical cross-cutting actions to achieve the vision.

Identified needs	VISION	Actions
1. A central framework for the GHS workforce	All-hazard preparedness One health approach	Implement expert guidance. Create a joint definition of GHS workforce
2. Robust data and security systems	Centralized data systems with distributed data	Adopt data-sharing procedures. Offer dissemination platforms
3. Coordinated training using technical innovations	No siloed thinking Input from workers and stakeholders on priorities	Adapt competencies for systems thinking and multisectoral collaboration. Research and evaluation to identify effective interventions and performance monitoring
4. Careers and capacity in GHS	Flexible workforce architectures at all levels Eliminate disparities in access to workforce development, Retention and recruitment Careers in GHS	Practices, policies, and legislation System design for diversity, equity, and inclusion Worker health and safety protections Support for capacity-building organizations

states. Like the Joint External Evaluation (JEE) of the GHSA, countries could be evaluated on the status of their health workforce development Action Plan and their capacity to respond to health threats.

COVID-19 revealed a need for interprofessional competencies that foster effective communication, coordination, and collaboration. The development and adoption of standardized training curricula and guidelines would raise baseline levels of preparedness and help to avoid confusion, mistrust, and lack of cooperation between stakeholders. Despite the benefits of standardized workforce competencies and strategies, there will always be a need to adapt training to the cultural context and level at which they are being implemented. Dedicated funding for training is also critical to develop a robust transdisciplinary GHS workforce, For example, the USAID EPT program funds the One Health Workforce Next Generation (OHW-NG) network, which provides collaboration opportunities across multiple universities in sub-Saharan Africa (AFROHUN) and Southeast Asia (SEAOHUN) (https://ohi.vetmed.ucdavis.edu/programs-projects/one-health-workforce-next-generation).

Ongoing knowledge assessments, opportunities to practice skills in simulated situations, and regular worker feedback will help to align training with workforce needs until a level of competency is achieved and performing a skill becomes more automatic, even in times of crisis and high-stress situations. At the start of the COVID-19 pandemic, it was observed that some health workers who had previously received training on the use of PPEs struggled to use them when it mattered most. Thus, there was a need to retrain frontline workers on how to don, doff and manage PPEs, a situation that adversely affected the response to the pandemic.[37] This makes the incorporation of simulation exercises in all training programs and routine preparedness drills important. Strategies, such as a train-the-trainer model can have an accelerator effect on the distribution of new ideas and processes. This makes the incorporation of simulation exercises in all training programs and routine preparedness drills important to achieve better outcomes.

Workforce development and planning should be considered a priority at the local, national, and international levels as each of these levels must be able to work together in a crisis to ensure GHS. It is key that not only organizational leadership address health workforce development, but that it also be considered by policymakers and governments to have robust political and financial support.

In times of emergency and surges, creative solutions are needed to meet the demand for qualified workers. The use of qualified, but currently unemployed healthcare providers, recent retirees, trainees that have nearly finished their training, and healthcare workers from the private sector or NGOs, can be considered for recruitment during times of need. The local or regional tracking of these individuals and the ability to contact them should be in place before they are needed. Strategies for workforce training must, therefore, consider all stakeholders.

System-level flexibility to respond to disasters can lead to resilience in healthcare systems. Building capacity and creating both flexibility and opportunity can maximize the roles and effectiveness of healthcare workers.[38] For example, when a pandemic or health emergency is recognized, policies should already be in place to create standard operating procedures, training, and guidelines quickly and collaboratively to scale up appropriately. This may involve creating regional training centers to pool educational resources, using e-learning methods to deliver content, and having credentialing systems that allow for fast-tracking for specific skills that can be rapidly acquired and implemented when needed. Several stakeholders, including Ministries of health, regulatory and credentialing boards, professional associations, healthcare worker training institutions, hospitals, employers, and healthcare workers themselves need to work together to implement task-sharing and/or task-shifting, and without preplanning, this can be slow and burdensome.

To ensure the availability of a prepared GHS workforce, current workers must be supported, both professionally and personally, to address their physical and mental health.[29] Workforce support should include safe working conditions, adequate PPE to protect themselves from contamination and illness, remuneration and incentives, opportunities for advancement and continuing education, and additional compensation for any work beyond their normal activities, including hazard pay when appropriate. Some have suggested compensation for the families of healthcare workers that lose their lives due to job-related illness or injury.[23]

System-level change can be enacted through policy and legislation. A recent bill in California (USA) state legislature (SB 1159) that expands the definition of worker injury to include, "illness or death resulting from the 2019 novel coronavirus disease (COVID-19)" is a model for codifying workforce protection and compensation.[39] Safe and healthy workplaces are important for supporting not only the current workforce but for new worker recruitment.

The COVID-19 pandemic has had catastrophic effects worldwide but has also opened a window to a path forward for developing a GHS workforce that is ready with the knowledge, skills, and capacity to work together to meet the next public health challenge.

CASE STUDY

The Africa CDC institute for workforce development program

Topic and problem

Global Health Security (GHS) and the ability to respond to and build resilience to epidemics requires a highly trained and confident workforce motivated to take timely and appropriate action at all levels. Between 1970 and 2019, there have been at least 1910 reported outbreaks in the African region including cholera, measles, dengue fever, Ebola, Marburg, Crimean-Congo hemorrhagic fever, Lassa fever, yellow fever, malaria, monkeypox, pertussis, Rift Valley fever, and polio.[40,41] In tandem with these outbreaks, there is also a chronic health workforce shortage. This is further worsened by the concentration of the health workforce mostly in dense urban areas, leaving many rural areas poorly staffed; the region has a ratio of 1.55 health workers (physicians, nurses, midwives) per 1000 people (WHO threshold density is 4.45 per 1000 people).[42] Undoubtedly, building GHS capacity requires health workforce development, mobilization, and deployment.

Background

In 2019, the Africa CDC Institute for Workforce Development (IWD) was established in partnership with the Rollins School of Public Health at Emory University (RSPH) to strengthen the public health workforce in Africa through training of public health professionals working at National Public Health Institutes, Ministries of Health, and other related institutes in African Union Member States. Courses such as Field Epidemiology, Public Health Surveillance, Antimicrobial

CASE STUDY (cont'd)

Resistance, and Leadership and Management were delivered online, in-person, or through mixed methods approach. With the emergence of SARS-CoV2, the IWD quickly pivoted to provide timely COVID-19 insights to healthcare professionals across Africa with the launch of the Africa CDC COVID-19 Clinical Community of Practice (CCoP). The CCoP took a three-pronged approach, hosting weekly webinars in both English and French and reinforcing messages through social media campaigns and messaging systems. Later, the project was expanded to provide risk communication training to journalists, community leaders, and youth groups.

What worked

- A digital-based capacity-building platform enhanced the reach of the project allowing for more widespread participation and bringing experts far and wide to the learner.
- A variety of platforms were leveraged to support workforce capacity building, including learning management systems (e.g., Canvas); conferencing (e.g., Zoom); and mobile apps (e.g., Facebook, Twitter, LinkedIn, and WhatsApp).
- The IWD developed and implemented robust monitoring and evaluation including key performance metrics. Data were collected across 30 indicators including attendance, time spent, and evaluation responses.

What did not work

- Despite being metric driven, the Africa CDC reached only 52 of the 54 African countries. The inclusion of Arabic in addition to English and French might have helped.
- The WhatsApp and Telegram groups comprising the Community of Practice needed constant monitoring to clean out unrelated solicitations.
- In any outbreak, the immediacy of clinical information is key; an initial attempt to set up a peer-to-peer teleconsultation service resulted in less than 3% utilization; the team realized that the best method of outreach would be one with the lowest barriers. Office Hours were launched where over 200 questions were asked.
- It was observed that frontline workers at the subregional levels under-patronized training sessions.

Outcomes

At the onset and during the COVID-19 pandemic, the program had already developed an existing platform for learning and communication. This platform facilitated engagement, knowledge, expertise, and best practices sharing among public health stakeholders and training participants across the African continent. The COVID-19 Community of Practice benefited from reinforcing messages; from the 46 webinars held, 10,800 health professionals attend to watch 37 expert panelists; 22 newsletters were sent out to 5958 emails, as well as social media posts across Facebook, LinkedIn, and Instagram. The Africa CDC IWD undertook an extensive review of e-Learning best practices as well as a bandwidth capacity assessment of target countries to understand the local context and needs of learners.

CASE STUDY *(cont'd)*

Lessons learned

- Learners expressed appreciation for the different platforms that allowed interaction with their peers and the instructors.
- Training programs are overpatronized by city and urban-dwelling health workforce.
- Community acceptance of new ways of training especially through digital platforms can be difficult initially.
- Funding for workforce development is an ongoing challenge.

Recommendations

- Workforce development should be guided by principles that include cultural humility, gender equity, ethical collaboration and knowledge sharing, transparent learning networks, multidisciplinary approaches, and perspectives.
- Countries should undertake a periodic assessment of capabilities and roles to ensure their national workforce is aligned with present and future needs.
- Countries should actively invest in employees and cultivate leaders that show promise. Offering pathways for professional growth will help with recruitment, retention, and culture.
- An impactful workforce is one that is built from diversity, equity, and inclusion. This approach will help build trust and foster support for global health security efforts and bridge disparities in prevention, detection, and response.
- Training programs should be decentralized to ensure maximum participation from health staff at the subnational levels.

References

1. World Health Organization. *Health Workforce Policy and Management in the Context of the COVID-19 Pandemic Response*; 2020. https://www.who.int/publications/i/item/WHO-2019-nCoV-health_workforce-2020.1.
2. Chaiechi T, Pryce J, Ciccotosto S, Billa L. State-wide effects of natural disasters on the labor market. In: *Economic Effects of Natural Disasters*. Elsevier; 2021:211–224. https://doi.org/10.1016/B978-0-12-817465-4.00014-5.
3. Willis K, Ezer P, Lewis S, Bismark M, Smallwood N. "Covid just amplified the cracks of the system": working as a frontline health worker during the COVID-19 pandemic. *Int J Environ Res Publ Health*. 2021;18(19), 10178. https://doi.org/10.3390/ijerph181910178.
4. Tulenko K, Vervoort D. Cracks in the system: the effects of the coronavirus pandemic on public health systems. *Am Rev Publ Adm*. 2020;50(6–7):455–466. https://doi.org/10.1177/0275074020941667.
5. World Health Organization. *WHO Guideline on Health Workforce Development, Attraction, Recruitment and Retention in Rural and Remote Areas*. World Health Organization; 2021. https://apps.who.int/iris/handle/10665/341139. Accessed March 19, 2022.
6. French AJ. Simulation and modeling applications in global health security. In: Masys AJ, Izurieta R, Reina Ortiz M, eds. Global Health Security. *Advanced Sciences and Technologies for Security Applications*. Springer International Publishing; 2020:307–340. https://doi.org/10.1007/978-3-030-23491-1_13.
7. World Health Organization. *Multisectoral Preparedness Coordination Framework: Best Practices, Case Studies and Key Elements of Advancing Multisectoral Coordination for Health Emergency Preparedness and Health Security*. World Health Organization; 2020. https://apps.who.int/iris/handle/10665/332220. Accessed March 19, 2022.

8. Lancet T. No health workforce, no global health security. *Lancet*. 2016;387(10033):2063. https://doi.org/10.1016/S0140-6736(16)30598-0.
9. Drehobl P, Stover BH, Koo D. On the road to a stronger public health workforce. *Am J Prev Med*. 2014;47(5):S280–S285. https://doi.org/10.1016/j.amepre.2014.07.013.
10. World Health Organization, Burton J. *WHO Healthy Workplace Framework and Model: Background and Supporting Literature and Practices*. World Health Organization; 2010. https://apps.who.int/iris/handle/10665/113144. Accessed March 19, 2022.
11. Rozhkov M, Cheung BCF, Tsui E. Workplace context and its effect on individual competencies and performance in work teams. *IJBPM*. 2017;18(1):49. https://doi.org/10.1504/IJBPM.2017.10001261.
12. World Health Organization. *Everybody's Business – Strengthening Health Systems to Improve Health Outcomes : WHO's Framework for Action*; 2007:44. Published online https://apps.who.int/iris/handle/10665/43918. Accessed April 29, 2022.
13. Kluge H, Martín-Moreno JM, Emiroglu N, et al. Strengthening global health security by embedding the International Health Regulations requirements into national health systems. *BMJ Glob Health*. 2018;3(suppl 1), e000656. https://doi.org/10.1136/bmjgh-2017-000656.
14. Kandel N, Chungong S, Omaar A, Xing J. Health security capacities in the context of COVID-19 outbreak: an analysis of International Health Regulations annual report data from 182 countries. *Lancet*. 2020;395(10229):1047–1053. https://doi.org/10.1016/S0140-6736(20)30553-5.
15. World Health Organization. *Global Strategy on Human Resources for Health: Workforce 2030*. World Health Organization; 2016. https://apps.who.int/iris/handle/10665/250368. Accessed March 19, 2022.
16. World Organisation for Animal Health. One Health Joint Plan of Action (2022-2026): Working Together for the Health of Humans, Naimals, Plants and Hte Environment. https://www.oie.int/app/uploads/2022/04/oh-jpa-14-march-22.pdf.
17. Martin R, Fall IS. Field Epidemiology Training Programs to accelerate public health workforce development and global health security. *Int J Infect Dis*. 2021;110(Suppl 1):S3–S5. https://doi.org/10.1016/j.ijid.2021.08.021.
18. Berezowski J, Akkina J, Del Rio Vilas VJ, et al. One health surveillance: perceived benefits and workforce motivations. *Rev Sci Tech*. 2019;38(1):251–260. https://doi.org/10.20506/rst.38.1.2957.
19. Streichert LC, Sepe LP, Jokelainen P, Stroud CM, Berezowski J, Del Rio Vilas VJ. Participation in one health networks and involvement in the COVID-19 pandemic response: a global study. *Front Public Health*. 2022;10, 830893. https://doi.org/10.3389/fpubh.2022.830893.
20. World Health Organization. *Digital Education for Building Health Workforce Capacity*. World Health Organization; 2020. https://apps.who.int/iris/handle/10665/331524. Accessed March 19, 2022.
21. World Health Organization, International Labour Organization. *Healthy and Safe Telework: Technical Brief*. World Health Organization; 2021. https://apps.who.int/iris/handle/10665/351182. Accessed March 25, 2022.
22. Penna AR, Hunter JC, Sanchez GV, et al. Evaluation of a virtual training to enhance public health capacity for COVID-19 infection prevention and control in nursing homes. *J Publ Health Manag Pract*. 2022;28(6):682–692. https://doi.org/10.1097/PHH.0000000000001600.
23. Adeyemo OO, Tu S, Keene D. How to lead health care workers during unprecedented crises: a qualitative study of the COVID-19 pandemic in Connecticut, USA. *PLoS ONE*. 2021;Vol 16(9), e0257423. https://doi.org/10.1371/journal.pone.0257423.
24. Nagesh S, Chakraborty S. Saving the frontline health workforce amidst the COVID-19 crisis: challenges and recommendations. *J Glob Health*. 2020;10(1), 010345. https://doi.org/10.7189/jogh-10-010345.
25. Hilgers L. *Nurses Have Finally Learned what They're Worth*. New Yourk Times; 2022 (Sunday Magazine) https://www.nytimes.com/2022/02/15/magazine/traveling-nurses.html.
26. Alsuliman T, Mouki A, Mohamad O. Prevalence of abuse against frontline health-care workers during the COVID-19 pandemic in low and middle-income countries. *East Mediterr Health J*. 2021;27(5):441–442. https://doi.org/10.26719/2021.27.5.441.
27. Bellizzi S, Pichierri G, Farina G, Cegolon L, Abdelbaki W. Violence against healthcare: a public health issue beyond conflict settings. *Am J Trop Med Hyg*. 2021;106(1):15–16. https://doi.org/10.4269/ajtmh.21-0979.
28. Chirico F, Nucera G, Magnavita N. COVID-19: protecting healthcare workers is a priority. *Infect Control Hosp Epidemiol*. 2020;41(9). https://doi.org/10.1017/ice.2020.148, 1117-1117.

29. Greenberg N, Docherty M, Gnanapragasam S, Wessely S. Managing mental health challenges faced by healthcare workers during Covid-19 pandemic. *BMJ*. March 26, 2020:m1211. https://doi.org/10.1136/bmj.m1211. Published online.
30. Osterhaus ADME, Vanlangendonck C, Barbeschi M, et al. Make science evolve into a One Health approach to improve health and security: a white paper. *One Health Outlook*. 2020;2(1):6. https://doi.org/10.1186/s42522-019-0009-7.
31. Lapinski MK, Funk JA, Moccia LT. Recommendations for the role of social science research in One Health. *Soc Sci Med*. 2015;129:51–60. https://doi.org/10.1016/j.socscimed.2014.09.048.
32. Michalon J. Accounting for One Health: insights from the social sciences. *Parasite*. 2020;27:56. https://doi.org/10.1051/parasite/2020056.
33. Szabo S, Nove A, Matthews Z, et al. Health workforce demography: a framework to improve understanding of the health workforce and support achievement of the Sustainable Development Goals. *Hum Resour Health*. 2020;18(1):7. https://doi.org/10.1186/s12960-020-0445-6.
34. Smith AC, Woerner J, Perera R, Haeny AM, Cox JM. An investigation of associations between race, ethnicity, and past experiences of discrimination with medical mistrust and COVID-19 protective strategies. *J Racial Ethn Health Disparities*. June 11, 2021. https://doi.org/10.1007/s40615-021-01080-x. Published online.
35. Bogaert K, Castrucci BC, Gould E, Rider N, Whang C, Corcoran E. Top training needs of the governmental public health workforce. *J Publ Health Manag Pract*. 2019;25(Suppl 2):S134–S144. https://doi.org/10.1097/PHH.0000000000000936. Public Health Workforce Interests and Needs Survey 2017.
36. Frenk J, Burke D, Spencer HC, et al. Improving global health education: development of a global health competency model. *Am J Trop Med Hyg*. 2014;90(3):560–565. https://doi.org/10.4269/ajtmh.13-0537.
37. Janson DJ, Clift BC, Dhokia V. PPE fit of healthcare workers during the COVID-19 pandemic. *Appl Ergon*. 2022;99, 103610. https://doi.org/10.1016/j.apergo.2021.103610.
38. Nuzzo JB, Meyer D, Snyder M, et al. What makes health systems resilient against infectious disease outbreaks and natural hazards? Results from a scoping review. *BMC Publ Health*. 2019;19(1):1310. https://doi.org/10.1186/s12889-019-7707-z.
39. SB 1159, Hill. *Workers' Compensation: COVID-19: Critical Workers*; 2020. https://leginfo.legislature.ca.gov/faces/billNavClient.xhtml?bill_id=201920200SB1159. Accessed May 4, 2022.
40. Mboussou F, Ndumbi P, Ngom R, et al. Infectious disease outbreaks in the African region: overview of events reported to the World Health Organization in 2018. *Epidemiol Infect*. 2019;147:e299. https://doi.org/10.1017/S0950268819001912.
41. Talisuna AO, Okiro EA, Yahaya AA, et al. Spatial and temporal distribution of infectious disease epidemics, disasters and other potential public health emergencies in the World Health Organisation Africa region, 2016-2018. *Glob Health*. 2020;16(1):9. https://doi.org/10.1186/s12992-019-0540-4.
42. Ahmat A, Okoroafor SC, Kazanga I, et al. The health workforce status in the WHO African Region: findings of a cross-sectional study. *BMJ Glob Health*. 2022;7(suppl 1), e008317. https://doi.org/10.1136/bmjgh-2021-008317.

CHAPTER

26

Health system preparedness and long-term benefits to achieve health security

Natalie Rhodes[1], *Garrett Wallace Brown*[1], *Luc Bertrand Tsachoua Choupe*[2], *Marc Ho*[3], *Stella Chungong*[4] *and Nirmal Kandel*[2]

[1]School of Politics and International Studies, University of Leeds, Leeds, United Kingdom; [2]Evidence and Analytics for Health Security, Geneva, Switzerland; [3]Ministry of Health, Singapore; [4]Health Security Preparedness Department, WHO Health Emergencies Programme, World Health Organization, Geneva, Switzerland

Introduction

All countries have been greatly impacted by the COVID-19 pandemic, yet there has been significant variability in readiness capacity,[1] disease burden,[2] and adopted responses.[3] It is this variability that offers a clear indication of the crucial relationship between health system capacities and effective health security response. In most countries, across high-, middle-, and low-income settings, health system deficiencies hampered effective response.[1–9] These included a lack of emergency planning and leadership,[1] poor health communication combined with limited population wide health literacy,[4] equipment shortages,[5,6] lack of sustainable health workforces,[7] disruptions to medical supply chains,[8] and incapacities to handle even limited increases in caseload frequency.[9] Other deficiencies have included poor information on health systems capacities and the ambiguous integration of private sector providers within national systems, resulting in often chaotic and ad hoc engagement with governments in national pandemic responses.[10] As such, the COVID-19 pandemic has underscored that there can be no genuine and resilient national or global health security without strong health systems characterized by essential capacities and resources present at all levels.

Modernizing Global Health Security to Prevent, Detect, and Respond
https://doi.org/10.1016/B978-0-323-90945-7.00001-4

However, previous rhetoric surrounding the need to strengthen health systems for health security has not produced commensurate investments and has noticeably tailed off after many of the epidemic events of the last 2 decades. Consequently, the COVID-19 pandemic is a pivotal moment and opportunity for the world to break the cycle of "panic-and-forget" (mobilized threat response followed by waning commitments) and secure the full commitment of global, national, and subnational stakeholders for long-term investments in Health Systems for Health Security through an all-of-government and all-of-society approach. Effective and coordinated strengthening of health systems will in turn contribute to the strengthening of health security for better prevention, detection, response to, and recovery from public health events and threats.

The World Health Organization (WHO) defines health security as relating to "the activities required, both proactive and reactive, to minimize vulnerability to acute public health events that endanger the collective health of populations living across geographical regions and international boundaries."[11] A health system is defined by the WHO as "all the organizations, institutions, resources and people whose primary purpose is to improve health."[12] In 2007, the WHO published the Health Systems Building Blocks framework, which presented health systems in the form of six blocks (these are service delivery, health workforce, health information systems, access to essential medicines, financing, and leadership/governance).[12] The main purpose of the building blocks framework was to improve understanding of what a health system is by defining desirable interlocking essential health system attributes and what constitutes as health systems strengthening. Although the building blocks do not provide a comprehensive description of the real-life complexity of health systems, community health or interactions between and among the blocks,[12–14] the model remains to be widely used as a useful heuristic and is largely accessible with its ability to succinctly describe key elements of a health system.[15]

Despite a nascent evidence base and policy discourse highlighting the inextricable linkages between health systems and public health security, the conceptual and policy links remain largely under explored. A 2021 review of the existing global peer-reviewed literature linking health systems and health security found that only 56.8% of the research made explicit links between health systems and health security, suggesting a limited evidence base and insufficient logic for preparedness and health security policy.[16] Moreover, the review found a tendency within the literature to treat health security as a "state of exception" rather than immersed in routine health policy, and a tendency for security policy to be siloed and reactive, focusing mainly on acute health emergencies and response (33.82%) versus upstream determinants (1.17%) and system preparedness (5.54%). There have been some efforts to strengthen and invest in health systems toward one that is reliable and sustainable and achieves Universal Health Coverage (UHC) and also improves national and global health security,[17–20] yet these efforts have remained insufficient and need to be strengthened and built upon.

This chapter draws heavily upon a scoping review conducted by the University of Leeds and the WHO Evidence and Analytics for Health Security team as part of an ongoing collaboration on health systems for health security.[21] The chapter sets out to analyze the linkages and gaps made between health systems and health security within the peer reviewed literature alongside suggested explanations and implications. In response to identified gaps, the chapter introduces the new WHO Health Systems for Health Security framework (HSforHS) where health security and health system strengthening are inextricably linked, and where practical steps that stakeholders can take to put the framework into practice are articulated.

Literature review and gap analysis

Although there is a promising, yet underdeveloped, research base demonstrating how specific improvements and investments in health systems and health system building blocks can enhance long-term health security, there remains a key lacuna to the extent that health systems and health security are often treated as distinct areas of research, with health security often separated from overarching health and development discourses.

The prioritization of health security over health systems

There is a propensity in the literature for authors to favor a security framework over the examination of how health systems could complement health security strategies or vice versa, reflecting a tendency to treat systems and security largely as separate fields of study or to treat linkages as self-evident or implicit. Moreover, despite the interconnectedness of the building blocks, they are rarely applied as a systems' lens to investigate the promotion of health security.[16]

The failure to substantively express their mutual inclusivity and co-dependency is surprising, given that previous governance failures raised questions about the sustainability of disease-specific responses and their impact on health and development outcomes.[22] In fact, it has been suggested that the historically recognized poor HIV/AIDS response caused "a rhetorical shift away from the silo-based approaches to development aid ... calling for 'investing in (other areas of) development for health."[23] Similar sentiments were iterated in the aftermath of the West African Ebola epidemic, where there were calls for renewed commitments to international public health arrangements and commitments, namely, via the implementation of the International Health Regulations (IHR).[24] It is this post-Ebola "wake up call" where the substantial beginnings of a health systems-health security nexus emerged,[25] with the idea that, "the most effective systems are those that are in use every day and can be scaled up in an emergency."[18] Yet, despite this emerging discourse, health systems and health security have remained distinct areas of research, with health security often separated from overarching health and development discourses.[26]

This siloed way health systems and health security have been approached poses several challenges for national and global health governance and our thinking about financial investment. Central to these challenges is the difficulty of presenting a holistic long-term strategy that moves beyond ad hoc reactive agendas.[27] Indeed, the need to take a broader systems approach toward longer-term health security investment is made more urgent when considering the evolving costs and risks of COVID-19 and the failures in national and global health governance preparedness and response.[28–31] Health systems saw the importance of being able to meet surge demands, having flexibly to deploy resources to areas of greatest need and capacity to reduce disruptions to essential services. Moreover, countries demonstrated different needs in mobilizing resources effectively. For instance, low-and-middle income countries (LMICs) often needed support to quickly scale-up skilled and specialized workforces during the emergency.

On the other hand, in high-income countries (HICs) with advanced health systems, the main challenge was a capacity to manage a huge surge in demand for health services in a short period of time.[32] Lastly, almost all countries, irrespective of income, faced challenges in the procurement of medical logistics to meet the demand for such supplies during the pandemic.[32] In terms of efficiency, wider health system investments necessary to help countries fully meet their IHR capacities would have represented a fraction of the cost of COVID-19 response, while also helping to deliver better day-to-day public health outcomes.

The tendency to treat health security as preparation for exceptional situations

Health security is often discussed in terms of being an exceptional form of response versus a concept embedded within a wider public health continuum and/or health system strengthening approach. This "exceptionalism" means that recommendations for improving security capacities are often isolated to a single building block, a sub-set of a building block, or ignoring health systems altogether, with a preference to focus on the use of emergency powers.[33–35] It has been argued that existing approaches to health security are anathema to a health system approach, highlighting how traditional health security paradigms "continue to speak in the terms of costly 'countermeasures' versus prevention and health system strengthening."[27]

Even on occasions where health security is discussed in relation to an individual building block or to broader system capacities, exceptionalism is still present. In relation to access to essential medicines, Elbe's work on the "pharmaceuticalization" of health security was critical of how pandemics have the capacity to disrupt every aspect of society.[36] Yet, in addressing such health security crises, vaccines, and in particular their stockpiling, facilitated "a new logic of managing infectious diseases," that seeks to implement measures to control a pathogen's disruptive impact, rather than substantively address the cause of disease.[37] As a result, decision-makers often pursue policies that minimize disruptions and costs, versus long-term initiatives aimed to improve overall health system capacity to prevent, prepare, and respond to emerging threats.[38,39]

Exceptionalism is also reflected through a strong focus in the literature on public health surveillance, where a "strengthened" health system is often measured against its immediate capacity to detect, alert with early warning, respond to, and recover from emerging risks as opposed to measurements based on the system's capacity for disease risk management. Njeru et al.[40] argued that health security often creates siloed improvements mainly in public health public health surveillance systems, diminishing wider system benefits that could be gained via the implementation of new technology across a health system. With regards to finance and investment, an inability of security approaches to appropriately engage with context-specific obstacles and moderators has been identified, with such approaches overlooking particular system and service delivery-related factors such as geographic remoteness,[41] difficulties in the transfer of subsidies[42] or in the case of the West Africa Ebola outbreak, the lack of "a smooth transition from short-term humanitarian relief to longer-term development-oriented programs," indicating wider problems in leadership and governance.[18]

This understanding of health security as a "state of exception," framed too by the broader contexts as described in previous reports,[11] could explain why a significant proportion of the existing literature focuses on acute infectious disease or biohazards (as discussed further in next section). This also may explain inadequate COVID-19 readiness and response, particularly in HICs, where notions of health security took the form of a "once in a century" exceptionalism, with the assumption that threats would mostly affect the LMICs and that HICs capacities were sufficiently prepared, creating a condition of complacency.[43]

Tendency to focus on acute health emergencies

There is a history of health "securitizations," where a particular disease threat gains rapid prominence and coverage, only to be replaced by the next infectious disease risk—for example, the securitization of HIV/AIDS, pandemic influenza[39] the zika virus,[44] and Ebola.[45] Research in health security often focuses on acute health emergencies and much less on preparedness[46,47] and often makes only general arguments for the need to support and strengthen fragile health systems as well as to enhance global mechanisms to respond to public health events of international concern.[48] Where more detailed health system recommendations have been discussed to enhance health security, they tend to focus on one or two subsets of a WHO building block(s). For example, in terms of service delivery, health information and communication (HIC) systems during acute emergencies via the delivery of public health surveillance services are often discussed but not general issues related to the delivery of healthcare services during the emergency. In relation to emergency capacities, there is a large focus on appropriate information infrastructure, targeting regulations and capacities needed to collect, transfer, and analyze data effectively. Information functions are considered to be critical for public health preparedness,[49] namely for the prompt detection and containment of epidemic-prone diseases, yet the review of the literature found that broader day-to-day health system benefits are less mentioned.[16]

A policy implication of this disease-specific "securitization" (declaring a disease an existential security threat requiring advanced resource mobilization to combat that threat) is that it can draw attention away from the important role of primary care and other what might be considered "mundane" attributes of health system performance in the provision of a strengthened health system.[18,50] In reality, UHC and public health security are complementary goals. Additionally, the focus on acute health emergencies results in the underrepresentation of NCDs. Reis and Cipolla[51] have stressed the imbalance existing between the priority given to epidemics, bioterrorism, and climate change by international health security initiatives and the fact that today the leading causes of mortality worldwide are noninfectious, inviting authorities to address broader determinants of health. Human health security therefore requires a more holistic and systems approach that reflects both acute and long-term health risks.

Implications of these gaps in literature

The scoping review revealed that health systems and health security are often treated as distinctive and separate domains, instead of being two sides of the same coin. Although core health system capacities to enhance health security are often singled out, such as laboratory capacity and surveillance, these investigations often remained siloed within the literature, diminishing complementary interconnections to wider health system functions or goals. In cases where connections are made in the literature, there is a tendency to prioritize security goals, or to focus on the examination of individual security elements, giving far less coverage to the examination of overall health system improvements as being a crucial part of a health security approach.[16]

As a result, there exists a need to fill evidentiary gaps about how investments in health system capacity building can generate returns on better long-term health security. For example, research should seek to explore what kinds of investments and/or programs can reliably lead to maximal gains from both a health system and a health security perspective. These should be accompanied by investments in health emergency preparedness, supported by investment cases, and the incorporation of lessons from the pandemic, such as the important role of community resilience and preparedness, increasing access to healthcare, and gender mainstreaming in a spirit of continual improvement and building back better for the future.

As a means to address the gaps identified, and in the wake of recent major outbreaks, including the COVID-19 pandemic, the increasing interest for "health system for health security" related-matters is an opportunity that can be built-on to support further research in this area. In particular, there is promising scope to better employ the WHO building blocks as a heuristic for strengthening health systems toward a more holistic, whole-of-society, and sustainable condition of health security. Yet, as this chapter argues, the evidence required to fully operationalize such a framework remains nascent, thus exposing key research and policy gaps that require additional attention. It is in response to this need that the WHO and partners have designed and launched the new Health Systems for Health Security framework and its associated benchmarks and toolkits outlined below.

The need for such a framework and its potential benefits as part of a preparedness versus reactive strategy can be demonstrated by The Kingdom of Thailand's health emergency activities described in the case study below and COVID-19 response. This is because Thailand's response to COVID-19 underscores three crucial aspects of the health system/health security nexus. First, the case study demonstrates how prior preparedness efforts involving Village Health Volunteers (VHV) and their training and integration into the health system allowed improved health emergency response. Second, having other health security responses embedded within the health system more widely supported a rapid deployment of public health responses. Third, Thailand's reliance on VHVs for critical frontline response highlighted key intersections between health financing, human resources, and program sustainability, since the pandemic exposed recruiting and renumeration challenges in maintaining VHV personnel. As a result, Thailand's experience helps to demonstrate why the new WHO HSforHS framework fills an important policy gap by making the interconnections between health systems and health security preparedness more explicit in addition to outlining what is required to strengthen these links for better preparedness policies.

CASE STUDY

The Kingdom of Thailand's response to the COVID-19 pandemic:

Problem and background

- Since January 1, 2020, there have been over 761 million confirmed cases of COVID-19 globally (WHO COVID-19 dashboard). The pandemic has affected every country in the world, either directly or indirectly. Health systems have been pushed beyond their capacity limits, with the pandemic exposing gaps in public health responses and the provision of routine clinical care.[52]
- The Kingdom of Thailand was the first country outside of China to detect a case of COVID-19 in early January 2020. Over 3 years later, Thailand sits well below the global average of 9764 cumulative cases of COVID-19 per 100,000 population with 6774 cases. Countries such as the United Kingdom, for example, have a cumulative count of over 36,000 cases per 100,000 population.
- Village health volunteers (VHV) have long played an essential role in primary health care in Thailand, through maintaining health records, performing disease prevention campaigns, and providing health education to their local communities.[53]

Observations and key issues

- There are over one million village health volunteers in Thailand, who assist subdistrict health centers to provide primary healthcare in over 75,000 villages across the country.[54] VHVs have a long history, being well established for over 40 years in the country, and are managed and financed through the Ministry of Public Health. VHVs are volunteers from the community they serve and therefore act as liaisons between the community and the health system.
- In March 2020, the Thai Government closed all schools and all public and business areas in Bangkok.[55] Travel within the country was allowed, along with the return of Thai citizens from abroad, for workers to return to their hometowns. This resulted in a large movement of people, with estimates of 70,000–80,000 people traveling per day at the peak of this time.
- As March 2020 was still very early in the pandemic, with no vaccine yet developed, public health and social measures were the foundation of pandemic response.

Interventions

- Early in the pandemic VHVs were provided with additional training on the COVID-19 virus and provided with personal protective equipment allowing them to provide COVID-19 screening and contact tracing by going door-to-door in their communities.[56,57] They were able to provide community surveillance, COVID-19 education, home quarantine support, and assist with the management of chronic health conditions in the community.
- In anticipation of the large movement of people in response to public health and social measures in Bangkok, the Thai

CASE STUDY (cont'd)

Government formed surveillance teams of VHVs to monitor returnees for signs of infection and to encourage self-quarantine.

- In the first 18 months of the pandemic, almost 600,000 VHVs participated in COVID-19 responses.[54]

Lessons learned

What worked

- VHVs were in a distinct position to rapidly assist with the COVID-19 response through being connected to their communities and already embedded with the health system. This allowed for a rapid dissemination of COVID-19 public health and social measures that could be aptly adapted to local community dynamics and shared through trusted sources.
- VHVs are trained on health education, promotion, disease prevention, and basic medical treatment. Since contributing to community surveillance in response to Avian Influenza in 2004, basic infectious disease surveillance has also been included in VHV training.[58]
- Having health security responses embedded within the health system supported a rapid deployment of public health responses.

What did not work?

- Over 50% of VHVs in Thailand are over the age of 50 years,[54] considerations for their health resulted in just over half of all VHVs participating in COVID-19 responses.
- Renumeration for VHVs was increased during the pandemic in recognition of critical involvement; this included an increase in the monthly salary alongside nonfinancial incentives such as national awards and recognition.[59] This was an additional cost to the government and suggests consideration for better renumeration going forward for essential VHVs.

Recommendations

- In Thailand, village health workers within the primary healthcare system have been credited with playing a crucial role in the country's COVID-19 response.
- Increased renumerations for VHVs should be investigated.
- Other countries with Village Health Workers should consider training and incorporating them into disease surveillance and control activities.
- This demonstrates the role that an established, functioning health system can play in a health emergency response with workforce, training. and community care rapidly able to be deployed.
- This example clearly illustrates the imperative role health systems play in health security, and how health security is reliant on resilient health systems.

Vision for the future

As stated by Dr. Tedros Adhanom Ghebreyesus, the Director-General of WHO, "quality health systems not only improve health outcomes in "peacetime," they are also a bulwark against outbreaks and other public health emergencies. UHC and health security are two sides of the same coin."[60]

The key elements of public health security and health systems are well-established, what is needed, as demonstrated in the example of Thailand above, is the elucidation of the linkages between the two. Building and enhancing these linkages involve a complex set of conceptual and practical issues for countries, WHO, and partners. There is need for a clear common narrative and well-defined framework to build resilient and responsive health systems that can surge and transform to meet the demands imposed by health emergencies. While this is no insignificant task, many of the tools needed to do this already exist.

There are a number of key aspects that can be drawn upon to construct such a framework, including lessons learned from case studies such as Thailand. These include: the IHR capacities; additional components from health systems; and components from other sectors that form critical dependencies with health and that strengthen health systems for health security.[61] These categories and their interlinkages are illustrated in Fig. 26.1.

In addition to the building block model introduced earlier in the chapter, a number of key notions must be taken into account to have a full and comprehensive view of health systems

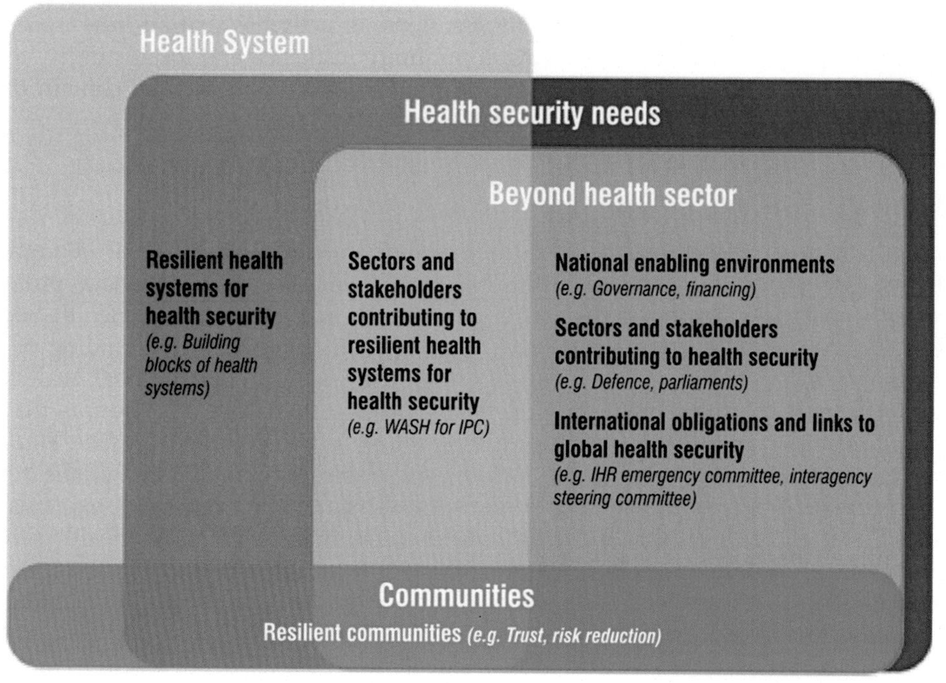

FIGURE 26.1 Components of the health systems for health security approach and interlinkages between one another.[32]

capacities, notably: Common Goods for Health (CGH),[62–64] Essential public health functions (EPHFs),[65] and Primary Health Care (PHC).[66,67] In particular, adopting a PHC approach is key to building strong and resilient health systems for health security. Although prevention, detection, response to, and recovery from health emergencies involves all levels of health system, it fundamentally begins and involves local communities. A PHC orientation of health systems and the systematic integration of emergency risk management within it can provide the essential foundations for both UHC and health security.

Beyond components described above, and as illustrated by the Thailand case study, additional capacities from other sectors are required to ensure a true whole-of-society approach for global health security. This imperative has been demonstrated by the COVID-19 pandemic, which principally hit vulnerable persons arising from preventable risk factors, economics, and social determinants.[68] Societies form a complex adaptive system, with all interconnected parts and change in any part of the system has reverberations across the system as a whole.[69] To fully engage in sustainable health security, there is a need to go beyond the health sector[70] and into the full scope of upstream determinants and actions needed to sustainably provide health systems for health security. This includes the involvement of other sectors that support health systems, in particular service delivery and adequate workforce,[71] and those which play a key role in human disease outbreaks interfacing with the agricultural and environmental sector in the context of a One Health approach[72] and the development of interventions such as the pharmaceutical sector in research and development but also in supply management and ensuring global access (particularly relevant in the case of the COVID-19 pandemic).[73] Finally, community engagement is essential for health security. Inclusive participation of local people in projects, interventions, or activities that address issues that affect their well-being is critical to building community resilience and local capacity to prevent, detect, respond to, and recover from health emergencies and thereby contain threats at their source.[74]

WHO Health Systems for Health Security framework

Through harnessing the tools and features outlined above with the goal of improved health security, responsive and resilient health systems, and social and financial protection with improved efficiency, the WHO has created the Health Systems for Health Security framework (HSforHS). The framework aims to promote a common understanding of what health systems for health security entails and how it contributes to better global health security, while delineating the essential components of health systems and other sectors that play an important role in meeting the demands imposed by health emergencies (See Fig. 26.2). Further, alongside a set of existing and new benchmarks and toolkits (e.g., the revised IHRs and a new Dynamic Preparedness Matrix Dashboard), the framework explains how countries can define, prioritize, and monitor actions and investments in health security, health systems, and other sectors for multisectoral and multidisciplinary management of health emergencies toward better global health security. This in turn will help partners and donors better support countries in strengthening health security by identifying where more investment in health systems is most needed, how best to do so, and how financing can be sustained. While the framework itself may seem simple, it is necessary to recognize

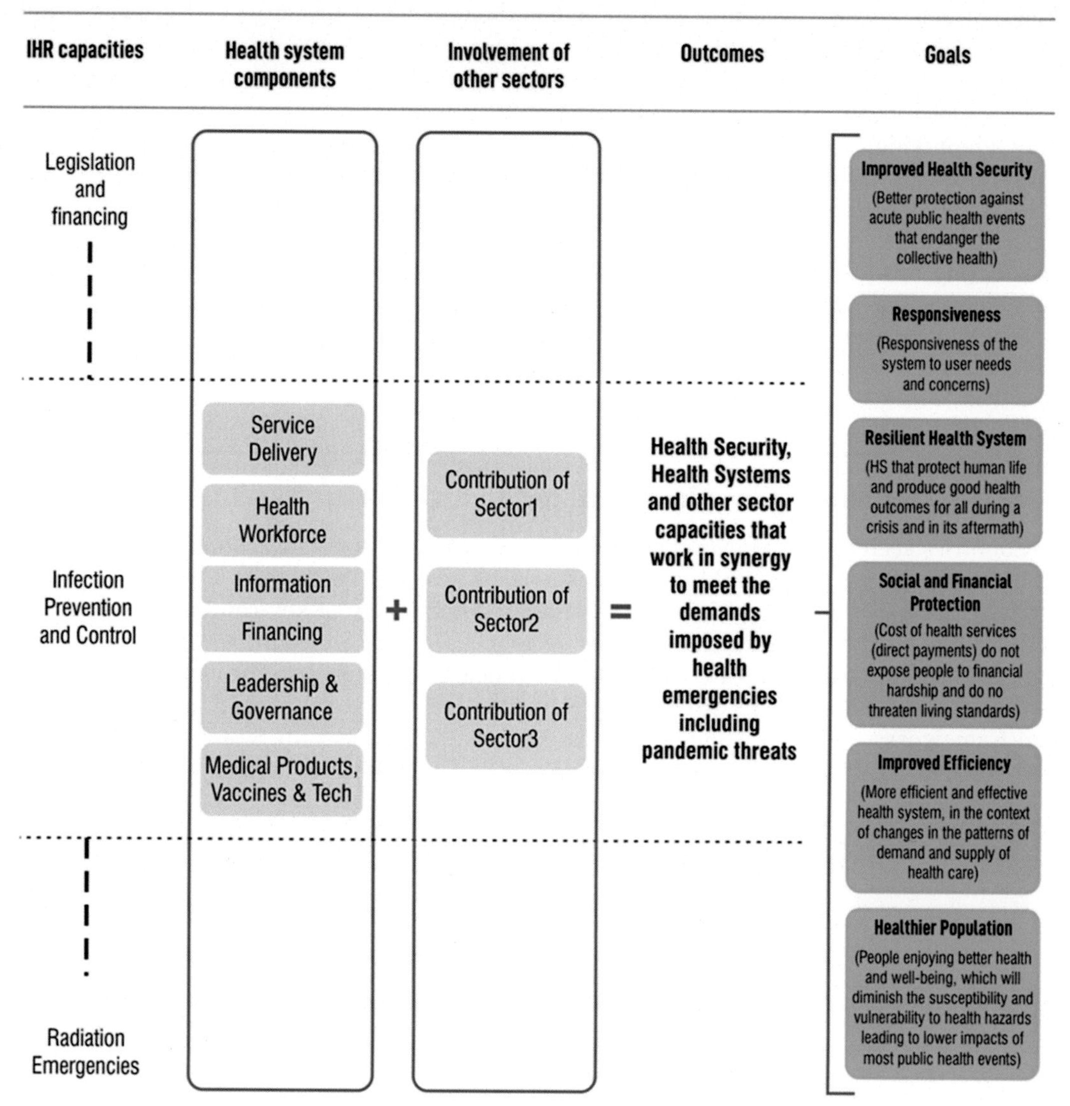

FIGURE 26.2 Building health systems for health security capacities to meet the demands imposed by health emergencies.[32]

the complexities and highlight challenges related to implementation of health systems for health security thus allowing more effective implementation of the framework.

As a result, there will be greater awareness of the importance of building health systems for health security. More synergistic working relationships will exist between health security, health systems, and other sectors for multisectoral and multidisciplinary management of health emergencies. Importantly, there will be increased investment in health systems for both the day-to-day service delivery (thus achieving UHC) as well as long-term health security by preventing, detecting, and quickly mitigating the occurrence and impact of health emergencies. Leveraging health systems for health security consist of developing, strengthening, and maintaining IHR capacities, components of health system, as well as from other sectors, which health systems are dependent on, to advance health security.

By identifying capacities in health systems and other sectors that contribute to health security, this framework and its subsequent products would help countries and other stakeholders better understand the research and policy gaps identified earlier in the chapter and thus more effectively invest in health systems for health security. The returns on investment of adopting this all-hazards, multisectoral, and multidisciplinary prevention and preparedness approach are aimed to have wider routine benefits across all sectors of society as well as facilitate long-term health security.

Actions

Four steps for building health systems for health security

Countries keen to move beyond a conceptual approach to concrete actions for Health Systems for Health Security should take four key steps (See Fig. 26.3).

First, assess existing capacities for IHR and the current state of key components in health systems (the six building blocks) and other sectors. This will help countries to identify existing gaps, which may hamper the management of health emergencies. The assessment of IHR capacities and health systems can be done using the IHR MEF tools, as well as the health system frameworks and their suite of associated tools such as Health Systems Assessments. Moreover, the WHO is currently designing a new Dynamic Preparedness Matrix (DPM),

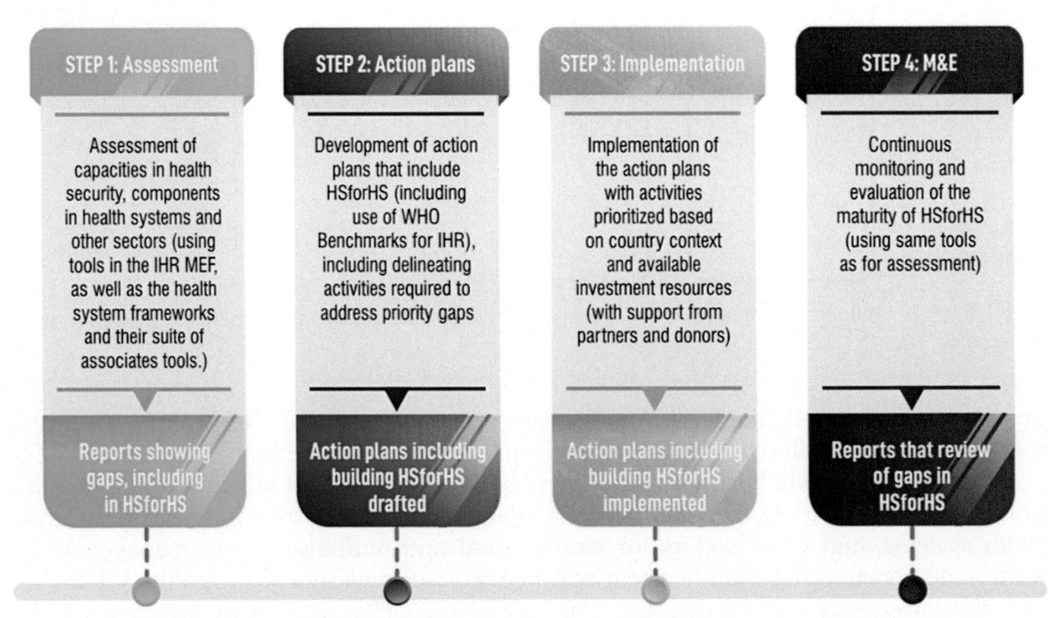

FIGURE 26.3 Four steps for building health systems for health security.[32]

an online dashboard that allows countries to routinely upload data to help assess current hazards, vulnerabilities, and capacities. The matrix allows for the generation of real-time gap scores, which can assist in the identification of preparedness priorities.

Second, the shortcomings identified should be rectified by developing comprehensive action plans that address gaps in health systems for health security, including through actions plans for IHR or health security and National Health Sector Strategic Plans. The plans should delineate actions and activities required to address essential missing components of health security, health systems, and other sectors through appropriate resources, capacities, and organizational systems that can work synergistically (rather than in parallel) to meet the demands imposed by health emergencies. A key component of the HSforHS framework and its associated materials will be the provision of trainings and related guidance tools to assist countries to locate appropriate measures to best integrate HSforHS principles into contextualized long-term health preparedness efforts.

Third, countries should implement planned activities for development of HSforHS capacities, resources, and organizational systems, while addressing gaps identified. Activities should be prioritized based on each country's context and available resources for investment and can be gleaned from suggested actions in the WHO Benchmarks for IHR. Additionally, partner agencies and donors should be engaged to support countries in implementation including allocating funds where more investment is most needed. As part of the support materials associated with HSforHS, guidance on identifying investment cases will be available, again assisting countries to capture sufficient health security returns on investment via system and multisectoral strengthening. In addition, the HSforHS framework stresses the need to locate crucial interconnections between sector partners and to pre-identify linkages required for sufficient health emergency preparedness and response. This includes the permanent adoption of sectoral focal points, establishing bi-annual multisectoral preparedness committees, and applying the whole-of-society lessons learned from the recent pandemic.

Finally, with implementation, the maturation of HSforHS over time should be continuously monitored and evaluated and be aligned with the tools used for gap assessment and any identified challenges in implementation. This is done with the aim of ever-improving efficient and effective management of health emergencies. To be clear, the longer-term reforms associated with the HSforHS framework do not, and should not, preclude more immediate health security measures. They are not mutually exclusive. Instead, the ultimate goal of HSforHS is to fundamentally challenge existing health security paradigms as identified in the first half of this chapter. In other words, to alter our thinking about what strengthening efforts are required for sustainable and sufficient health security. As COVID-19 painfully exposed, current policies were inadequately prepared, resulting in substantial social, economic, and human costs. In this context, the HSforHS framework provides an important response.

Prioritizing investment using a maturity model in the WHO benchmarks for IHR

The development of capacities for strengthening of health systems for health security while addressing the challenges identified (through assessment process) should be guided by a maturity model. This offers countries a conceptual representation of graduated actions to be implemented for scaling up health emergencies management capacities, starting from their current state. The maturity model for HSforHS is aligned with that presented in the WHO Benchmarks for IHR capacities.[60] The benchmark document describes key actions and attributes from all 22 IHR capacities technical areas and provides a roadmap of suggested actions that can be applied to build and strengthen IHR capacities, strong and resilient systems components, and other sector capacities.

These benchmarks serve three primary purposes in terms of strengthening health systems for health security and expanding investments in them. First, they provide a definition of desirable attributes—the actions required, in health security, health systems, and other sectors for health security at each level of the benchmark. Next, they provide a way of defining health systems for health security priorities for countries, development partners, and the WHO. Finally, they provide a useful way of clarifying essential actions that require a more integrated response and recognize the interdependence of each action in the benchmarks.

WHO is updating the WHO Benchmarks for IHR Capacities, along with experts from partner agencies and academics. The second edition of the benchmarks builds on lessons learned from recent and ongoing health emergencies including the COVID-19 pandemic (such as illustrated in the Thailand case study) and will follow the same approach as described in the HSforHS framework. The updated benchmarks will include suggested actions for strengthening capacities related to the IHR 2005, as well as components in health systems and other sectors.

Implementing health systems for health security at different levels in a country and across sectors

The maturity of the health system and its contributions to health security can vary within the same country, with health systems at different administrative, geographical, or federal levels showing different levels of maturity. As such, performance at the national level may not reflect that of remote communities or regions with very poor capacities to manage public health emergencies. Furthermore, urban settings, especially capital cities, often hold the highest capacities of health systems for health security in a country, supporting surrounding peri-urban and rural regions. It is thus essential to account for geographical and community disparities. Planning for strengthening of HSforHS therefore needs to be done not just at national level but also at subnational and supranational levels, with relevant priority actions selected to address different types of gaps at each level (See Fig. 26.4).

A safer, healthier, sustainable way forward

By identifying components in health systems and other sectors that contribute to health security, this framework and its subsequent products (including scientific papers, crash-courses and training, dataset, etc.) would help countries and other stakeholders to better understand and more effectively invest in health systems for health security. The returns on investment of adopting this all-hazards, multisectoral, and multidisciplinary prevention and preparedness approach will have wide benefits across all sectors of society while promoting sustainable health security.

Supranational Level

1. Global and regional framework, guidance and standards
2. Global and regional coordination mechanism for preparedness and response
3. External support, strategic partnership and collaboration
4. Provision of knowledge, skill and resources

National Level

1. Legislation, policies and strategies
2. All sectoral functional coordination and partnerships
3. Defines priorities, developing plans and resource mobilization
4. Contingency planning and resource allocation for emergencies
5. Specialized care, training of health care workers and distribution
6. Development of risk communication strategies and dissemination
7. Logistic management and distribution

Intermediate Level

1. Trained health workers (surveillance with access to specialized care and facilities) and their training
2. Multisectoral coordination and resources/information sharing
3. Information management and dissemination
4. Laboratory testing, facilitation and referral
5. Development and access to risk comm. materials, training and dissemination
6. Logistic management and distribution (vaccine, drugs, equipment)

Community Level

1. A trained health worker (surveillance guidelines, case management of priority diseases and/or referral)
2. Access to reporting (early warning and IT tools)
3. Specimen collection and referral (access to outbreak investigation kits and transportation)
4. Risk communication to community (social mobilization, IEC materials, community engagement)
5. Access to minimum WASH, IPC provision and logistics
6. Availability of vaccines and drugs for local endemic diseases

FIGURE 26.4 Levels of application of health systems for health security.[32]

References

1. Kandel N, Chungong S, Omaar A, Xing J. Health security capacities in the context of COVID-19 outbreak: an analysis of International Health Regulations annual report data from 182 countries. *Lancet Lond Engl.* 2020;395(10229):1047–1053. https://doi.org/10.1016/S0140-6736(20)30553-5.
2. Assefa Y, Gilks CF, Reid S, et al. Analysis of the COVID-19 pandemic: lessons towards a more effective response to public health emergencies. *Glob Health.* 2022;18:10. https://doi.org/10.1186/s12992-022-00805-9.
3. Sturmberg J, Paul E, Van Damme W, Ridde V, Brown GW, Kalk A. The danger of the single storyline obfuscating the complexities of managing SARS-CoV-2/COVID-19. *J Eval Clin Pract.* 2021:1–14. https://doi.org/10.1111/jep.13640.
4. Paakkari L, Okan O. COVID-19: health literacy is an underestimated problem. *Lancet Publ Health.* 2020;5(5):e249–e250. https://doi.org/10.1016/S2468-2667(20)30086-4.
5. Doctors Say Shortage of Protective Gear Is Dire During Coronavirus Pandemic - The New York Times. Accessed September 21, 2021. https://www.nytimes.com/2020/03/19/health/coronavirus-masks-shortage.html.
6. Ranney ML, Griffeth V, Jha AK. Critical supply shortages — the need for ventilators and personal protective equipment during the Covid-19 pandemic. *N Engl J Med.* 2020;382(18):e41. https://doi.org/10.1056/NEJMp2006141.
7. Bourgeault IL, Maier CB, Dieleman M, et al. The COVID-19 pandemic presents an opportunity to develop more sustainable health workforces. *Hum Resour Health.* December 2020;18(1):1–8.
8. COVID-19 update: coronavirus and the pharmaceutical supply chain. Accessed September 21, 2021. https://www.europeanpharmaceuticalreview.com/article/116145/covid-19-update-coronavirus-and-the-pharmaceutical-supply-chain/.
9. Healthcare resource statistics - beds. Accessed September 21, 2021. https://ec.europa.eu/eurostat/statistics-explained/index.php?title=Healthcare_resource_statistics_-_beds.
10. Williams OD. COVID-19 and private health: market and governance failure. *Development.* 2020;63(2):181–190. https://doi.org/10.1057/s41301-020-00273-x.
11. World Health Organization. *The World Health Report 2007 : A Safer Future : Global Public Health Security in the 21st Century.* World Health Organization; 2007. https://apps.who.int/iris/handle/10665/43713. Accessed September 21, 2021.
12. World Health Organization. *Everybody's Business — Strengthening Health Systems to Improve Health Outcomes : WHO's Framework for Action.* World Health Organization; 2007. https://apps.who.int/iris/handle/10665/43918. Accessed September 21, 2021.
13. Huff-Rousselle M. Reflections on the frameworks we use to capture complex and dynamic health sector issues. *Int J Health Plann Manag.* 2013;28(1):95–101.
14. Sacks E, Morrow M, Story WT, et al. Beyond the building blocks: integrating community roles into health systems frameworks to achieve health for all. *BMJ Glob Health.* 2019;3(Suppl 3):e001384. https://doi.org/10.1136/bmjgh-2018-001384.
15. Mounier-Jack S, Griffiths UK, Closser S, Burchett H, Marchal B. Measuring the health systems impact of disease control programmes: a critical reflection on the WHO building blocks framework. *BMC Publ Health.* 2014;14:278. https://doi.org/10.1186/1471-2458-14-278.
16. Brown GW, Bridge G, Martini J, et al. The role of health systems for health security: a scoping review revealing the need for improved conceptual and practical linkages. *Glob Health.* 2022;18(1):1–17.
17. Kluge H, Martín-Moreno JM, Emiroglu N, et al. Strengthening global health security by embedding the International Health Regulations requirements into national health systems. *BMJ Glob Health.* 2018;3(Suppl 1):e000656. https://doi.org/10.1136/bmjgh-2017-000656.
18. Heymann DL, Chen L, Takemi K, et al. Global health security: the wider lessons from the West African Ebola virus disease epidemic. *Lancet.* 2015;385(9980):1884–1901. https://doi.org/10.1016/S0140-6736(15)60858-3.
19. Ayanore MA, Amuna N, Aviisah M, et al. Towards resilient health systems in sub-saharan Africa: a systematic review of the English language literature on health workforce, surveillance, and health governance issues for health systems strengthening. *Ann Glob Health.* 2019;85(1):113. https://doi.org/10.5334/aogh.2514.
20. Erondu NA, Martin J, Marten R, Ooms G, Yates R, Heymann DL. Building the case for embedding global health security into universal health coverage: a proposal for a unified health system that includes public health. *Lancet Lond Engl.* 2018;392(10156):1482–1486. https://doi.org/10.1016/S0140-6736(18)32332-8.

21. University of Leeds. Leeds chosen by WHO to help create global health strategy. Accessed February 17, 2022. https://www.leeds.ac.uk/news-global/news/article/4820/leeds-chosen-by-who-to-help-create-global-health-strategy.
22. Schneider K, Garrett L. The end of the era of generosity? Global health amid economic crisis. *Philos Ethics Humanit Med*. 2009;4(1):1–7.
23. Woodling M, Williams OD, Rushton S. New life in old frames: HIV, development and the "AIDS plus MDGs" approach. *Global Publ Health*. 2012;7(sup2):S144–S158. https://doi.org/10.1080/17441692.2012.728238.
24. Fidler D. *Epic Failure of Ebola and Global Health Security. 21 Brown J World Aff 179 SprSum*; 2015. Published online January 1, 2015 https://www.repository.law.indiana.edu/facpub/2139.
25. Brown GW. *The 2015 G7 Summit: A Missed Opportunity for Global Health Leadership*. 2015:9. Published online.
26. Mackintosh M, Mugwagwa J, Banda G, et al. Health-industry linkages for local health: reframing policies for African health system strengthening. *Health Pol Plann*. 2018;33(4):602–610. https://doi.org/10.1093/heapol/czy022.
27. Paul E, Brown GW, Ridde V. COVID-19: time for paradigm shift in the nexus between local, national and global health. *BMJ Glob Health*. 2020;5(4):e002622.
28. Global Economic Prospects. World Bank. Accessed January 10, 2021. https://www.worldbank.org/en/publication/global-economic-prospects.
29. A World in Disorder. Accessed September 21, 2021. https://www.gpmb.org/annual-reports/overview/item/2020-a-world-in-disorder.
30. Global economic bailout is running at $19.5 trillion. It will go higher - CNN. Accessed January 11, 2021. https://edition.cnn.com/2020/11/17/economy/global-economy-coronavirus-bailout-imf-annual-report/index.html.
31. Weekly epidemiological update on COVID-19-27 July 2021. Accessed August 16, 2021. https://www.who.int/publications/m/item/weekly-epidemiological-update-on-covid-19—-27-july-2021.
32. World Health Organization. Health Systems for Health Security: A Framework for Developing Capacities for International Health Regulations, and Components in Health Systems and Other Sectors That Work in Synergy to Meet the Demands Imposed by Health Emergencies.; 2021. Accessed September 21, 2021. https://www.who.int/publications/i/item/9789240029682.
33. Aginam O. Globalization of health insecurity: the world health organization and the new international health regulations. *Med Law*. 2006;25(4):663–672.
34. Ipe M, Raghu TS, Vinze A. Information intermediaries for emergency preparedness and response: a case study from public health. *Inf Syst Front*. 2010;12(1):67–79. https://doi.org/10.1007/s10796-009-9162-3.
35. Rushton S. Global health security: security for whom? security from what? *Polit Stud*. 2011;59(4):779–796. https://doi.org/10.1111/j.1467-9248.2011.00919.x.
36. Elbe S. The pharmaceuticalisation of security: molecular biomedicine, antiviral stockpiles, and global health security. *Rev Int Stud*. 2014;40(5):919–938. https://doi.org/10.1017/S0260210514000151.
37. Elbe S, Roemer-Mahler A, Long C. Medical countermeasures for national security: a new government role in the pharmaceuticalization of society. *Soc Sci Med*. 2015;131:263–271.
38. Hameiri S. Avian influenza, "viral sovereignty", and the politics of health security in Indonesia. *Pac Rev*. 2014;27(3):333–356. https://doi.org/10.1080/09512748.2014.909523.
39. Kamradt-Scott A, McInnes C. The securitisation of pandemic influenza: framing, security and public policy. *Global Publ Health*. 2012;7(Suppl 2):S95–S110. https://doi.org/10.1080/17441692.2012.725752.
40. Njeru I, Kareko D, Kisangau N, et al. Use of technology for public health public health surveillancereporting: opportunities, challenges and lessons learnt from Kenya. *BMC Publ Health*. 2020;20(1):1–11.
41. Garcia-Subirats I, Vargas I, Mogollón-Pérez AS, et al. Barriers in access to healthcare in countries with different health systems. A cross-sectional study in municipalities of Central Colombia and North-Eastern Brazil. *Soc Sci Med*. 2014;106:204–213. https://doi.org/10.1016/j.socscimed.2014.01.054.
42. Kolie D, Delamou A, van de PR, et al. "Never let a crisis go to waste": post-Ebola agenda-setting for health system strengthening in Guinea. *BMJ Glob Health*. 2019;4(6):e001925. https://doi.org/10.1136/bmjgh-2019-001925.
43. Lincoln M. Study the role of hubris in nations' COVID-19 response. *Nature*. 2020;585(7825):325. https://doi.org/10.1038/d41586-020-02596-8.
44. Gostin LO, Hodge JG. Zika virus and global health security. *Lancet Infect Dis*. 2016;16(10):1099–1100. https://doi.org/10.1016/S1473-3099(16)30332-2.

45. Honigsbaum M. Between securitisation and neglect: managing Ebola at the borders of global health. *Med Hist*. 2017;61(2):270–294. https://doi.org/10.1017/mdh.2017.6.
46. Khan Y, O'Sullivan T, Brown A, et al. Public health emergency preparedness: a framework to promote resilience. *BMC Publ Health*. 2018;18(1):1344. https://doi.org/10.1186/s12889-018-6250-7.
47. Ravi SJ, Meyer D, Cameron E, Nalabandian M, Pervaiz B, Nuzzo JB. Establishing a theoretical foundation for measuring global health security: a scoping review. *BMC Publ Health*. 2019;19(1):954. https://doi.org/10.1186/s12889-019-7216-0.
48. Gostin LO. Ebola: towards an international health systems fund. *Lancet*. 2014;384(9951):e49–e51. https://doi.org/10.1016/S0140-6736(14)61345-3.
49. Stolka KB, Ngoyi BF, Grimes KEL, et al. Assessing the public health surveillanceSystem for priority zoonotic diseases in the democratic republic of the Congo, 2017. *Health Secur*. 2018;16(S1):S44–S53. https://doi.org/10.1089/hs.2018.0060.
50. Howell A. The global politics of medicine: beyond global health, against securitisation theory. *Rev Int Stud*. 2014;40(5):961–987.
51. Reis N, Cipolla J. The impact of systems of care on international health security. In: contemporary developments and perspectives in international health security-volume 1. *Intech*. 2020;1:47–68.
52. Lal A, Erondu NA, Heymann DL, Gitahi G, Yates R. Fragmented health systems in COVID-19: rectifying the misalignment between global health security and universal health coverage. *Lancet*. 2021;397:61–67.
53. World Health Organization. *Thailand's 1 Million Village Health Volunteers – "Unsung Heroes" – Are Helping Guard Communities Nationwide from COVID-19*; 2020. https://www.who.int/thailand/news/feature-stories/detail/thailands-1-million-village-health-volunteers-unsung-heroes-are-helping-guard-communities-nationwide-from-covid-19.
54. Krassanairawiwong T, Suvannit C, Pongpirul K, Tungsanga K. Roles of subdistrict health personnel and village health volunteers in Thailand during the COVID-19 pandemic. *BMJ Case Rep*. 2021;14:e244765.
55. Kaweenuttayanon N, Pattanarattanamolee R, Sorncha N, Nakahara S. Community Surveillance of COVID-19 by village health volunteers. *Thailand. Bull World Health Organ*. May 1, 2021;99(5):393–397.
56. Tejativaddhana P, Suriyawongpaisal W, Kasemsup V, Suksaroj T. The roles of village health volunteers: COVID-19 prevention and control in Thailand. *Asia Pac J Health Manag*. 2020;15(3):i477.
57. Bezbaruah S, Wallace P, Zakoji M, Perera WLSP, Kato M. Roles of community health workers in advancing health security and resilient health systems: emerging lessons from the COVID-19 response in South-East Asia Region. *WHO South-East Asia J Public Health*. 2021;10(suppl 1).
58. Nittayasoot N, Suphanchairmat R, Namwat C, Dejburum P, Tangcharoensathien V. Public health policies and health-care workers' response to the COVID-19 pandemic. *Thailand. Bull World Health Organ*. April 1, 2021;99(4):312–318.
59. Tangcharoensathien V, Vandelaer J, Brown R, Suphanchaimat R, Boonsuk P, Patcharanarumol W. Learning from pandemic responses: informating a resilient and equitable health system recovery in Thailand. Front. *Publ Health*. 2023:11.
60. Ghebreyesus TA. How could health care be anything other than high quality? *Lancet Glob Health*. 2018;6(11):e1140–e1141. https://doi.org/10.1016/S2214-109X(18)30394-2.
61. World Health Organization. *WHO Benchmarks for International Health Regulations (IHR) Capacities*. World Health Organization; 2019. https://apps.who.int/iris/handle/10665/311158. Accessed September 21, 2021.
62. Lo S, Gaudin S, Corvalan C, et al. The case for public financing of environmental common Goods for health. *Health Syst Reform*. 2019;5(4):366–381. https://doi.org/10.1080/23288604.2019.1669948.
63. Soucat A. Financing common Goods for health: fundamental for health, the foundation for UHC. *Health Syst Reform*. 2019;5(4):263–267. https://doi.org/10.1080/23288604.2019.1671125.
64. Gaudin S, Smith PC, Soucat A, Yazbeck AS. Common Goods for health: economic rationale and tools for prioritization. *Health Syst Reform*. 2019;5(4):280–292. https://doi.org/10.1080/23288604.2019.1656028.
65. World Health Organization. *Essential Public Health Functions, Health Systems and Health Security: Developing Conceptual Clarity and a WHO Roadmap for Action*. World Health Organization; 2018. https://apps.who.int/iris/handle/10665/272597. Accessed September 21, 2021.
66. World Health Organization. *Primary Health Care and Health Emergencies*. 2018.
67. Chan M. Primary health care as a route to health security. *Lancet*. 2009;373(9675):1586–1587. https://doi.org/10.1016/S0140-6736(09)60003-9.

68. Williamson EJ, Walker AJ, Bhaskaran K, et al. Factors associated with Covid-19-related death using OpenSAFELY. *Nature*. 2020;584(7821):430–436. https://doi.org/10.1038/s41586-020-2521-4.
69. Sturmberg JP, Martin CM. Covid-19 – how a pandemic reveals that everything is connected to everything else. *J Eval Clin Pract*. 2020;26(5):1361–1367. https://doi.org/10.1111/jep.13419.
70. WHO Commission on Macroeconomics and Health. *Working Group 2, Organization WH. Global Public Goods for Health : The Report of Working Group 2 of the Commission on Macroeconomics and Health*. World Health Organization; 2002. https://apps.who.int/iris/handle/10665/42518. Accessed September 21, 2021.
71. Jones DS, Dicker RC, Fontaine RE, et al. Building global epidemiology and response capacity with field epidemiology training programs. *Emerg Infect Dis*. 2017;23(Suppl 1):S158–S165. https://doi.org/10.3201/eid2313.170509.
72. Douphrate DI. Animal agriculture and the one health approach. *J Agromed*. 2021;26(1):85–87. https://doi.org/10.1080/1059924X.2021.1849136.
73. Wouters OJ, Shadlen KC, Salcher-Konrad M, et al. Challenges in ensuring global access to COVID-19 vaccines: production, affordability, allocation, and deployment. *Lancet*. 2021;397(10278):1023–1034. https://doi.org/10.1016/S0140-6736(21)00306-8.
74. Armstrong-Mensah EA, Ndiaye SM. Global health security agenda implementation: a case for community engagement. *Health Secur*. 2018;16(4):217–223. https://doi.org/10.1089/hs.2017.0097.

Further reading

1. World Health Organization. *International Health Regulations (2005): IHR Monitoring and Evaluation Framework*; 2018. https://www.who.int/publications-detail-redirect/international-health-regulations-(-2005)-ihr-monitoring-and-evaluation-framework. Accessed September 21, 2021.
2. World Health Organization. *NAPHS for All: A Country Implementation Guide for National Action Plan for Health Security (NAPHS)*. World Health Organization; 2019. https://apps.who.int/iris/handle/10665/312220. Accessed September 21, 2021.
3. IHR-PVS National Bridging Workshop | Strategic Partnership for Health Security and Emergency Preparedness (SPH) Portal. Accessed September 21, 2021. https://extranet.who.int/sph/ihr-pvs-bridging-workshop.

CHAPTER

27

Measuring progress of public health response and preparedness

Parker Choplin[1], Wondimagegnehu Alemu[2,3], Nomita Divi[4], Ngozi Erondu[5,6], Peter Mala[7] and Ann Marie Kimball[5,8]

[1]Rollins School of Public Health, Emory University, Atlanta, GA, United States; [2]International Health Consultancy, LLC, Decatur, GA, United States; [3]Rollins School of Public Health, Emory University, Decatur, GA, United States; [4]Ending Pandemics, San Francisco, CA, United States; [5]Chatham House, London, United Kingdom; [6]O'Neil Institute, Georgetown University, Washington, DC, United States; [7]World Health Organization, Geneva, Switzerland; [8]Department of Epidemiology, University of Washington, Bainbridge Island, WA, United States

Introduction

The increased frequency of severe emerging and re-emerging infectious pathogens such as severe acute respiratory syndrome (SARS)-2003,[1] Ebola West Africa and Democratic Republic of Congo,[2,3] Middle East Respiratory Syndrome (MERS),[4] Zika Virus,[5] and now the COVID-19 pandemic present substantial global public health risks. The broad global response included adopting the revised International Health Regulations (2005) following the 2003 SARS outbreak and the amendment of the IHR (2005) following the West African Ebola epidemic.[6,7] With the COVID-19 pandemic still surging through countries, and new variants posing greater threats,[8] public health professionals and global leaders are once again looking to revise the IHR (2005) and codify that within a broader "Pandemic Framework."[9]

According to the IHR (2005), countries are expected to build their national core capacities to be able to prevent, detect early, and respond timely to public health emergencies of international concerns (PHEIC).[10] Ideally infection should be contained within national borders. If a developing country requests, the World Health Assembly (WHA) 58.3 resolution urges countries to actively collaborate with each other and provide support to developing countries in building, strengthening, and maintaining these capacities. The WHA resolution also calls upon the World Health Organization (WHO) Director General to collaborate in resource

Modernizing Global Health Security to Prevent, Detect, and Respond
https://doi.org/10.1016/B978-0-323-90945-7.00011-7

mobilization to support developing countries with achieving the IHR core capacities, in addition to requiring countries to actively evaluate and report their progress toward these capacities to the WHA.[7]

Further, evaluating readiness of countries for early detection and rapid response to public health risks is an integral part of preparedness and response and a requirement of the IHR (2005).[10] Evaluations of preparedness and response capacities offer evidence-based information to identify strengths, limitations, gaps, and challenges as well as document progress over time.[11] These findings can be used to guide revision of national preparedness plans and readjust resource allocation with intention of enhancing prevention from, detection of, and response to public health emergencies both at national and sub-national levels.[12] Member States use a variety of tools such as self-assessment of IHR core capacities, simulation exercises, joint external evaluations, and after-action reviews as part of the evaluation of preparedness and response.[13]

Currently, countries have access to a handful of public health preparedness and response evaluation tools/indexes including the following: the Global Health Security Index (GHSI); the Joint External Evaluation (JEE); After Action Reviews (AAR); Intra-Action Reviews (IAR); Simulation Exercises (SimEx); the IHR State Party Annual Reporting (SPAR) tool; Timeliness Metrics; and RAND Corporation's Infectious Disease Vulnerability Index.

In earlier chapters the origin and implementation of the IHR Monitoring and Evaluation Framework (IHR MEF) by the WHO and partners have been detailed. At the heart of the International Health Regulations is the balance between interrupting traffic (travel and trade) and population safety during public health events. Following the revision in 2005, the WHO set up an ambitious monitoring framework to assist Member States in measuring and strengthening their health security to address pandemic threats. The IHR MEF consists of four assessment tools, three voluntary (JEE, AAR, and simulation exercises) and one mandatory (SPAR).[14] The JEE and SPAR tools, along with the Global Health Security Index, and Infectious Disease Vulnerability Index, measure public health preparedness and response capacities to ensure global health security or their progress toward IHR core capacities. Tools such as Simulation Exercises, After Action Review, Intra-Action Review, and Timeliness Metrics assess functionality and allow countries to measure performance of capacities in between public health emergencies, during and after public health events, like an outbreak, and effectively identify gaps and weaknesses in performance. These performance measurement tools and metrics are also useful for assessment of subnational capacities. However, as COVID-19 illustrated, countries can feel prepared, and even rank high in preparedness capacities, yet such novel pathogens, or unusual major public health emergencies can present characteristics that exploit gaps in fairly well-developed preparedness and response capacities. Under such circumstances a country's level of preparedness may not guarantee the same level of performance.

The global community is suitably alarmed by the course of the COVID-19 pandemic.[15] COVID-19 touched every inhabited continent of the planet within a year. Variants identified in one continent leapt to other continents within weeks.[16] While it is evident that the world was not adequately prepared for the pandemic, and the toll for that lack of preparedness has been disastrous; it is less apparent why initial detection of COVID-19 was not timely enough. Furthermore, to varying degrees, neither preventive nor response measures have been effective. As a result, rapid containment of the pandemic has failed in nearly every country including those with high preparedness and response capacity scores based on current metrics.

Literature review

The table below provides a snapshot of the eight preparedness tools discussed in this chapter. These tools are globally utilized by government and nongovernment organizations to assess a region or country's present ability to contain an epidemic and/or pandemic. For each tool, listed are the number of indicators, evaluation level, strengths, limitations, and utilization (Table 27.1).

The World Health Organization (WHO) developed the IHR Monitoring and Evaluation Framework to provide an overview of five approaches that countries can utilize to assess its core public health capacities.[14] The framework includes IHR State Party Self-Assessment Annual Report (SPAR), Joint External Evaluation, After Action Review, and Simulation Exercises, plus the new addition of the Intra-Action Reports.

Electronic State Party Annual Reporting tool (e-SPAR)

Under the IHR (2005), State Parties are required to report progress made toward the IHR core capacities.[21] Since 2010, this reporting has been done on an annual basis and is completed by the IHR national focal point (NFP). The "national focal point" is usually a person or an entity, who sits in the national public health institute or ministry of health. As one result of IHR Monitoring and Evaluation Framework development, reporting was converted to an online platform in 2019. This was a significant improvement in access. Through the online portal, State Parties of the WHO can report annual updates on their capacity requirements to the World Health Assembly (WHA). SPAR evaluates 35 indicators for 15 IHR capacities ranging from zoonotic events to human resources.[17] Ideally, countries should use a multisectoral approach, but because the IHR focal point sits in a health institute, this is not always the case. Most member states report to WHO on the progress of their core capacities.

Prior to COVID-19, the WHO reported that across all countries, there was progress in disease surveillance, laboratory capacity, and IHR coordination, and a need for more improvement in chemical events, capacities at points of entry, and radiation emergencies. The Joint External Evaluation (JEE) used SPAR as the baseline tool, though there is evidence that not all the indicators were well aligned.

Joint External Evaluation (JEE)

The JEE is a voluntary process a country can use to evaluate its capacity to "prevent, detect, and rapidly respond to public health threats"[22] and is conducted every 2–3 years using a peer-to-peer approach.[23] The tool measures a country's specific level and headway in achieving preparedness targets through four major domains: prevention, detection, response,

TABLE 27.1 Overview of preparedness and response preparedness evaluation tools. This table provides an overview of the tools discussed at length in the following section.

Evaluation tool	Indicators (#)	Evaluation level	Strengths	Limitations	Utilization
Electronic State Party Annual Reporting Tool (e-SPAR)	35[17]	National	Health systems resilience, cross-border coordination, subnational preparedness	Self-assessment and self-reporting	171 Member states
Joint External Evaluation	56	National	Reviews publicly available, peer-to-peer evaluation and has a multisectoral approach.	Findings are not standardized or weighted leaving potential for bias during the analysis.[18] Smaller states and federal government systems have had complications implementing all JEE processes.[18] Requires a lot of human resources.[18]	113 Member states
After Action Reviews	Not relevant	National or subnational, inter-country	Systematically reviews capabilities following an event. Inform national or regional preparedness and readiness plans.	Voluntary; not always performed or performed too long after an event occurs.	65 Member states
Intraction Reports	Not available	National and/sub-national	Adaptable toolkit, made communication between experts easier.	Not available	84 countries, generating 129 reports[19]
Simulation Exercises	Not available	Not available	Not available	Not available	169 total, 13 countries in 2021
Infectious Disease Vulnerability Index	7	National	Algorithm can be used to isolate a particular vulnerabilities or areas of greatest concern.[20]	Solely focuses on infectious diseases that have the potential for transnational spread. Not all data available.	Not available
Global Health Security Index	34	National	Scores for each sub-indicator are weighted.[13] adds three new competencies not assessed by JEE (health system resilience, compliance with international norms, risk environment).[13]	Uses open-source data for evaluation.[13] Potential issues with relevance and adaptability of certain indices.[13] Scoring across indicators is lacking consistency.[13]	195 countries

TABLE 27.1 Overview of preparedness and response preparedness evaluation tools. This table provides an overview of the tools discussed at length in the following section.—cont'd

Evaluation tool	Indicators (#)	Evaluation level	Strengths	Limitations	Utilization
Timeliness Metrics	11	National or subnational	Can be self- conducted. Performance metric. Timeliness milestones can be incorporated into surveillance forms to allow for regular capture of timeliness. Enables multisectoral engagement.	Not meant for cross country comparison.	Retrospectively used in 20 countries. Being built into surveillance to be used prospectively in 5 countries to date.

and "other IHR-related hazards and Points of Entry (PoE)."[22] As of September 2021, 113 countries implemented the first round of the JEE.[22] The first round of evaluations provided a baseline for countries' capacities and capabilities.[22] The second round and other future JEEs will assess progress in capacity development and sustainability.[22]

Advantages of the JEE include external experts' objectivity, along with paired learning and experience sharing that takes place between external and local experts during the exercise. However, this tool does not come without challenges. For instance, some countries are willing to use the JEE but have inadequate human resources to prioritize a voluntary review over other primary concerns.[18] Additionally, there are potential issues with understanding and implementing the JEE process until more subject experts join the country's team.[18] Further, researchers experienced complications with validating field visits in countries experiencing instability and conflict.[18] There is a potential need for a revised JEE tool that is applicable in small states or countries run by federal government systems in addition to the probability for subjective and biased JEE indices. Because of the way JEE findings are analyzed and presented, some researchers suggest incorporating a weighing application.[18] Finally, the score is applied at the national level and does not capture within-country variability of capacity.

After Action Review (AAR)

An AAR is both a voluntary qualitative review of actions taken after a public health event of concern and a means for documentation of challenges and best practices. Based on the context, it may follow four main formats: debriefing, key informant interviews, working group format, and mixed methods; each depends on the nature of events, local, and cultural context of the country conducting the AAR, the number of areas under review and the location of stakeholders. The AAR provides an opportunity to review the functional capacity of public health and emergency response systems for the identification of areas for improvement. Reviewers can translate the report findings into plans and actionable roadmaps for improvement. The WHO typically recommends that countries conduct an after-action review immediately or up to 3 months after the national declaration of the end of a public health event, to draw from the fresh memory and ensure an accurate and detailed portrayal of response actions. This timing builds off the momentum from ending the crisis. Outcomes of AARs are useful for updating National Action Plans for development of preparedness capacities and readiness. The AAR is the only tool or metric that systematically reviews functional capacities and capabilities following a real-life event. Unfortunately, a substantial portion of these reviews are conducted in low-middle income countries that lack the funding and resources to implement key recommendations.

Intra-Action Review (IAR)

The country COVID-19 Intra-action Review (IAR) is a process developed by WHO that was modeled after the WHO after-action review. Developed in April 2020, the IAR is used to guide countries to conduct periodic reviews of their national and subnational COVID-19 response in real time. These reviews enable countries to evaluate and learn from their current COVID-19 experience to allow for improvement in their response to COVID-19 in their country. The IAR is a country-led process that brings together various individuals with knowledge of the public health response pillars that are being assessed such as testing, intersectoral collaboration, public communication, and surveillance. The IAR allows for collective learning where experiences are shared, and challenges and strengths are identified. The WHO has created a toolkit consisting of various ready-to-use and customizable tools for easy adaptation during the intra-action review. The IAR can be conducted as a face-to-face format or via an online format, which is recommended particularly during high levels of community transmission. During the fourth and fifth meeting of the Emergency Committee under the International Health Regulations, convened by the WHO Director-General respectively on July 31st and October 30th′ 2020, countries were encouraged to share their best practices and conduct and share the results from their IARs with WHO.[24] AARs and IARs can strengthen readiness and resiliency to future challenges from COVID-19 and future outbreaks. As of August 15, 2022, 84 countries have conducted 129 COVID-19 IARs using or adapting the methodology proposed by WHO. Some countries have conducted multiple IARs.[19]

Simulation Exercises (SimEx)

Simulation exercises (SimEx) mimic real-life emergencies in an effort to enhance and validate preparedness and response techniques, plans, and systems.[25] SimEx comes in many varieties, from table-top exercises to drills and field/full scale events. Simulation exercises or "table top" exercises were extensively deployed during the Avian Influenza threat in South East Asia by the Mekong Basin Disease Surveillance network,[26] their utility led to their global adoption.

Infectious disease vulnerability index

RAND Corporation constructed the Infectious Disease Vulnerability Index in 2015. RAND uses an algorithm that stems from an extensive literature review analyzing infectious disease outbreaks, which ultimately produced seven domains: "demographic, health care, public health, disease dynamics, political-domestic, political-international, and economics."[20] The purpose of this assessment is to identify a country's ability to prevent and contain an infectious disease outbreak so that technology and funding can be more efficiently allocated to prevent future outbreaks.[20] Originally produced for use by the U.S. Department of Defense, Health and Human Services (HHS) and the international community, the algorithm runs on open-source data, primarily collected from the WHO or World Bank, and produces a vulnerability score.[20] Some countries do not have available data for all of the measurements in which case, the mean value of that particular measurement for a similar subgrouping of countries was used.[20] The RAND team uses a weight system to produce final countries' scores between 0 and 1.0.

The tool is currently not publicly available due to the method used to handle missing values, which resulted in the tool being less user-friendly.[20] But, for now, individuals can request permission from the authors and can change the weight system based on their own priorities and assumptions. An additional precaution, the results depend on the data quality, the strengths of publications used to generate the algorithm, and the authors' judgments.[20]

Global Health Security Index (GHSI)

In 2019, experts at the Nuclear Threat Initiative, Johns Hopkins Center for Health Security, and the Economist Intelligence Unit developed the GHS Index because of the increased probability of emerging pathogens and accidental or deliberate releases of one.[27] The GHS Index assesses a country's capacity to cope with infectious disease outbreaks and evaluates six core concepts: prevention, detection and reporting, response, health systems, compliance with norms, and risk of infectious disease outbreaks.[27] The GHS Index pulls data from publicly available reports and studies under the approach that a country is safer and more secure when its citizens and other countries are able to freely access information on a country's preparedness capacities to better prepare themselves.[27] Additionally, there are concerns regarding the tool's adaptability to low- and middle-income countries due to varying priorities.[13]

Timeliness Metrics

While the SPAR and JEE tools have proven valuable in identifying gaps and driving preparedness planning and investments, they focus on assessing country capabilities rather than performance. Timeliness, long recognized as a key element in successful prevention and control is not directly addressed in these efforts. A framework for assessing the timeliness of disease surveillance would complement such efforts by providing a set of quantitative performance measures for tracking progress routinely, consistently, and in real time.[28]

Timeliness is an important criterion for evaluating performance of infectious disease surveillance systems. Through the use of 11 clearly defined outbreak milestones, timeliness metrics can capture the speed of outbreak detection, verification, response, and other key actions across the timeline of an outbreak and evaluate progress over time.

National surveillance programs, international agencies, and donor organizations can use timeliness metrics to identify gaps in surveillance performance and track progress toward improved global health security. The metrics have been incorporated into WHOs GPW-13, Simulation guidance and After-Action Guidance and are also being used by various non-profits and national governments.

Observed disconnect between existing metrics and experience with COVID-19

The pandemic revealed some major gaps in our current regimen. While it is too early to do a complete "after action" review, given the global COVID-19 pandemic is ongoing, some critical questions are being raised about the validity of existing metrics and relevance of the IHR (2005) in the face of a fast moving and exponentially spreading virus such as COVID-19. According to the interim report of the IHR Review Committee (IHR-RC) and the report of the Independent Oversight and Advisory Committee for the WHO Health Emergencies Program (IOAC), JEE and other existing tools for measuring preparedness should be reviewed based on the lessons learned during the COVID-19 pandemic. The independent review panel found that[29] the legally binding IHR in their current form are conservative instruments that constrain rather than facilitate action. To address these gaps, proposed solutions from the independent panel include the need for the WHA to give the WHO explicit authority to publish information on outbreaks with pandemic potential immediately without requiring prior approval of national governments to improve detection; and the need for Universal Periodic Peer Review (UPPR) to strengthen measurement of leadership and accountability (i.e., governance) dimensions of preparedness and response to public health emergencies, and the broader preparedness and response capacities.[26] This is key going forward, given the observed contribution of leadership and governance in shaping response to the ongoing COVID-19 pandemic in most countries.

Ongoing efforts to improve the assessment tools, as suggested by the IHR-RC and IAOC, commenced in March 2021. A global virtual technical consultative meeting was convened by WHO from March 9 to 10, 2021. Subsequently, a technical working group (TWG) was formed, as advised by the consultative meeting, to address the recommendations of the meeting. This recommendation included specific improvements that should be considered for JEE and SPAR based on lessons from COVID-19. The TWG, made up of over 180 experts, met virtually from July 13th to September 30th' 2021, and made specific changes to JEE and SPAR tools.

The efforts to improve assessment tools are informed by the fact that solutions or any changes to existing tools need to be designed to fit the problem. In other words, what was measured in the past, how did those measures "work" in terms of identifying gaps and contributing to success, especially during the COVID-19 pandemic, and how could/should systems of measurement and evaluation change to enhance global health security?

While the ongoing efforts to improve assessment of preparedness capacities is commendable, the current disconnect between capacity assessment and capacity development due to inadequate financing for preparedness remains a major gap in global health security.

Improvement of the tools must go hand in hand with optimization of necessary resources for capacity development to address the gaps identified. In a nutshell, capacity building and health system strengthening for enhanced health security have not received desired attention from both domestic and external financing.

There has been a general appreciation of operational shortfall. The profound and continued scale of the pandemic emergency attests to this fact. Four major themes are emerging: 1) The need for stronger political leadership and intersectoral collaboration in response to public health emergencies; 2) the major challenge and risk presented by the ongoing urbanization of global populations; 3) the lack of historic attention to supply lines and clinical surge capacity in metrics; and 4) the uneven access to lifesaving tools such as testing kits, oxygen supplies, clinical therapeutics, and, most notably, vaccines especially in public health emergencies of vaccine preventable diseases.

The modern world has evolved a new epidemiologic space for diffusion of infection—global travel. This has been described as the "Global Express" for microbial transmission.[30] The IHR is currently missing metrics measuring travel and border closures which serves as an example of why the reach of WHO guidance and evaluation may not be adequate for this pandemic, and why considering alternate or complementary structural placement for global emergency governance might be useful. Specifically, the issue of the mobility of humans and goods illuminates a critical gap in intersectoral planning and implementation, which needs to be addressed to prevent future pandemics.

In dealing with this global emergency, many countries used a "stop traffic" approach. For travel this continues to date in the form of blocking commercial and shipping access. The WHO has continuously eschewed closing borders as a strategy during pandemic threats. According to the RC, this is largely due to the inherent tension between the IHR's aim to protect health and the need to protect economies by avoiding travel and trade restrictions. Countries have continued to embrace this strategy, a dichotomy in the official narrative between cooperation and unilateral actions such as travel restrictions emerged. At least one analysis has suggested this strategy may have reduced risk, especially when quarantine on arrival was included.[31]

Many countries engaged in trade policy activism beginning in February and March of 2020. At this time, such limitations continue to be pursued to restrict exports of critical medical equipment such as personal protective equipment (PPE), or vaccines, or to limit imports of potential harmful goods (such as frozen foods into prevention, retention, and contingency (PRC)).[32]

In sum, there is little doubt that the evaluations through the JEE, SPAR, simulation exercises, AAR, and the new IAR represent a strong value addition to the mix. However, the International Health Regulations and their implementation have demonstrated gaps in the face of major epidemics including the ongoing COVID 19 pandemic. This is partly due to the inherent disconnect between the current provisions of the IHR and the operational realities of optimal response to public health emergencies. In the example of global mobility of people and goods, the core balancing act of the historic genesis of the IHR, it has proven to be a critical and ongoing gap.

CASE STUDY

Topic and problem identified: Public health preparedness score does not equal the ability to respond to an emergency.

Background

While measuring preparedness is a complex effort, various efforts such as the Joint External Evaluation (JEE) and Global Health Security Index (GHSI) have attempted to quantify preparedness at the national level. Countries across the world have used preparedness evaluation tools like the Joint External Evaluation and Global Health Security Index to measure their ability to prevent, detect, and respond to an emergency.[33] Observations from COVID-19 have made it clear that quantifying preparedness is even more complex than realized and that current JEE and GHSI score did not predict preparedness and performance during COVID-19. Current measurement approaches are not sufficient to predict performance during an emergency (Table 27.2).

TABLE 27.2 Cases of COVID-19 per 10,000 people by WHO region.

WHO region	COVID-19 case counts (country level)	JEE score	GHSI
Africa (AFRO)	High	3.2	31.9
	Medium	2.0	34.2
	Low	2.4	32.2
Americas (PAHO)[a]	High	4.1	83.5
	Medium	–	–
	Low	4.5	75.3
Southeast Asia	High	2.8	33.8
	Medium	3.0	56.6
	Low	2.7	40.3
Europe (EURO)	High	3.5	52
	Medium	3.5	62.9
	Low	3.9	49.3
Eastern Mediterranean (EMRO)	High	4	39.4
	Medium	2.4	25.7
	Low	2.9	26.2
Western Pacific	High	3.5	49.5
	Medium	3.1	43.1
	Low	4.3	54

[a] *The PAHO region only had two countries with published JEE scores at the time of publication.*

CASE STUDY (cont'd)

Observations

COVID-19 cases were pulled from WHO website on December 12, 2021. Total country population was taken from the World Bank; population was last updated in 2020. Inclusion criteria consist of having a public JEE and GHSI score available, recorded COVID-19 cases with the WHO, and recorded population by the World Bank. Countries were then separated based off of WHO region. Three countries were then selected based on case count per region, 1 with the highest, lowest, and then the middle country was determined by the median. Countries' names are not listed to keep them anonymous and prevent any potential adverse reactions to where a country ranks in the table.

Key issues

- The validity of current assessment tools for preparedness was drawn into question as COVID-19 mortality and morbidity were highest in countries with strong health systems, especially in the early face of the pandemic
- Other dimensions such as leadership or lack thereof are not currently captured by current tools; however, they played an important role in shaping outcomes of COVID-19 pandemic in many countries

New interventions

- A focus on the need for ongoing readiness exercises has become apparent and is currently being encouraged to prevent countries from becoming complacent
- Ongoing readiness tabletop exercises are being encouraged as part of the JEE evaluations, so that JEEs are not a one-time snapshot[34]
- The Dynamic Preparedness Metric (DPM) is the World Health Organization's newest idea to decrease the gap between preparedness and response.[35] This metric includes the dimensions of hazard, vulnerability, and capacity. The aim of the DPM index is not to rank countries but rather to facilitate the identification of countries' specific limitations in preparedness and possible mitigation strategies[35]

Lessons learned

- Preparedness capacity scores are not good predictors of performance; lack on congruence between metrics and outcomes; current metrics being formally measured did not correlate with a country's success during COVID-19 pandemic.
- High capacity may not translate into performance in times of need, especially in the absence of motivation of leaders and decision-maker to act optimally, according to expectancy theory.
- Overconfidence appears to be one of the greatest weaknesses in response management.[36]
- As COVID demonstrated, no amount of pandemic preparedness can overcome poor or failed leadership and/or erroneous policies.[37]
- Local inequalities and structural racism within countries exacerbated the effects of the pandemic.[38]
- Experience from COVID-19 shows that application possessed capacity, those who utilized the limited capacity rather than delay enforcement actions resulted in better outcomes.
- The accuracy of any pandemic preparedness score depends not only on a countries' capabilities but also on what

CASE STUDY *(cont'd)*

governments will (and will not) do in any given situation.

Recommendations

- There is need to better understand and improve assessment of dimensions such as leadership and political will for optimal action in public health emergencies that are not addressed by current tools.
- Need for a better understanding of determinants of effective utilization, or underutilization, of possessed capacity.
- No one single preparedness assessment tool can capture all the dimensions of the preparedness capacities. There is need for introduction of new instruments to complement existing tools such as the Universal Health and Preparedness Review (UHPR) and DPM.
- Review and improvement of existing assessment based on lessons from COVID-19.
- A diverse team of experts should be used to assess the complex set of factors that shape a country's capacity to respond,[38] ideas, and inputs should be gathered from outside of the public health world.
- Determine if limited capacity can be implemented as a dimension.
- Measurement should include qualitative assessment of a country's political leadership's willingness to accept scientific information and recommendations.
- Existing measures of corruption and trust should be used in future indices.
- Change in score over time for each individual country may be a more useful way of tracking progress instead of comparing countries against each other.[13]

Vision for the future

Capacity to monitor and evaluate IHR core capacity building

Global health security is only as strong as the weakest, least prepared territory. Regular monitoring and evaluation of country readiness to respond to public health threats are essential. Enhanced proven tools that allow better prediction of preparedness status at national, regional, and global levels for pandemics and epidemics are needed. However, the current pandemic has shown limitations in the capacity to measure and predict performance using current ranking approaches. The lessons learned from programs such as Polio Eradication Initiative (PEI) indicate that key indicators, applied to regularly monitor program performance, using a rigorously developed system help ensure success. In most African countries, the lessons learned from PEI surveillance were applied to monitor performance Integrated

Disease Surveillance and Response (IDSR) using core indicators. By advancing early warning systems through integration of multiple surveillance systems, countries have improved their decision-making capacity to respond to public health events.[39] Monitoring performance at the lowest possible administrative units complemented by information generated from community-based data worked well in an adequately resourced setting. The same approach could be used to measure the IHR Core Capacities building performance level at district level. The findings could be aggregated, reviewed, and monitored and actioned at the provincial and national levels. This requires adapting methods to fit the requirements of IHR core capacity building, fostering a viable multisectoral and multistakeholder coordination mechanism and strong willed staff and resourced IHR focal structure.

One issue which runs through both the preparedness of a nation and the measurement of that preparedness is engaging across sectors. WHO has begun to test a new assessment tool, the IHPR, which attempts to bridge this gap. Generally resourcing the health sector is central as presented above; however, the allocation of resources across sectors relies on high-level policy development. The new tool attempts to capture this dynamic. Testing is in the early stages in Thailand. It is not clear at this point that this methodology can accomplish the task; however, the pandemic has made clear that evaluation of factors beyond the health sector itself will be critical to success.

Investing in artificial intelligence and public health informatics

Innovative approaches including web-based sources of data have been extensively used in surveillance, detection, and prediction of public health threats. Mavragani et al. (2020) write that information that could not be accessed using the traditional data collection methods could be obtained by using web-based sources on time.[40] This will improve availability of the data in real time for improved prediction and preventive action by further improving existing monitoring tools.

Countries have developed their National Action Plan for Health Security (NAPHS) based on the findings of the joint external evaluation. However, accessing resources to support implementation of the plan is often challenging. Although governments are expected to mobilize resources from domestic and external sources to support implementation of the NAPHS, performance is sub-optimal. It is not unusual to find most planned activities were not implemented. Building pandemic and epidemic preparedness, particularly in low- and middle-income countries, requires purposeful and deliberate investment of resources (human, material, and time).

National IHR core capacities to prevent, detect, and respond to public health emergencies of international concern (PHEIC) is ideally built through engagement of multiple stakeholders (i.e., comprising government sectors, academia, development partners, private sector, civil society organizations (CSOs), and other bodies). The government should take the leadership to ensure that the coordination mechanism is put at supra-ministerial level, preferably at the head of government level. The public health leadership should continue their advocacy to ensure high-level political commitment, popular support, and community engagement.

Actions

Countries should prioritize and further strengthen monitoring of preparedness for improved predictable performance using key performance indicators. This requires development of a comprehensive monitoring and evaluation framework with milestones, timeline, and with assigned responsible structure along with delegation of authority for action.

Intersectoral breadth of measurement and evaluation is critical. This is particularly true, given that the political aspects of government leadership have proven one aspect of national response, which was not at all addressed in current tools for measurement. It is a cross cutting need, which also applies to assurance of supply lines and issues of travel and trade. Partnership with private sector industry is critical. Reinforcing the intersectoral reach of the IHR will be crucial at all levels, global, national, and subnational.

Urbanization and travel and trade are critical areas for additional innovation in metrics. Some countries include densely populated megacities, while others do not. Some countries are hub travel centers and include major shipping ports, while others do not. Thus, global metrics must be tailored and tested in those areas where they are most pertinent to successful health security. Metrics on operations and performance are already baked into city operations and into travel and trade systems. Repurposing of this extant information could be useful in defining the overall monitoring and evaluation framework. Careful consideration of potential information bias is always central to such efforts.

Ongoing and timely performance measurement of key indicators requires updating policy and guidance on the use of contemporary innovations in data science applications such as AI for surveillance and public health informatics including use of Internet-based information. Although the use of web-based sources including social media has been used extensively, these sources were considered as unofficial and therefore could not be used without prior authorization by countries. However, updating the policy and guidance to make use of the nontraditional sources of data such as the Internet and social media will further enhance the existing system.

Updating policy and guidance on the use of AI and public health informatics in monitoring and evaluation of preparedness and response entails investment. Government should commit resources to finance critical components of the program from its own government budget line. In addition, the government should be committed to take the leadership to mobilize resources from domestic sources as well as external partners to ensure sustainable financing for implementation of recommended actions. While looking for more money, the government may have to increase transparency and efficiency to obtain more for the existing money.

Regular monitoring of performance of IHR core capacity building using key performance indicators requires a designated structure within the IHR focal point, which is well resourced and delegation of authority and power. Using lessons learned from successful programs, the structure should be represented at district and provincial levels to allow prompt localization and containment of emergent threats. This structure should also be empowered by providing adequate resources for supporting capacity building at district and provincial levels.

Retrospective analyses are critical to efforts at monitoring and evaluation going forward. In line with the discussion of supra-ministerial methods above, a recently published study of 177 countries has further outlined the critical need for enhanced guidance in metrics.[41] The finding that trust in government and among individuals was the strongest correlate of success suggests that metrics for measurement need to broaden. The source of the measurement of trust in this work was survey data from the World Values survey, the Wellcome Global Monitor, and a Gallup World poll. Such diverse sources of measurement could well prove important in broadening the lens of evaluation to assure what we are measuring is truly indicative of what can succeed during a pandemic. Beyond polling, incorporation of operational metrics from other sectors may also enhance our understanding.

Equity in access to information, products and materials, and financing is not a reality across the globe currently. While the vaccine access issue has been profound, similar access issues in personal protective gear, testing kits, genetic sequencing capacity, and oxygen and medicines have all compromised success. Countries were largely on their own to identify emergency funding to address their varied needs during the pandemic. The development banks played a large role in this scramble. When the IHR were adopted, the appeal to developed countries to assist less developed countries in building capacity was met through funding of international organizations and scattershot bilateral assistance. All of the recommended enhancements discussed above will require financing. Operationalizing national plans which address the many gaps of current tools will also require funding. Assuring some level of coordination of resource management, both emergency and longer term, requires either vigorous interagency work by the WHO or an "uber WHO" council within the UN system fit to purpose.

Finally, all measurement and improvement will require transparency. Creating reliable transparency across measurement, evaluation, and operationalization is critical to ensure compliance and cooperation. The COVID-19 pandemic has shown that as pandemics evolve, and transmission patterns shift measurement regimes need to be tailored in time to capture what works when and how. Shifts in how and what is measured will inevitably occur, and such shifts need to be fully understandable by those who use them for guidance at national, provincial, and local level. National sensitivities and sovereignty concerns need to be balanced with the global health security requirement to accurately track capacity and resources in pandemic times. When the COVID-19 crisis is past, the interpandemic period should be exploited to frame stronger population protection systematically and measure progress accurately.

References

1. Lew TWK, Kwek T-K, Tai D, et al. Acute respiratory distress syndrome in critically ill patients with severe acute respiratory syndrome. *JAMA*. 2003;290(3):374–380.
2. Changula K, Kajihara M, Mweene AS, Takada A. Ebola and Marburg virus diseases in Africa: increased risk of outbreaks in previously unaffected areas? *Microbiol Immunol*. 2014;58(9):483–491.
3. Team EOE. Outbreak of Ebola virus disease in the democratic republic of the Congo, April-May, 2018: an epidemiological study. *Lancet*. 2018;392(10143):213–221.
4. Arabi YM, Balkhy HH, Hayden FG, et al. Middle east respiratory syndrome. *N Engl J Med*. 2017;376(6):584–594.
5. Wikan N, Smith DR. Zika virus: history of a newly emerging arbovirus. *Lancet Infect Dis*. 2016;16(7):e119–e126.
6. World Health Organization. Fifty-Eighth World Health Assembly. Geneva; 2005:58:159.
7. World Health Organization. Sixty-Seventh World Health Assembly. Geneva; 2014:67:207.
8. Berg S. *What Doctors Wish Patients Knew about the Dangerous Delta Variant*. Public Health; 2021. American Medical Association2021.

9. Taylor AL, Habibi R, Burci GL, et al. Solidarity in the wake of COVID-19: reimagining the international health regulations. *Lancet*. 2020;396(10244):82–83.
10. World Health Organization. International health Regulations (2005). Headquarters W; 2008:86.
11. Larsson A. *A Framework for Evaluating Emergency Preparedness Plans and Responses Strategies*. International Institute of Applied Systems Analysis; 2008 (Interim Report).
12. Oppenheim B, Gallivan M, Madhav NK, et al. Assessing global preparedness for the next pandemic: development and application of an epidemic preparedness index. *BMJ Glob Health*. 2019;4(1), e001157.
13. Razavi A, Erondu N, Okereke E. The global health security index: what value does it add? *BMJ Glob Health*. 2020;5(4), e002477.
14. World Health Organization. International Health Regulations Monitoring and Evaluation Framework (IHR MEF). https://www.who.int/emergencies/operations/international-health-regulations-monitoring-evaluation-framework. n.d.
15. World Health Organization. *WHO Coronavirus (COVID-19) Dashboard*; 2021. Published https://covid19.who.int/.
16. Prevention CfDCa. *About Variants of the Virus that Causes COVID-19*; 2021. Published https://www.cdc.gov/coronavirus/2019-ncov/variants/variant.html.
17. World Health Organization. *State Party Self-Assessment Annual Reporting Tool*. 2nd ed. World Health Organization; 2021.
18. Talisuna A, Yahaya AA, Rajatonirina SC, et al. Joint external evaluation of the International Health Regulation (2005) capacities: current status and lessons learnt in the WHO African region. *BMJ Glob Health*. 2019;4(6). e001312-e001312.
19. World Health Organization. *Intra-Action Review*; 2022. Published https://extranet.who.int/sph/intra-action-review?region=All.
20. Moore M, Gelfeld B, Okunogbe AT, Paul C. *Identifying Future Disease Hot Spots: Infectious Disease Vulnerability Index*. Santa Monica, CA: RAND Corporation; 2016.
21. Razavi A, Collins S, Wilson A, Okereke E. Evaluating implementation of international health regulations core capacities: using the electronic states parties self-assessment annual reporting tool (e-SPAR) to monitor progress with joint external evaluation indicators. *Glob Health*. 2021;17(1), 69-69.
22. World Health Organization. *Joint External Evaluation Tool: International Health Regulations*. 2nd ed. 2005. Geneva2018.
23. World Health Organization. *Joint External Evaluation Tool: International Health Regulations*. 3rd ed. 2005.
24. World Health Organization. *Statement on the Fifth Meeting of the International Health Regulations*. Emergency Committee regarding the coronavirus disease (COVID-19) pandemic [press release]. 2005.
25. World Health Organization. *Simulation Exercises*; 2021. Published https://www.who.int/emergencies/operations/simulation-exercises.
26. Kimball AM, Moore M, French HM, et al. Regional infectious disease surveillance networks and their potential to facilitate the implementation of the international health regulations. *Med Clin N Am*. 2008:92.
27. Index GHS. About. https://www.ghsindex.org/about/. Published n.d. Accessed April 1, 2021.
28. Crawley AW, Divi N, Smolinski MS. Using timeliness metrics to track progress and identify gaps in disease surveillance. *Mary Ann Liebert*. 2021;19(3):309–317.
29. The Independent Panel for Pandemic Preparedness and Response. COVID-19: make it the last pandemic. The Independent Panel for Pandemic Preparedness and Response2021.
30. Kimball AM. *Risky Trade: Infectious Disease in the Era of Global Trade*. Aldershot UK: Ashgate Press; 2006.
31. Koopmans R. *A Virus that Knows No Borders? Exposure to and Restrictions of International Travel and the Global Diffusion of COVID-19*. Research Unit: Migration, Integration. Transnationalization; 2020.
32. Evenett S, Fiorini M, Fritz J, et al. Trade policy responses to the COVID-19 pandemic crisis: evidence from a new data set. In: *Policy Research Working Paper2020*. 2022.
33. Prevention CfDCa. *Joint External Evaluations (JEE) for improved health security*; 2019. Published https://www.cdc.gov/globalhealth/security/ghsareport/2018/jee.html.
34. Asghar RJ, Kimball AM, Khan AS. Global health security: rethinking joint external evaluations to ensure readiness? *Mary Ann Liebert*. 2019;17.
35. Kandel N, Chungong S. Dynamic preparedness metric: a paradigm shift to measure and act and act on preparedness. *Lancet*. 2022;10(5):e615–e616.
36. Bollyky TJ, Kickbusch I. Preparing democracies for pandemics. *BMJ*. 2020:371.

37. Goldschmidt P. The global health security index: another look. *Front Epidemiol.* 2022;2.
38. Baum F, Freeman T, Musolino C, et al. Explaining Covid-19 performance: what factors might predict national responses? *BMJ.* 2021:372.
39. Jian SW, Chen CM, Lee CY, Liu DP. Real-time surveillance of infectious diseases: Taiwan's experience. *Health Security.* 2017;15(2).
40. Mavragani A. Tracking COVID-19 in Europe: infodemiology approach. *JMIR Public Health Surveill.* 2020;6(2).
41. Collaborators CNP. Pandemic Preparedness and COVID-19: an exploratory analysis of infection and fatality rates, and contextual factors associated with preparedness in 177 countries from January 1, 2020 to September 20 2021. *Lancet.* 2022;399(10334).

Index

'*Note:* Page numbers followed by "f" indicate figures, "t" indicate tables and "b" indicate boxes.'

H

V

W

Y

Z

Printed in the United States
by Baker & Taylor Publisher Services